E-maintenance

Kenneth Holmberg · Adam Adgar · Aitor Arnaiz
Erkki Jantunen · Julien Mascolo · Samir Mekid
Editors

E-maintenance

Springer

Kenneth Holmberg, Prof.
VTT Technical Research Centre of Finland
Metallimiehenkuja 6–8
02044 VTT, Espoo
Finland
kenneth.holmberg@vtt.fi

Adam Adgar, Dr.
Teesside University
School of Science and Engineering
Borough Road
Middlesbrough, Tees Valley TS1 3BA
UK
a.adgar@tees.ac.uk

Aitor Arnaiz, Dr.
Fundación Tekniker
Avda. Otaola, 20
20600 Eibar, Guipúzcoa
Spain
aarnaiz@tekniker.es

Erkki Jantunen, Dr.
VTT Technical Research Centre of Finland
Metallimiehenkuja 6–8
02044 VTT, Espoo
Finland
erkki.jantunen@vtt.fi

Julien Mascolo, Dr.
Centro Ricerche Fiat S.C.p.A
Strada Torino, 50
10043 Orbassano, Torino
Italy
julien.mascolo@crf.it

Samir Mekid, Dr.
King Fahd University Petroleum & Minerals
Department of Mechanical Engineering
Dhahran 31261
KSA
smekid@kfupm.edu.sa

ISBN 978-1-4471-6005-2 ISBN 978-1-84996-205-6 (eBook)
DOI 10.1007/978-1-84996-205-6
Springer London Dordrecht Heidelberg New York

British Library Cataloguing in Publication Data
A catalogue record for this book is available from the British Library

Softcover re-print of the Hardcover 1st edition 2010

Cover design: eStudioCalamar, Figueres/Berlin

Printed on acid-free paper

Springer is part of Springer Science+Business Media (www.springer.com)

Preface

This is the first book to present the topic of *e-maintenance*, which has appeared in the scientific and technological discussions at conferences and meetings during the last decade. E-maintenance is a synthesis of two large trends in our society: on the one hand the growing importance of maintenance as a key technology to keep machines running properly, efficiently and safely in industry and transportation, and on the other hand, the very rapid development of information and communication technology (ICT). This has opened the way to completely new concepts and solutions with more detailed equipment for health information and more effective diagnostic and prognostic tools and user interfaces to ensure good reliability and availability of plants and vehicles remotely worldwide.

The authors of the book are European top experts on ICT and maintenance technology both from academia and industry. They have worked very intensively together for the last four years, starting in 2005 within the European Commission funded research and development project DYNAMITE – Dynamic Decisions in Maintenance. The R&D group consisted of about 50 experts altogether from nine European countries: Estonia, Finland, France, Germany, Greece, Italy, Spain, Sweden and UK.

This book presents an overview of the subject of e-maintenance including trends, scenarios and needs in industry and advanced ICT technologies and future solutions to global and mobile industrial maintenance needs. The pioneering e-maintenance concept DynaWeb is presented, and the group of experts that were involved in its development describe the detailed technologies, their development and experiences gained with this R&D process, as well as future perspectives.

The book is divided into 16 chapters, which include the new integrated e-maintenance concept, intelligent, wireless, MEMS, and lubricating oil sensors, smart tags, mobile devices and services, semantic web services, strategies for e-maintenance and related cost effective decisions, industrial demonstrations as examples of e-maintenance, as well as related e-training.

The book is intended for engineers and qualified technicians working in the fields of maintenance, systems management, and shop floor production lines maintenance. It constitutes a good tool for the further development of e-maintenance in both current and new industrial sites.

It is the hope of the authors that this book will open new views and ideas to researchers and industry on how to proceed in the direction of a sustainable and environmentally stabile society.

Europe
October 2009

The authors

Acknowledgements

The authors gratefully acknowledge the support of the European Commission Sixth Framework Programme for Research and Technological Development. This book summarises work performed as part of FP6 Integrated project IP017498 DYNAMITE "Dynamic Decisions in Maintenance".

The authors are grateful for the support and encouragement received from the European Commission Scientific Officers Andrea Gentili, Philipp Dreiss and Barry Robertson. We also wish to thank the project reviewers appointed by the Commission, Flavio Testi and Christoph Hanisch, for their advice and guidance during the R&D work.

The excellent help and assistance from a great number of colleagues and staff members, as well as the encouragement and financial support from all organisations participating in DYNAMITE is gratefully acknowledged:

- VTT Technical Research Centre, Finland
- Fundación Tekniker, Spain
- University of Sunderland, UK
- University of Manchester, UK
- Université Henri Poincaré, France
- Linnaeus University, Sweden
- Zenon S.A. Robotics & Informatics, Greece
- FIAT Research Centre, Italy
- Volvo Technology, Sweden
- Goratu Maquinas Herramienta, Spain
- Wyselec, Finland
- Martechnic, Germany
- Engineering Statistical Solutions, UK
- Diagnostic Solutions, UK
- Prisma Electronics, Greece
- IB Krates, Estonia

The financial support from the following national funding agency is gratefully acknowledged:

- Spanish Ministry of Science and Innovation (grant no. DPI2007-29958-E)

The authors also wish to thank Ms Christina Vähävaara for the skilful and meticulous editing of the manuscript.

Contents

Contributors

Addison, Dale
University of Sunderland, UK
E-mail: dale.addison@sunderland.ac.uk
Web: www.sunderland.ac.uk

Adgar, Adam
University of Teesside, UK
E-mail: a.adgar@tees.ac.uk
Web: www.tees.ac.uk

Albarbar, Alhussein
Manchester Metropolitan University, UK
E-mail: a.albarbar@mmu.ac.uk
Web: www.mmu.ac.uk

Al-Najjar, Basim
Linnaeus University, Sweden
E-mail: basim.al-najjar@lnu.se
Web: www.lnu.se

Arnaiz, Aitor
Fundación Tekniker, Spain
E-mail: aarnaiz@tekniker.es
Web: www.tekniker.es

Baglee, David
University of Sunderland, UK
E-mail: david.baglee@sunderland.ac.uk
Web: www.sunderland.ac.uk

Bellew, Jim
Martechnic Gmbh, Germany
E-mail: jim.bellew@martechnic.com
Web: www.martechnic.com

Eberhagen, Niclas
Linnaeus University, Sweden
E-mail: niclas.eberhagen@lnu.se
Web: www.lnu.se

Emmanouilidis, Christos
CETI/Athena Research & Innovation Centre, Greece
E-mail: chrisem@ceti.gr
Web: www.ceti.gr

Garramiola, Fernando
Goratu Maquinas Herramienta S.A., Spain
E-mail: fgarramiola@goratu.com
Web: www.goratu.com

Gilabert, Eduardo
Fundación Tekniker, Spain
E-mail: egilabert@tekniker.es
Web: www.tekniker.es

Giordamlis, Christos
Prisma Electronics, Greece
E-mail: christos@prisma.gr
Web: www.prisma.gr

Gorritxategi, Eneko
Fundación Tekniker, Spain
E-mail: egorritxategi@tekniker.es
Web: www.tekniker.es

Halme, Jari
VTT Technical Research Centre, Finland
E-mail: jari.halme@vtt.fi
Web: www.vtt.fi

Holmberg, Kenneth
VTT Technical Research Centre, Finland
E-mail: kenneth.holmberg@vtt.fi
Web: www.vtt.fi

Iung, Benoit
Université Henri Poincaré, France
E-mail: Benoit.Iung@cran.uhp-nancy.fr
Web: www.cran.uhp-nancy.fr

Jantunen, Erkki
VTT Technical Research Centre, Finland
E-mail: erkki.jantunen@vtt.fi
Web: www.vtt.fi

Katsikas, Serafim
Prisma Electronics, Greece
E-mail: sekat2004@yahoo.gr
Web: www.prisma.gr

Krommenacker, Nicolas
Université Henri Poincaré, France
E-mail: nicolas.krommenacker@cran.uhp-nancy.fr
Web: www.cran.uhp-nancy.fr

Lecuire, Vincent
Université Henri Poincaré, France
E-mail: vincent.lecuire@cran.uhp-nancy.fr
Web: www.cran.uhp-nancy.fr

Levrat, Eric
Université Henri Poincaré, France
E-mail: eric.levrat@cran.uhp-nancy.fr
Web: www.cran.uhp-nancy.fr

Mascolo, Julien
FIAT Research Center, Italy
E-mail: julienetienne.mascolo@crf.it
Web: www.crf.it

Mekid, Samir
The University of Manchester, UK
King Fahd University Petroleum & Minerals, KSA
E-mail: smekid@kfupm.edu.sa
Web: www.manchester.ac.uk; www.kfupm.edu.sa

Naks, Tonu
IB Krates OÜ, Estonia
E-mail: tonu@krates.ee
Web: www.krates.ee

Nilsson, Per
Volvo Technology AB, Sweden
E-mail: Per.Henrik.Nilsson@volvo.com
Web: www.volvo.com

Pietruszkiewicz, Robert
The University of Manchester, UK
E-mail: rpietrus@gmail.com
Web: www.manchester.ac.uk

Salles, Nicolas
Université Henri Poincaré, France
E-mail: nicolas.salles@cran.uhp-nancy.fr
Web: www.cran.uhp-nancy.fr

Spais, Vasilis
Zenon S.A. Automation Technologies, Greece
E-mail: v_spa@zenon.gr
Web: www.zenon.gr

Starr, Andrew
The University of Hertfordshire, UK
E-mail: a.starr@herts.ac.uk
Web: www.herts.ac.uk/csc

Tohver, Avo
IB Krates OÜ, Estonia
E-mail: avo@krates.ee
Web: www.krates.ee

Tommingas, Toomas
IB Krates OÜ, Estonia
E-mail: toomas@krates.ee
Web: www.krates.ee

Voisin, Alexandre
Université Henri Poincaré, France
E-mail: alexandre.voisin@cran.uhp-nancy.fr
Web: www.cran.uhp-nancy.fr

Yau, Alan
University of Sunderland, UK
E-mail: alan.yau@sunderland.ac.uk
Web: www.sunderland.ac.uk

Zhu, Zhenhuan
The University of Manchester, UK
E-mail: Zhenhuan.Zhu@manchester.ac.uk
Web: www.manchester.ac.uk

Abbreviations

ACK	Acknowledgment
ACL	Asynchronous Connectionless Link
ADC	Analogue-to-digital Converter
AE	Acoustic Emission
AI	Artificial Intelligence
AmI	Ambient Intelligence
ANN	Artificial Neural Networks
AoA	Angle of Arrival
AP	Access Point
API	Application Programming Interface
AR	Augmented reality
ASIC	Application-Specific Integrated Circuit
BDM	Breakdown Maintenance
BN	Bayesian Networks
BN	Base Number
BP	Back Propagation
BSS	Basic Service Set
CAD	Computer-aided Design
CAM	Content Aggregation Model
CAP	Contention Access Period
CBM	Condition Based Maintenance
CBR	Case Based Reasoning
CCD	Charge Couple Device
CCK	Complementary Code Keying
CEO	Chief Executive Officer
CFP	Contention-Free Period
CNC	Computer Numerical Controlled
CM	Condition Monitoring
CMI	Computer Managed Instruction
CMMS	Computerised Maintenance Management Systems

CMOpS	Computer Maintenance Operational System
CMOS	Complementary Metal-oxide-semiconductor
COTS	Commercial off-the-shelf
CPT	Conditional Probability Table
CRC	Cyclic Redundancy Check
CRIS	Common Relation Interface Schema
CSMA	Carrier Sense Multiple Access
CUSUM	Cumulative Sum
DAG	Directed Acyclic Graph
DCF	Distributed Coordination Function
DIFS	Distributed Inter-Frame Spacing
DPSK	Differential Phase Shift Keying
DSP	Digital Signal Processing
DS	Distribution System
DSSS	Direct Spread Sequence Shifting
Dynamite	Dynamic Decisions in Maintenance
ECU	Electronic Control Units
EEPROM	Electrically Erasable Programmable Read-only Memory
EMC	Electromagnetic Compatibility
ESD	Electrostatic Discharge
ESS	Extended Service Set
EP	Extreme Pressure
ERP	Enterprise Resource Planning
ES	Expert System
FFD	Fully Functional Devices
FFT	Fast Fourier Transform
FMEA	Failure Mode and Effect Analysis
FSO	Full Scale Output
FTA	Fault Tree Analysis
GFSK	Gaussian Frequency Shift Keying
GPS	Global Positioning System
GTS	Guaranteed Time Slots
GTTT	Generalised Total Test on Time
HDD	Hard Disk Drive
HMD	Head Mounted Displays
HR/DSSS	High Rate/Direct Sequence Spread Spectrum
HSI	Human System Interface
HTML	Hyper Text Markup Language
HTTP	Hypertext Transfer Protocol
IBSS	Independent Basic Service Set
IC	Integrated Circuit
IP	Internet Protocol
ICP	Integrated Circuit Piezoelectric
ICT	Information and Communications Technologies

IEEE	Institute of Electrical and Electronics Engineers
ISM	Industrial, Scientific and Medical
ISO	International Standards Organization
IT	Information Technology
ITS	Intelligent Tutoring Systems
ITU-T	International Telecommunication Union – Telecommunication
KBS	Knowledge Based System
KPI	Key Performance Indicators
LAN	Local Area Network
LCI	Life Cycle Income
LCP	Life Cycle Profit
LCC	Life Cycle Cost
LCMS	Learning Content Management Systems
LED	Light-emitting Diode
LIP	Learner Information Package
LMS	Learning Management Systems
LO	Learning Objects
LOM	Learning Object Metadata
LQI	Link Quality Indicators
LRD	Light Receiving Device
MAC	Medium Access Control
MDAQ	Machine Data Acquisition
MDSS	Maintenance Decision Support System
MEMS	Microelectromechanical Systems
MES	Maintenance Execution System
MIL	Matrox Imaging Library
MIMO	Multiple Input – Multiple Output
MIMOSA	Machinery Information Management Open Systems Alliance
MPDU	MAC Protocol Data Unit
MMME	Man Machine Maintenance Economy
MTBD	Mean Time Between Degradation
MTBF	Mean Time Between Failures
NC	Numerically Controlled
NIR	Near Infrared
NIRS	Near Infrared Spectroscopy
NDT	Non-destructive Testing
OEE	Overall Equipment Effectiveness
OEM	Original Equipment Manufacturer
OFDM	Orthogonal Frequency Division Multiplexing
O&M	Operations and Maintenance
OPD	Optical Particle Detector
OSA-CBM	Open Systems Architecture for Condition Based Maintenance
OSA-EAI	Open Systems Architecture for Enterprise Application Integration

OTAP	Over The Air Programming
OWL	Ontology Web Language
PAN	Private Area Network
PAPI	Personal and Private Information
PC	Personal Computer
PCF	Point Coordination Function
PDA	Personal Digital Assistant
PdM	Predictive Maintenance
PHM	Proportional Hazard Modelling
PHY	Physical Layer
PIFS	Priority Inter-Frame Spacing
PLC	Programmable Logic Controller
PLL	Phase Locked Loops
PM	Preventive Maintenance
PPDU	Physical Protocol Data Unit
PPM	Planned Preventive Maintenance
P2P	Person-to-Person
RCM	Reliability Centred Maintenance
RF	Radio Frequency
RFD	Radio Frequency Device
RFD	Reduced Functional Devices
RFID	Radio Frequency Identification
RISC	Reduced Instruction Set Computer
RMS	Root Mean Square
ROCOF	Rate of Occurrence of Failures
ROIIM	Return on Investment in Maintenance
RPC	Remote Procedure Call
RSSI	Received Signal Strength Indication
RTE	Run-Time Environment
RTLS	Real-Time Location System
RUL	Remaining Useful Life
SCADA	Supervisory Control and Data Acquisition
SCO	Synchronous Connection-Oriented
SCO	Sharable Content Objects
SCORM	Sharable Content Object Reference Model
SHM	Structural Health Monitoring
SIFS	Short Inter-Frame Spacing
SLED	Super Light-Emitting Diode
SME	Small-to-Medium sized Enterprise
SN	Sequencing and Navigation
SNR	Signal-to-Noise Ratio
SOA	Service-Oriented Architecture
SOAP	Simple Object Access Protocol
SoC	System on Chip

SQL	Structured Query Language
SSID	Service Set Identifier
SW	Semantic Web
TAN	Total Acid Number
TBN	Total Base Number
TCP	Transmission Control Protocol
TDIDT	Top Down Induction of Decision Trees
TDoA	Time Difference of Arrival
TDMA	Time Division Multiple Access
ToA	Time of Arrival
TPM	Total Productive Maintenance
TQM	Total Quality Maintenance
TQMain	Total Quality Maintenance
TTT	Total Time on Test
UART	Universal Asynchronous Receiver-Transmitter
UCD	Use Case Diagrams
UML	Unified Modelling Language
URI	Uniform Resource Identifier
USB	Universal Serial Bus
UWB	Ultra Wire Band
XML	Extensible Markup Language
XSD	XML Schema Definition
VBM	Vibration-Based Maintenance
VET	Vocational Education and Training
VR	Virtual Reality
WEP	Wired Equivalent Policy
WINS	Wireless Intelligent Network Sensors
WILE	Web-based Intelligent Learning Environments
WIP	Work-In-Progress
WLAN	Wireless Local Area Network
WMAN	Wireless Metropolitan Area Network
WORM	Write–Once, Read–Many
WPAN	Wireless Personal Area Network
WSN	Wireless Sensor Network
WWAN	Wireless Wide Area Network

Chapter 1
Introduction

Kenneth Holmberg

Maintenance is a field of technology that consists of technical skills, techniques, methods and theories that all aim at "keeping the wheels in our society rolling properly". The purpose is to find both technical and organisational solutions for large assets like factories, power plants, transportation vehicles and building technology equipment, as well as for smaller assets such as household machines, hobby devices and consumer products, to function properly, in a cost-effective way, with low energy consumption, without polluting the environment and in a safe, controlled and predictable way.

The huge costs and risks related to improper maintenance have been both observed and documented in the industry. Poorly functioning production machines and unreliable products are not good for a company's business. Maintenance is directly linked to competitiveness and profitability and thus to the future of a company (Pehrsson and Al-Najjar 2005).

In the last decades several organisational approaches to arrange the maintenance work as efficiently as possible have been developed. Such methods are, *e.g.*, total productive maintenance (TPM), reliability-centred maintenance (RCM) and condition-based maintenance (CBM) (Campbell and Jardine 2001, Márquez 2007). These methods have been implemented in the industry with mainly very good results.

At the same time people have realised that the strategy to wait to repair equipment until it fails is often not a good solution. The break down may come at an inconvenient time and the sudden and unexpected stoppage can be very expensive. The breakdown may even become a source of problem for nearby equipment (secondary damage), the environment (pollution) and may even pose health and safety problems to nearby personnel. One solution is to use scheduled maintenance, stopping the equipment regularly for checking and service. The problem with this approach is that the equipment is stopped also in unnecessary cases, and sometimes the stop and unnecessary service action may introduce new problems.

The optimal solution is to know continuously the condition of the asset and its components and take repair and service actions only when really needed. It is, of course, a big challenge to have complete control over the asset condition and also know what the optimal maintenance decisions are each time. However, current technological development offers new and advanced techniques and methods to support this approach.

Currently, there is an improved understanding of the physical, mechanical and electrical phenomena initiating and triggering disturbances and failures. There is the potential to develop low cost micro size integrated sensors for observing the behaviour of a device. There are high capacity and advanced methods for condition data collection, signal analysis, data mining, reasoning and decision making. There are methods for computer based diagnostics and prognostics of plant conditions. New wireless techniques and the internet offer the possibility of using mobile hand-held computers (PDA, personal digital assistant) to have access to large information globally and on line (Holmberg and Helle 2008).

This development opens a new possibility in asset maintenance. It is called *e-maintenance* and has been defined as "The network that integrates and synchronises the various maintenance and reliability applications to gather and deliver asset information *where it is needed*" (Baldwin 2001). The e-maintenance solutions typically offer answers to the following:

- What: which equipment needs maintenance?
- When: when is the maintenance needed?
- Who: computerised maintenance management systems.
- How: manuals, spare part availability.

The concept of e-maintenance integrates existing telemetric maintenance principles with web services and modern e-collaboration methods. Collaboration allows us to share and exchange not only information but also knowledge and e-intelligence (Han and Yang 2006, Muller *et al.* 2008).

In this book we present a flavour of advanced techniques and methods that form the basis of an integrated e-maintenance approach, including solutions such as advanced micro sensors, smart tags (RFID, radio frequency identification), online oil sensors, PDA maintenance applications, ontology based diagnostic and prognostic methods, wireless communication, semantic web service for distributed intelligence, dynamic cost effectiveness based decision making tools and a holistic e-maintenance concept.

In this book the development of such techniques and methods is reported and the state-of-the-art is reviewed. Moreover, experiences both from laboratory testing as well as the use of e-maintenance in industrial environments are reported. The reported cases are demonstrations on the global level, with milling machines, machine tools, foundry hydraulics, maritime lubrication systems and automatic stamping machines. An e-training package for implementing successful e-maintenance applications is presented.

The development work and industrial demonstrations were carried out in the European Commission 6th Framework Programme project "Dynamite" (Dynamic Decisions in Maintenance) by 17 academic and industrial partners in Europe. It is our hope that this book will help the reader to understand the different advanced techniques that e-maintenance is based on and how e-maintenance as a concept can offer new and optimal solutions for asset management in a modern net-based information environment for globally active enterprises.

References

Baldwin RC (2001) Enabling an e-Maintenance infrastructure. Maintenance Technology 12, available at www.mt-online.com/articles/1201_mimosa.cfm

Campbell JD, Jardine AKS (2001) Maintenance excellence – Optimizing equipment life-cycle decisions. Marcel Dekker, New York

Han T, Yang BS (2006) Development of an e-maintenance system integrating advanced techniques. Computers in Industry 57:569–580

Holmberg K, Helle A (2008) Tribology as basis for machinery condition diagnostics and prognostics. International Journal of Performability Engineering 4:255–269

Márquez AC (2007) The maintenance management framework. Springer, London

Muller A, Márquez AC, Iung B (2008) On the concept of e-maintenance: Review and current research. Reliability Engineering and System Safety 93:1165–1187

Pehrsson A, Al-Najjar B (eds) (2005) Creation of industrial competitiveness. Acta Wexionensia No 69/2005, Växjö University Press, Sweden, ISBN: 91-7636-467-4, ISSN: 1404-4307

Chapter 2
Maintenance Today and Future Trends

Andrew Starr, Basim Al-Najjar, Kenneth Holmberg, Erkki Jantunen, Jim Bellew and Alhussein Albarbar

Abstract. This chapter describes the state of the art in maintenance and its future trends. The key areas that have influenced maintenance in the last 40 years are *management* of people and assets, and *technological* capability. These areas are important because they aim to take the best advantage of expensive resources, whether that advantage be profit, or to provide the best possible service with limited resources. The chapter first sets out the current range of maintenance in industrial practice. It is recognised that many businesses do not undertake the full extent of the work reported here, but it is our purpose to survey the state of the art. The chapter then continues to survey the influences of nascent technologies and ideas, before making some predictions about the future. Indeed, some of the most advanced condition-based maintenance effectively aims to predict the future. However, here we do not offer a crystal ball calibrated to international standards; we will constrain ourselves to an informed, independent opinion.

2.1 State of the Art in Management

Maintenance today contributes to the aim of sustainable development in society, including environmental and energy saving aspects, safety aspects and economical aspects. Advanced maintenance has a critical role to play in improving companies' competitiveness. Technology will not be effective without excellent management. The reliability and availability of machines and instruments are crucial factors of competitiveness, particularly in applications where safety and availability are important. Automation and integrated production have resulted in larger technical systems, which are more difficult to control, and more sensitive and vulnerable to diverse consequential effects because of breakdowns.

Reliability, availability and lifetime planning first advanced in the nuclear energy industry. The aerospace industry quickly followed, developing methods to assure reliability by distributing and duplicating the crucial features. Safety and risk analyses have been developed and adapted not only in the chemical industry but to some extent in most industrial fields.

However, existing methods are not always so easily applicable to conventional power plants, or to the process and metal industries, where availability is often a more important criterion than reliability. In other words, the downtime is more important than a small probability of failure. A failure can be acceptable if the repair and restarting times are short. Maintainability and maintenance support performance are therefore most important in such cases.

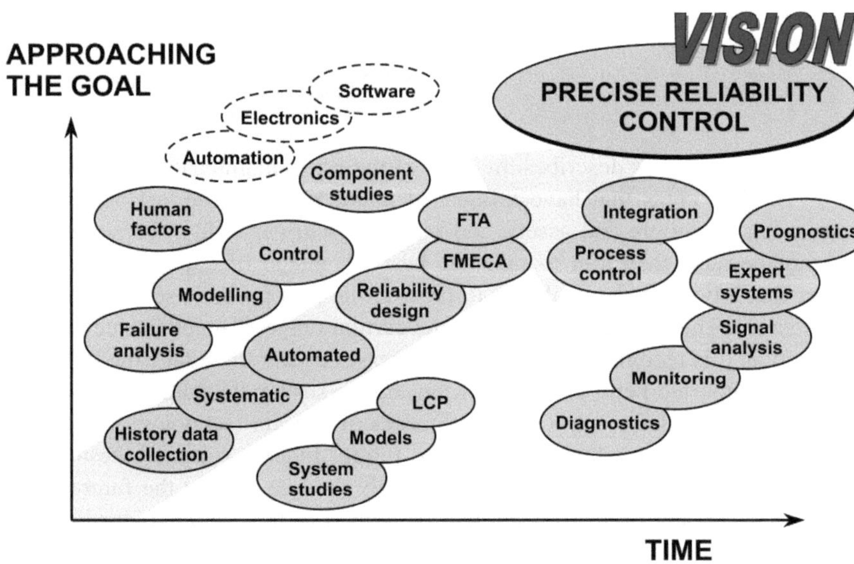

Figure 2.1 The fusion and advance of maintenance technologies

Traditionally, the manufacturer guaranteed the faultless action of a product for a certain warranty period. Nowadays, life cycle profit (LCP) planning is gaining popularity and it is based on the reliability of a product during its whole lifetime. Statistically-defined failure frequency, availability, and the lifetime of the product can now be used as a competitiveness argument. This will also give a reliable basis for recycling a product.

Higher reliability of industrial plants and machines means fewer risks, both personal and environmental, and better control, as well as energy conservation and lower expenses during the operating lifetime. The international competitiveness of the industry can be improved by developing new techniques and methods to specify and control the product reliability more precisely and convincingly. This is a very important sales argument in a situation where the gap between different products, in terms of performance and functional features, diminishes as a result of extremely advanced product development driven by competition.

Today's product design methods are mainly based on optimising the performance of the products and little attention is given to reliability and lifetime estimations. Few design tools emphasise reliability and availability.

This fusion of technologies is illustrated by Figure 2.1, in which the influence of a wide range of technological advances is considered over the last two decades. Because of the great variety of different techniques, based on expert knowledge in several fields of technology involved, there is a need to approach the reliability and maintainability problems from a general, holistic point of view, starting from the problem of the customer and ending with the satisfied user. The Technical Research Centre of Finland (VTT) has developed a systematic approach (Holmberg 2001, Holmberg and Helle. 2008). This is aimed at improving the synergistic interactions between the different fields of expertise by showing a logical and comprehensive structure, where each expert can find his place and see the connections to experts from other fields, all working with the same aim of a satisfied end user, as shown in Figure 2.2.

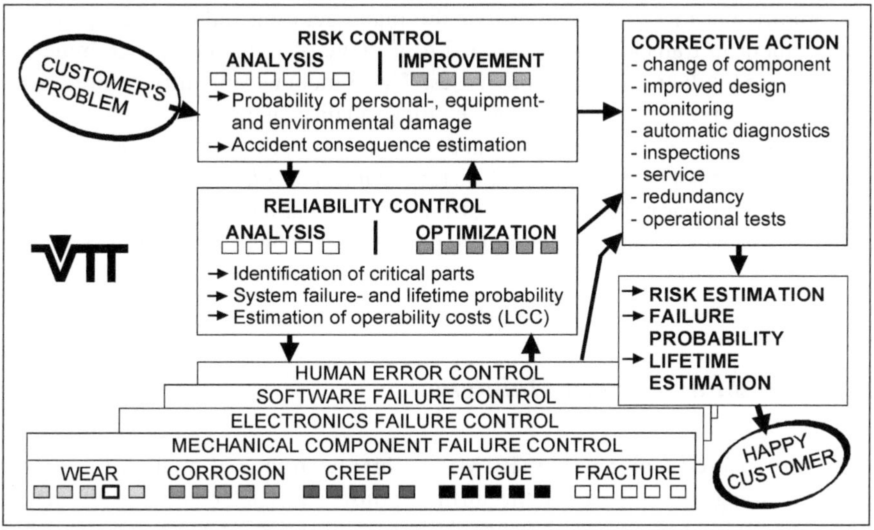

Figure 2.2 Holistic approach to maintenance integration

The probability of personnel, equipment and environmental damage can be analysed and the accident consequences estimated by systematic methods of risk control. The critical parts are identified, the probability of system failure and lifetime are calculated, and the operability costs are estimated by statistically based techniques of reliability control. When the critical parts of the production system that need improvement have been identified, the right techniques and tools for improvement actions are found in the fields of mechanical component failure control, electronics failure control, software failure control or human error control.

When a critical function is identified, such as the wear endurance life of a certain component, a component operability analysis is carried out; this includes an analysis of the old solution, a robust lifetime design approach to recommended improvements, an analysis of the new solution and, as a result, the improvement actions with estimated improved failure probability and probable lifetime.

The recommended measures to be taken can be a change of component, redundancy, improved design, extended monitoring, automatic diagnosis, inspections, operational tests or service instructions. The output of the holistic approach is recommendations for improvements together with estimations of their effects on the risks, the probability of failure and the lifetime.

2.2 Integrated Programmes and Planning Processes

The holistic maintenance concept has been developed extensively during the last decades. As noted above, a significant driver was the reliability of nuclear power and aircraft, but the competiveness of the manufacturing industry, coupled with improvement in transport areas, has led to a range of integrated programme philosophies and holistic planning processes. These include reliability centred maintenance, total productive maintenance, total quality maintenance, lean maintenance and many others. Data oriented techniques such as proportional hazard modelling offer some integration of history (Jardine *et al.* 1998). A good deal of proprietary know-how is also being offered to the market. The aim of these integrated approaches is to offer a complete plan, usually compiled from the strategies in Section 2.3 below.

2.2.1 Reliability-centred Maintenance

Reliability centred-maintenance (RCM) is a highly structured method for maintenance planning, developed for the airline industry and later adopted by several other industries and military branches (Moubray 2001). RCM partitions a machine in a systematic way to analyse its construction by using failure mode and effect analysis (FMEA) in order to identify significant components and failure modes. It then selects the appropriate maintenance action for each of these components using structured criteria, with the key aim of reducing or eliminating failure. In order to implement RCM, a bank of failure data is preferred.

As a structured method, RCM has some strong features, including a good audit trail and consistent decision-making. However, there are some drawbacks:

- Failure data is not easy to obtain because equipment and components are usually replaced before failures to avoid high consequential costs especially in the process and chemical industries.
- Reliability may not be the main focus – manufacturing plants typically focus on availability.
- The RCM structure is not concerned with the outcome of monitoring, *e.g.*, it does not make full provision for the use of condition monitoring techniques, so that the development of potential failures is not followed until just before failure.

2.2.2 Total Productive Maintenance

Total productive maintenance (TPM) aims to maximise equipment effectiveness. It consists of a range of methods that are known from maintenance management experience to be effective in improving reliability, quality and production. TPM tries to improve a company through improving personnel and plant, and changing the corporate culture. Cultural change at a plant is a difficult task to perform and it involves working in small groups, a strong role for machine operators in the maintenance program, and support from the maintenance department (Willmott and McCarthy 2000).

One of the essential forces driving total quality management (TQM), TPM is an improvement cycle or Deming cycle, *i.e.*, plan-do-check-act. While this cycle has been used when a failure occurs, it is more economical to control the machine condition and to prevent failure or manufacture of defective items. For example, monitoring the vibration related to the product quality or machine damage may help to detect quality deviation before manufacture of defective items or further damage.

TPM requires operators to take over some of the maintenance staff tasks, *e.g.*, cleaning, lubrication, tightening fasteners, adjustment and reporting of observations of changes in the machine condition. All these tasks are important and useful to stop some failure causes but they cannot stop all failure modes. More detailed information concerning the machine condition is of great importance for supporting this operator maintenance, and instruments assist the operators in searching for abnormalities in the equipment.

2.2.3 Total Quality Maintenance

One of the essential forces driving TQM and TPM is the improvement cycle. The action can be interpreted so that action is started at an early stage, *i.e.*, as soon as a significant deviation in the equipment/process condition is observed. About 99%

of the mechanical failures are preceded by some detectable indications of condition change (Al-Najjar 2001). A range of condition monitoring techniques is considered in Section 2.5.2 below.

TQM is a means to maintain and improve continuously the technical and economic effectiveness of the production process and its elements (Al-Najjar 2001). It is not just a tool to serve or repair failed machines; rather it is a means to maintain the quality of all the elements involved in the production process. Thus, the role of TQM is the following:

- Monitoring and controlling deviations in a process, working conditions, product quality and production cost.
- Detecting damage causes, their developing mechanisms and potential failures in order to interfere (when possible) to stop or reduce the machine deterioration rate before the production process and product characteristics are intolerably affected.
- Performing the required action to restore the machine/process or a particular part of it to as good as new.

All these should be performed at a continuously reducing cost *per* unit of good quality product. Here, failure is defined as *a termination of a component's ability, to perform its required function, which can be defined on basis of the machine function, capability, production rate, production cost, product quality or personnel/machine safety.*

2.3 Strategies

This section reviews tried-and-tested maintenance strategies and the 21st century variants. Maintenance is a key part of any business activity, since its principal objective is to preserve the availability of the assets that are used for the business. In formulating a maintenance plan, the aim is to minimise the combined cost of operating the business and maintaining the plant. The organisation of maintenance activity is based around the continuous process of the business, plant breakdowns, availability of personnel and spares. Planning and administration is required to match the resources (men, spares and equipment) to the expected maintenance workload. A range of strategies is available to the maintenance manager. The state-of-the-art policy uses a combination of run-to-failure, time-based maintenance, design out, condition based maintenance and opportunity maintenance. Traditionally, the trigger for initiating maintenance has been either failure of the plant or a time-based preventive plan. Condition based maintenance (CBM) is an improved method of preventing failures, based on detection of machine deterioration.

2.3.1 Run-to-failure

In run-to-failure, also known as maintain-on-failure or breakdown maintenance (BDM), the plant item is allowed to fail before maintenance is initiated; this is appropriate if the consequences of failure are small, *e.g.*, a light bulb. It is only appropriate to run to failure if it does not matter whether the machine fails, or how long the repair will take or how much it will cost. Sometimes a failure is not predictable using any instrument or analysis, and only checking for failure will detect the fault. Unfortunately the strategy is widely used in inappropriate situations. At failure, the task of the repair team is to restore the machine to a state in which it can perform the required function as quickly as possible.

The strategy has some advantages:

- Planning is simple – the organisation need only adapt to match the failure rate.
- Work is not scheduled until it is really needed.

However, it has major disadvantages:

- Failure can, and probably will, occur at an inconvenient time, *e.g.*, when the plant is at full load, or while it is starting.
- A component fault may go unnoticed, leading to expensive consequential damage, *e.g.*, bearing seizure causes damage to a shaft.

Box 2.1 Run-to-failure in the maritime industry
In high risk environments, *e.g.*, maritime, the option to run to failure is dangerous and extremely expensive. That being said, there are many instances where such practices occur through negligence, under-resourced operators and as a consequence of poor management. An owner of an asset, looking for a short term cost reduction may decide to cut maintenance costs to a minimum. In such circumstances "run-to-failure" occurs in an industry least able to handle the consequences.

- Dangerous and/or expensive failure consequences should be expected.
- No data are available regarding the past, present and possible future state of the machine.
- A large breakdown crew may need to be available on standby. All the required expertise should be either within the plant or easily accessed from external resources, which is almost always costly, or a longer waiting time should be expected.
- A large spares inventory is necessary to ensure quick repair.
- Failures exceeding the capacity of the repair team lead to "fire-fighting".

2.3.2 Time-based Maintenance

In planned (or time-based) preventive maintenance (PPM) maintenance is scheduled in advance to prevent failure. This concept was developed through the mid 20th century, focusing on preventing failures through replacing components at particular times. It is assumed that the machine/component life is predictable, and maintenance is based on hours run or calendar time elapsed. Individual or block replacement is made. This is suitable for repeatable degradation modes, *e.g.*, wear processes or constant rate corrosion.

The strategy has some advantages:

- A more effective use of time.
- Spares are only ordered as required.

However, it has disadvantages:

- The plant may not fail according to a fixed time period (calendar or run hours) – this is likely in complex plant.
- Failures may still occur.
- The method depends on statistical analysis; in many cases suitable and correct failure data are not present.
- The plant may not need maintaining – spares and labour are used unnecessarily, and the plant is unavailable during maintenance.
- Unnecessary strip down and bearing changes may cause problems.

PPM advocates replacement or repair at a fixed time after installation, independent of its condition. The time period used to construct a maintenance schedule can be either calendar time or component running time. A component is replaced at a fixed time T, or at failure, whichever occurs first.

The timing of maintenance activity in a PPM programme is calculated to minimise overall costs. Many applications of maintenance optimisation models exist, and analysis of the role of these models in maintenance is given. The often interrelated model assumptions and characteristics are divided into four groups concerning equipment, maintenance situation, production–demand situation and type of model and solution procedure.

PPM works well provided that it is acknowledged that some failures will occur. The majority of the failures will be pre-empted, but some will still occur because of uncertainty about the underlying failure distribution of the plant/component life, which is occasionally shorter than the maintenance interval. The most effective use of time-based PPM will be in equipment that has a very predictable life, *e.g.*, components that are designed to wear. Typical PPM activities are shown in Table 2.1.

Table 2.1 Typical PPM activities

Visual and aural inspection for leaks, noise, looseness and cleanliness
Lubrication of bearings and slides
Adjustment of belts and couplings
Checking electrical connections
Checking performance
Cleaning filters and strainers
Replacing parts at intervals: belts, seals, bearings, *etc.*

Often the simplest, low-cost methods of inspection and cleaning are not done properly because the job is uninteresting and considered unimportant. Many inspection tasks, however, do not assess the real performance of a component, *e.g.*, an electrical connection may look sound, but have a high resistance oxide coating on its contact surfaces.

Box 2.2 PPM in the maritime industry
Across the marine industry time-based maintenance is the norm. Usually described as planned or preventative maintenance, the process includes stopping and inspecting machinery based on time in use and replacing components at specific periods. The industry is regulated by schedules enforced through classification societies and supported by standards recognised by insurance companies. It is normal for a vessel to be taken out of service for a total survey after a predetermined time. This will include dry docking of the ship, cleaning and inspecting the hull and attending to all of the below water equipment. As this is a hugely expensive undertaking, all other internal surveys and refits are scheduled to coincide. Because of the cost and dislocation of this process, operators strive for longer and longer periods between dry docking. These periods are dependent upon the type of vessel, the trade routes in which it is engaged and even the coatings applied to the hull. However, the internal machinery usually requires maintenance between dry dockings and the challenge is always to keep the vessel in service. Therefore, a considerable amount of redundancy is built into the shipboard systems with reserve or duplicate equipment, spares carried on board or stocked at ports along the ship's itinerary. In the case of tramp ships (vessels without a scheduled itinerary) it is common to air freight large components, such as pistons and cylinder liners, around the world to maintain the service.

Some failures occur despite programmed PPM, for a number of reasons. Sometimes the maintenance action is inappropriate, *e.g.*, regular bearing changes can replace good bearings, with plenty of remaining life, with a poor bearing of short life. Some maintenance actions cause damage by disrupting seals and bearings, and allowing dirt to contaminate clean components.

2.3.3 Opportunity Maintenance

An important extension of time-based maintenance or PPM is the planning of maintenance around the opportunity for access. The principal problems arise from plants that physically move, *e.g.*, vehicles such as trains, and those that run almost continuously and therefore must be deliberately shut down in order to maintain them, such as steel making, chemical plants and nuclear power plants. A great deal of planning is necessary to prepare what can be done before the shutdown, what must be done during the shutdown, and what may be done after the shutdown and the plant operating again. Statistical data is useful to establish whether repairs are necessary now or at the next shutdown; such techniques find sophisticated application in aircraft maintenance. Turnaround management focuses on the planning and execution of opportunity maintenance (Lenahan 2005).

Planned opportunity maintenance can also arise from an unscheduled event. If work that was planned for a shutdown can be undertaken during an unscheduled repair period, then it is possible to extend an operating period or delay a scheduled survey.

2.3.4 Design Out

"Design out" as a maintenance strategy means that a failure is addressed by a new or updated design process, with the intention of reducing or preventing future failures. It is pertinent to enquire why design out is required at all:

- A machine or process may be working beyond its original design specification in speed or capacity.
- Legacy equipment may lack sufficient information to make informed judgement about capacity.
- Despite best efforts in previous design, the specified properties of a component or system may not match its actual behaviour.

This strategy is an inclusive philosophy: many maintenance operators undertake redesign of repeat failures. Strictly speaking, it should prevent further breakdown maintenance but could still result in PPM or condition-based maintenance. It is sensible to undertake design out on cost grounds, and indeed reliability centred maintenance favours design out if it is technically feasible and worth doing, on the grounds that it has the potential to eradicate a risk.

2.3.5 Condition-based Maintenance

Advanced maintenance plans avoid failure by detecting early deterioration, spotting hidden or potential failures. Condition-based maintenance (CBM) initiates maintenance when deterioration in machine condition occurs. The component or equipment is usually replaced or repaired as soon as the monitoring level value exceeds the normal. CBM combines the advantages of other strategies, with the following benefits:

- Better planning of repairs is possible, *i.e.*, out of production time.
- Inconvenient breakdowns and expensive consequential damage are avoided.
- The failure rate is reduced, thus improving plant availability and reliability.
- A reduced spares inventory is required.
- Unnecessary work is avoided, keeping the repair team small but highly skilled.

It is possible to prevent unnecessary strip-down of machines and replacement of parts. Manufacturers' recommendations for overhaul do not always take into account the machine loading and conditions of use. Disassembly may cause damage, and bearing replacement, in particular, may lead to premature failure. The time between such replacements is necessarily shorter than the estimated machine life.

The trigger for CBM activity is a measured parameter that is indicative of the machine condition. This may be a performance indicator, or a diagnostic measurement that gives early warning of deterioration, and is termed condition monitoring (CM). Additional information is available from control and monitoring systems that offer performance data from existing sensors or extra sensors, chosen to detect machine condition. Condition monitoring techniques are generally developments of established diagnostic methods. There are three methods that may be regarded as general-purpose methods in that they detect incipient failures in a wide range of machine components: thermal, lubricant and vibration monitoring. Many other techniques are effective for particular fault indicators; they are described in Section 2.3.2.

It is important that CBM is applied to appropriate plants, rather than as an overall policy. Some techniques are expensive and it would not be cost effective to use them everywhere. It is also crucial that the CM techniques are selected to suit the problem – it is all too common that a technique is assumed to be the panacea for all the problems. It is therefore important to evaluate the criticality of plant before beginning the process. In some companies CBM is often only applied to a "critical" plant.

A plant can be critical on three grounds:

1. safety;
2. capital value; and
3. potential for production/service loss.

Analysis of failure history identifies a plant that has shown down time and unreliability. A thorough audit of current maintenance activities will also identify which areas are currently spending the most on maintenance – whether it is justified or not. CBM has the best potential for cost reduction in critical units, because it is there that catastrophic incidents can occur, valuable assets can be destroyed and production can be lost.

If the monitoring is imperfect, a sudden increment in a measured variable can cause an inappropriate maintenance decision. If a replacement is made well before actual failure, most of the advantages of CBM are negated.

It is possible to extend CBM, not only to track incipient failures, but to consider the root cause of failures, trace defect initiation and developing mechanisms and follow defect development. This way of monitoring components and equipment increases the possibility of detecting deviations in both the machine condition and product quality at an early stage. This works best when the correlation between spectrum constituents of the CM parameter and the deviations in the machine state and product quality are defined clearly.

The concept can be described on two working levels.

- The first level can be called *proactive maintenance*: detection and correction of defect causes such as unsuitable lubricant, misuse, faulty construction, faulty bearing installation, pollution in lubricants, bent or thermally hogged shafts and high operating or environmental temperature. (see also design out above).
- The second level may be called *predictive maintenance* (PdM): monitoring of symptomatic conditions. This is necessary when the failure process is active and when it was not possible to correct defect causes or when it exceeded a predetermined level.

The use of CBM may lead to appreciable reductions in production cost and capital investments and increments in the quality rate, profits and market share, which in turn put maintenance as a contributor in company profit.

2.3.6 Summary

This section has reviewed the range of industrial maintenance strategies and their limitations. The best maintenance plan incorporates all the strategies, playing to their strengths. CBM has been defined in detail and has been shown to be an advanced strategy, which is aimed especially at high criticality problems that are not adequately treated by the traditional strategies of run-to-failure and time based preventive maintenance. It is important to emphasise, however, that each strategy is part of a planned approach, used where appropriate.

Industries with remote, critical assets, such as the maritime sector, theoretically have the most to gain from condition based maintenance, and the leading operators are seriously exploring opportunities. There are, however, many obstacles to widespread adoption, not the least of which is the culture of caution and conservatism that, for good reasons, permeates many industries. Change occurs slowly in industries that have long investment cycles and heavily bureaucratised systems.

2.4 Maintenance Information and Control Systems

2.4.1 Features of the Typical Maintenance System: from SME to Global Enterprises

In this section we consider the maintenance system as a management tool for information and control. Computational tools are considered below. The basic tools included in most maintenance systems are summarised in Table 2.2.

Maintenance systems started as simple planning exercises with wall charts and card indexes, and such devices are still in use because they are visual and virtually foolproof. In larger plants, computerised models were used very early on to manage and control large inventories. Job control and history data followed. CBM has a special need for high resolution data storage and also for sophisticated tools for information extraction and decision-making. Early computerised systems tended to focus on job scheduling, resource management, and inventory, but CBM systems rapidly developed the sophistication in rich data storage and interpretation, *e.g.*, vibration analysis. The integration of systems including both technical and management data led to important advances in the optimisation of maintenance programmes incorporating CBM, and achieved significant savings for users, while radically improving reliability and safety.

Table 2.2 Typical maintenance system features with integrated condition-based maintenance

Feature	Basic features	Extensions for CBM
Modular structure	✓	
Plant inventory/asset register	✓	
Job catalogue	✓	
Work planning	✓	
Stores	✓	
Report generation	✓	
Plant history	✓	
Inventory structure down to monitoring points		✓
Defined monitoring parameters		✓
Data collection		✓
Communications		✓
Diagnosis and trending		✓

The standardisation of maintenance data was significantly advanced by MIMOSA, the Machinery Information Management Open Systems Alliance. MIMOSA standards are compliant with, and formed the informative reference to, ISO 13374-1 for machinery diagnostic systems (ISO 2002). Led by software house Entek and many condition monitoring technique providers, the interchange of data between proprietary systems became commonplace. The US Department of Defense took up MIMOSA standards in 2006, and large systems providers such as Invensys and SAP have worked with MIMOSA.

A wide range of software systems exist. Small-to-medium sized enterprises (SMEs) tend to implement simpler systems, but may benefit from a simplicity and clarity that eludes larger systems.

2.4.2 Limitations to the Penetration of Integrated Systems

It might have been expected that integrated maintenance systems would have almost universal appeal, to all but the smallest customers, but some important problems remain. The huge effort by MIMOSA started over 10 years ago to standardise interconnection with significant uptake by some specialist suppliers. Some key concepts, and connectivity between major system elements, are still the preserve of specialist sectors and providers. For example, computer-based systems starting from the perspective of enterprise resource planning (ERP) have only recently started to integrate the results of technical monitoring programmes for event initiation.

Clearly, the integration of large-scale systems is not trivial and takes a long time to achieve. The recognition of the important role of maintenance is still poor today, in many industries the focus is on new methods of production. An integrated maintenance system is technologically very complex and is not easy to integrate. However, often there is also a desire to keep maintenance systems separate from production (*e.g.*, control) systems, which is understandable because it may avoid risk to production, but on the other hand it may be a hindrance to efficient maintenance and reliability, which are prerequisites for production. It is possible, therefore, that management structures that treat maintenance as a separate activity may also treat its systems separately, to the extent that it is not possible to integrate the systems structures. In such an organisation, the technical and managerial systems and data will, therefore, be permanently disconnected from the company's key business systems.

The forward strides in connectivity, facilitated by the internet and mobile telephone networks, have offered almost limitless scope to systems integrators. However, users must continue to demand standardised systems interconnection, or specialist suppliers will tie the user into proprietary arrangements, which ultimately fail to maximise the benefit to either party.

All software and data in complex systems carry their own maintenance cost. The effort involved with updating the software and data is a huge and demanding task. It is necessary to keep up with changes in the hardware, such as mass storage devices, systems software, such as new operating systems, and to apply normal changes to the system associated with, for example, changes to the plant supported by the maintenance system, such as a new piece of equipment.

2.5 State of the Art in Technology

2.5.1 Computing Tools

The use of computer systems is critical for the organisation of systems pertaining to maintenance. Section 2.2.3 considered the structure of maintenance software systems, but it is worth discussing briefly the hardware and firmware associated with practical maintenance systems. Low cost, pervasive computing tools, from the internet to the data stick, are readily adopted as part of systems solutions in maintenance. Maintenance practitioners have rapidly adopted highly mobile devices, with mobile internet connections, so that they can communicate with databases while on the move, either around a large plant or in many plants. Different solutions are adopted by employees of large companies and service providers to those companies. Local (wired) area networks are often inaccessible to maintenance staff because they belong to the host's production function, so independent wireless communications are a distinct advantage. Maintenance practitioners also need access to normal business systems, *e.g.*, email and internet, so they select the best proprietary solutions for both business and technical functions.

A key feature of maintenance systems is compatibility. In the past, mistakes were made in the selection of both hardware and software, leading to clumsy or impossible connections between technical systems and management systems. It is now a prerequisite that hardware and software not only be compatible now, but that legacy and future systems will connect as smoothly as possible. In maintaining engineering capital equipment, we expect a long life; therefore it is important that the design and installation data, the spare parts data, the service history and condition monitoring data are kept accessible, possibly for decades. We know that such systems will outlive computing devices and software versions. The current generation of lightweight PC notebooks, coupled with USB interfaces for hardware, and WiFi or 3G modem for internet, is capable of local processing and remote database access, but we can be sure that it will be superseded.

A major advantage of interconnecting, standard systems is that the specialist providers, of software, hardware and analysis, can concentrate on their core functions. For example, it is no longer necessary for a provider of multi-channel vibration analysis to specify and build the data acquisition hardware and computer platform. Similarly it is not necessary to be concerned about the communication of actions arising from the vibration analysis, for example to raise a job to correct the source of the vibration. New systems provide functions, which were formerly only available in desktop computing, on a hand-held PDA (personal digital assistant).

Networking is particularly influential in the design of the state of the art for e-maintenance. The ubiquitous nature of the internet has changed our perception of how a network functions; it is no longer a proprietary connection for a specific purpose, tied to hardware, but rather it is used for all our information needs and is available by wireless almost anywhere. The question of security is answered by a range of standard technologies, for example encryption and secure server functions, used in credit card transactions. Many maintenance software systems host all their data on a web site to maximise accessibility; secure access is a straightforward solution.

2.5.2 Measurement Tools and Services

Condition monitoring uses a range of methods to estimate the health of machines and processes, with the aim of confirming health or scheduling maintenance action prior to failure. This section briefly reviews some of the popular techniques available, but there are many more techniques for specific measurement problems.

Monitoring of process parameters – sometimes the best indicator of the condition of a machine is its performance, *e.g.*, pressure, flow rate and energy consumption of a pump. Unfortunately, robust machines can deliver normal performance up to a point very close to catastrophic failure, so specific measured parameters such as those below are adopted.

Vibration analysis measures the acceleration, velocity or displacement of moving mechanical components, sometimes directly, but more often at an available surface that gives an indication of internal events, without disrupting processes or containment. The vibration may lead to audible noise, excessive stresses and subsequent failures. Its measurement and analysis can detect a very wide range of common machine faults. The raw vibration data is typically processed into the following:

- Overall vibration levels and levels of selected bands, consistent with known fault types.
- Frequency spectra and other analysis techniques seeking insight into many specific faults; many specialised techniques are available for sophisticated diagnostics, *e.g.*, in turbines.
- High frequency emission for rolling element bearings, sometimes processed using event counting and thresholding, and sometimes simply banded.

Acoustic emissions (AE) measure the elastic waves passing through structures, and sometimes through the air, arising from events and continuous processes such as friction. Analysis of the strength and pattern of continuous and burst AE gives an early indication of faults in mechanical components and structures, with some important capabilities for low speed machines, *e.g.*, in bearings and gears at speeds below the capabilities of vibration analysis and large, loaded structures such as bridges and cranes.

Temperature monitoring – sensors of varying sophistication and cost can detect the temperature of electrical equipment, coolants, lubricants and mechanical components. Simple sensors measure specific points in, or on the surface of, a system.

Infrared analysis – emissions in the infrared region of light are indicative of high temperatures. The equipment available for measuring the infrared emission usually works in real time, so the dynamic changes can be examined immediately. The method is useful for locating "hot spots" over a wide physical area and can be used for the exterior of buildings, pipe work and ducting, mechanical systems and electrical/electronic systems. Sophisticated diagnostics can be conducted with thermal imaging over a wide area. Thermal imaging has made significant penetration into new markets over the last ten years, with lower cost and with increasing expertise.

Lubricant analysis – the additives, contaminants and debris reveal the service condition of the lubricant itself, avoiding unnecessary oil changes, and also the level of wear particles from rubbing surfaces in the machinery. The technique is effective for slow moving and reciprocating machinery where vibration techniques are sometimes less effective.

Leak detection – methods from "soap and water" to ultrasonic and tracer gas techniques can detect minute leaks.

Corrosion monitoring – electrical resistance and potential techniques, hydrogen detection, sacrificial coupon and bore holes can be used to measure corrosion and its varying rates of progress during production runs.

Crack detection – many non-destructive testing (NDT) methods may be used for crack detection, *e.g.*, dye penetrants, flux testing, and ultrasonics.

2.5.3 Portable Instruments

Most measurement processes benefit from real-time access to their measured environment, as opposed to sampling followed by laboratory measurement. In industrial environments, the access for large instrumentation and power supplies has always been limited, but "portable" implies that a trolley is no longer required. The benefit of small, self powered instruments is clear. However, portability is a relative measure: a device that is too heavy to carry may still gain huge advantage by being packaged so that it can be moved out of a laboratory. Devices that are too large and heavy for aircraft may be quite acceptable on a ship.

Portable instruments generally carry a range of basic functions. Transducers have to be small enough to handle and must be powered by batteries for some hours. The simplest instruments have a very simple display, such as a meter or LED indicator, but typically we expect some local analogue and/or digital processing, followed by data storage and ability to transmit the data to a PC. Communications can include wired and wireless upload and download, in a variety of standards. Some instruments are equipped with significant power in digital signal processing (DSP). The versatility of such platforms means they may find application in many fields in science and medicine, as well as engineering.

Process parameters tend to be connected to existing monitoring systems, so they only need portable instrumentation in unusual circumstances. Most condition monitoring parameters are not connected to permanent systems, so they need portable data collection.

Vibration analysis tends to use piezoelectric accelerometers, although other transducers are available. Input analogue electronics includes filters and amplifiers. Sampling is usually conducted at a high rate (up to about 50 kHz), prior to a range of DSP, but typically the fast Fourier transform, to produce a frequency spectrum. A range of post-processing can be applied for diagnostics. The raw signals and spectra, and a range of other derived measurements, are stored. Display is typically a graphical array, up to full PC screen resolution, but often smaller, to save space. PDA devices, fitted with input electronics, can now offer similar capability.

AE measurements use some of the same power supply, processing, communication and display devices but need specific input electronics. Portable devices use analogue processing, in simple terms, to resolve burst events from background continuous emission and then to measure overall amplitude and severity.

Simple temperature devices are common. The time constants or lags associated with engineering equipment mean that slow measurements (in the order of 1 s) are sufficient. Hence the data collection and display is also simple. Thermal imaging requires much more sophisticated instrumentation. The camera is a typically a cooled charge couple device (CCD) array, which produces a high resolution image at a high data rate. The resolution tends to lag the typical television camera a little in terms of development, but the image can be at least 640×480 pixels, at a rate of 25 frames *per* second. Input electronics allow real-time contrast mapping to stretch or compress the display resolution across the temperature range of interest, achieving temperature sensitivity of up to 0.1°C. Frame and video sequence storage is then possible on typical video storage devices. The technology has benefited directly from advances in domestic portable video devices, including its reduction in size.

Lubricant and wear debris analysis are still emerging from the laboratory at the time of writing this. Portable devices are possible, but tend to require relatively large transducers, including a wide range of fluid handling mechanisms and optics. There is no single technology that provides a universal measure for lubricant characteristics or wear debris features, so the benefits of mass production have less impact. The more popular measures, *e.g.*, water in lubricant or fuel, have seen greater reduction in size of instrument. Larger instruments, *e.g.*, particles counters, are "transportable", their weight arises from high pressure sampling valves, vessels and pipe work. Sampling of fluids and particulates using magnets, filters and bottles is an important complementary alternative.

2.5.4 Laboratory-based Services

Many practitioners use the services of experts and laboratories. It is not always possible to develop low cost and portable instruments. The cost of the development is amortised only if sufficient goods can be sold. Some instruments are fundamentally large, *e.g.*, scanning electron microscopes or mass spectrometers. Laboratory-based services offer maintenance practitioners an industrial service using scientific principles. A very wide range of measurements is available, but as an example, the detail available in laboratory-based tribological analysis of wear debris far exceeds that available from portable methods

The cost of undertaking routine laboratory-based services can be quite reasonable compared to owning instruments and hiring staff. The sampling process has to be carefully managed, for example the data collection must avoid corruption, such as may occur in bottle sampling of fluids.

The return of measured data, with diagnostic and prognostic reports, has migrated to electronic format, and service organisations can return data in the correct format for direct uploading to the client's own maintenance management system. MIMOSA compliance has been particularly important in this respect.

2.6 New Paradigms: Customisation and Sustainability

An important development in maintenance thinking is the business model of ownership. In moving towards shared risk, the ownership of capital equipment can change. In order to produce a product or service, it is not necessary to own the equipment – it may be paid for in a range of methods and agreements. There is an increasing tendency for the manufacturers of equipment to retain ownership of their own products. Traditionally, manufacturers made money from selling spare parts. As we have seen from our examination of maintenance strategy above, replacement of spare parts can be effective, but we would prefer to move to proactive and predictive maintenance, which will minimise the use of spare parts. If manufacturers own their own products, then they also wish to minimise the use of spare parts. The focus moves to the maintenance of the service provided by the product. For example, the owner of a fleet of cars requires a specified transport service at a suitable cost. The car manufacturer is principally interested in selling new cars, and as a secondary concern, spare parts. However, if the manufacturer continues to own the cars, then the interest in spares is to minimise their use and cost.

Power plant manufacturers have been innovative in this regard. The complexity of diesel or gas turbine power plant is of little concern to the user. The current challenge is to make hybrid systems like combined heat and power, or increasingly renewable energy sourced power, not only as reliable as simpler power plant, but with the same level of convenience. Remote condition monitoring has ensured that a reliable service can be offered, without the need for the user to acquire specialist skills. Continued ownership of the plant has increased the innovation in supporting products in the field and underlined the importance of maintenance as an essential part of ownership.

The sub-contract of maintenance services has a similar model. Some of the most difficult condition monitoring problems are well served by this approach: integrated services for rotating machines, for example, now include provision of specialist bearings, lubrication and monitoring services. This means that a paper producer, for example, can concentrate on making paper, instead of managing the replacement of bearings and lubricants. The rollers in the paper machine are still required to be located with minimum friction, so the integrated service can concentrate on providing that function.

Life cycle costing is not yet universal, and it has been observed that the reason is that the average lifetime of capital equipment considerably exceeds the average period-of-office of a CEO! Notwithstanding this truism, the cost of maintenance is now regarded as a major input to investment decisions. The costs of energy and maintenance far outweigh the initial purchase price in many situations. Hence the maintenance of equipment and the monitoring of its operating efficiency will be major contributors to sustainability.

The idea of a standard product is also changing. The biggest problems in maintenance do not arise from mass-produced items, but from bespoke ones. In the maritime sector, for example, where preventive maintenance is critical, the nature of small production runs and mobile assets means that spare parts are a special problem. They are frequently heavy and bulky, and parts must be held in strategic ports worldwide or on board the ship. Unlike vehicles that are manufactured by a handful of companies in long production runs with standardised components, ships are built to many specialised designs in small numbers by shipyards all over the world with little commonality. In the commercial marine sector alone the Lloyd's Register/Fairplay database lists 160,000 vessels above 100 gross tonnes. The average age of the world's fleet is around 18 years. Consequently, the availability of spare parts is a constant global challenge.

The mobility of expertise, and the sub-contract or outsourcing of skills, is considerably enhanced by e-maintenance. Clearly, the initial building of good relationships is still fostered by face-to-face meetings, but electronic communication allows the rapid transfer of documents, data, images, video, software updates and training. This means that expertise can be mobile across continents and time zones, and can have global impact. With highly mobile assets such as ships and land vehicles, which may have relatively low skilled crews, the outsourcing of maintenance and repair activity is common. The work is undertaken by local repair companies, specialist maintainers and technologists, and so-called "flying crew", highly-skilled specialists who are temporarily transported to the site.

A major future influence will be the contribution that maintenance makes to sustainability. The concept of life extension, and maximising reliability at minimum cost, is the bread and butter of maintenance, and such efforts add genuine competitiveness to businesses. However, the new challenge of our time is carbon reduction. The carbon cost of engineered systems is high during their lifetimes, comprising the embedded carbon of manufacturing and the carbon used during operation. Maintenance has significant contributions to make, by avoiding unnecessary replacement and by supporting efficient operation (Starr *et al.* 2007).

2.7 New Developments in Decision Making

Decisions in industrial maintenance involve risks to equipment, expenditure and personnel, and yet they are based on inaccurate data. Technical inputs such as estimates of machine health, using sophisticated measurements and signal processing, must be combined with vague information about risk and cost, in a dynamic assessment aimed at an optimal long term outcome such as maximum reliability or minimum cost. The decision is typically in the hands of a human, who must weigh technical risk against limited resources, taking into account physical limitations, the requirements of the law, and the "political" implications of the conflicting decisions. Research work aims to provide automated assistance in the decision making process by using data fusion methodology to combine dissimilar quantities and information, and by coping with missing data. The computed recommendations are aimed at the human, who must remain in control of the process (Esteban *et al.* 2005).

The problem is a top-level decision, which draws on lower-level data fusion problems, converting data to information and then to decisions, and which might combine measured data and knowledge. The problem has several areas of uncertainty:

- The technical alert has varying degrees of confidence and urgency, usually unknown in the field.
- The cost data may be difficult to obtain or may be out of date.
- The criticality or risk data may be difficult to obtain, out of date or invalid for the current equipment arrangement.
- Any of the input data may be unsynchronised or absent.

Combination of the three input streams, based on cost-based criticality, produces a ranking based on the risk of expenditure. The ranking is optimised to minimise consequential cost. The method provides an audit trail for decision-making, which is important in many industrial sectors (Moore and Starr. 2006).

2.8 New Developments in Technological Tools

Maintenance benefits from a wide range of technologies, which have often been developed for a different purpose. A number of new off-the-shelf tools are described below, which are likely to create significant advances, including wireless sensors, miniaturisation and MEMS (micro electro-mechanical systems), a range of new disruptive technologies, and pervasive sensing and intelligence.

2.8.1 Wireless Sensors

Wireless communication is experiencing explosive growth in many areas of electronics, and it is clear that there are some fundamental advantages in shedding both the cables and the plugs associated with conventional communications. The cost of devices for consumer electronics, such as mobile telephones, toys and computer peripherals, has plummeted. Distributed wireless monitoring is now a reality and it is developing fast.

The systems architecture of a wireless sensor is the starting point. There are parallels with, and distinct differences to, older portable devices used for monitoring and diagnostics, as shown in Table 2.3.

Low cost changes the thinking in design (Albarbar *et al.* 2007). Multiple measurands are possible and desirable in a wireless sensor. Connectivity is critical, as part of the wider e-maintenance network. Commercial off-the-shelf (COTS) radio platforms offer high functionality, integration, and low cost, exploiting radio standards such as Bluetooth (IEEE 802.15.1) and Zigbee, an extension to IEEE 802.15.4, which is specially intended for sensors (Pietruszkiewicz *et al.* 2007). Size, processing and power supplies are developing rapidly. The sensor's internal processing capacity is capable of accommodating basic diagnostic functions, trending, prognosis and decision making, and communicating health status indications.

The software for performing remote, automated, distributed monitoring, in a robust fashion without recourse to human intervention, remains a challenge for the success of such a system. The embedding of knowledge and procedure will be important for the penetration of the new devices into the unmanned monitoring of future applications.

Table 2.3 Characteristics of portable instruments and wireless sensors

Characteristic	Portable instrument	Wireless sensor
Transducer	High specification sensors. Sampling generates rich but massive data. The instrument diagnoses as well as detects a fault. Most transducers have a wired connection to the instrument.	Low cost miniature devices embedded in the instrument. Permanently fixed, so can sample frequently. Lower specification. Wired connections and terminations are not required.
Signal conditioning and interfacing	The transducer requires the right voltage(s) input and arrangement of its output to match with subsequent processing.	Necessary, but tailored to the permanent transducer connections
Hardware filter electronics	Important for extraction of features from the rich signal or removing unwanted parts of a signal	Filters may still be necessary for some transducers
Analogue to digital conversion (ADC)	In most instruments the signal is sampled and digitised for further processing. The ADC has to run fast: in some applications this could go up to MHz.	Lower specification, tailored to the application.
Digital signal processing (DSP)	Most instruments modify the signal to extract interesting features, enabling better resolution of faults. A wide range of algorithms may be provided for the skilled operator to apply.	Processing is defined in advance. The DSP can be tailored to a minimum required for the fault resolution. For some applications it will not be required.
Computer platform	Micro processor and local memory; fairly powerful. Human supervision takes care of power management and error states.	More limited capacity, but does not have to respond to a human operator – "real time" is permissibly slower. Power management is important. Automatic software must handle start-up, monitoring, messages, and potential error states.
Communications	Links to host computer systems: wired and wireless links *e.g.*, RS232, infra-red and Bluetooth	Wireless standard is important, especially in the context of power consumption and longevity. Some miniaturised wired connections might be included for programming.
Display	Large colour screen with graphical user interface	No display required. Some devices include LEDs for checking status.

Table 2.3 (continued)

Characteristic	Portable instrument	Wireless sensor
Packaging	A substantial box suitable for industrial conditions; strong terminals for the electronic connections	The device only needs simple protection from the environment. Contact of the transducers with surfaces is important in some cases.
Battery	Rechargeable or long-life cells, housed within the package. Recharging or replacement is practical.	Power management is critical. Wired recharging is not practical. Renewable sources are a practical proposition.
Operating conditions	Human operator makes most instructions.	Operates autonomously most of the time.
Diagnosis requirements	Detailed information	Simple fault detection, monitoring and alerting
Cost	Up to €10k depending on package	Up to €100, but tending to reduce – "disposable"

2.8.2 Miniaturisation, Cost Reduction and MEMS

Miniaturisation can be simply defined as smaller than the last attempt; it is a relative measure. The reduction in size, with corresponding power requirements, is familiar in many walks of life. Computing, for example, has reduced from a roomfull, to a box, to a laptop, and now to a pocket-sized device. One manufacturer has promised a holographic keyboard, and USB data sticks have reduced to the size of a postage stamp. Almost every device used in maintenance has reduced in size; there are, however, some human limitations. Display devices, such as screen and meters, need to be big enough to read, key pads need to be large enough for fingers, sometimes wearing gloves. In some industries, the favoured package for sensors has taken impact and other abuse into account, even if the transduction element is tiny. The reliable electrical connection has depended on relatively large connectors and cables, even for tiny signal currents. The use of wireless devices has made considerable advance in package size; if signal connections are no longer necessary, a large contribution to mass and volume is removed. The next challenge is the reduction in size, and removal of connections, for replaceable batteries, with local "power harvesting" as the alternative.

A critical part of size reduction is also cost reduction. Some materials and processes only become viable in a large quantity, and the maintenance sector has often benefited from a technology advance, when it has had a more popular application. Distinctly different decisions are made between capital equipment, even at the low cost end, and consumable items. Devices that are disposable may have very different expectations in life, target applications, packaging, selection of materials and technology choice. However, the acceptable price of a disposable item can be surprisingly high in some industries. This means that the threshold for budgeting influences the strategic decisions for sensor architecture in many applications. If the price is acceptably low, short life is acceptable; light weight packaging, coupled with battery power, can allow access to measurement problems that would be inaccessible to expensive, longer life solutions. Multiple low cost sensors can allow access to highly-resolved measurements over a wide area, with local communication between sensors. Damage to, or loss of, a sensor is of little consequence, so higher risks are acceptable to such a measurement system.

MEMS have seen a great deal of development in recent years, and many products are now on the marketplace (Nexus 2009). The key advances are in the miniaturisation of the transduction elements, in the measurement of, *e.g.*, strain, pressure, acceleration, angular rate, displacement, force, ultrasonics, flow, temperature, optoelectronic photonic properties and magnetic properties, amongst others. Some packages of MEMS sensors include multiple transducer elements, for different parameters, on the same substrate. The advantages of MEMS lie in the size and unit costs of mass-produced items; for example, the ADXL105 accelerometer measures about $10 \times 8 \times 4$ mm and can be configured to work up to about 10 kHz with sensitivities of 250 to 1000 mV/g and temperature measurement, at the cost of a few Euros. The extra functionality offered by local processing, either in the package or in associated circuitry adds a real boost to "smart" sensing. However, the power and communications requirements still provide some constraints. Transducers are generally already small; it is the packages and connectors that are large. Reducing the size of the transducer alone is not sufficient to reduce the size of the whole package.

Overall, size reduction is a relative measure: what is large in a car or an aircraft may be very small on a ship. We can be certain, however, that technological developments in related fields will continue to offer further size reduction in maintenance devices.

Albarbar reported the details of the transducer selection and signal conditioning (Albarbar *et al.* 2007). The work investigated a range of low cost sensing elements at component level, for integration with the COTS platform. The requirement for temperature, pressure and vibration has led to testing of a variety of piezo and MEMS devices. The mass production of such devices has offered high quality at low cost, but requires some expertise to interface the devices and to mount them in suitable packages. The wireless sensor does not need to be handled, so the package does not have to be excessively robust.

The basic construction of a piezoresistive pressure micro sensor, using a flexible silicon membrane as the sensing element, is shown in Figure 2.3. MEMS accelerometers are generally divided into two main types: piezoresistive and capacitive-based accelerometers. The schematic of a piezoresistive MEMS accelerometer is shown in Figure 2.4 (Plaza *et al.* 2002). These accelerometers generally consist of a proof-mass suspended by a "spring", which in MEMS is usually a cantilever or beam. When the device is subjected to acceleration, the inertia of the mass causes changes in the gap between it and the bulk of the device. Vibration sensors can operate using the same principle (Partridge *et al.* 2000). The mass may move out of the plane of the silicon wafer or in the plane (as is common in surface micro machined devices) (Xie *et al.* 2000). The piezoresistive accelerometer incorporates a piezoresistor on a cantilever beam structure, as shown in Figure 2.4. The electric signal generated from the piezoresistive patch and the bulk device due vibration is proportional to the acceleration of the vibrating object. Capacitive-based MEMS accelerometers measure changes of the capacitance between a proof mass and a fixed conductive electrode separated by a narrow gap.

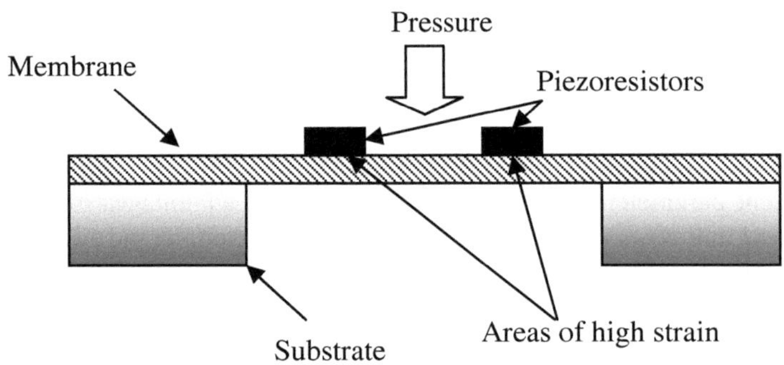

Figure 2.3 Piezoresistive pressure micro sensor based on membrane structure

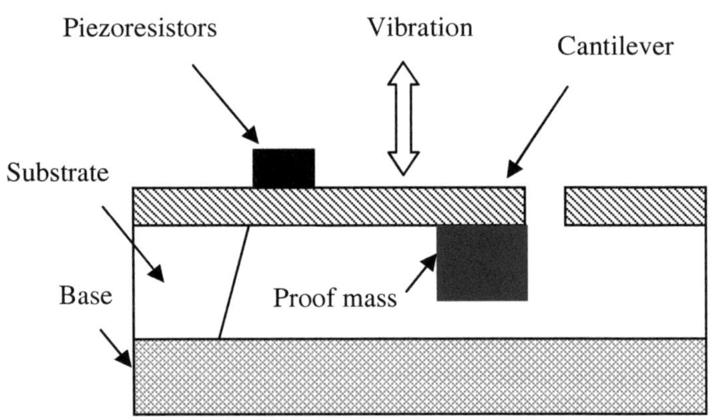

Figure 2.4 Typical piezoresistive micro accelerometer using the cantilever design

The results from a calibration test of the transfer function of a MEMS accelerometer and an integrated circuit piezoelectric (ICP) accelerometer are of interest. Figure 2.5 illustrates graphically the response from 0–12 kHz, under test on a shaker driven with white noise and compared to a conventional B&K piezoelectric accelerometer (Albarbar *et al.* 2007, 2009). Note the peaks at 3.7 kHz, probably associated with the mounting assembly, and at 10.5 kHz, a resonance above which the signal dies away fairly rapidly. Given the typical running speeds of electrical and mechanical machines, these bandwidths allow access to most of the features of interest in fault detection. Note that these bandwidths were achieved by stiff mounting direct to a metal surface screwed to the shaker.

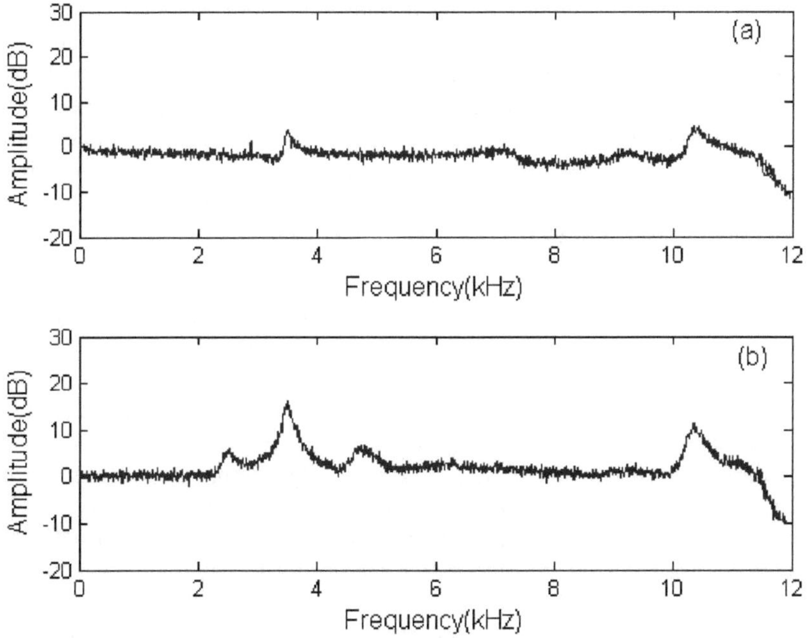

Figure 2.5 Transfer function for: (a) MEMS, and (b) ICP accelerometers

2.8.3 Disruptive Technologies and the Future

A disruptive technology is one that changes the game; for example, internet trading has changed the way we do business, forever. Maintenance management and technology has benefited hugely from several disruptive technologies, even if they were intended for completely different target audiences, *e.g.*, personal computing profoundly changed the nature of maintenance management software and its uptake.

The ability to predict the future is perhaps restricted to fortune-tellers, but the authors claim some knowledge of condition monitoring prognostics, which is in a similar vein! In this section it is our purpose to make some speculative remarks about the possible influencing technologies and ideas that may change the way in which we do maintenance in the future.

Nanotechnology has flooded the research literature and popular airwaves, with some questionable advances and speculative claims. Events at a nanometre scale are already of great interest to maintenance practitioners but they are difficult to measure: fracture mechanics, considering the sub-surface cracking in bearing surfaces; lubricant properties, such as metallic particulate and additives. Of particular interest is the spherically-structured Buckminster Fullerene C60, which has already been evaluated as a dry lubricant by NASA and academic laboratories. The carbon nanotube structure is also likely to lead to interesting lubricant properties. Nanostructures are also likely to provide structures, *e.g.*, for the delivery of additives to extreme conditions over a long period, with a strong connection with medical applications. Many biological structures are measured in nanometres, and the ability of using tiny structures as treatments and tell-tale evidence of wear, or of some other physical interference, *e.g.*, of mechanical seals, may prove successful.

Algorithms: The ability to judge a good measurement or decision, from a bad one, is a major challenge for maintenance practitioners. The raw data is influenced by a very wide range of interference and uncertainty. Researchers expend great effort to improve such numerical estimation and classification problems, but the solutions are hard to implement. Algorithms that have been demonstrated on supercomputing platforms in the past, *e.g.*, genetic algorithms were demonstrated by NASA in the 1950s, can now be executed on a typical PC, and with a bit of patience, on sensor and mobile telephone platforms. Moore's law suggests that processing power doubles every year; hence the ability to have whatever processing you require, in "real time" (*i.e.*, fast enough), will be achieved sooner or later (Schaller 1997). What remains is the choice of the *right* processing for the job: numerical problems do not all yield to neural networks, and classification problems do not all yield to knowledge-based systems. A good deal of learning will be necessary to implement the solutions required, but there will be no shortage of distributed processing capability. An analogous example is the JPEG algorithm for photographic compression; it is so good and so ubiquitous that we forget its presence and can no longer remember being without it (ISO 1992). In the future we may regard the Fourier spectrum as an ancient precursor of multi-dimensional data-mining displays, given that Fourier died in 1830 and the fast Fourier transform was built into hardware almost 50 years ago. On a smaller scale, smart sensors with automatic alarms are already with us. There are, of course, new risks in using methods with which we are not familiar; human supervisors of automated systems have to be wary of false alarms. Image processing algorithms will almost certainly make an impact in condition monitoring; many sensors already produce multi-dimensional data, some of which is a clear picture (*e.g.*, thermal images, wear debris particulate micrographs), some a reconstructed picture (*e.g.*, scans from non-destructive testing, such as ultrasonic thickness testing) and some of which is multi-dimensional mapping, not a picture (*e.g.*, waterfall plots or wavelet transforms). Most of these data forms will be difficult to interpret and will need significant intelligent automation to yield useful results in the field or in automatic processes.

Self healing and robustness: an inviting idea is that a device or component *knows* it is damaged, and instigates a healing process, rather like a wound on a finger. We have some limited experiences of such processes in action, for example the rolling action of balls in a bearing causes some surface treatment to occur where spalling has occurred. In redundant systems, "hot" switching keeps a system working. Extreme pressure (EP) additives in lubricants cause localised chemical treatment in response to damaging conditions in gears. However, biological processes completely repair a problem. We could conceive of structural repairs to composites, for examples, where biological agents penetrate delamination to make complete repair. Nanocomposites, passively suspended in fluids, could form the matrix for structures which rebuild damaged seals, retaining valuable lubricants. Automated systems with very low human supervision could arrange the replacement of robotic mechanisms or tools by other robotic agents. Other highly redundant approaches could give the impression of self-healing characteristics, *e.g.*, clusters of smart sensors for industrial or military purposes can already re-route their *ad-hoc* "mesh" networks (*e.g.*, Zigbee) if a node is lost. Robustness is the ability to give service or response under harsh conditions or partial failure. Physically, redundant systems are robust because they replace a critical component or system with another one. In control and monitoring, a robust system may have more than one approach to a single problem, *e.g.*, with a range of potential algorithms, and in some cases different hardware and software compilers to monitor a process. In some of the more notorious problems in condition monitoring, a robust approach is to use a range of solutions which raise alarms about all the potential faults which could occur. In helicopters, for example, a range of vibration features are coupled with a range of lubricant/wear debris features and many other usage and monitoring parameters. However, failures do still occur. New low cost, distributed hardware will allow many more parameters to be monitored in parallel, with sufficient processing power to monitor any features desired, using sophisticated voting and checking algorithms to reduce false alarms, the curse of automated systems.

2.8.4 Pervasive Sensing and Intelligence

Telecommunications in whatever form, data and information will be universal; it will not be a question of *whether* you can get data, but what you will do with it and how you will interpret its information, before rapid and substantial reduction or disposal. For example, a car built in 1990 had approximately 10 sensors on board; a similar model built in 2005 had approximately 50 sensors, with built-in diagnostic software.

Algorithms encompassing adaptability and intelligence are all around us. We do not have to instruct a mobile telephone to pass its connection from one antenna to the next, nor do we have to tell it to manage its battery power. We regard such packages of functions as normal, entry-level features. It is also normal that the telephone is equipped with music and other media recording and playback, and a camera, and that it fits easily in the pocket. The sensor package, and its processing platform, is *pervasive* in that it can be found everywhere.

The connections provided for data, whether computer, telephone services or other, are so commonplace and so competitively priced, that the fitting of hard-wired data networks will become as much a relic as a typewriter.

The host platforms of mobile sensor and telecommunications devices already carry a range of housekeeping functions. In the same way as mobile telephones can download utilities and games, sensor platforms can carry pre-processing algorithms. Standard processors and languages will allow specialist providers to offer a range of utilities relevant to maintenance providers. Many sensors need pre-processing, *e.g.*, for extraction of relevant features from rich data. The jury is probably still out on the benefit of retaining and transmitting *all* the available data. One school of thought is that only events need reporting and the other is that all the data needs transmitting to a *data warehouse* for later processing. The dilemma is that event data may not retain sufficient detail for diagnostics, but the data warehouse may become too large to use effectively. One can be sure, however, that smart sensors will become smarter; for example, a low-grade event such as a warning could automatically trigger more intensive measurement.

The march towards smarter, more integrated, pocket devices such as smart phones and PDAs (really small computers) is as inexorable as their increasing connectivity. Integrated packages already include remote access to records, maintenance instructions, personnel, drawings, and diagnostics. Users will become increasingly impatient if such devices offer anything less than they expect at their desk. Networks of sensing and PDA-type devices will allow a small number of people to monitor a wide area.

The architecture of large-scale systems has a number of unresolved issues. Most systems use a central computerised control host, often based on a database program. However, a schism exists between the users of distributed processing and the centralised approach. Distributed systems allow local handling of bulk data, with corresponding small data packets to the centre. Most of the routinely monitored data is then discarded, transmitting and retaining the chosen key features and events. The centralised approach streams all the available data to a data warehouse, retaining all of it for later inspection, hence losing nothing. Features and events are easily uncovered by central searches. The storage and transmission costs, however, are considerable overheads, and it is difficult to know what is worth keeping: simple parameters, large samples like AE data, or continuous video data? The moment we think of down-sampling (reducing the data in some way) the chief benefit of the centralised system is lost. The standardisation of data structures and their storage is certainly not fully resolved. Two major EU Framework 6 research projects (TATEM and DYNAMITE, for information see the Cordis website, European Commission 2009) used MIMOSA as their data standard (ISO 2002) with its common relational interface schema (CRIS), and many popular commercial software products adopted the standard in its fledgling form over ten years ago. However, it is clear that MIMOSA's penetration is far from universal, and not all user groups perceive it as a core requirement. Despite its low entry cost and open systems approach, some of the biggest system providers resolutely stick to proprietary standards.

2.9 Conclusions

Maintenance has a sophisticated approach to performance optimisation. In the context of the competitive industries that benefit from advanced maintenance, the availability and reliability of processes and systems are differentiating factors. These factors have promoted the concept of maintenance from one of the underpinning cost centres to become an essential part of leading profit centres.

Maintenance practitioners and systems providers quickly adopt new technologies to exceed their initially conceived value and extent, such as novel use of portable computing. Traditional approaches to technology exploitation required economies of scale, such as employed specialists for database management and use of instruments, but new business models have allowed low entry cost, coupled with high quality remote services.

Future implementation of maintenance systems will see greater integration of business and technical systems, with more intelligent use of collected data. They will protect users against change of personnel, with the inherent loss of their learning, and allow better informed choices for decision makers. Technological data collection, with its attendant signal processing for extraction of information from raw data, will embed an increasing amount of intelligent processing at source, while increasing the speed of communication through wearable computing and robust mesh networking. Sophisticated strategies are under development for mobile plant, vehicles and aircraft, to allow independent local processing with intermittent communication to a central system for parts ordering and work scheduling.

Limitations to progress include the standardisation of system and communication components, and training. Certain hardware, *e.g.*, mobile computing, has a shelf life considerably shorter than high capital engineering equipment. It becomes rapidly obsolete and cannot be replaced without upgrading software as well as hardware. Much of the required communications networks exist, but they are not universal, and the business models needed to access them are still under development, *e.g.*, access to cellular telephone and WiFi hotspots. Some businesses are sensitive to security arrangements for public networks. The use of such wide-ranging systems, from detailed technical programming of smart sensors through to management of information leading to business-critical maintenance decisions, requires some exceptional people to run it, their very mobility will cause them to change employer and job function, so the capture of their knowledge and the training of new people will continue to be essential for the exploitation of advanced maintenance. The data from embedded, smart sensors is likely to grow – whether we want it or not – and the problems of managing the data will also grow.

The exciting potential of condition-based maintenance in high-risk environments, *e.g.*, aircraft and maritime industries, probably offers greater benefits than in any other industry. Pilot projects currently underway will certainly enhance the understanding and build the confidence to extend the strategy and techniques further into those industries. The benefits will need to be achieved in safety and security as much as economics in order to persuade high-risk operators to expand applications. The focus and direction will tend to be seen on new equipment and will certainly be carried out by the high-tech, sophisticated operators.

References

Albarbar A, Pietruszkiewicz R, Starr A (2007) Towards the implementation of integrated multi-measurand wireless monitoring system. Proceedings 2nd World Congress on Engineering Asset Management (WCEAM) June 2007, Harrogate, ISBN 978-1-901-892-22-2

Albarbar A, Sinha J, Starr A (2009) Performance evaluation of MEMS accelerometers. Measurement 42:790–795

Al-Najjar B, Wang W (2001) A conceptual model for fault detection and decision making for rolling element bearings in paper mills. J Quality Maintenance Engg 7:192–206, ISSN 1355-2511

Esteban J, Starr AG, Willetts R, Hannah P, Bryanston-Cross P (2005) A review of data fusion models and architectures: towards engineering guidelines. Neural Computing and Applications, Springer, London, 14:273–281, ISSN 0941-0643 (Paper) 1433-3058 (online)

European Commission (2009) http://cordis.europa.eu/search/, last accessed September 2009

Holmberg K (2001) New techniques for competitive reliability. Int. J. COMADEM 4:41-46

Holmberg K, Helle A (2008) Tribology as basis for machinery condition diagnostics and prognostics. Int J Perform. Engg 4:255–269

ISO (1992) ISO/IEC IS 10918-1 – Information technology – Digital compression and coding of continuous-tone still images – Requirements and guidelines

ISO (2002) ISO 13374-1 – Condition monitoring and diagnostics of machines – Data processing, communication and presentation – Part 1: General guidelines

Jardine A, Makis V, Banjevic D, Braticevic D, Ennis M (1998) Decision optimization model for condition-based maintenance. J Quality Maintenance Engg 4:115–121, ISSN 1355-2511

Lenahan T (2005) Turnaround, shutdown and outage management: effective planning and step-by-step execution of planned maintenance operations. Butterworth-Heinemann, London, ISBN 0750667877

Moore WJ, Starr AG (2006) An intelligent maintenance system for continuous cost-based prioritisation of maintenance activities. Comput Ind 57:595–606, ISSN 0166-3615

Moubray J (2001) Reliability-centered maintenance. Industrial Press, New York, ISBN 0831131462

Nexus (2009) http://www.enablingmnt.com/html/nexus_market_report.html, last accessed Sept 2009

Partridge A, Reynolds J, Chui B, Chow E, Fitzgerald A, Zhang L, Maluf N, Kenny T (2000) A high-performance planar piezoresistive accelerometer. J Microelectromech Syst. 9:58–66
Pietruszkiewicz R, Starr A, Albarbar A, Tiplady K (2007) Development of the wireless intelligent sensors for condition monitoring systems. Proceedings 2nd World Congress on Engineering Asset Management (WCEAM) June 2007, Harrogate, ISBN 978-1-901-892-22-2
Plaza J, Collado A, Cabruja E, Esteve J (2002) Piezoresistive accelerometers for MCM package. J Microelectromech Syst 11:794–801
Schaller RR (1997) Moore's law: past, present and future. Spectrum, IEEE, 34:52–59
Starr A, Albarbar A, Pietruszkiewicz R, Mekid S (2007) Developments in wireless sensing for condition monitoring. Condition Monitor, Coxmoor, ISSN 0268-8050
Starr A, Bevis K (2009) The role of education in maintenance: the pathway to a sustainable future. Proc. WCEAM Athens (in press)
Willmott P, McCarthy D (2000) TPM: A route to world class performance. Butterworth-Heinemann, Oxford, ISBN 0750644478
Xie H, Fedder G (2000) CMOS z-axis capacitive accelerometer with comb-finger sensing. Proc. IEEE Micro Electro Mechanical Systems (MEMS), 496–501

Chapter 3
Information and Communication Technologies Within E-maintenance

Aitor Arnaiz, Benoit Iung, Adam Adgar, Tonu Naks, Avo Tohver, Toomas Tommingas and Eric Levrat

Abstract. This chapter describes the state of the art in information and communication technologies (ICT) related to maintenance and its future trends. Several topics apply, from pure technological advances in acquisition, communication and storage of information, to the identification of advanced information standards for systems interoperability, such as MIMOSA and open system architecture for condition based maintenance (OSA-CBM) (Bengtsson 2004).

The first section introduces the concept of e-maintenance and follows this with a broad review of the state of the art on some ICT technologies. This serves as an introduction to the Dynamite approach presented in the next chapter. A global framework between different systems that forms a "plug & play" basic mode of operation is outlined. The chapter serves as an introduction to forthcoming chapters dealing with specific technologies that have been converted into "capabilities", such as wireless communications, intelligent web services and smart PDAs.

3.1 Introduction

The chapter is set out as follows, Firstly, an introduction to the e-maintenance concept is made, stating the need for e-maintenance solutions, This need is driven by the increased complexity of the information related to the maintenance tasks, which is even more important in CBM and PdM strategies. An identification of the meaning of e-maintenance in this book follows, with special focus on the link between maintenance technologies and services.

Secondly, a review of the state of the art on ICT technologies is provided. Among these, some technologies are pointed out as enabling technologies, in the sense they may provoke sharp changes in the maintenance processes, as well as in the strategies regarding operation and maintenance.

In conclusion, this chapter serves as an introduction to forthcoming chapters dealing with specific technologies that have been converted into "capabilities", such as wireless communications, intelligent web services and smart PDAs.

3.2 Introduction to E-maintenance

With today's growing demands on system productivity, availability and safety, product quality, customer satisfaction and the decrease of profit margins, the importance of the maintenance function has increased (Al-Najjar and Alsyouf 2003, Crespo and Gupta 2006). Indeed the maintenance function plays a critical role in a company's ability to compete on the basis of cost, quality and delivery performance. For example, Westkämpfer (2003) in defining the new paradigm of life cycle management explains that as about 5 to 6% of the product price is spent on maintenance and service yearly, the main demands for maintenance in industrial manufacturing are preventive maintenance and short reaction, low cost of maintenance, upgrading of software and control, and guarantee of output rates and quality. Moreover it was highlighted that for securing a high quality product at a competitive price, an effective maintenance policy is needed to globally enhance the performance *efficiency* of the production process, while fewer failures and better control of the production plant would help minimise pollution and fulfil society's demands.

Thus, in countries where modern maintenance practices have yet to be well adopted by the industry, the potential savings from modern maintenance are massive. These modern and efficient maintenances imply identification of, at least, the root-cause of component failures, reduction of the failures of production systems, elimination of costly unscheduled shutdown maintenances, and improvement of productivity as well as quality. To support this role, the maintenance concept has undergone several major developments leading to proactive considerations, which require changes in transforming traditional "fail and fix" maintenance practices to "predict and prevent" e-maintenance methodology (Lee 2001, Ben-Daya *et al.* 2009, Muller *et al.* 2008). This includes, for instance, potential impact on the service to customer, product quality and cost reduction. The advantage of the latter is that maintenance is performed only when a certain level of equipment deterioration occurs rather than after a specified period of time or usage, from current mean time between failure (MTBF) practices to mean time between degradation (MTBD) technologies.

3.2.1 Maintenance Today: What Are the Main Issues?

As previously summarised, the maintenance function is currently critical for a manufacturing organisation to be able to maintain its competitiveness. Maintenance is changing from a cost centre to a profit centre to check the performances (Al-Najjar and Alsyouf 2003). Indeed, without well-maintained equipment, a plant will be at a disadvantage in a market that requires low-cost products of high quality to be delivered quickly. This means that changes in the production environment have made the maintenance task increasingly complex (Swanson 2003). Higher levels of automation can make diagnosis and repair of equipment more difficult. The high level of capital intensity associated with automated equipment also places greater pressure on the maintenance function to rapidly repair equipment and to prevent failures from occurring.

The complexity for maintenance can thus be related to a lot of factors within the production environment (Swanson 2003):

- Manufacturing diversity (variability of demand patterns and the complexity of the products being produced).
- Process diversity determined by the characteristics of process technology.
- Accessibility to the site of the components or the unsafe situation related to the type of process (*e.g.*, nuclear, aeronautics, space, offshore).
- Growth of information and communication technologies to implement innovative solutions for improving operation and maintenance practice. Complexity has a direct effect on an organisation's information-processing needs. An innovative solution consists of a set of specific components (hardware, software, hybrid) and resources (*e.g.*, applications, services) forming the IT infrastructures for supporting enterprise automation as a whole. Each infrastructure is composed of one or several networks with the servers, workstations, applications, databases but also smart sensors, PDA, *etc.* It is also characterised by its operating principles (wireless infrastructure, highly fault-tolerant, secured, *etc.*) and the concrete implementation of a technological interoperability consisting in deploying the right ICT related to the standards to present, store, exchange, process, communicate, data, information, knowledge, intelligence.
- Maintenance mission accomplishment should be in phase with production environment performances. In that way, maintenance requires the cooperation of, and association with, virtually every department (production, procurement, engineering, accounting, human resources, *etc.*) in the plant, and especially with production. Thus maintenance has to be seen as a major element of a system that will be developed in association with the prime elements of the system-of-interest and as part of the overall system engineering process (see System Engineering Initiative INCOSE, http://www.incose.org).
- The number of maintenance actors (not only conventional actors but also advanced ones) involved in all a life-cycle management oriented approach. Indeed, faced with sustainability aspects, maintenance has to be considered not

only in the production or operation phases but also in all other phases of the life cycle maintenance process (Takata *et al.* 2004). Some of these actors are human and this implies labour diversity with, for example, the use of decentralised maintenance crews. Some of these actors are automated (CMMS, sensors, PLCs, *etc.*). These actors are representative of the strategic level (ERP, maintenance experts), tactical level (CMMS, MES, SCADA) and operational level (PDA, maintenance operator, MEMS, *etc.*).
- Heterogeneity and complexity of the actions supported by each actor. For example, from an observation resulting in general from condition-monitoring done by a sensor or by an operator, it is necessary to analyse what occurred for identification of the failure origin. For this activity various materials, such as models, experience feedback or various documents, are needed.

These factors are today taken more or less well into account in the maintenance strategies implemented in companies. In fact, this strategy choice must result in an optimisation, or better a compromise, between the direct maintenance cost and the indirect maintenance cost resulting from the strategy deployment. This optimisation results in solving more or less adequately each factor.

Swanson (2001) explains that there are three types of maintenance strategies: the reactive strategy (breakdown maintenance), the proactive strategy (preventive and predictive maintenance) and the aggressive strategy (TPM). This synthetic view was extended by Wang (2002) to develop a survey of maintenance policies of deteriorating systems. Traditionally, many companies employed a reactive strategy for maintenance, "fixing machines only when they stopped working". More recently, ICT emergence and the increased sophistication of maintenance personnel have led some companies to replace this type of reactive approach. A proactive strategy for maintenance utilises preventive and predictive maintenance activities that prevent equipment failures. An *aggressive* strategy, like total productive maintenance (TPM), focuses on actually improving the function and design of the production equipment.

CBM is a practice illustrating the predictive strategy and concerns making decisions and performing necessary maintenance tasks based on the detection and monitoring of selected equipment parameters, the interpretation of readings, the reporting of deterioration and the vital warnings of impending failure. Thus this type of strategy is well in phase with most of the complexity factors that require *dynamics* in the decision to face with most of the diversities. For example, scheduled preventive maintenance strategies carried out some time too late in relation to the current status of the potential failure are not easily compatible with this maintenance vision.

CBM is the first step toward e-maintenance practice: In addition to the dynamics it integrates the possibility to make the different remote maintenance actors work as a whole to form a network-based maintenance infrastructure. Indeed, this new philosophy allows the fulfilment of the maintenance global objective depending on a mandatory *collaboration* of knowledge between human and/or automated actors all along the system life cycle.

3.2.2 E-maintenance: Towards a Consensus or a Lot of Different Definitions?

Within the framework of the previously expressed production-maintenance environment, the term e-maintenance emerged in early 2000 and is now a very common term in maintenance related literature. However, it is not yet consistently defined in current maintenance theory and practice as shown by the following different e-maintenance definitions:

"The network that integrates and synchronises the various maintenance and reliability applications to gather and deliver asset information where it is needed. E-maintenance is a subset of e-manufacturing and e-business" (http://www.mtonline.com/articles/1201_mimosa.cfm).

"The ability to monitor plant floor assets, link the production and maintenance operations systems, collect feedbacks from remote customer sites, and integrate it to upper level enterprise applications"(www.imscenter.net).

"Transformation system that enables the manufacturing operations to achieve predictive near-zero-downtime performance as well as to synchronise with the business systems through the use of web-enabled and tether-free (*i.e.*, wireless, web, *etc.*) infotronics technologies" (Lee *et al.* 2006).

"E-maintenance as (the "e" in e-maintenance means) = *excellent* maintenance = *efficient* maintenance (do more with fewer people and less money) + *effective* maintenance (improve RAMS metrics) + *enterprise* maintenance (contribute directly to enterprise performance)" (http://www.mt-online.com/newarticles2/04-00uptime.cfm).

"Maintenance management concept whereby assets are monitored and managed over the Internet. It introduces an unprecedented level of transparency and efficiency into the entire industry" (http://www.devicesworld.net/iscada_applications_maintenance.html).

"E-maintenance is integrating the principles already implemented by telemaintenance (Ben-Daya *et al.* 2009) which are added to the web-services and collaboration principles (Iung *et al.* 2009) to support pro-activity while keeping maintenance as an enterprise process (holistic approach) – integration concept (*i.e.*, IEC/ISO 62264) for optimising performances".

What is the definition most used by the engineers and scientists working in the e-maintenance area? Perhaps the last one, but this is not sure because some initiatives have shown the difficulties to converge towards a unified way of understanding e-maintenance (Iung *et al.* 2004).

Moreover, in addition to these conceptual definitions, some e-maintenance contributors pragmatically consider e-maintenance more as a maintenance strategy (*i.e.*, a management method), maintenance plan (*i.e.*, a structured set of tasks), a maintenance type (such as CBM, RCM, TPM, corrective, preventive, predictive, or proactive) or a maintenance support (*i.e.*, resources, services to carry out maintenance). Some results of these contributions have already been published at least in (a) a special issue on e-maintenance (Iung and Crespo 2006), (b) a review on e-maintenance (Muller *et al.* 2008) and (c) a proposal of an e-maintenance framework (Levrat *et al.* 2008a).

3.2.3 E-maintenance: a Symbiosis Between Maintenance Services and Maintenance Technologies

At the end of so many different definitions of e-maintenance, one important fact is emphasised. E-maintenance emergence is linked with two main factors:

- The appearance of e-technologies allows an increase of maintenance efficiency to optimise maintenance related workflow (*i.e.*, infotronics technologies). e-maintenance support is globally made-up with Intra-Net, Extra-Net and Inter-Net parts. These parts are built from many different e-technologies such as web technology, new sensors, wireless communications, mobile components (*e.g.*, PDA) (Arnaiz *et al.* 2006).
- The need to integrate business performance, which imposes the following requirements on the maintenance area: openness, integration and collaboration with other services of the e-enterprise and introduces a new way of thinking regarding maintenance. This leads one to consider the e-maintenance value chain composed not only with conventional maintenance processes (which are not upgraded because they are re-used in the same way for e-maintenance) and upgraded conventional maintenance processes (from CMMS to e-CMMS, from documentation to e-documentation) but also new processes (new services) that are emerging from e-maintenance requirements such as the business process of prognosis degradation (Jardine *et al.* 2006), or opportunistic maintenance (Levrat *et al.* 2008b).

E-maintenance is "scientifically and technologically" more than a mosaic of models, technologies and standards. It has to be considered as a "system"[1] and the development and integration of such systems (as systems-of-systems) needs numerous interoperations with other systems and objects.

In relation to this system view, through the e of e-maintenance, the pertinent data *vs.* information *vs.* knowledge *vs.* intelligence become available and usable at the right place, at the right time for making the best anticipated maintenance decision all along the product life cycle: the concept of 3R as proposed by Lee *et al.* (2006). Thus, e-maintenance transforms manufacturing companies to a business service to support all customers anywhere and anytime

[1] An integrated set of elements that accomplish a defined objective. These elements include products (hardware, software, firmware), processes, people, information, techniques, facilities, services and other support elements.

Collaboration allows us not only to share and exchange data and information but also knowledge and (e)-intelligence and this happens among all the actors (human, units, departments), as well as along the entire product life cycle. Thus an academic challenge is now to structure e-maintenance knowledge in order to define a new framework and more precisely a new scientific discipline devoted to "e-maintenance".

Finally, it could be considered that the emergence is based on the prevalence of the need (the new way of thinking about maintenance, the new maintenance needs) compared with the e-technologies. Indeed, ICT is perceived mainly as a means required for e-maintenance development but it is not sufficient to contribute to the e-maintenance added value in terms of know-how and services according to the maintenance statement and not the technology statement. In that way e-maintenance has to include also strategic vision, business processes, organisations and approaches to perform all the factors identified in the previous section. However, the advance of technologies also contributes to a great extent to the development of this vision, as described by Muller *et al.* (2008) and Campos (2009). This is also the main aim of Dynamite and gives substance to the rest of the chapter, where the state of the art on ICT technologies is confronted with the needs and activities carried out with at the Dynamite project.

3.3 ICT for E-maintenance

ICT is a very broad subject, but it is worthwhile mentioning the major areas of development during last years and what has been incorporated into everyday maintenance activities. It is also relevant to point out what is yet to be incorporated into the daily maintenance and operations processes.

There are basically two technology sources that can be tagged as having been disruptive over the preceding years:

- First, the use of miniaturised devices is increasing the ways data can be acquired or "sensed". This also applies to the appearance of mobile systems, because communication has released many modes of communication.
- Second, the extension of communication technologies (including wireless) has ultimately boosted the usage of the Internet as a main distributed platform for business operation.

Spreading from the above, there are a number of technologies worth mentioning. As it is possible to make many different classifications (*e.g.*, according to the type of technology involved, according to historical appearance), this chapter makes the identification according to functionality: data acquisition, data processing and conversion and finally communication of data to humans and machines.

3.3.1 Miniaturisation Technologies for Data Acquisition

The vast reduction in the cost of electronics and the development of the new micro-technologies has opened up new possibilities in sensor technology. Although this applies to many different fields, some examples of new possibilities are the on-line sensorisation of lubricants, the use of RFID tags, the irruption of smart mobile devices, and the inclusion of all miniaturised devices within ubiquitous computing systems.

3.3.1.1 New Sensor Systems

Concerning lubricant sensors, thick film technology enables sensor heads to be made smaller and more accurately. What is more, at the top of the technological development, optical micro-sensors are being developed for measuring visible and infrared wavelengths that can be correlated to many different fluid properties, providing reliable readings for many parameters that nowadays can only be analysed with laboratory equipment. With appropriate communications with central intelligence systems, smart sensors are able to run unattended, performing self-tuning and auto-calibration, *etc.* (Aranzabe *et al.* 2004).

What are the main advantages of using these kinds of sensors, that is, using diffraction gratings, miniaturised systems, or micro-optical systems?

- The most important advantage is, of course, the achievement of much reduced sensor sizes, which could even rival those of vibration sensors. This can allow the introduction of laboratory-like detection systems in reduced machinery. This spans from most of machine-tools to cars and compressors.
- On the other hand, even though sensor prototyping has a cost, a low cost micro-fabrication is also foreseen, when using silicon-based materials replicated on polymers.

At Dynamite, several new sensors have been developed and tested. On the one hand, current MEMS systems have been expanded, with a specific focus on self-powered systems (see Chapter 6). On the other hand, new optics devices have been developed into full prototype sensors able to identify machinery lubricant conditions (see Chapter 7).

RFID is a technology that involves tags that emit radio signals and devices called readers that pick up the signal. These smart tags are the basis of the technology, which is rapidly emerging as the replacement for the barcode. In fact, RFID systems are beginning to make an impact on manufacturing and logistics operations, and it is believed that advantages may also be gained soon in the maintenance field. Although cost is a major factor limiting the uptake of this technology by companies at the present time, it is recognised that this will become less of a factor as micro-manufacturing costs decrease and operating efficiencies are squeezed to higher and higher requirement levels. The ability of businesses to plan accordingly and, in emergencies, react quickly is one key advantage of these new techniques (Adgar *et al.* 2007).

On of the motivations in Dynamite has been the expectation of smart tags as fully tailored products for e-maintenance applications based on commercial hardware. The tags should be able to store and communicate identity and historical information. Hence the use of active and passive solutions with write and read capability will be required. The possible benefits to maintenance activities are only just beginning to become clear. Users of such technology would enjoy immediate access to information including machinery data, sensor identification, audit trails of maintenance activities, spare part information and use of maintenance tools. This topic is addressed in detail in Chapter 8.

3.3.1.2 Smart PDAs and Mobile Devices

Since their inception (Chess *et al.* 1995), mobile agents have been used in a wide variety of applications. There are several advantages in employing mobile computing compared to conventional wired computer applications. Among other things, mobile computing offers the flexibility to initiate applications at flexible locations in unstructured networked environments, to quickly and efficiently search for and retrieve relevant information from heterogeneous data sources, to perform tasks while utilising limited or intermittent connectivity and to provide asynchronous services to client requests (Samaras 2004). Adding the ease and flexibility of carrying a handheld wireless device, mobile computing has the potential to transform the way a range of industrial management, monitoring and control tasks are performed (Buse and Wu 2004). This potential is still largely unexplored in maintenance management.

Although the usage of wireless devices within an e-maintenance framework has been suggested in the past (Lee 2001), integrated maintenance management solutions based on combined usage of wireless sensing, RFID tags, hand-held devices and central or remote server-side computing and data-offices (Legner and Thiesse 2006) are still in their infancy. Part of the difficulty is attributed to the challenge of integrating equipment, devices and computing resources and code from very heterogeneous sources (Bartelt *et al.* 2005) but also to the great complexity of optimising the management of maintenance in modern industry.

Today there is a huge amount of potential PDA hardware available, which can be divided into four principal subgroups:

1. regular consumer PDAs;
2. retail/logistics PDAs;
3. smart phone PDAs; and
4. custom reference platforms.

However, there are important features that are still hard to find off the shelf, such as short range wireless, RFID readers or PDA expansion slot availability. Such features are deemed important for mobile solutions tailored to serving industrial maintenance management needs.

Within Dynamite, the usage of PDA devices plays a key role in bringing mobile maintenance management closer to daily practice on the shop floor. PDAs are used in synergy with intelligent sensing devices and smart tags on the lower-end of the data processing architecture, but also with central server's databases and data processing and remote access applications at the higher-end of the architecture. PDA is expected to become a ubiquitous expert advisor and, at the same time, a flexible data collector. A complete description of the work done in this area is included in Chapter 9.

3.3.1.3 Ubiquitous Computing

A third field where miniaturisation of technologies plays an important role is the appearance of multiple sources of computing. Firstly, this makes possible to have the computing power at the operator's hand. The PDAs and portable devices described in the previous section are initial examples of a "wearable" computing power. Secondly, and of more interest to this section, devices may be mimicked with the area surrounding the operator (Arnaiz *et al.* 2004).

Defined by the EC Information Society Technologies Advisory Group ambient intelligence (AmI) emphasises greater user-friendliness, more efficient services support, user-empowerment, and support for human interactions. In this vision, people will be surrounded by intelligent and intuitive interfaces embedded in everyday objects around them and an environment recognising and responding to the presence of individuals in an invisible way. This vision of ambient intelligence places the user at the centre of future development. Therefore, technology should be designed for people rather than making people adapt to technology (Friedewald and Da Costa 2003).

Maintenance tasks tend to be difficult because they require expert technicians. Maintenance working conditions are characterised by information overload (manuals, forms, video, real-time data), collaboration with suppliers and operators, integration of different sources of data (drawsings, components, models, historical data, reparation activities). Ambient intelligence provides a new working environment to maintenance technicians; it offer access to ubiquitous and up-to-date information about the equipment wherever the equipment or the operator is (enabling remote maintenance and life-cycle management) and user friendly and intelligent interfaces (context-aware applications). Advantages provided by the use of AmI in maintenance environment come from (Ducatel *et al.* 2000):

- simplifying distributed computing, better distributed knowledge management;
- intelligent resource management;

- overcoming user interface problems;
- overcoming data exchange and communication problems; and
- personalisation, adaptation to the user.

Ubiquitous computing is related to the integration of micro-processors into everyday objects like furniture, clothing, white goods, toys, even paint. Microsystems and electronics play an important part in the ambient intelligence (AmI) environment. An early example of ubiquitous computing applications is locating people and objects.

Ubiquitous computing also needs communications enabling these objects to communicate with each other and the user by means of *ad-hoc* and wireless networking, as well as intelligent user interfaces enabling the inhabitants of the AmI environment to control and interact with the environment in a natural (voice, gestures) and personalised way (preferences, context). Agent technology is also providing new distributed architectures and better communication strategies for the applications, making the information exchange easier and allowing integration of new modules like sensors or diagnosis algorithms with less effort from the point of view of customers and machine tool builders.

Finally, standards are indispensable in the AmI scenario to support interoperability and interactivity between heterogeneous environments. The knowledge shared over the network needs standards to allow knowledge acquisition, validation, management, dissemination and reuse.

3.3.2 Standards for Data and Information Communication

Technological advances such as those referred to in the previous section would be difficult to apply if there were not any adequate standards. In this sense, new standards in last years have allowed many important advances, both in the wireless communication area, easing connectivity of many miniaturised systems, and in the logical communication and architecture of maintenance processes. The related standards are briefly outlined in the following section.

3.3.2.1 Wireless Standards and Technologies

The wireless network enables communication of information without wires. Many types of wireless communication systems exist and for classification purposes several parameters could be considered, such as cost, frequency, capacity, *etc.* Wireless networks can be also divided according to the size of the physical area that they cover, resulting in the following categories:

- wireless personal Area Network (WPAN)
- wireless local area network (WLAN)

- wireless metropolitan area network (WMAN)
- wireless wide area network (WWAN).

Dynamite work is mostly interested in the advantages that wireless LAN and PAN technologies may bring within the factory. Among these, the most important options are as follows.

IEEE 802.11 standards are already broadly used and are commercially available under the references 802.11a, 802.11b and 802.11g. IEEE 802.11 specifies physical and medium access layers. For the physical layers, spread spectrum techniques are used on the 2.4 GHz ISM[2] frequency band (802.11b/g) and 5 GHz frequency band (802.11a), offering various data rates between 1 Mbps and 54 Mbps. For wireless communications systems, the transmission range with 802.11 standards depends on several factors, such as data rate, transmit power and radio frequency. With 802.11b technology, the typical indoor range is 30 m at 11 Mbit/s up to 90 m at 1 Mbit/s. 802.11x transmission is by design relevant to transmission of large data files, compared to 802.15.x transmission. As such, 802.11x performance drops considerably when the data traffic is primarily related to large numbers of small packets.

Bluetooth (IEEE 802.15.1) is a specification for wireless personal area networks, which was originally meant to eliminate cables between devices like mobile phones, personal digital assistants, laptops and their accessories. All devices can be easily interconnected to coordinate and exchange information using an infrastructure-less short-range wireless connection. Bluetooth also uses a 2.4 GHz ISM band, and the transmission range is usually around from 10 to 100 m, with high-power Bluetooth devices.

ZigBee (IEEE 802.15.4) is a wireless standard developed by the ZigBee alliance[3]. The ZigBee consortium defines a Zigbee stack with network and application layers above the IEEE 802.15.4 stack, which offers physical and MAC layers. The IEEE 802.15.4 itself is a specification investigating low data rate wireless solutions with very low complexity and very low power consumption (years with standard batteries) [802.15.4]. Contrary to Bluetooth, Zigbee supports very large numbers of nodes (using 64-bit address space) within star, mesh and cluster tree networks. Nevertheless, to achieve good energy efficiency on physical and MAC layers, Zigbee is optimised for a short range, typically 10 m with a maximum data rate of 250 Kbps.

A comparison between these three main technologies is shown in Table 3.1.

---

[2] Radio bands for Industrial, Scientific and Medical purposes

[3] zigbee alliance http://www.zigbee.org/

Table 3.1 Comparison between Wi-Fi, Bluetooth and Zigbee

Feature	WiFi (IEEE 802.11b)	Bluetooth (IEEE 802.15.1)	ZigBee (IEEE 802.15.4)
Radio	DSSS[a]	FHSS[b]	DSSS
Data rate	11 Mbps	1 Mbps	250 kbps
Nodes *per* master	32	7	64,000
Slave enumeration latency	Up to 3 s	Up to 10 s	30 ms
Data type	Video, audio, graphics, pictures, files	Audio, graphics, pictures, files	Small data packet
Range (m)	100	10	70
Extendability	Roaming possible	No	Yes
Battery life	Hours	1 week	>1 year
Bill of material (US$)	9	6	3
Complexity	Complex	Very complex	Simple

Other technologies exist for wireless communications such as those shown in Figure 3.1. Among these, WiMedia is one of most promising. Indeed, ultra wide band (UWB) technology offers great opportunities for short-range wireless multimedia networking. WiMedia-based UWB specifications have been architected and optimised for wireless personal-area networks delivering high-speed (480 Mbps and beyond), low-power multimedia capabilities for the PC, CE, mobile and automotive market segments.

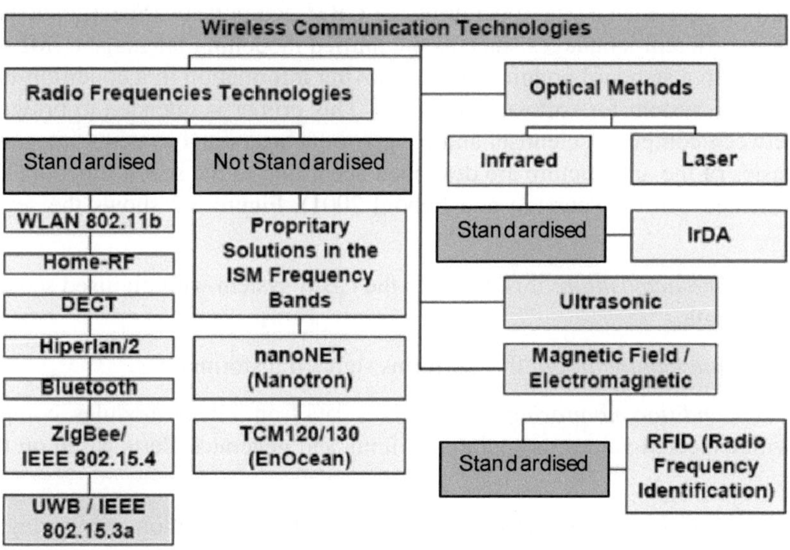

Figure 3.1 Wireless communication technologies

3.3.2.2 OSA-CBM Architecture

The implementation of a CBM system usually requires the integration of a variety of hardware and software components. Therefore, a complete CBM system may be composed of a number of functional blocks or capabilities: sensing and data acquisition, data manipulation, condition monitoring, health assessment/diagnostics, prognostics and decision reasoning. In addition, some form of human system interface (HSI) is required to provide a means of displaying vital information and provide user access to the system. Thus, there is a broad range of system level requirements that include communication and integration with legacy systems, protection of proprietary data and algorithms, the need for upgradeability, reduction of engineering design time and costs.

With these requirements in mind, OSA-CBM (open system architecture for condition based maintenance, www.mimosa.org) is designed as an open non-proprietary CBM communications framework to provide a functional platform flexible enough to suit a broad range of applications. The standard is maintained by the operations and maintenance information open systems alliance (MIMOSA). MIMOSA™ is an alliance of operations and maintenance (O&M) solution providers and end-user companies who are focused on developing consensus-driven open data standards to enable open standards-based O&M interoperability (MIMOSA). Standardisation of a networking protocol within the community of CBM developers and users will, ideally, drive CBM suppliers to produce interchangeable hardware and software components.

The goal of OSA-CBM is the development of architecture and data exchange conventions that enables interoperability of CBM components. Specifications are written in different languages, such as the unified modelling language (UML) and correspond to a standard architecture for moving information in a condition-based maintenance system for software engineers. This primer is intended to bridge the gap between computer scientists and program managers and systems integrators. The basics of the architecture are described according to the seven functional layers presented below (Thurston and Lebold 2001). Figure 3.2 shows the seventh layer.

Layer 1 – data acquisition: this provides the CBM system with digitised sensor or transducer data.

Layer 2 – data manipulation: this performs signal transformations.

Layer 3 – condition monitoring: this receives data from sensor modules, compares data with expected values or operations limits and generates alerts based on these limits.

Layer 4 – health assessment: this receives data from condition monitoring and prescribes if the health in the monitoring component, sub-system or system is degraded. Moreover, it is able to generate, diagnostic based upon trends in health history, operational status and loading and maintenance history, and also propose fault possibilities.

Layer 5 – prognosis: this plans the health state of equipment into the future or estimates the remaining useful life, taking into account estimates of future usage profiles.

Layer 6 – decision support: this generates recommended actions, related with maintenance or how to run the asset until the current mission is completed without occurrence of breakdown, and alternatives. It takes into account operational history, current and future mission profile, high-level unit objectives and resource constraints.

Layer 7 – presentation layer

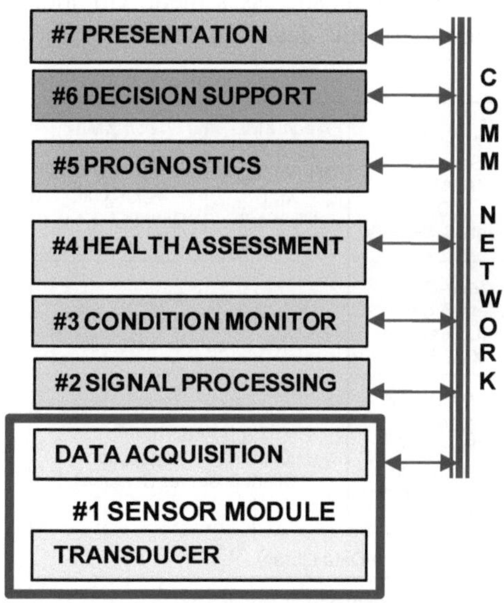

Figure 3.2 OSA-CBM layers (Thurston and Lebold 2001)

3.3.2.3 MIMOSA Protocols and OSA-EAI Architecture

The tasks in the field of operations and maintenance are manifold. Asset management, monitoring, diagnostics, maintenance task management, decision support, *etc.*, usually cannot be supported by a single computer system. Commonly several interacting computer systems are used, possibly made by different suppliers. To allow the interaction between different systems the overlapping data entities must be well identified and each system must provide a suitable interface for the others to use the necessary information. Both tasks are far from trivial. Composing a data model that satisfies different applications is a challenge on its own. When creating interfaces between different applications there are basically two choices (a) develop dedicated set of interfaces for each application or (b) use standard bridge for exchanging the data. Dedicated interfaces are unquestionably more effective in terms of performance and use of resources. However, they lead to tight coupling between applications and work well only in the case when the interfaces systems come from the same supplier or suppliers cooperate closely. A standard interface provides much more flexibility with the price of performance and on the assumption that all suppliers agree to follow the same standard.

Open systems architecture for enterprise application integration (OSA-EAI) is another MIMOSA-coordinated standard (Figure 3.3). It was created to solve/remedy the problem of application integration. It is an open data exchange standard in several key asset management areas: asset register management; work management, diagnostic and prognostic assessment, vibration and sound data, oil, fluid and gas data, thermographic data and reliability information (MIMOSA 2004).

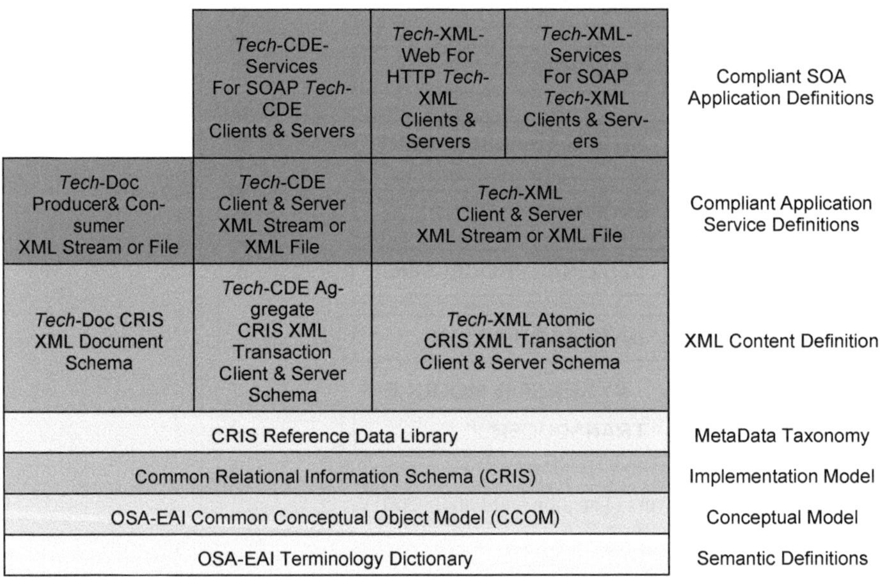

Figure 3.3 MIMOSA OSA-EAI architecture diagram (OSA-EAI Tech Summary 2007)

OSA-EAI is composed of several layers defining the data model contents, relations and interfaces. The common relational informational schema (CRIS) provides common implementation schema for the conceptual model. The primary representation of CRIS is XML schema (XSD), which defines the common format that all data sources must be able to translate. To ease the creation of CRIS-compatible data sources Oracle and Microsoft SQL table creation scripts are provided. CRIS reference data library provides mechanisms for maintaining and referencing classification taxonomy for all items (enterprises, sites, assets, agents, measurement locations, engineering units, *etc.*) MIMOSA also maintains industry-standard taxonomies and codes for many of these classifiers. The next layers define data exchange formats for services and definition of SOAP web service transport (OSA-EAI Tech Summary 2007).

To support data exchange between different applications and even enterprises, OSA-EAI specification contains mandatory unique identification methodology. This methodology allows integration of all items and agents identification nomenclature. The hierarchy of data elements is very flexible (OSA-EAI Tech Summary 2007), but obviously such flexibility does not come for free. Complex referencing mechanisms and extreme normalisation may cause performance degradation. Also, while the support for multiple enterprises and sites is essential for successful data exchange, inside a single application operating on a fixed site it is a source of considerable extra complexity.

The mentioned problems become apparent only when the application uses the CRIS data model for physical storage. In fact, the OSA-EAI standard does not even address the persistence of CRIS – the standard is oriented on data exchange and service-level compatibility. This is also a reason why most applications implement their proprietary database format and implement the OSA-EAI-compliant interfaces layer. Later in this chapter we will describe an approach taken in the Dynamite project to achieve data persistency.

3.3.3 Data and Information Processing and the Impact of Machine Learning Systems

Another series of technological drivers for the application of new technologies have much to do with the actual status of automation systems and "computational intelligence", and the tools readily available to help to model a maintenance tasks.

Most existing commercially available products only perform inferential steps. This is because learning – any change in diagnostic knowledge – is very difficult to be encoded, if at all possible. However, changes concerning most of the monitored machinery may appear everywhere, so the ability to modify the inferential steps is a must. In fact, learning abilities are really what make us consider a system "intelligent". It is not possible to consider a system "intelligent" when it keeps on making the same mistake forever (Arnaiz and Gilabert 2004).

Learning can come in two different ways: As a fully data-dependant batch process, where a model is constructed out of a data system, and as an incremental approach, where an existing model is slightly modified by new data or expertise. Here, we concentrate on adaptive systems and leave the data-driven batch model construction for the next section.

One interesting approach is *case based reasoning (CBR).* Usually IA approaches are based on a general knowledge about a domain of a problem, setting associations through a set of general relationships between problems and conclusions. However, CBR uses specific knowledge of previous experiences in particular situations. A problem is resolved finding a similar uncertain situation (case) in the past and reusing its solution in the new situation. INRECA (Althoff 1996) is a system development methodology of CBR. Nowadays it is able to facilitate the design of systems starting from predefined templates. The most interesting thing is the existence of templates specifically designed for maintenance applications. The system has been continuously updated, and has served as a base to tools like NBCWorks and applications like the following ones, which integrate decision trees as form of organisation of the memory:

Another model for knowledge adaptation is *Bayesian networks (BN)*. A Bayesian network is a model that represents the states of some parts of the world that we are modelling. It describes how these states are related through conditional probabilities. A BN must represent all the possible states that can exist in our world. A machine-tool can be running normally or giving a failure. In a medical diagnosis, a man can be sick or healthy. That is a causal system where some states tend to occur more frequently when previous states are present. BN are very useful because they are adaptable. It is possible to start building a network with a delimited knowledge in our domain and grow them as more information becomes known. Furthermore, it is possible to provide feedback to the network if the solution given is not right, adapting the probabilities between the states. Finally, it can handle the uncertainty. As a consequence, graphical probabilistic models (Figure 3.4), and more specifically BN, are becoming popular (Arnaiz and Arzamendi 2003).

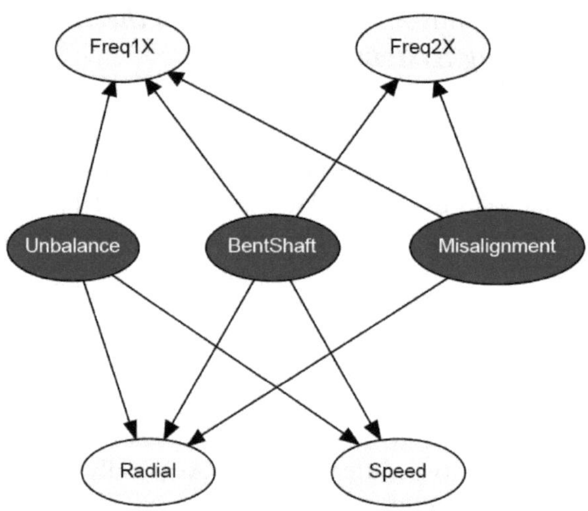

Figure 3.4 Excerpt of a machinery diagnosis graph model

We can also mention *data mining* systems as a final range of algorithms that are employed in the search of a model for maintenance task automation; we have to mention the algorithms that perform batch processing of the available data in order to model the solution. These algorithms are very close to the paradigms in the previous paragraphs, as they fall in the same family of machine learning algorithms. However, in this case, model construction usually starts from no prior knowledge and in this sense, if new knowledge appears it is used to reconstruct the model again. These systems are very close to statistical models, as statistical models also perform a parametric or non-parametric modelling base (Esbensen 2002).

From a computational perspective, *neural networks* is a technology of particular interest for pattern classification and functional synthesis, with literally dozens of applications in maintenance (*e.g.*, Emmanoulidis *et al.* 2006), where the most interesting areas are pattern classification and functional synthesis consisting of establishing relationships between several continuous-valued inputs and one or more continuous-valued outputs. This involves data interpretation through filtering noise, forecasting/prediction problems, *etc.* A very interesting related approach is neuro-fuzzy systems (McIntyre and McGarry 1999)

Another important family of data mining systems is represented by *induction algorithms*. Here learning algorithms are based on the presentation, on the part of an actor, of positive and negative examples of a concept. This information, along with a "general" knowledge or "background", must serve for the system as learning to "recognise" the concept so that new examples are classified in the right way on the basis of the learned concept. The process has similarity with the learning that makes artificial neural networks (ANN), although in this case comprehensible symbolic structures are generated, with the capacity to explain the results. In order to get this learning, two groups of methods are distinguished: inductive and deductive. Inductive algorithms are based on finding a hypothesis (structure) that represents the concept to learn on a sufficiently great set of examples, which will serve to represent the concept in unobserved (test) examples. Most well known methods are decision trees (*e.g.*, top down induction of decision trees – TDIDT) with many examples of application in the area (Arnaiz *et al.* 2005).

Intelligent Web Services

Web services represent a development of the use of the web. Initially the web was used to transport pages of Hyper Text Markup Language (HTML) from a files system somewhere on the internet to a browser that would render it and display it to a user. With respect to communications, web services are an extension of remote procedure call (RPC) in the same veneer as DCOM/COM+, CORBA and RMI. What is actually novel is the use of a plain text format for the exchange of messages as well as a standard Internet protocol such as HTTP/TCP for message transport. This guarantees that any machine connected to the net will be able to participate in a web service exchange since HTTP is usually always open on even the strictest firewall configurations (Cerami 2002).

A web service is a software system identified by a URI, whose public interfaces and bindings are defined and described using XML. Additionally its definition can be discovered by other software systems. These systems may then interact with the web service in a manner prescribed by its definition, using XML based messages conveyed by internet protocols (www.v3.org).

At Dynamite one of the objectives has been to push the inclusion of web services capable of performing intelligent actions, and especially taking advantage of artificial intelligence technologies such as those outlined previously. This will create a series of intelligent web services on demand that can be accessed whenever necessary, through the appropriate communication mechanisms stemming from standards such as MIMOSA (CRIS XML), or even higher "semantic" additions. Thus, many information processing algorithms explained above, such as Bayesian networks based diagnosis systems, have been "framed" onto a web-services architecture. A detailed view of this outcome is given in Chapter 11.

3.4 Conclusions

This chapter has put together the most relevant information that, regarding information and communication technologies has driven the research within Dynamite.

The first section included a description of the e-maintenance concept. This concept was important to focus the motivation of the "dynamic" approach to the consecution of a flexible maintenance framework.

Second, a number of ICT technologies were clearly marked as starting points for the research. This involved from smart tags up to intelligent automation systems. The main goal behind the use of these technologies was the conversion for general purpose technologies into "capabilities" easy to use within a general maintenance scenario.

Last, this revision leads to the presentation of the global concept (DynaWeb) that will be presented in next chapter. Web services, MEMS sensors and smart tags, information storage, smart PDAs and wireless communication systems are all examples of developed capabilities that will be also shown in next chapters.

References

Adgar A, Addison JFD, Yau CY (2007) Applications of RFID technology in maintenance systems. In Proc 2nd World Congress on Engineering Asset Management, Harrogate, UK, 11–14 Jun 2007. Coxmoor, Oxford

Al-Najjar B, Alsyouf I (2003) Selecting the most efficient maintenance approach using fuzzy multiple criteria decision making. International Journal of Production Economics 84:85–100

Althoff KD (1996) Evaluating case-based reasoning systems: The INRECA case study. Thesis work, University of Kaiserslautern

Aranzabe A, Terradillos J, Arnaiz A, Merino S, Gómez D (2004) Application of micro-technologies in on-line condition monitoring of lubricants. Proc 14th Int Colloquium Tribology, Tribology and Lubrication Engineering TAE–Technische Akademie Esslingen, Germany, 269–278

Arnaiz A, Arzamendi J (2003) Adaptive diagnostic systems by means of Bayesian networks. Proc 16th Int Congress on Condition Monitoring and Diagnostic Engineering Management COMADEM, Växjo, Sweden, September 2003, 155–164. Växjö University Press

Arnaiz A, Aranburu I, Terradillos J, Gorritxategi E, Ciria JI (2005) Intelligent data analysis for condition monitoring. An application to oil status prediction. Proc 18th COMADEM, Augt 2005, Cranfield, UK, 515–524. Cranfield University Press

Arnaiz A, Emmanoulidis C, Iung B, Jantunen E (2006) Mobile maintenance management. Journal of International Technology and Information Management15:11–22

Arnaiz A, Gilabert E (2004) Learning approaches in maintenance and monitoring. Proc 17th Int Congress COMADEM, Cambridge, UK, 509–518

Bartelt C, Fischer T, Niebuhr D, Rausch A, Seidl F, Trapp M (2005) Dynamic integration of heterogeneous mobile devices. Proceedings of DEAS 2005, First Workshop on Designing and Evolution of Autonomic Application Software, May 21, 2005, St Louis, MO

Ben-Daya M, Duffuaa SO, Raouf A, Knezevic J, Ait-Kadi D (2009) (Eds) Handbook of maintenance Management and Engineering – Part Integrated e-Maintenance and intelligent maintenance systems. Springer, Berlin, ISBN 978-1-84882-471-3

Bengtsson M (2004) Condition based maintenance system technology. Where is development heading? Proc 17th European Maintenance Congress (Euromaintenance), Barcelona, Spain, 2004, 147–156. AEM. Puntex Publicaciones

Buse DP, Wu QH (2004) Mobile agents for remote control of distributed systems. IEEE Transactions on Industrial Electronics, 51/6, December

Campos J (2009) Development in the application of ICT in condition monitoring and maintenance. Computers in Industry 60:1–20

Cerami E (2002) Web services essentials: distributed applications with XML-RPC, SOAP, UDDI & WSDL. O'Reilly, Farnham, UK (ISBN 0-596-00224-6)

Chess D, Grosof B, Harrison C, Levine D, Parris C, Tsudik G (1995) Itinerant agents for mobile computing. Journal IEEE Personal Communications, 2/5, October

Ducatel K, Bogdanowicz M, Scapolo F, Leijten J, Burgelma, JC (2000) Scenarios for ambient intelligence in 2010. ISTAG 2001 Final Report, IPTS, Seville

Emmanouilidis C, Jantunen E, MacIntyre J (2006) Flexible software for condition monitoring, incorporating novelty detection and diagnostics. Computers in Industry 57:516–527

Esbensen K (2002) Multivariate data analysis in practice. CAMO Process AS. 5th edition

Friedewald M, Da Costa O (2003) Science and technology roadmapping: Ambient intelligence in everyday life. JRC/IPTS–ESTO Study, June 2003

Iung B, Al Najjar B, Arnaiz A (2004) New Model and technologies to select and improve a cost-effective condition-based maintenance policy practically. Report of the Idea Factory Session, IMS-FORUM2004-Como-Italy-19 May, 2004

Iung B, Crespo Marquez A (2006) Special issue on e-Maintenance. Computers in Industry, 57:473–606

Iung B, Levrat E, Crespo-Marquez A, Erbe H (2009) Conceptual framework for e-maintenance: illustration by e-maintenance technologies and platform. Annual Review in Control, Issue 3, 2009

Jardine A, Lin D, Banjevic D (2006) A review on machinery diagnostics and prognostics implementing condition-based maintenance. Mechanical Systems and Signal Processing, 20:1483–1510

Lee J (2001) A framework for web-enabled e-maintenance systems. Proceedings 2nd International Symposium on Environmentally Conscious Design and Inverse Manufacturing, EcoDesign'01

Lee J, Ni J, Djurdjanovic D, Qiu H, Liao H (2006), Intelligent prognostics tools and e-maintenance. Computers in Industry, Special issue on e-Maintenance, 57:476–489

Legner C, Thiesse F (2006) RFID based maintenance in Frankfurt Airport. IEEE Pervasive Computing

Levrat E, Iung B, Crespo Marquez A (2008a) e-maintenance: review and conceptual framework. Production Planning and Control, 19:408–429

Levrat E, Thomas E, Iung B (2008b) Odds-based decision-making tool for opportunistic production-maintenance synchronisation. Journal IJPR, 46:5263 – 5287

MIMOSA (2004) MIMOSA Brochure, http://www.mimosa.org/downloads/13/whitepapers/index.aspx, accessed 04/2009

Muller A, Crespo Marquez A, Iung B (2008) On the concept of e-maintenance. Review and current research. Reliability Engineering and System Safety, 93:1165–1187

OSA-EAI Tech Summary (2007) MIMOSA's open system architecture for enterprise application integration (OSA-EAI) Version 3.2 Technical Architecture Summary. http://www.mimosa.org/downloads/44/specifications/index.aspx, accessed 04/2009

Samaras G (2004) Mobile agents: what about them? Did they deliver what they promised? Are they here to stay? Proceedings of the 2004 IEEE International Conference on Mobile Data Management (MDM'04)

Swanson L (2001) Linking maintenance strategies to performance. IJPE 70:233–244

Swanson L (2003) An information-processing model of maintenance management. International Journal of Production Economics 83:45–64

Takata S, Kimura F, van Houten FJAM, Westkämper E, Shpitalni M,Ceglarek D, Lee J (2004) Maintenance: changing role in life cycle management. Annals of the CIRP, 53:643–656

Thurston M, Lebold M (2001) Standards development for condition-based maintenance systems. New frontiers in integrated diagnostics and prognostics. 55th Meeting of the Society for Machinery Failure Prevention Technology, MFPT (2001)

Wang H (2002) A survey of maintenance policies of deteriorating systems. European Journal of Operational Research, 139:469–489

Westkämpfer, E (2003) Assembly and disassembly processes in product life cycle perspectives. Keynote paper, Annals of CIRP, 52/2

Chapter 4
A New Integrated E-maintenance Concept

Aitor Arnaiz, Benoit Iung, Basim Al-Najjar, Erkki Jantunen, Kenneth Holmberg, Tonu Naks and David Baglee

Abstract. This chapter outlines the work done in Dynamite project (http://DYNAMITE.vtt.fi), as well as the resulting concept nicknamed as DynaWeb. DynaWeb represents the link between Dynamite and the e-maintenance technologies described in previous chapters and results in a global framework where all technologies can participate within an advanced maintenance solution.

This chapter serves as an introduction to the rest of the chapters dealing with specific technologies that have been converted into 'capabilities', such as intelligent sensors, wireless communications, intelligent web services or smart PDAs, as well as to the final demonstrations.

4.1 Introduction

The Dynamite vision aims at promoting a major change in the focus of condition based maintenance, essentially taking full advantage of recent advanced of information technologies related to hardware, software and semantic information modelling. Special attention is also given to the identification of cost-effectiveness related to the upgraded condition based maintenance (CBM) strategies, as well as to the inclusion of innovative technologies within CBM processes. It is expected that the combination of the use of new technologies together with a clear indication of cost-benefit trade-off will facilitate the upgrade into CBM. This expectation is thought to be particularly relevant in many cases where non-critical machinery exists, and especially for the vast majority of SME companies where the distance between planned and condition based maintenance is too wide.

However, it is difficult to find a single solution that fits for all concerning the maintenance needs. Arnaiz *et al.* (2007b) clearly showed that it is not possible to find such a system, as existing strategies, machinery and legacy systems differ between companies, as well as perceived technical problems and economical motivations.

As a consequence, one of the main goals in Dynamite is to bring together a group of technologies that can be integrated in a structured way, yet which is flexible enough to allow the selection of a particular subset of the technologies, depending on the final scenario of application. This has lead to the development of DynaWeb concept (Arnaiz *et al.* 2007a).

4.2 E-maintenance Scenario Analysis

The design of a flexible structure for Dynamite is supported by the assumption of the existence of numerous companies that can benefit from a subset of the technologies addressed in the project, providing customised plug and play to the desired upgrades with respect to each company's existing maintenance activities. It is also understood that there is no single 'upgrade' solution that fits for all concerning the maintenance needs.

Given this, one of the first activities performed in the project, part of a conventional study of requirements for ICT and sensor technologies, has been the study of the use cases involved in the project, with the aim to identify clear separate scenarios for demonstration of new technologies, which should also facilitate the implementation of cost-effective maintenance solutions. Scenarios are extracted out of initial use cases as likely representative examples of a wider group of companies having similar objectives in the maintenance process, sharing similar technology status or sharing a need with respect to maintenance technologies needed. As a result, different groups have been separated, including large companies with de-centralised production, OEM suppliers, small companies with few dozens of machining systems and third party consultants.

Table 4.1, which was compiled from end user analysis during the first stages of Dynamite, clearly shows that the initial assumption is true, and that it is not possible to find a single system for a global upgrade of existing maintenance systems. The existing strategies, legacy systems and other issues differ very much, as do perceived technical problems and economical motivations. However, if this table is taken as reference, it allows generalising scenarios that can go beyond a particular use case and thus provide an entry point to the technologies for companies that share similarities with one of the scenarios (roles, operational contexts, applications, components, preferred upgrades, etc).

Table 4.1 Summary of end-user scenarios in DYNAMITE project (Arnaiz *et al.* 2007b)

Providers	Plant operators	OEM manufacturers	Consulting	Transport (OEM + consulting.)
Context	Single location (Manufacturing plant). Multiple machines	Technical Assistance Services. Guarantees. Multiple locations	Specialised services (*e.g.*, lube analysis) for multiple locations	Specific machinery on movement
Application (Components)	Milling, drilling and high speed machine tools (Hydraulic systems, gearbox, spindle)			Motors (Marine, Automotive)
Current strategy	New PM (10–20% CBM)	BDM	PM	–
Current economic motivations	Overall economic impact to the company for different maintenance strategies not always known	Uneven workload. Enforce/surveil remote maintenance procedures on guaranteed machinery	Decrease downtime, repair and maintenance costs	Plan new cost-effective e-maintenance for new equipment
Current technical problem(s)	Evaluation of machine condition depending on expert knowledge (subjective)	Improve diagnosis Communication sensors to OEM (bypass CNC)	Lack of experienced diagnosis and decisions over existing parameters	Lack of proper knowledge
Interesting technologies	Include advanced sensors Wireless communication Smart PDAs Include cost-effectiveness	Upgrade to CBM Use remote monitoring E-maintenance Wireless gateways Cost effectiveness	Use e-Maintenance to remotely assess expert and communicate to operators. Training systems.	Initiate predictive maintenance
Likely CBM/PM parameters (sensors)	Temperature, voltage, current, Oil level, oil quality, vibration, pressure, wear debris			

4.3 DynaWeb Integrated Solution

A new concept named DynaWeb has been developed. This concept is best described as an information and communication platform that provides operational interaction between 'plug-in' technologies in the framework of a distributed information scenario, where technologies of interest may vary from one maintenance use case to another (Jantunen *et al.* 2008).

In order to develop this platform, a study of likely actors associated to future DynaWeb activities was made. The synthesis appears in Table 4.2, which identifies the main role identified in each case, the data expected and their expected involvement in OSA-CBM layered information processing steps.

Table 4.2 Main characteristics of Dynamite actors

Actor	Role	Expected Data	Associated OSA-CBM levels
Maintenance Expert	Strategic decisions on maintenance in accordance with Enterprise policy	Policies	
CMMS/ERP	Manage life cycle of the maintenance work-orders in accordance with selected maintenance strategy	Data on spares, work orders, events, indicators	Operational decision support
Computer Maintenance Operational System (CMOpS)	Support maintenance dynamic processes for selecting the best maintenance work-order and supporting it	Historical/trend data Reliability data	Prognosis Diagnosis
PDA	To assist the maintenance operator in carrying out everyday tasks. To embed CMOpS functions	Operator data	Condition monitoring Signal processing
Sensors	To deliver data and information on machine status	Status data	Condition monitoring – Signal processing – Data acquisition
Smart Tags	To automate machine/part identification and deliver historical information on machine state	Identification data	Signal processing

Even though some other actors may also participate in maintenance activities, such as MES or ERP systems, Dynamite technologies are focused on the activities related to the above defined actors.

Once actors were identified, they were framed into a flexible architecture concerning communication channels between the actors. The graphical layout indicates typical communication architecture with respect to a company where information and actors are distributed.

Figure 4.1 provides a schematic overview of the complete system concept depicted for information and communication technologies that are considered within the Dynamite project. This view identifies the existence of three layers (on the right-hand side of the figure) with the location of actors with respect to the company and also states the interoperability of these actors with different technologies.

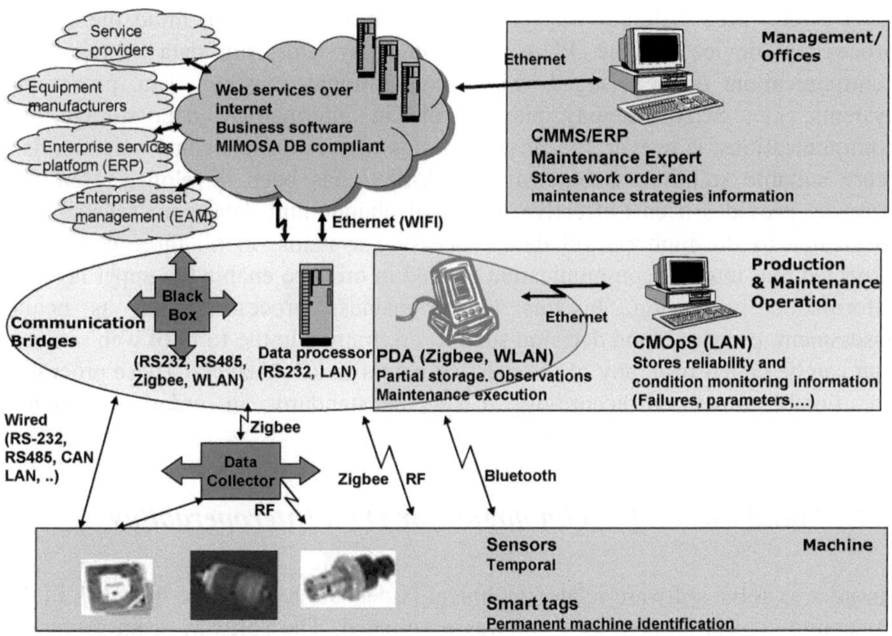

Figure 4.1 DynaWeb ICT structure (Arnaiz *et al.* 2009)

The first level corresponds to the machine and identifies sensors and smart tags as associated to this level of interoperation. It is also expected that sensors hold temporal information concerning current condition values, with little or no historical information attached. The second level corresponds to the production shop floor and identifies two main actors: The PDA and the computer and maintenance operational support (CMOpS). It is argued that these can both hold temporal information concerning operator activities and input values and that CMOpS will hold historical records on selected condition information. The third level corresponds to headquarters and management staff, where both tactical and strategic decisions are made. CMMS as well as maintenance expert agents are located at this level, together with information concerning scheduled operations and maintenance strategies (Holmberg *et al.* 2005).

In all three levels, there is a special effort within Dynamite to provide technologies able to provide a flexible information and communication infrastructure, primarily based on the use of wireless systems. Thus, at the first level it is expected that most of the new sensors (some developed within Dynamite) will be able to communicate directly to the upper level, for instance, one of the novel options is an USB connection to a PDA, *i.e.*, the PDA acting as a flexible, powerful and portable data logger. However, for those with wired connections, a data collection system is developed so that it is possible to 'plug' conventional communications (*e.g.*, RS232) to be converted into a Zigbee output. The second level offers three different ways of interoperation. Having in mind the central processing device of the PDA, two other systems for data storage and communication are envisaged. First, conventional existing data processing systems (*i.e.*, SCADA) may play a role in intermediate data storage and communications. However, for those scenarios where wireless communication is a more suitable solution, a Gateway 'black-box' has been developed. Here the gateway provides a cost-effective means of channelling data from sensors in a local area to the higher level data processing systems. In all three cases it is expected that Internet communication is used in order to enable the upper layer of information processing. In this layer, business processes such as health assessment, prognosis and decision support are framed in the form of web services that can be called from any of the existing actors at lower layers. These processes are finally structured according to several standards in order to enhance interoperability (Arnaiz *et al.* 2009).

4.3.1 Standards and Technologies for Data Interoperability

In order to solve software related technical issues and to integrate the work in the Dynamite project a software team was nominated. The software team discussed issues related to the programming tools and techniques used in the project and tried to unify the programming work and support the communication between various modules developed by various partners in Dynamite.

The software team members created a document/deliverable that describes the programming techniques used in programming the software modules of Dynamite project "D7.6 A Short Tutorial on How to Build Web Services, Agents and PDA Software". Even though the title suggests that the document is not big in size, in the end it became a 130 pages long document containing the most important aspects and guidelines for programming Dynamite modules. During this process of definition of the basic rules for programming it was realised that the challenge of how to organise the communication between various modules of Dynamite had not been solved in the project plan. After long discussions in a separate meeting the decision was made to rely on a common database when integrating the various software modules. The decision was not easy to make especially because many of the partners already had some maintenance related software together with various database formats. However, it was realised that here lay the key to success of Dynamite, *i.e.*, it could only succeed if all the modules could communicate together, and therefore a common ground was needed. In fact, the whole idea in e-maintenance is to be able to pass information to where ever it is needed so the correct maintenance action can be taken at the right time using effective methodology.

When the decision had been made to rely on a common database for data exchange between the Dynamite software modules it was a logical decision that the database should be MIMOSA. Why to choose MIMOSA? MIMOSA organisation gives the following definition "MIMOSA is a not-for-profit trade association dedicated to developing and encouraging the adoption of open information standards for operations and maintenance in manufacturing, fleet, and facility environments. MIMOSA's open standards enable collaborative asset lifecycle management in both commercial and military applications." Clearly this is the optimal strategy for an international project aiming for collaborative work and integration of results.

As such MIMOSA is relatively big containing hundreds of tables. MIMOSA definition covers issues related to measurements, condition monitoring, diagnosis, prognosis and management of maintenance work orders, *etc.* Rather soon after the adaption of MIMOSA it became clear that it was not an easy step for such partners who were not used to working with relational databases. Even though MIMOSA is well documented and is easy to download and install to run, *e.g.*, under a SQL server, it is not an easy step to start using a database in a logical way. In fact quite a lot of effort was spent in discussing how MIMOSA should be used in order it to be an effective tool.

The common MIMOSA database was installed in the project server. A clever user interface was built to help the use of MIMOSA and especially the manual input of data into MIMOSA. After using MIMOSA for data exchange by the end of the project all partners agreed that the decision to go for MIMOSA was the right decision to make and that in fact no other solution had been seen to have had a similar effect in supporting the integration within Dynamite. Figure 4.2 shows the communication within DynaWeb in simplified flowchart format. As can be seen the MIMOSA database is the central point where most of the data goes and where it can be read from. Moreover, there is also communication between various modules on a module to module basis following the same common data format.

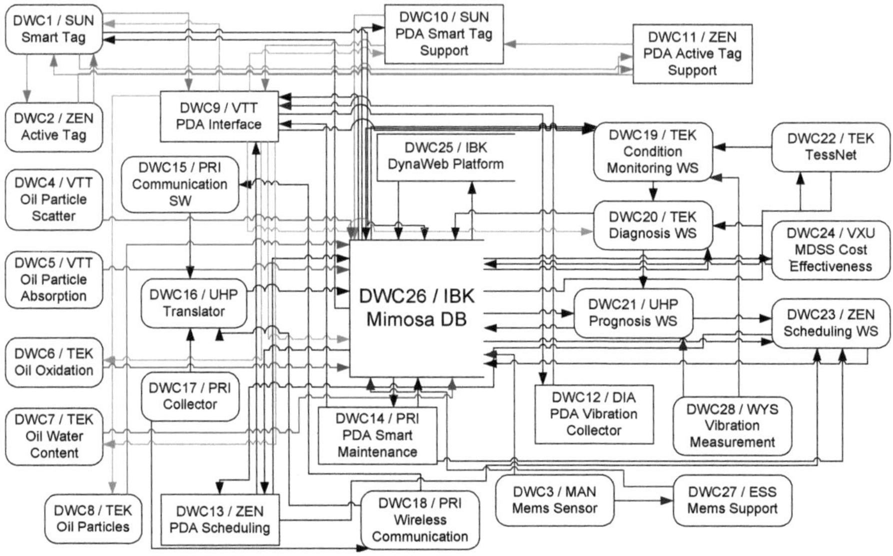

Figure 4.2 Simplified flowchart of the communication within DynaWeb through the MIMOSA database

4.3.2 Implementing the Solution

Given the fact, that the DynaWeb platform is composed of numerous components, the choice of OSA-CBM as the layered architecture of business processes, and MIMOSA as the information exchange standard was quite obvious. For the data persistence layer the first temptation was to go to the direction many MIMOSA compliant applications choose (*e.g.*, Rockwell Emonitor®, http://www.rockwellautomation.com) – each component/application uses its native data storage and implements an interface to provide necessary data for the others. There are dedicated middleware systems on the market that support such interoperability (*e.g.*, Mtelligence MIMOSA Interop Server, http://www.mtelligence.net). However, this would have reduced the effect of component synergy. Different components were designed to analyse overlapping data from different angles and duplicating the overlapping part in different applications did not seem the best way to proceed. Based on this consideration, the decision was made to keep all data in the central DynaWeb database server and implement CRIS as close to the standard as possible.

When implementing this decision a few problems where discovered; nevertheless, the result was positive. The first problem was the documentation; it was not that easy to identify proper location for all the data needed by the various web services. However, the final credit goes to MIMOSA; after some investigation it was possible to accommodate all Dynamite data within the CRIS data model

(OSA-EAI CRIS 2008) without modifying the database schema or misusing existing columns. Also, it should be mentioned that documentation in the latest release 3.2.1 of OSA-EAI is significantly improved (OSA-EAI v3.2.1 2008).

The second, more serious, problem was in the data model architecture. It was mentioned already earlier that, due to its high flexibility, OSA-EAI uses multi-level identification for data elements and this means multi-column keys. In worst cases it may mean up to seven columns that participate in a foreign key relationship (Figure 4.3). Such a referencing mechanism makes the data access layer in applications complicated and expensive to develop. Even worse, during the development and in R&D projects it would be useful to access and modify test data through a general-purpose user interface such as Microsoft Access. In the case of a fully normalised database with multi-column filters this is rather difficult, not to say impractical. The usual solution is to simplify the database schema to a single key and derive the missing relations in the interface layer (*e.g.*, Mathew 2006). As our intention was to use an unmodified CRIS database schema and make the service layer thinner by avoiding data conversions, a different approach was chosen. A middle layer (named MIMOSA Views) in the database was created (see Figure 4.4), which hides multi-level primary keys behind one derived key, takes care of assigning current values for primary keys in the case of creating new items and updates the modification date-time fields. Provided that this extra layer is implemented directly in the database server, it successfully hides the complexity of references in CRIS data model from all higher layers of applications and allows significant reductions in development time.

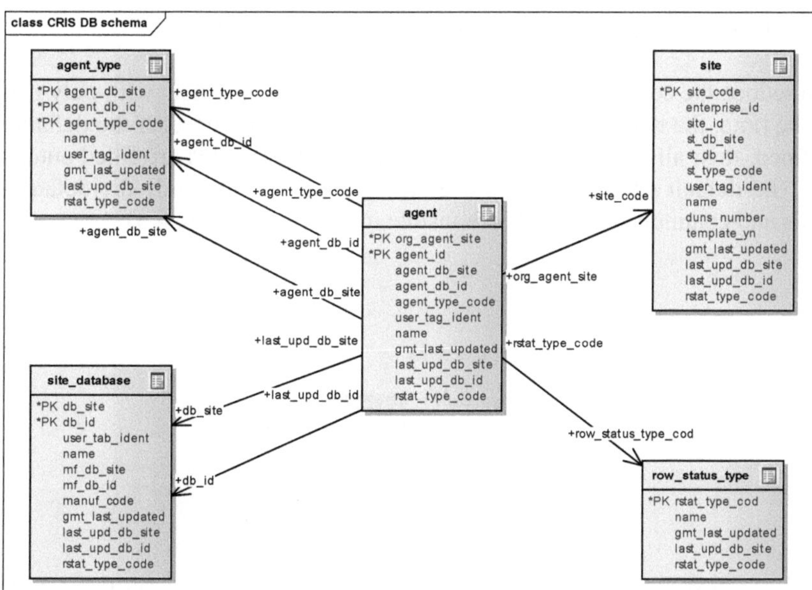

Figure 4.3 Example of table relationships in CRIS DB schema (OSA-EAI CRIS 2008)

Interacting with database through MIMOSA Views means 80% less fields in primary keys and 57% less relation joins, thus reducing greatly the development effort. This approach enabled the creation of Common MIMOSA UI for DYNAMITE with a reasonable development effort. Common MIMOSA UI for DYNAMITE serves as a universal front-end for editing the data in database and simplifies managing and testing the database. All relationships are taken care of by the application.

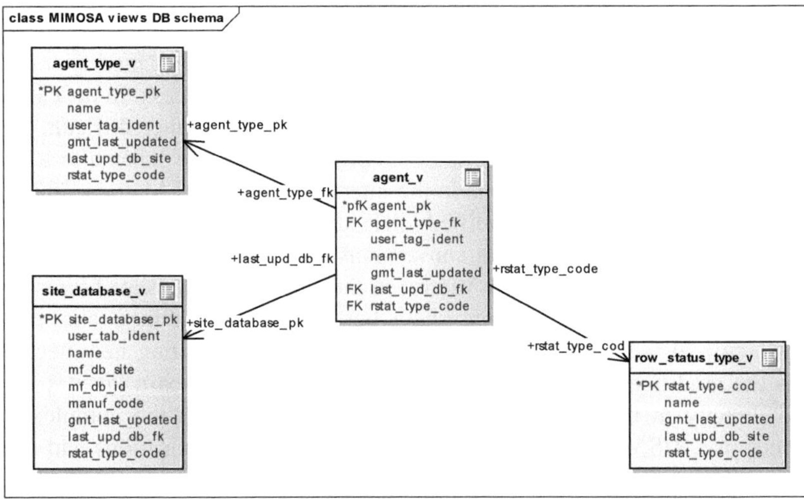

Figure 4.4 Simplified database schema using MIMOSA views

To reduce the complexity of data entry and filtering operating with only one site at the time, the default site and database feature is introduced. If the default site is defined, then all entered records of site and database fields are filled automatically. The default site can also be used as a filter, and then only the default site records and common site records are visible in the view (Figure 4.5).

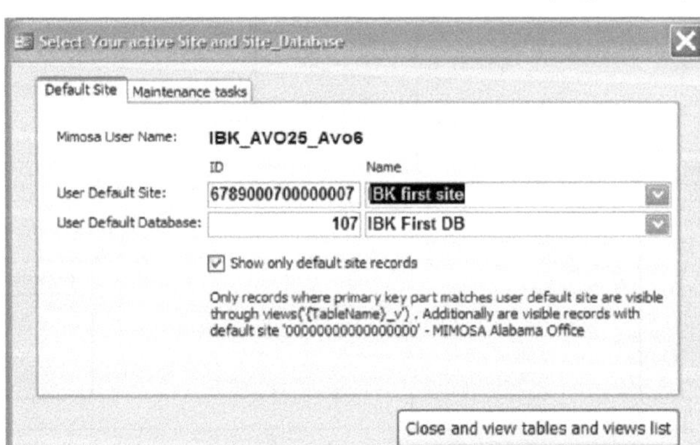

Figure 4.5 Default site and filtering options

The Dynamite solution for data exchange is to use one common database and for each component to interact directly or use web services with database built by the CRIS DB schema.

4.4 Intelligent Sensors

The development of intelligent sensor is fundamental to support different diagnostic, prognostic and maintenance activities. In the Dynamite project, three categories of intelligent sensors are focused including smart tags, micro sensors and lube sensors.

The three categories of intelligent sensors have different purposes. The investigation of smart tags concentrates on utilising existing radio-frequency identification (RFID) technology to improve asset maintenance and management. Different to the smart tags, the role of investigating micro sensors is to develop a more powerful and self-powered wireless microelectromechanical systems (MEMS) sensor for advanced condition monitoring. Finally, the investigation of lube sensors is targeted on various sensor techniques for analysing and detecting different lubrication features.

For smart tags, both passive RFID and active RFID technology are considered. Passive RFID is suggested as a replacement of the barcode system and it is specifically recommended for asset identification and inventory purposes, including machines, spare parts and tools. Moreover, passive RFID and PDA can be used together as a perfect maintenance tool. It can effectively reduce improper asset identification in order to prevent a series of inappropriate maintenance activities.

Alternatively, active RFID is perfect for the real-time location system (RTLS) of mobile assets. It can be applied for security and mobile asset tracking purposes to detect any unauthorised people getting into a protected area, and also search and reserve any shared mobile resources like vehicles.

For micro sensors, a powerful wireless MEMS sensor for advanced condition monitoring has been developed. It combines and integrates three sensors including vibration, temperature and pressure together and a Zigbee wireless communication module into a single MEMS sensor. By using this, multiple sensor values can be collected and transmitted directly to the wireless sensor network for processing at the same time.

Also, in order to support the idea of wireless senor technology continuously working for a long time, a self-powered electronic power management module has been developed to bridge the gap between the wireless sensor and the harvesting devices including solar-based devices and vibration-based devices for recharging batteries.

In the Dynamite project, four types of lubrication sensors are focused in the lubrication system: fibre optic laser and super light-emitting diode (SLED) absorption and scatter sensors for solid contaminants, and particle, oxidation and water sensors. The various sensor techniques described here are important in analysing and detecting different lubrication features

- Firstly, a low cost fibre optic laser and SLED absorption and scatter sensors for a solid contaminants sensor uses optical fibres as data and energy channels to give a cleanliness index for solid particle content in the measured lubrication.
- Secondly, a particle sensor is developed to measure particle content of lubricating oil. Through using a CCD camera together with an illumination system, particles even smaller than 1 micron size can also be detected.
- Thirdly, an oxidation sensor is used to measure oxidation level of lubricating oil. Through measuring the transmittance of the light in the visible range (380–780 nm) of the light spectra, the degradation status of the lubricating oil can be calculated by the correlation of absorption of light and the oxidation level of lubricating oil.
- Fourthly, a water sensor can measure the water content of lubricating oil. Similar to the oxidation sensor, the transmittance of light in the near infrared (NIR) range (about 1400 nm) of the light spectra can be measured and checked with the correlation of the absorption of light and the water content of lubricating oil.

Figure 4.6 illustrates a complete information flow of a lubrication system. Various lube sensors are connected to the target system for data collection. Then a PC or a PDA can be used as a basic data collector to extract features and critical information and send the results to the MIMOSA database for storage and further processing.

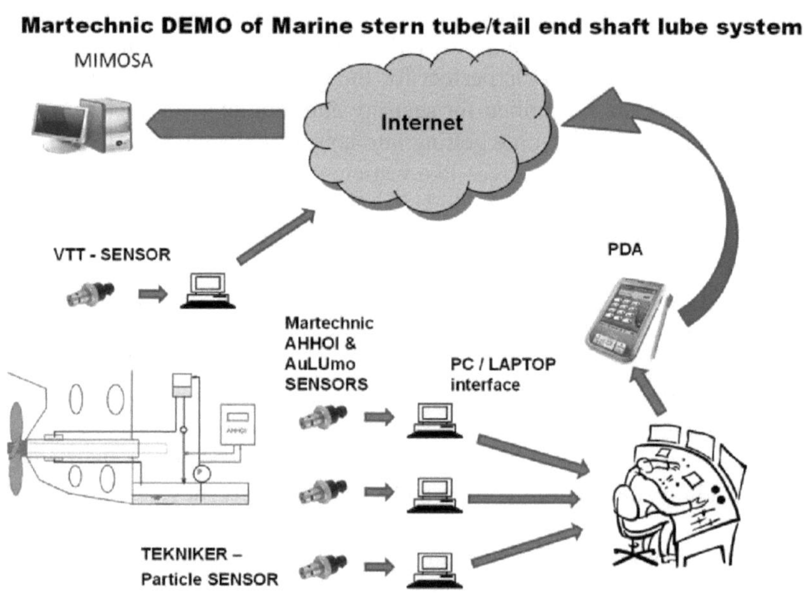

Figure 4.6 The complete information flow from lube sensors to MIMOSA database

In order to support the manipulation of sensor data in the MIMOSA database, a complete set of web-services are designed and developed to support asset information querying, monitoring, diagnostics and prognostics. Those web-based services and sensor techniques are detailed as different DynaWeb components.

Finally, it is necessary to note that all techniques described in this section including smart tags, wireless MEMS sensors and lube sensor systems have been successfully tested in the laboratory and through demonstration. They are fully supported through the connection to the MIMOSA database for storage and retrieval of information. Based on this, engineers can call the different web services provided to read up-to-date data anywhere by using their PC, PDA or mobile phone through accessing WiFi and the mobile internet.

4.5 Information and Communication Infrastructure

Actual industrial maintenance activities are mainly driven by traditional strategies where events are normally time-based. Although it has been pointed out that more advanced condition-based strategies can provide clear savings in many maintenance activities, their application is normally prevented by different causes, such as the need to manage, both physically and logically, an increasing volume of data and information. At Dynamite, one of the main developments has been related to the development of a flexible architecture concept to provide flexible data and information management.

On the one hand, a platform of web services to provide intelligent processing capabilities has been designed. This platform is logically structured according to OSA-CBM decision layers, from condition monitoring to decision support, but also to the existing operators (sensors, PDA, CMMS, etc).

In DynaWeb, in order to provide the most convenient analysis flow, information processing is understood as a distributed and collaborative system, where there are different levels of entities that can undertake intelligence tasks. Given this, with the help of use case diagrams (UCD) using the standard unified modelling language (http://www.uml.org), a system architecture has been defined to identify the interactions between actors and the required functions. Of particular importance is the UCD definition for operation, evaluation and execution of tasks (see Figure 4.6). The specification of this UCD includes four layers that correspond to the central information processing layers of OSA-CBM standard (Thurston and Lebold 2001):

- *Condition monitoring:* Condition monitoring receives data from the sensor modules and the signal processing modules. Its primary focus is to compare data with expected values. The condition monitoring layer should also be able to generate alerts based on preset operational limits or changes in the trend.

- *Health assessment:* This receives data from different condition monitoring sources or from heath assessment modules. The primary focus of the health assessment module is to prescribe if the health of the monitored component, subsystem or system has degraded. The health assessment layer generates diagnosis records and proposes fault possibilities. The diagnosis is based upon trends in the health history, operational status and loading and maintenance history.
- *Prognostics:* This module takes into account data from all the prior layers. The primary focus of the prognostic module is to calculate the future health of an asset, with account taken of the future usage profiles. The module reports the failure health status of a specified time or the remaining useful life.
- *Decision support:* In this context this is related to "schedule work orders". CMMS (computerised maintenance management system) schedules work orders based on component predictions. After that it distributes work orders to different operators' PDAs. The PDAs need to read the smart tags in order to learn about the components (Adgar *et al.* 2007).

All four layers, together with the different actors that can access the information, are represented in Figure 4.7.

The result, as indicated previously, is a three level framework (machine, plant, company) that provides a flexible configuration that allows the system to be used 'on-demand' and to grow according to the needs (new sensors and functionality), together with a flexible communications infrastructure, where a generic wireless 'gateway' device is being developed, in order to complement existing communications options (wired or wireless) between sensors and company decision areas when other communication options are not available (such as SCADA, PDAs, *etc.*). This framework has also been enriched with different ICT components to facilitate this web distributed e-maintenance solution

A mobile handheld device has been developed, which includes wireless access to smart tags and sensors and centralised databases within the e-maintenance infrastructure, including application software for analysis of monitoring data and early stage diagnosis of faulty conditions. Specific developed modules that can be pointed out include the interfaces for managing the information and communicating with operators, the infrastructure and agents for interoperation with remote web services, and the inclusion of specialised models to retrieve information from smart tags and optimise the maintenance scheduling

A dual system for wireless communication across different operators has been developed. This system is composed of a physical gateway for communication between machine and plant, plus a data collector to facilitate communication between wired systems and wireless gateways. The protocols used are Zigbee and Wifi. This physical system is complemented by a translator developed to assist the interpretation of SQL queries into the MIMOSA database structure.

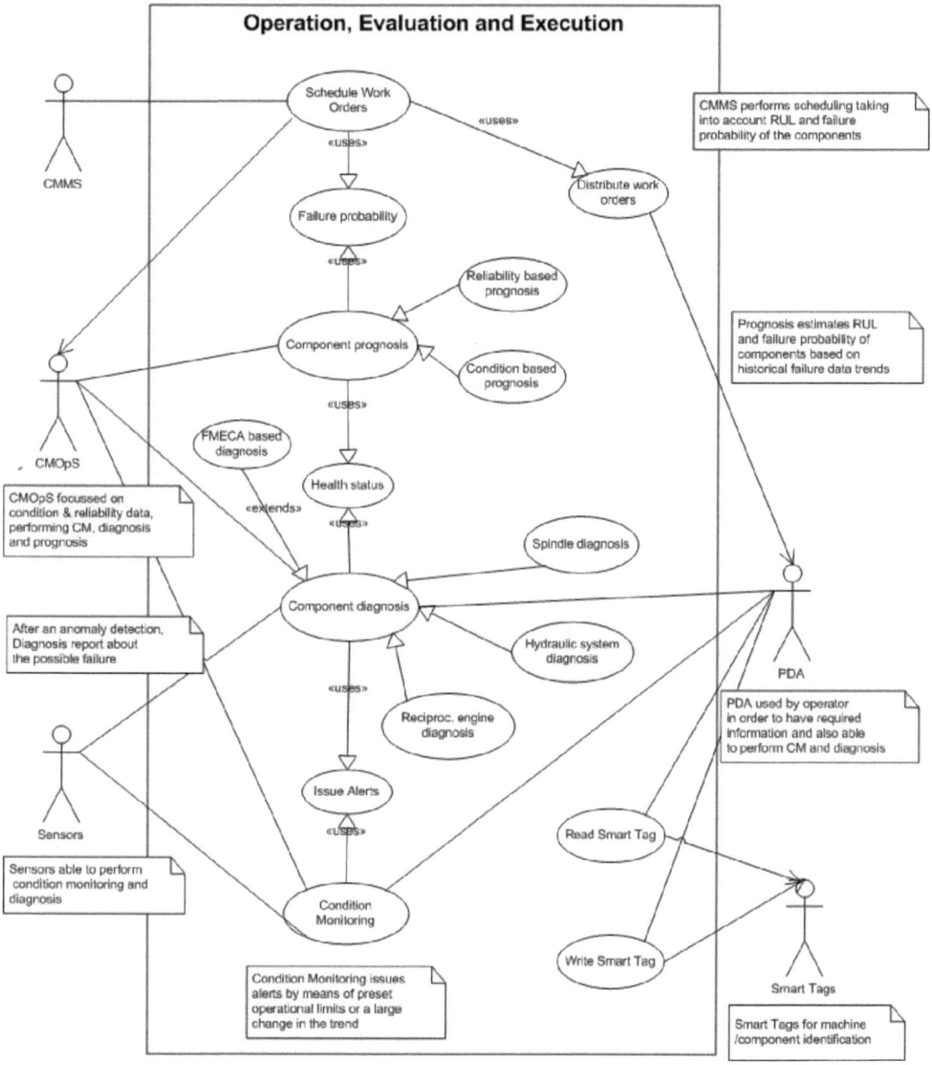

Figure 4.7 Use case diagram for operation, evaluation and execution

A complete set of web services have been designed and developed, covering the complete OSA-CBM information process layered structure. These services can operate from any type of location (sensors, PDAs, PCs) and provide a standardised means to access to information located in any machine, without the need to change existing legacy systems, with a minimum need for adaptation.

Finally, it can also be stated that it was decided to use the MIMOSA architecture as a central part of our development process. This has allowed different positive outcomes:

- Interoperability between all developed components has been secured not just across the entire project
- Many components developed can now be used in two different ways. Straight to MIMOSA databases, or *via* XML message passing to non-MIMOSA databases, also using MIMOSA architecture in the protocols. In this way, the web services can also be standardised, no matter where the information is stored. This is illustrated in Figure 4.8.
- The use of MIMOSA standard, even though not much extended yet at the beginning of the project, seems to be noticeable in many different areas, and this now turns into a positive selling argument, as the compliance of any system with MIMOSA allows interoperability among an increasing number of maintenance software systems.

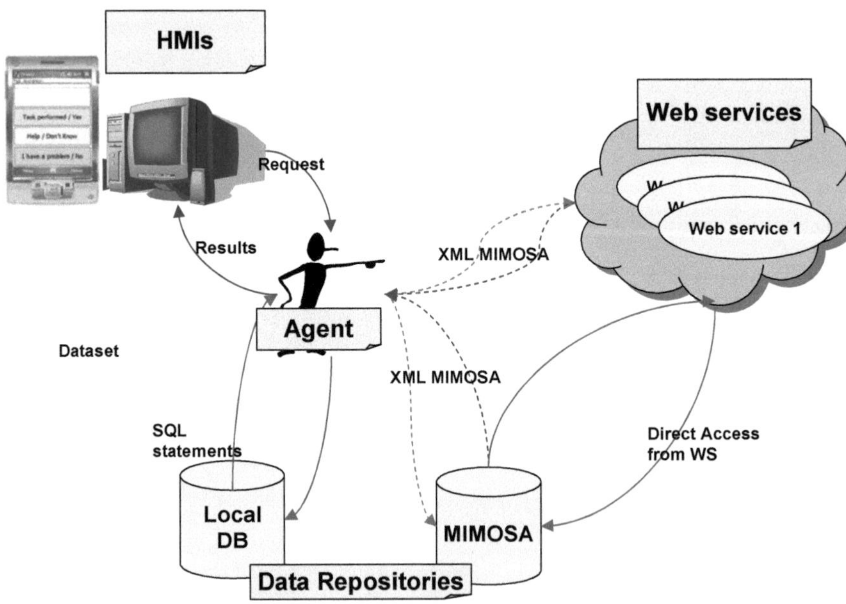

Figure 4.8 Communication options between HMI, agent and web services, depending on existing database characteristics

All the results achieved have been detailed in a list of DynaWeb components. Concerning ICT communications the available components resulting from Dynamite are shown below

For direct operation of the PDA

2. Active smart tag, asset tracking	Zenon
9. Mobile maintenance PDA user interface	VTT
10. Smart tag PDA support	Sunderland University
11. Active smart tag PDA support	Zenon

12. Hand held PDA vibration data collector	Diagnostics Solutions
13. PDA scheduling support	Zenon
14. Smart PDA maintenance user interface	Prisma technologies
For communication across different operators (sensors, PDA, PC), MIMOSA database and web services	
15. Communication SW module	Prisma technologies
16. MIMOSA translator	University Henri Poincaré
17. Collector (=Gateway)	Prisma technologies
18. Wireless communication system for e-maintenance	Prisma technologies
Web services development	
19. Condition monitoring web service	Fundación Tekniker
20. Diagnosis web service	Fundación Tekniker
21. Prognosis web service	University Henri Poincaré
22. DynaWeb e-maintenance platform (TESSNet)	Fundación Tekniker
23. Scheduling web service	Zenon
27. MEMS SW support module	Diagnostics Solutions
28. Vibration measurement system	Wyselec
Structures for MIMOSA	
25. DynaWeb platform	IB Krates
26. MIMOSA database	IB Krates

4.6 Cost-effectiveness Based Decision Support System

In general, decisions of when and why to stop a producing machine and whether it is cost-effective or not are crucial for production profitability especially in companies of intensive capital investments, *e.g.*, the process industry, shipping and engineering manufacturing, where stoppage time is very expensive. It is vital to have a system that provides the reliable data required to achieve cost-effective and dynamic maintenance decisions for maintaining and improving company profitability and competitiveness.

A novel maintenance decision support system (MDSS) has been developed, which offers three different strategies for cost-effectiveness in production and maintenance processes that can be applied in an integrated manner or separately (Figure 4.9). MDSS can help companies to reduce economic losses through mapping the situation of production and maintenance processes and enhance maintenance performance. It allows us to follow up maintenance performance measures more frequently and thereby be able to react quicker in the case of disturbances and thus avoid unnecessary costs. It also facilitates tracing the causes behind deviations.

MDSS consists of three toolsets, where every toolset consists of one to three tools with different functions, see Table 4.3. MDSS provides services that existing systems cannot. It helps to identify and prioritise problems, suggests the most beneficial areas for future investments in maintenance, follow up and control investments results. The applicability and usefulness of MDSS have been tested successfully by personnel from FIAT/CRF (Italy) and Goratu (Spain), and it has been installed at Fiat/CRF, Italy for testing since 16 January 2009. The final conclusion of the test and demonstration of MDSS is that it is user friendly and can be used successfully for analysis of data and achievement of maintenance dynamics and cost-effective decisions. For more details see Chapters 12 and 13.

Table 4.3 Functions of the toolset and tools included in MDSS

Toolsets	Tools	Features and function
MDSS		Easy to use, effective and low cost
Toolset 1		
to enhance the accuracy of maintenance decisions	PreVib (prediction of vibration level)	To predict the vibration level of a component/equipment in the next planned maintenance action or measuring moment for avoiding sudden and dramatic changes and catastrophic failures.
	ProFail (probability of failure)	To assess the probability of failure of a component (using machine past data) at need or when its vibration level is significantly high.
	ResLife (residual lifetime)	To assess the residual life of a component for avoiding failures and delivery delays. It can be used to control whether or not it is possible for the production process to proceed according to the production schedule.
Toolset 2		
simulate and select the most cost-effective maintenance solution	AltSim (alternative simulation)	To simulate technically applicable alternative solutions suggested for a particular problem and to select the most cost-effective maintenance solution using an intelligent motor.
Toolset 3		
to identify & prioritise problem areas	MMME (man-machine-maintenance-economy)	To identify and prioritise problem areas and to assess the losses in the production time.
to map, follow up, analysis and assess the cost-effectiveness of maintenance	MainSave (maintenance savings)	To monitor, map, analyse, follow up and assess maintenance cost-effectiveness, *i.e.*, maintenance contribution in company profit.

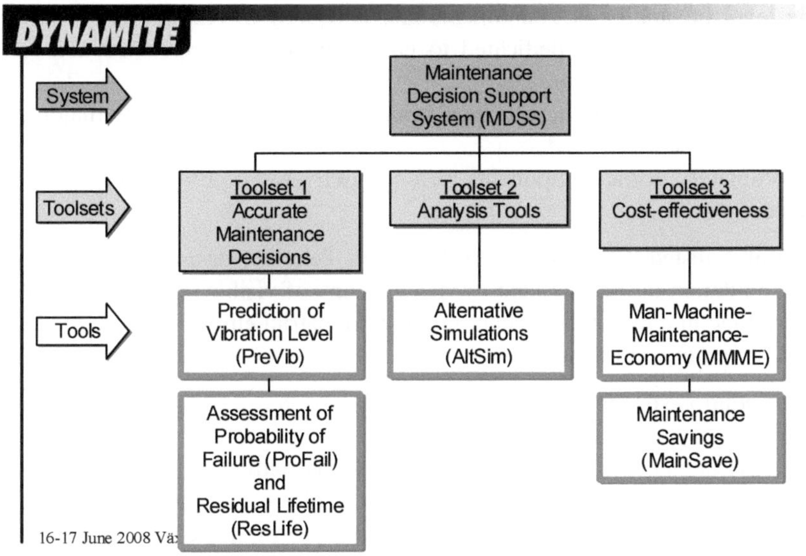

Figure 4.9 MDSS strategies for cost-effectiveness

4.7 DynaWeb Demonstrations

DynaWeb demonstrations were carried out in an industrial environment on a global level, with a milling machine, machine tools, foundry hydraulics and a maritime lubrication system.

The functionality of DynaWeb and its components were tested and demonstrated in the following way:

- Demonstrations were performed at four different test sites at Fiat, Volvo, Goratu and Martechnic.
- Technical and economical evaluations of these demonstrators were carried out.
- Recommendations for further implementation, development and industrialisation were made.

The following results were achieved at the four demonstration sites:

(1) Fiat tested and demonstrated the integration between 25 DynaWeb hardware, software components and services. The demonstration was done in an industrial machining centre similar to what is used in car production. Detailed results are available in the technical reports. As a summary of the demonstration:

- The overall results are extremely positive, with technical and economical feasibility proven.

- The level of quality of components and adequacy to requirements was high, with people extremely dedicated to enhancing their components and testing them on the demonstrator.
- As expected, integration was not straightforward and required major effort from all partners involved.
- Unfortunately, some components were not delivered on time and thus not integrated.

(2) Volvo tested the oil sensor system from Tekniker designed to measure the level of oxidation of the lubricant by spectroscopy of visible light. The demonstration was done in a hydraulic system in a real industrial environment (a production line in the foundry). In conclusion:

- The oxidation sensor hardware and software worked well in the foundry installation.
- The environment in the foundry at Volvo was extremely dirty, which was a good test for the sensor but made it impossible to have the computer at the same location. The sensor required a continuous low speed oil flow without air bubbles and at a low oil pressure. The sensor signal jumped up and down depending on, *e.g.*, irregular oil flow, air bubbles, *etc.*, which made the interpretation more difficult and not straightforward.
- Volvo IT policy made it almost impossible to demonstrate communication with the MIMOSA database at the external server but a one-way web service communication to store data in the MIMOSA database was created by Tekniker and included in the software and tested.

(3) Goratu tested several DynaWeb components and their communication to the MIMOSA database.

- The VTT particle scatter lube sensor for hydraulic system, the Tekniker water content lube sensor for cooling system and the Wyselec vibration measurement system for spindle vibration were implemented at a Goratu machine, and the data collected where sent to the MIMOSA database located at the IBK server. All the data collected provided good information for Goratu, who did not have any kind of information related to these issues. Apart from this, the web services provided an important tool for machine reliability. Web services allowed Goratu to implement the online diagnosis and condition monitoring, which had until then been impossible.
- The hand held vibration unit and the PDA maintenance user interface were also tested. These allowed the insertion of new assets into the database and taking measurements using a PDA, which due to its size is very comfortable for the user.
- This demonstration gave new feedback to Goratu, who will use this information for machine improvements and new utilities for the customers. The innovation is very important, but some improvements are needed for a full imple-

mentation in an industrial environment, higher flows and pressure for sensors and better filters for vibration system.

(4) At Martechnic the demonstration consisted of a simulated application of a stern tube bearing/tail end shaft assembly from an 8000TEU container ship. For logistical and security reasons this demonstration could not take place on board the ship. Specially designed test rig conditions on board a ship were replicated and the cycled lube oil was progressively contaminated with water and particulate matter.

- The demonstration, which ran for nine days, evaluated four sensors (one from Tekniker was not enabled for DynaWeb communication). The other three (two from Martechnic and one from VTT) performed satisfactorily, communicating their results *via* two separate routes to the MIMOSA database.
- The demonstration was deemed a success and the economic scenario surrounding this application clearly demonstrated considerable benefits resulting from the application of the DYNAMITE concepts.

4.8 Conclusions

This chapter outlined the motivations, activities, technologies and results related to the Dynamite project. Even tough the specifics related to each topic are described in detail in the following chapters, this one serves as an introduction to the rest of the book.

Firstly, it must be pointed out that a pioneering e-maintenance solution named DynaWeb has been developed. It is based on scenario analysis of future industrial needs and trends for plant operators, OEM manufacturers, transportation and consulting companies. DynaWeb is a flexible web distributed ICT structure capable of multi-level condition monitoring and maintenance data treatment with common MINOSA structure, internet web services, training services and decision support based on technical and economical considerations. Lastly, DynaWeb consists of 28 integrated hardware and software components. They include smart MEMS sensors with energy harvesting, on-line lubrication sensors, smart tags for identification and location of components, maintenance actions supporting handheld mobile computers (PDAs), wireless communication and a strategic and economical decision support system.

In addition, a condition monitoring data and statistically based decision support system MDSS has been developed. It includes toolsets for accurate maintenance decisions, maintenance analysis and cost-effectiveness.

Finally, DynaWeb components and the integrated structure have been successfully tested and demonstrated on a global level, with a milling machine, machine tools, foundry hydraulics, a maritime lubrication system and automatic stamping machine industrial installations.

References

Adgar A, Addison JFD, Yau C-Y (2007) Applications of RFID technology in maintenance systems. In Proc 2nd World Congress on Engineering Asset Management, Harrogate, UK, 11-14 June 2007. Coxmoor, Oxford

Arnaiz A, Gilabert E, Jantunen E, Adgar A (2009) Ubiquitous computing for dynamic condition based maintenance. Journal of Quality in Maintenance Engineering (JQME), 15:151–166

Arnaiz A, Iung B, Jantunen E, Levrat E, Gilabert E (2007a) DYNAWeb. A web platform for flexible provision of e-maintenance services. Int Congress on Enterprise Asset Management and Condition Monitoring, Harrogate, UK, June 2007. Coxmoor, Oxford

Arnaiz A, Levrat E, Mascolo J, Gorritxategi E (2007b) Scenarios for development and demonstration of dynamic maintenance strategies. ESReDA (European Safety, Reliability & Data Analysis) 31st Seminar, Sardinia, Italy, May 2007

Holmberg K, Helle A, Halme J (2005) Prognostics for industrial machinery availability. Maintenance, Condition Monitoring and Diagnostics – International Seminar, Oulu, Finland, POHTO, 17–29

OSA-EAI CRIS (2008) Common relational information schema (CRIS), Version 3.2.1 Specification, 31/12/2008, http://www.MIMOSA.org/downloads/44/specifications/index.aspx, accessed 04/2009

OSA-EAI v2.3.1 (2008) OSA-EAI specification v2.3.1. http://www.MIMOSA.org/downloads/44/specifications/index.aspx, accessed 04/2009

Thurston M, Lebold M (2001) Standards development for condition-based maintenance systems. New frontiers in integrated diagnostics and prognostics. 55th Meeting of the Society for Machinery Failure Prevention Technology, MFPT

Chapter 5
Intelligent Wireless Sensors

Samir Mekid, Andrew Starr and Robert Pietruszkiewicz

Abstract. This chapter summarises the latest trends in the use of intelligent sensors in engineering applications. New approaches and computation methods used in different research areas are discussed here. Materials used for this report were selected from a broad range of academic and public sources. They emphasise the scientific motivation that brings technological development in that area.

The key points discussed are:

- Parameters and types of sensors currently used.
- What makes the science world interested in intelligent sensors?
- What are the currently developed applications in academia?
- Benefits from using sensors.
- Processing capacities offered by intelligent sensors.

The key targets for intelligent sensors are to research and utilise novel technologies that can perform the required functions robustly, inexpensively and at extremely low power.

5.1 Introduction

5.1.1 Fundamental Definitions

5.1.1.1 Definition of an Intelligent Sensor or Smart Transducer

There are many products available in the market using the term "intelligent", stating that a product is more advanced than the competition (Figure 5.1). To differentiate between these products a definition is required. There are several definitions available describing intelligent sensors:

- "A sensor that is capable of modifying its internal behaviour to optimize the collection of data from the external world" (White 1997).
- "A device that combines a sensing element and a signal processor on a single integrated circuit" (Powner and Yalcinkaya. 1995).
- "A smart sensor provides various functions beyond those necessary to generate better decision making or better controlled quantity. The intelligence aspect is improved in a networked environment" (Mekid 2006)

5.1.1.2 Effectiveness of Conventional Sensors

Existing conventional sensors have a very limited functionality. They were designed for specific applications to acquire a specific type of measurement and pass the collected data to the higher level monitoring system. Conventional sensors do not have any data calculation capabilities. This task is related to the centralised monitoring system and results in the following problems:

- *Complexity:* a limited number of sensors may be installed in each system, imposed by the level of complexity that human designers can deal with.
- *Cost:* the system is composed of a small number of highly specialised, relatively expensive sensors.
- *Flexibility:* the resulting system cannot be easily expanded, modified, maintained or repaired. Highly trained personnel are required for these functions.

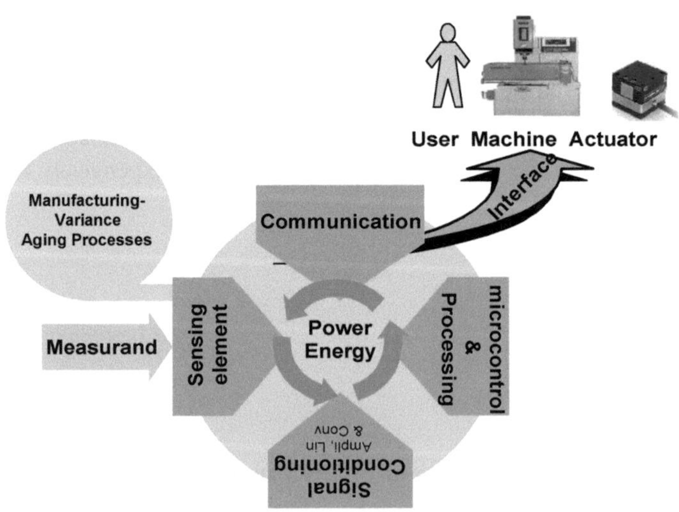

Figure 5.1 Intelligent sensor architecture

Intelligent sensors have more to offer than just the data collection task. These sensors are implemented with data processing capabilities and can perform additional functions such as:

- *Compensation:* self-diagnostics, self-calibration, adaptation.
- *Computation:* signal conditioning, data reduction, detection of trigger events.
- *Communications:* network protocol standardisation, communication with other sensors
- *Integration:* Coupling of sensing and computation at the chip level, *e.g.*, MEMS (micro-electro-mechanical systems)
- *Others:* Multi-modal, multi-dimensional, multi-layer, active, autonomous sensing.

5.1.2 Benefits of Using Intelligent Sensors

Intelligent sensors are able to operate with effective data collection techniques. They enable the development and application of more flexible sensor networks that efficiently utilise and coordinate the limited resources of each individual sensor. By focusing resources according to the state of the surrounding environment and on the immediate task, more efficient operation of the sensor is ensured. The following are some other benefits:

- *Accuracy:* an intelligent sensor will incorporate features that enable it to compensate for systematic errors, system drift and random errors produced due to system parameters or the characteristics of the sensor. Self calibration is one of the most required characteristics.
- *Reliability:* the incorporation of data and sensor validation techniques to detect corrupted data, self-testing of network path connections and sensor operation, as well as calibration of sensor drift, provides yet another level of system reliability in addition to techniques already applied in the network design.
- *Adaptability:* the processing parameters of an intelligent sensor system should be determined automatically and adopted by a higher level in the system architecture. This enables the optimisation of the measuring and processing operations, as well as enabling the sensor to adequately respond to changing environmental conditions.
- *Bandwidth reduction:* in the case where the number of sensors in a system is expanding, it might cause severe data processing bottlenecks. By localising signal processing and reduction, the data communication bandwidth can also be reduced.
- *Advanced data processing:* enabling the "intelligence" to be implemented to the sensing strategy. This is the main advantage of using intelligent sensors above conventional sensors.

5.1.3 Businesses Driven Development of Intelligent Sensors

Intelligent sensor technologies attract a lot of interest from various industries, looking for implementation of this technology. Currently the most influencing markets include:

- Industrial automation and process control, unattended sensors and real-time monitoring over wide areas, utilities such as automated meter reading, building automation, heating and cooling patterns to provide optimised control. In a level of declining industrial automation marketplace, big growth can be expected (Shen *et al.* 2004).
- Medical industry, especially micro-endoscopy for which the ability to navigate micro or nanostructures through the human body has the potential to make a significant impact on modern medicine.
- Condition monitoring for critical and vital equipment. The working level must be kept at its highest level, *e.g.*, avionics, space research, manufacturing and military applications.
- Automotive applications where modern vehicles are designed to be more comfortable and safe, thus they are packed with sensors. The development of new sensors becomes a very important part of automotive industry.

It is being predicted by automation experts that within the next few years, these technology developments will impact industrial and commercial markets, bringing new opportunities. The forecast encompasses more popularity for technology growth, *e.g.*, distributed sensing and computing will be present almost everywhere: homes, offices, factories, cars, shopping centres, super-markets, farms, forests, rivers and lakes. The trend will impact many aspects of life. Smart, wireless networked sensors will soon be everywhere around us, collecting and processing huge amounts data from air quality and traffic conditions, to weather conditions and tidal flows. And this means not only monitoring a few isolated sensors, but literally tens of thousands of intelligent sensor nodes, which will provide not only local measurements, but overall patterns of change.

With the advances in nanotechnology, atomic-scale sensors will emerge on the scene. Soon, MEMS and nanotechnology will yield tiny, low-cost, low-power sensors. The tiny aspect is important because they can be scattered around unobtrusively to measure just about everything that can be imagined. Low power means they will not need to carry large batteries and may even be solar or source powered. Low cost is also an advantage key as the required numbers will be enormous.

Several new companies are already producing ultra low-power, postage stamp-sized smart sensors gathering good results in a variety of applications. However, thumbnail size sensors are still only an interim stage. Soon, integrated sensors and silicon will yield microscopic components that can be scattered around like “smart dust”.

5.2 State-of-the-art Intelligent Sensors

Over the past decade or so, intelligent sensors have risen from being an academic pipe dream into practical and commercial devices. The main reason for their continued success is largely as a result of the major advances in the area of microelectronic technologies. Limitations of conventional sensors can be compensated by the use or the advantages brought by the capabilities offered by the intelligent, miniaturised and wireless sensors.

Like any sensor, an intelligent sensor is primarily aimed at providing estimations of system variables, to be used in control loops, or in decision making for control, maintenance or management of the system. However, intelligent sensors are not components that are just interconnected to the rest of the application. An intelligent sensor that conforms to this definition should autonomously perform advanced processing functions such as self-validation, self re-calibration, fault detection, and sensory data filtering and feature extraction.

The presented functional view of intelligent sensors shows that features of interest from the scientific point of view are the growth in processing capabilities such as: estimation, characterisation, validation and fault tolerance. On the contrary, due to the extra functions they implement, they fully participate in the architecture of complex distributed control systems by offering services at the supervision level.

One of the prime issues with wireless sensor networks is the power consumption of numerous sensor nodes and the requirements to provide periodic maintenance including battery replacement. This issue can be resolved by the implementation of energy harvesting methods. Their use can eliminate the need for battery replacement, making sensors even more independent and their functionality even more distributed.

The main area of improvement comes from the possibility of using the advanced data processing directly in the sensor. The true distributed data processing opens brand new possibilities and suggests a future direction from the technological point of view. This shift from the conventional centralised data processing and moving the monitoring task direct to its source will be the main challenge for the future generations of WINS (wireless intelligent network sensors).

The first effort to apply the wireless sensing technology to a structural health monitoring system, for example, was in civil engineering structures applications (Straser *et al.* 2003). Based on the work by Straser, Lynch *et al.* (2003) demonstrated a model of a wireless sensor using standard integrated circuit components. Other researchers presented more models of wireless sensor networks (Maser *et al.* 1997, Mitchell *et al.* 1999, Liu *et al.* 2001). Moreover, in 2000, the "Smart Dust" project was funded by US Defense Advanced Research Projects Agency (DARPA), in which the ultimate goal was to develop low-cost, small and high-reliable wireless sensing systems (Spencer 2003).

5.2.1 Several Functions Within One Platform

The main objective for the design specification of an intelligent sensor is to identify the state-of-the-art in technology design and match the requirements to the need of the condition monitoring sector. The specification should suggest available directions for the design of a multi-measuring device for the proposed sensors range. The specification includes general requirements for the functioning of the device as well as the functional specification for possible hardware solutions and possible alternatives.

The second aspect of the specification is to investigate commercial solutions for wireless transmission mediums. Three main contesters were selected: Bluetooth, ZigBee and WiFi. The device should receive as well as transmit, so that it can be programmed. Wireless standards have been reviewed in the context of power requirements, bandwidth, range, and commercial outlook for the applications.

Figure 5.2 illustrates the conceptual dependencies of the main parts of the designed system.

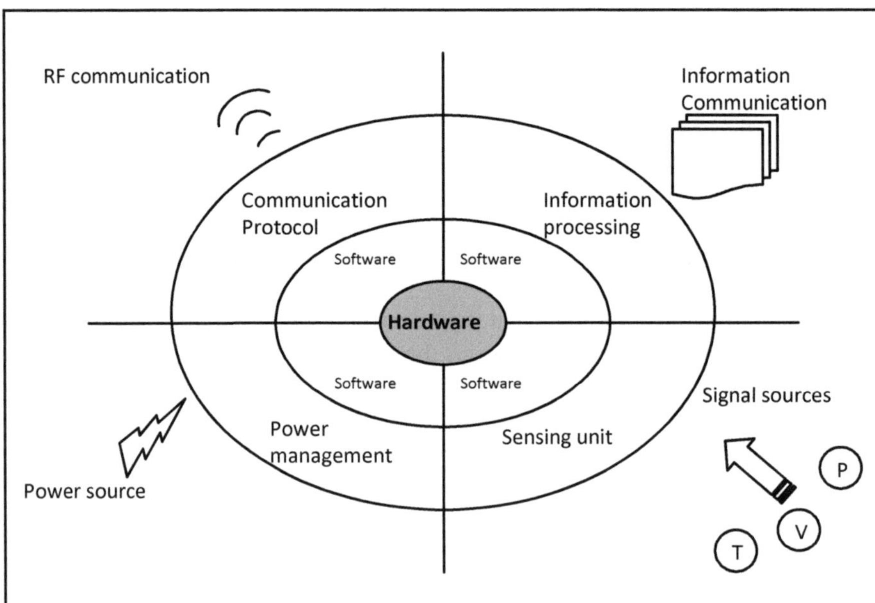

Figure 5.2 Environment influencing the hardware specification

There are four design considerations that will obviously influence the design of the sensor:

- signal sources – sensing unit or transducers
- information – information processing unit

- power source – power management in the platform
- radio frequency communication – communication protocol.

Figure 5.1 shows a conceptual configuration of an intelligent sensor. The designed intelligent sensor was foreseen as a self contained monitoring system. By applying the internal pre-processing in the sensor itself, it should be able to test data and automatically decide if the system is functioning normally. If the results from the data suggest an abnormality, a decision about the severity of the fault should follow. In the case of the detection of a faulty situation, the sensor should be able to make a decision on what to do next. This decision is based on automated reasoning. Intelligence is also necessary to perform self diagnostics to ensure the micro-sensor is working properly. A validation of the data acquisition unit therefore needs to be incorporated into the communicated diagnostic information. Decision making and reasoning will be the part of the advanced capabilities of the intelligent sensor designed to operate in a network system with a minimum of unnecessary traffic. These functions are an essential part of the intelligent sensor and the intelligent sensors based system.

Depending on the communication type between the model and the other elements in the system (communication strategies) there might be various requirements on the hardware and possible applications of the sensor. Considering the different possible levels of distribution within a condition monitoring system, three profiles of communication are being offered. These communication strategies demonstrate the use of distributed processing methods and will provide experience and results from the tests. The strategies represent the range of distributed processing, from the minimal distribution of intelligence as in a conventional data acquisition system, to the other extreme, where distribution of the data processing offers elements virtually independent from the supervisory system. The middle profile offers a balanced solution with distributed processing capability and elements cooperating and being controlled by the central supervisory stations. These three different strategies are designed to illustrate the use of different levels of distribution and show how the communication strategies might influence required hardware involved in the intelligent sensors.

5.2.2 Hardware

Today's low-end sensors use low cost reduced instruction set computer (RISC) microcontrollers with a small program (about 100 kb) and data memory size. An external flash memory with large access times may be optional to provide secondary storage and to alleviate the application size constraints imposed by the chip memory. Common on-board I/O buses and devices include serial lines such as the universal asynchronous receiver–transmitter (UART), analogue to digital converters and timers.

Two approaches have been adopted for the design of transducer equipment. The most general and expandable approach, as pioneered by Crossbow (2006), consists of developing transducer boards that can be attached, and possibly stacked one on top of the other, to the main microcontroller board through an expansion bus. A typical transducer board from Crossbow provides light, temperature, microphone, sounder, tone detector, two-axis accelerometer and two-axis magnetometer devices. Alternatives include economical versions that provide a reduced set of transducers or more expensive versions that boast GPS, for instance. Special boards are also available that carry no transducers but provide I/O connectors that custom developers can use to connect their own devices to the Crossbow sensors.

The second approach is to put transducers directly on the microcontroller board, a solution also followed by Moteiv (Kahn *et al.* 1999).

Transducers are soldered or can be mounted if needed, but the available options are very limited and generality and expandability is affected. On the other hand, these on-board transducers can reduce production costs and are more robust than standalone transducer boards, which may detach from the microcontroller board in harsh environments.

By means of the transceiver circuitry a sensor unit communicates with nearby units. Although early projects considered using optical transmissions (Moteiv website and the SmartDust program), current sensor hardware relies on RF communication.

Optical communication is cheaper, easier to construct and consumes less power than RF but requires visibility and directionality, which are extremely hard to provide in a sensor network. RF communication suffers a high path loss and requires complex hardware, but is a more flexible and understood technology.

Currently available sensors employ one of two types of radios. The simplest and cheaper alternative offers a basic carrier sense multiple access (CSMA) medium access control (MAC) protocol, operates in a license free band (315/433/868/916 MHz) and has a bandwidth in the range 20–50 Kbps. Such radios usually offer a simple byte oriented interface that allows software implementations of arbitrary energy efficient MAC protocols.

Newer models support an 802.15.4 radio operating in the 2.4 GHz band and offering a 250 Kbps bandwidth. The latter offers the possibility of using an internal, *i.e.*, on-board, antenna, which makes sensors more manageable and self-contained with respect to an external whip antenna. The radio range varies with a maximum of about 300 m (outdoor) for the first radio type and 125 m for the 802.15.4 radios (Baronti *et al.* 2007).

5.2.3 Wireless RF Standards

Wireless sensors systems present novel requirements for low cost, low power, short range, and low bit rate RF communication. In contrast to previous emphasis in wireless networks for data communication, distributed sensors and embedded microcontrollers raise new requirements while relaxing the requirements on latency and throughput. The sensor's communication module becomes an embedded radio with a system that may be added to compact micro-devices without significantly impacting cost, form factor or power. However, in contrast to previously developed simple, low power RF modems, the WINS device must fully support networking capability. In addition, the WINS radio should be compatible with compact packaging. Communication and networking protocols for the embedded radio are a topic of project. However, simulation and experimental verification in the field indicate that the embedded radio network must include spread spectrum signalling, channel coding, and time division multiple access (TDMA) network protocols.

The operating bands for the embedded radio are most conveniently the unlicensed bands at 902–928 MHz and near 2.4 GHz. These bands provide a compromise between the power cost associated with high frequency operation and the penalty in antenna gain reduction with decreasing frequency for compact antennas. The currently available prototype, operational, wireless sensors networks are implemented with a self-assembling, multi-hop TDMA network protocol (Asada *et al.* 1998).

Well known challenges accompany the development of RF systems in CMOS technology (Abidi 1995). Of particular importance to the embedded radio are the problems associated with low transistor trans-conductance and the limitations of integrated passive RF components. In addition, WINS embedded radio design must address the peak current limitation of typical battery sources, of 1 mA. This requires implementation of RF circuits that require one to two orders of magnitude lower peak power than conventional systems. Due to short range and low bit rate characteristics, however, the requirements for input noise figure may be relaxed. In addition, channel spacing for the embedded radio system may be increased relative to that of conventional RF modems, relaxing further the requirements on selectivity. Constraints on operating requirements must consider, however, resistance to interference by conventional spread spectrum radios occupying the same unlicensed bands (Abidi 1995).

Of the three domains, a sensor node expends maximum energy in data communication. This involves both data transmission and reception. It can be shown that for short-range communication with low radiation power (about 0 dbm), transmission and reception energy costs are nearly the same. Mixers, frequency synthesisers, voltage control oscillators, phase locked loops (PLL) and power amplifiers all consume valuable power in the transceiver circuitry. It is important that in this computation we not only consider the active power but also the start-up power consumption in the transceiver circuitry. The start-up time, being of the order of hundreds of micro-seconds, makes the start-up power non-negligible. This high value for the start-up time can be attributed to the lock time of the PLL. As the transmission packet size is reduced, the start-up power consumption starts to dominate the active power consumption. As a result, it is inefficient in turning the transceiver on and off, because a large amount of power is spent in turning the transceiver back on each time.

Narrow-band radios are finally disappearing, having not been entirely successful in every implementation. The competition between direct sequence and frequency hopping still continues. The newest contender, ultra wide band, is becoming known in the marketplace and could be a serious option in many applications. It has been suggested that IEEE 802.11 will have a big impact in industrial markets. It has not happened so far and Bluetooth technology still is an alternative and a common industrial communication standard. This has caused a number of vendors to abandon their own radio systems and eliminate their development efforts to provide in-house radios in favour of the huge cost-and-effort saving potential offered by the new wireless standards. Table 5.1 shows the existing communication technologies.

Table 5.1 Wireless standards for data application (EZURIO 2006)

Wireless standards for data application					
Source		Mature	New	Trial	Development
ETSI	2G	GSM GPRS	EDGE		EDGE Ph2
	3GPP		UMTS/WCDMA TD-CDMA	HSDPA	HSUPA
	3GPP2	CDMA	1xRTT	1xEV-DO	1XEV-DV
IEEE	802.11	Wi-Fi.11b .11a .11g			.11n
	802.15	802.15.3 – UWB			MBOA/DS-UWB
		802.15.1 v1.1	v1.2	v2.0+DER.15.4(ZigBee)	
	802.16	WiMAX	.16°	.16d	.16e
	802.20	MAN			

ETSI standard including: *2G, 3GPP, 3GPP2* are the technologies used in mobile phones.
IEEE standard containing the range *802* are the low range wireless.

It is still uncertain whether Bluetooth will be able to find the solution to all problems it has with its industrial applications. Bluetooth has been examined by some vendors and it was decided that it was not appropriate for their particular market and applications. In some cases, vendors may have been discouraged by the tradeoffs made to control cost and improve throughput at the expense of reliability in Bluetooth. Certainly, Bluetooth as a technology should not be totally discounted just because it is not ideal, it may still be successful. There are two spin off technologies originating from the Bluetooth standard. They are Bluetooth EDR with a better bandwidth and a new protocol in industrial communication,

ZigBee. A number of companies, including several new on the market, are actively bringing new wireless technologies to market.

A short comparison of the most popular technologies and their parameters is presented in the Table 5.2 and Figure 5.3.

Table 5.2 Comparison of wireless communications protocols (Allen 2005)

Parameter	Wi-Fi (IEEE 802.11)	Bluetooth (IEEE 802.15.1)	ZigBee (IEEE 802.15.4)
Range	About 50 m	About 10 to100 m	About 10 m
Bandwidth/throughput	868 MHz/20 kbits/s 915 MHz/40 kbits/s 2.4 GHz/250 kbits/s	2.4 GHz/1 Mbit/s	868 MHz/20 kbits/s 915 MHz/40 kbits/s 2.4 GHz/250 kbits/s
Power consumption	400 mA (TX on) 20 mA (standby)	40 mA (TX on) 0.2 mA (standby)	30 mA (TX on) 1 μA (standby)
Protocol stack size	100 kbytes	100 kbytes	32 kbytes
Battery life	Minutes-hours	Hours-days	Days-years
Relative node physical size	Large	Medium	Small
Relative cost/ complexity	High	Moderate	Low

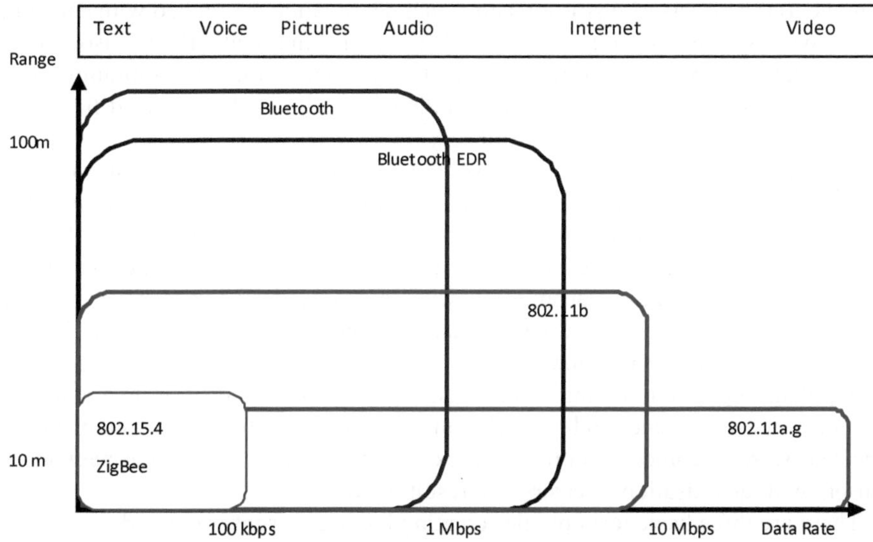

Figure 5.3 Range cooperation at similar power levels (EZURIO 2006)

5.2.4 Intelligent Sensor Networks

The categorisation of a sensor as intelligent implies that the sensor incorporates more functionalities than merely providing an output measurement as introduced previously. There are some discrepancies governing what makes a given sensor intelligent. There are varying levels of sophistication used by sensors, which have claimed to be intelligent or smart, ranging from merely incorporating an operational amplifier for the output signals, to advanced data modelling techniques for condition monitoring. Hence, a smart sensor is defined as a sensor that provides functions beyond those necessary for generating a correct representation of a sensed or controlled quantity and having on board advanced processing capability. The intelligent sensors discussed in this book have built-in wireless communication capability. This function typically simplifies the integration of the transducer into applications in a networked environment.

Intelligent sensors were designed to work and interact as a complex system, unifying the sensors into a network. Intelligent sensor networks are used in applications where a number of various sensors are needed or where the sensor devices are distributed geographically.

The initial aim was the simplification of the wiring required for signal transmission. Additionally, the digital nature of networked signals brings robustness and reliability to the system. The digital transmission is relatively immune to the effects of distortion and signal degradation associated with carrying an analogue signal over long distances. This implies that networked sensors have ADC (analogue-to-digital converter) capabilities. The ability to communicate a much wider range of information in both directions allows the expected functions from intelligent sensors to be fully utilised.

Another aspect is that the networked sensors typically contain a local microprocessor that handles sensor signals and their transmission. This gives the opportunity not to limit the microprocessor to transmission functions only, but also to use the calculation capability to perform additional calibration or signal corrections. Using digital transmission sensors can be designed to have multiple sensing functions. Each signal can be handled and transmitted separately by the sensor without extra connections.

A potential problem arising from organisation of communication in a complex system is network bandwidth, which can cause unreliability of communication. The hardware becomes a more complex circuitry compared to non-networked sensors with quantisation of errors as a result of ADC.

There are three basic technologies creating intelligent sensor networks:

1. Micro miniature, ultra-low-power sensors. Currently, these are usually MEMS structures that are fabricated identical to silicon integrated circuits.
2. Embedded silicon chips, wireless transceivers and firmware for P2P communications and self-organising systems. While the individual nodes are

relatively fragile and communicate over only small distances, the complete networks are robust, with communication through multiple redundant paths.
3. Software for communications, control and optimisation for thousands of nodes.

Together, these technologies bring intelligent wireless sensor networks that can be used over wide areas. The main ability is to monitor the plant in the real-time and possibly analyse data before it will be transferred.

The prognoses say that the wireless sensor networks will soon become as important as the internet. Just as the latter allows access to digital information anywhere, sensor networks will provide vast arrays of real-time, remote interaction with the physical world. The industrial automation business will be generating significant growth in this new arena.

5.3 Expected Features and Design of Intelligent Sensors

Prior to the discussion of the features and design of intelligent sensors, conventional sensors are introduced as background information for intelligent sensors because for many of them they constitute the sensing component.

5.3.1 Conventional Sensors

Conventional sensors are basically composed of the sensing element that will measure a certain phenomenon. The data are acquired by an external system to this sensor and analysed separately.

The outlook on the applications of conventional sensors is necessary to prepare a background for intelligent sensors. These are listed below and include the following types of sensors:

- Mechanical sensors such as metallic, thin-film, thick film and bulk strain gauges, pressure sensors, accelerometers, angular rate sensors, displacement transducers, force sensors, bulk and surface acoustic wave sensors, ultrasonic sensors, flow meters and flow controllers.
- Electromechanical sensors of all ranges from macro to micro, on any substrates, such as metal, plastic or silicon.
- Thermal sensors such as platinum resistors, thermistors, diode and transistor temperature sensors, thermocouples, thermopiles, pyroelectric and piezoelectric thermometers, calorimeters and bolometers.
- Optoelectronic/photonic sensors such as photovoltaic diodes, photoconductors, photodiodes, phototransistors, position-sensitive photodetectors, photodiode arrays, charge-coupled devices, light-emitting diodes, diode lasers, other quantum devices and liquid-crystal displays.

- Ionising radiation sensors such as gamma ray, charged particle and neutron detectors.
- Integrated optics/fibre optical devices such as those based on photometry, fluorimetry, interferometry and ellipsometry.
- Microwave/millimetre wave sensors measurement of the micro radio signals.
- Magnetic sensors such as: magneto resistors, Hall-effect devices, magnetometers, magnetic-field sensors, solid-state read and write heads.
- Chemical and biological sensors, with emphasis on the electronics and physics aspects of transducing chemical and biological signals into information about chemical and biological agents.
- Mass-sensitive devices such as quartz crystal microbalances and surface acoustic wave devices.
- Other sensors such with some degree of intelligence, used for applications such as on-line monitoring, process control, and test kits; sensor signal processing and fusion; thin-film and thick-film gas sensors, humidity sensors, specific ion sensors (such as pH sensors), radon sensors, carbon monoxide sensors, viscosity sensors, density sensors, acoustic velocity sensors, proximity sensors, altimeters and barometers.

5.3.2 Examples of Application of Conventional Sensors

A number of applications are introduced here.

- Sensor phenomena and characterisation (sensitivity, selectivity, noise, ageing, hysteresis, dynamic range, interfering effects, *etc.*).
- Sensor systems and applications such as multiple-sensor systems, sensor arrays and "electronic nose" technology, sensor buses, sensor networks, voting systems, telemetering; combined sensors (*e.g.* electrical and mechanical), automotive, medical, environmental monitoring and control, consumer, alarm and security, military, nautical, aeronautical and space sensor systems, and robotics and automation applications.
- Sensor arrays: large and high density sensor arrays, distributed sensor networks, sensitive skin systems, intelligent sensor arrays.
- Sensor-actuators, including integrated sensor-actuators, smart sensor-actuators and network able sensors-actuators.

Some of the above mentioned sensors were designed as intelligent, remotely accessible devices. The development of technologies might bring a broader need for incorporation of intelligence and remote access. It is expected that in the near future most conventional sensors will be available with upgraded intelligence built into their structure. Such a framework is generic, with the intention of being able to apply the upgraded intelligence to a variety of sensor types and across civil and military sensor platforms.

5.3.2 Expected Features of Intelligent Sensors

The main interest for the use of wireless intelligent sensors comes from the advanced functionality offered by this type of sensors. Unlike the conventional sensor they are designed to create a complex architecture enabling implementation of advanced measurement strategies, a facility which is not present or difficult to establish with conventional sensors. Intelligent sensors do not measure new parameters, their use allows us to measure the same parameters but in different scale. Applications of WINS are the new sophistication in feature extraction, allowing better monitoring of the phenomena than was possible before. Additionally, the use of the distributed intelligence in form of local data processing allows faster sampling and reduction of the retained data.

Sensor networks represent a significant improvement over traditional sensors, which are deployed in the following two ways:

- Sensors can be positioned far from the actual phenomenon, for example something known as sense perception. In this approach, large sensors that use some complex techniques to distinguish the targets from environmental noise are required. This is in the case where the specific measurement is not the main point of concern, but the effects are. For example, in building automation, the wind force will have an effect on the stability of the building.
- Several sensors that perform only sensing can be deployed. The positions of the sensors and communications topology are carefully engineered. They transmit time series of the sensed phenomenon to the central nodes where computations are performed and data are fused. This is especially in the case of complex machinery or large objects requiring multiple points of measurement.

An intelligent sensor approach can also be used to improve the higher-level sensor management's confidence in the reliability of sensory data. Onboard condition monitoring and fault detection techniques can be used in preference to reliance on sensor redundancy for ensuring robust measurements. Advanced data based modelling techniques can be used to model non-linear and time-variant sensor systems, avoiding the limitations of linear physical sensor models, and allowing for reconfiguration of the sensor to correct possible errors.

Other functions of an intelligent sensor will be the capability of being autonomous, adaptive to changes in its environment and self-adjusting to effects caused by the environment and other faults. The principal characteristic of such an intelligent sensor is that it is capable of communicating reliable and self-validated signals or features to higher-level supervision systems, for purposes such as information fusion, tracking and estimation. Poor sensory data can be identified by the sensor itself and flagged for quality problems, together with estimates of the likely cause to enable attempts at sensor reconfiguration.

5.3.2.1 Applications in Engineering Areas

There are many research projects undertaken that utilise the benefits of the implementation of intelligent sensors. Applications where wireless sensors have been used range from vibration monitoring to systems that can detect the presence of micro-organisms in food products by sensing the temperature rise of the products, or capacitive displacement transducers that can accurately sense position, speed and acceleration by measuring the (trans) capacitances in multi-electrode structures. Neural networks are already finding many uses, particularly in multi-element sensor arrays such as those found in the so-called electronic nose.

The list of applications where wireless sensors have found their use is very long. A few examples from the literature survey are presented below:

- Wireless sensors network for vibration measurements. Structural health monitoring (SHM) systems are applied for condition monitoring of machines and structures, structural integrity assessment, damage detection and structural failure prediction. Measurement data acquired by many sensors are essential for SHM; application of many sensors located on mechanical structures without wiring makes the monitoring process more efficient (Boiko 2005).
- Passive wireless strain and temperature sensors. Approaches to wireless strain and temperature measurements that employ passive sensors based on two types of surface acoustic wave devices, reflective delay lines and resonators (Shrena *et al.* 2003).
- Surface acoustic wave devices based wireless measurement platform for sensors. In some applications, a wireless readout is necessary because of the difficulty of fixed connection between sensors and signal processing unit. Wireless surface acoustic wave sensors are being used for sensing some physical and chemical phenomena passively (Han and Shi 2001).
- Wireless sensors to measure gaps efficiently. A network of small, wireless sensors helps measure seal gaps in real time (Danowski *et al.* 2003).
- Nano-based resonator gas sensors for wireless sensing systems. Microwave carbon nano-tube resonator sensors for gas sensing applications.
- Wireless sensors for damage detection and correlation-based localisation. Damage detection and correlation-based localisation demonstration using wireless sensors (Patra *et al.* 2000).
- Noise reduction in RF cavity wireless strain sensors. In this research the noise reduction techniques for this new type of wireless sensor for use in monitoring strain in civil structures is analysed.
- Passive wireless SAW sensors based on Fourier transform. Application of surface acoustic wave resonators as sensor elements for different physical parameters such as temperature, pressure and force (Bazuin 2003).
- Wireless sensors for structural health monitoring. It is "smart" in that it contains a chloride sensor and a RFID chip that can be queried remotely both to identify it and to indicate chloride concentration levels.

- Integrated wireless piezoelectric sensors. Piezoelectric sensor arrays and sensor networks have been suggested as a means to monitor the integrity of composite structures throughout the service life, for instance of an aircraft (Zaglauer 2000).
- Wireless network of sensors for continuous monitoring of vital bio-signals. The concept of a wireless integrated network of sensors can provide an advanced monitor and control medium for healthcare services (Prentza *et al.* 2004).
- Wireless sensors in agriculture and food industry. Wireless sensors and sensor networks are being applied in agriculture and food production for environmental monitoring, precision agriculture, man to machine based machine and process control, building and facility automation and RFID-based traceability systems are given (Zhang *et al.* 2006).
- Wireless microwave based moisture sensors. Microwave moisture sensors, battery powered and capable of communication with the host control system *via* spread spectrum wireless communications (Moschler and Hanson 2004).
- Wireless sensors applied to model analysis. Approaches to wireless hardware and software are suggested that could parallel calculations and thus reduce calculation time and improve data quality by elimination of wires (Kiefer *et al.* 2003).
- Smart microphone, suitable for outdoor acoustic surveillance on robotic vehicles. This smart microphone will incorporate MEMS sensors for acoustic sensing, wind noise flow turbulence sensing, platform vibration sensing (Asada *et al.* 1998).

5.3.2.2 Future Directions for Intelligent Sensors

There are many applications waiting to benefit from the use of intelligent sensors. There is a general perception that it will become a new trend in technology to use intelligent devices. The intelligent sensors are expected to use distributed processing to give additional benefits.

As with every new technological development, popularisation is not possible if the technology is kept commercially protected. Any system without interoperability and exchangeability will always be limited. The solution to this problem will be standardisation of transducer interfaces, *e.g.*, the electrical and mechanical connections. It is also crucial to have an open communications protocol available to other vendors, allowing them to join efforts and contribute to the popularisation process. The expectation is that the addition of communication capabilities will contribute to a general usability of the technology and not be limited to a specialised high-tech application. In the ideal world the intelligent sensors are plug-and-play, autonomous, distributed, re-configurable, self-calibrated, inexpensive (compared to conventional systems) and most importantly are the obvious choice to use.

5.3.3 Processing Capacity Offered by the Use of Intelligent Sensors

The main advantage of intelligent sensors comes from the fact of using the advanced processing methods adding the "intelligence" factor to the equation. The evolution of microprocessors used in the smart devices is just one aspect of increased performance. Digital signal processors (DSPs) are running faster and have improved development tools, making it simpler for designers to employ them in applications that need faster math processing than general-purpose microprocessors can provide. Many smart devices are starting to move to DSPs, which are able to extract more data from sensors. DSPs also make it possible to run more diagnostics. The new process sensors would be able to run the same normal controls and diagnostic algorithms that would normally be run on the control system (Chuang and Thomson 2005).

The second advantage of intelligent sensors in comparison to conventional sensors is the possibility of working together in large numbers and communicating the results. The large number of processors in the array may be used to enhance reliability by allowing for redundancy. This is ideal for automatic detection of faults and allocation of tasks to adjacent sensors. Different analysis tasks may be distributed to different processors within the array, allowing a single system to perform several analyses simultaneously. Recognition and analysis algorithms might be distributed to the array as a whole, using the complete system as a 'pattern recognition' device. Such patterns are often diagnostics of particular mechanical faults and may be detected by such means. Systems are likely to use all three of these possibilities, providing systems with specialised processing and an element of redundancy, providing powerful diagnostic procedures with graceful degradation in the case of individual sensor failure (Esteban *et al.* 2005).

The advanced processing methods employed by researchers constitute their strengthraises the intelligent sensors above conventional devices. These methods include following:

- *Data pre-processing*. The physical sensing element produces a response to the environment that it is placed in, according to the sensor's transduction process. The pre-processing stage converts the physical sensing element's response from the sensor modality, which, for example, may be acoustic intensity or temperature, into a more useful engineering unit that is representative of the raw environmental parameter, such as electrical current or voltage. Pre-processing includes software for calibration of the sensory data, which for an intelligent sensor system should be adaptive to compensate for long-term bias and ageing effects in the sensor. Depending on application, the calibration process may incorporate linearisation of the signal that can be implemented using look-up tables, removal of direct current bias effects using a normalisation approach and conditioning of the signal to correct for deviations caused by temperature effects (Boltryk *et al.* 2005).

- *Information processing*. This encompasses the data related processing that aims to enhance and interpret the collected data and maximise the efficiency of the system, through signal conditioning, data reduction, event detection and decision-making. This may involve a collection of filtering and other data manipulation techniques together with advanced learning techniques for feature extraction and classification in order to provide the most relevant data in an efficient representation to the communications interface (sensor networks).
- *Condition monitoring and fault detection.* In a classical implementation of condition monitoring, sensors are deployed to monitor the condition of a system to detect abnormal behaviour. For example, the characteristics of frequency spectra originating from vibration in machine bearings can be used as an indicator of progressive bearing wear. Together with expert knowledge about the system, the observation of certain spectral components can be used to detect the onset of specific failure mechanisms. However, condition monitoring of sensory data itself is conceptually different because the fault detection system has to be robust to genuine changes in the process variable (Boltryk *et al.* 2005).
- *Sensor modelling and uncertainty.* An analytical model of the sensor element for residual calculation is usually restricted to be linear and is often time invariant. Since deriving adequate mathematical models of complicated sensor systems can be intractable, a data-based, kernel representation is instead chosen for sensor modelling. A data modelling approach is naturally suited to non-linear systems, and since the model is derived based on example system data, it is not necessary to specify mathematical sensor models from first principles. A further advantage of data-based models is that they can be autonomously retrained using up to date data to accommodate deviations in the characteristics of the sensor caused by effects such as ageing. Estimates of measurement uncertainty are required by the sensor management for data fusion processes such as Kalman filtering; such a kernel representation can estimate prediction uncertainty directly (Boltryk *et al.* 2005).
- *Compensation.* This is the ability of the system to detect and respond to changes in the network environment through self-diagnostic routines, self-calibration and adaptation. An intelligent sensor must be able to evaluate the validity of collected data, compare it with that obtained by other sensors and confirm the accuracy of any following data variation. This process essentially encompasses the sensor configuration stage.
- *Communications component.* This of intelligent sensor systems incorporates the standardised network protocol that serves to link the distributed sensors in a coherent manner, enabling efficient communications and fault tolerance. Traditional task-specific sensor systems often contain a number of limitations in terms of complexity, cost and flexibility. Intelligent sensors aim to overcome these limitations through the utilisation of standardised transducer interfaces and communications protocols, resulting in autonomous, distributed, re-configurable sensors.

- *Qualitative characterisation.* Some applications need qualitative (symbolic) characterisation of the estimate, for example for rule-based control or diagnostics, where the variables are described by qualitative values; for example, large, small, increasing, or, decreasing. *ad hoc* algorithms exist for the quantitative to qualitative transformation, using either crisp or fuzzy sets. Data characterisation is often based on a set of successive estimations on a sliding time window. The size of that time window may be fixed as a parameter of the intelligent sensor, or it may be self-adjusted so as to optimise some given optimality criterion (Pottiea and Clareb. 2004).
- *Fault tolerance.* Each version of the estimation service is associated with the resources it needs to perform correctly. When resources do not behave nominally, the version will provide erroneous estimates. Fault tolerance issues in intelligent sensors can be considered at different levels. The first level is associated with the definition of real time strategies by which the system is able to continue its operation in spite of faults. Two such strategies can be used. Fault accommodation is the strategy by which the fault is compensated so as to avoid erroneous estimates. Sensor reconfiguration is the strategy by which another version of the estimation service, whose resources are not faulty (or, if faulty, whose faults can be compensated), is run as the current operating version. The second level is associated with the evaluation of the system fault tolerance, *i.e.* its ability to accept faults while being still able to run some version of the estimation service. The fault tolerance possibilities depend on the nature of the faulty resources and of the nature of the fault. Technological validation is associated with resources that are critical, *e.g.* the power supply, and thus fault tolerance with respect to such faults is rarely possible, unless hardware redundancy has been implemented for these resources.
- *Validation of sensors.*This is required to avoid the potential disastrous effects of the propagation of erroneous data. This is a different problem to overcoming individual sensor failure. A control system operating on decisions made on faulty data can lead to unpredictable behaviour or even complete system failure. The impact of such errors may be reduced through the use of a dense sensor network. The incorporation of data validation into intelligent sensors increases the overall reliability of the system. So an effective means for performing this function is required. Two approaches are analytical redundancy and hardware redundancy. Analytical redundancy utilises a mathematical model that compares the static and dynamic relationship between sensor measurements and effectively determines the expected sensor value. The computational expense of this approach can become prohibitive as the number of sensors and model complexity is increased. Hardware redundancy may involve the use of additional sensors and selection of data that appears similarly on the majority of sensors. This approach is not applicable, however, in cases where the presence of an excessive number of sensors has a detrimental effect on the given environment. Knowledge based systems are one alternative, where

an intelligent sensor incorporates expert systems that apply reason and infer the solution (Issnip website).

- *Data validation.* This is a very important concept as far as systems sensitive to malfunctions are considered. When included in such systems, intelligent sensors should also provide data that qualify or disqualify the estimation (and the associated parameters) that they produce, or at least evaluate the confidence level with which it can be associated (Pottiea and Clareb 2004).
- *Technological validation.* This is a (partial) data validation approach. Technological validation is concerned with the conditions under which the estimation procedure runs; namely it checks if some hardware resources of the sensor, which are in general common to all the versions, are in normal operation. Technological validation is concerned with the power supply of the sensor's transducers, with the checksum of the microprocessor's memory, with the connection to the network, *etc.* The technical validation process does not guarantee that the estimation produced by the sensor is correct, but only that the operating conditions were correct (Pottiea and Clareb 2004).
- *Reconfiguration of data validation.* Functional data validation rests on redundancy, which means that the result provided by the estimation version currently in use is checked against estimations provided by other versions (in the observer-based approach) or it is checked by computing residuals (in the analytical redundancy based approach). Fault isolation needs more information than fault detection, since in order to design structured residuals, at least two residuals must be available, which means at least three versions of the estimation service (the larger the number of resources to be isolated, the more versions are needed). Fault identification needs at least the same information as fault isolation, since when three versions of the estimation service are available, and the faulty one is isolated, the estimation of the fault signal follows directly (Pottiea and Clareb 2004).
- *Distribution of the sensing field.* Intelligent sensors make it possible to conceive applications that employ arrays of interacting micro-sensors, creating in effect spatially distributed sensory fields. To achieve this potential, however, it is essential that these sensors are coupled to signal conditioning and processing circuitry that can tolerate their inherent noise and environmental sensitivity without sacrificing the unique advantages of compactness and efficiency.

5.3.4 General Design Requirements for Intelligent Sensors

A list of general requirements according to the functional aspects of the wireless intelligent sensors will include criteria from the following two sections:

5.3.4.1 Quantifiable Requirements

Low-power modes and small physical size. Long-term operation of wireless sensors places a premium on power. Battery size is the greatest single size constraint for the sensor in many situations. Most applications require three to five years of battery life (Table 5.3). To achieve this level of performance, the software must execute all necessary functions quickly and then turn off the hardware and stay in the sleep mode till the next event.

Table 5.3 Example of battery characteristics

Parameter	Value	Units
Dimensions	50 × 30 × 10 *	mm
Voltage level	3.2	V
Battery storage capacity	500	A hours
Peak current	150	mAs

Robust and reliable performance. Most wireless sensor networks will consist of numerous devices that are largely unattended. The engineer will expect them to be operational most of the time. To that end, the operating system on a single node or sensor should not only be robust but also able to continue functioning when other devices on the network fail (Table 5.4). This will ensure that if one sensor or device should fail, the network or application is not jeopardised.

Table 5.4 Key characteristics within sensor networks

Characteristics	Description	In use
Effectiveness	Speed of transactions performed by the program	Must provide time reliability expected by the external system
	Response time (promptness)	Initial values tested on 500 ms
	Refreshing time (refreshment)	Initial values tested on 500 ms
Reliability	Probability of error during the performance	Minimal
	Frequency of errors	Minimal
	Mean time between errors	Minimal
	Accessibility (percentage of time the system available)	Minimal
Robustness	Reboot time after a crash	3 s.
	Probability of the data destruction after a crash	Low

Electromagnetic compatibility (EMC) resistance and electrostatic discharge (ESD). The devices will be working in very hostile environments. Protection from EMC effects has to be included in the early stages of the design. The devices will be working on limited RF power to reduce the amount of power, which will require optimisation of the transmission source. The power levels can be controlled and set to the required valued according to the distance and the environment that device is applied for (Table 5.5).

Table 5.5 EMC thresholds to be set

Parameter	Value	Units
EMC resistance	*	mW
EMC emission	*	mW

* Prototype tests required

5.3.4.2 Unquantifiable Requirements

Self-configuration. Long, complex installation procedures destroy the benefit of wireless sensors. Installing a sensor should be as easy as gluing the unit to the point of measurement. This can be provided by use of the ZigBee protocol with ability of self-configuring the network (White 1997).

Distributed processing requirements. A sensor network's primary mode of operation is to flow sensor information from place to place with some processing in between. There is always a limit on the volume of data transfer as well as storage or buffer capacity on a wireless sensor because of size, cost and power consumption considerations. Therefore, to reduce inbound and outbound traffic, especially in controlling radio communications, distributed data pre-processing needs to be highly efficient.

Diversity in design and use. Networked wireless sensors will tend to be application specific rather than general purpose, and because of cost and size considerations, they will carry only the hardware and software actually needed for the application. With the wide range of potential applications, the variation in sensor device requirements is likely to be great. It will be desirable to reduce the variability of physical hardware and embed as much system variation within the software components.

Integration in intelligent sensors involves the coupling of sensing and computation at the chip level. This can be implemented using micro electro-mechanical systems (MEMS), nano-technology.

A hierarchical structure can be used to describe the functionality of the system, where the lower layer performs the signal processing functions, the middle layer performs the information processing and the upper layer performs the knowledge processing and communications. This should be considered in the sensor design as well as in the system design.

Many important WINS applications require the detection of signal sources in the presence of environmental noise. Source signals decay in amplitude rapidly with radial distance from the source. To maximise detection range, sensor sensitivity must be optimised. In addition, due to the fundamental limits of background noise, a maximum detection range exists for any sensor. Thus, it is critical to obtain the greatest sensitivity and to develop compact sensors that may be widely distributed. Clearly, MEMS technology provides an ideal path for implementation of these highly distributed systems.

5.4 Hardware Requirements for Wireless Sensors

Identification of the hardware for the design of the intelligent wireless sensor is complicated due to many factors coming into play. This specification aims to bring and discuss all the possible options and aspects influencing the specification of the hardware. Environmental issues relating to sensor deployment will also be critical in determining sensor hardware – robustness, intrinsic safety, waterproofing, *etc.*

A sensor node is made up of four basic components (modules): a sensing unit, a processing unit, a transceiver unit and a power unit (Figure 5.4). Design and implementation aspects, as well as correlation between the units will be discussed in the next paragraphs, presenting the way of influencing the hardware and possible solutions available for the design. The modularity of the design is intentional and modules can be interchanged depending on the circumstances in which that sensor will be applied. Power generation initially assumes the use of batteries, but if possible a power harvesting module will be applied.

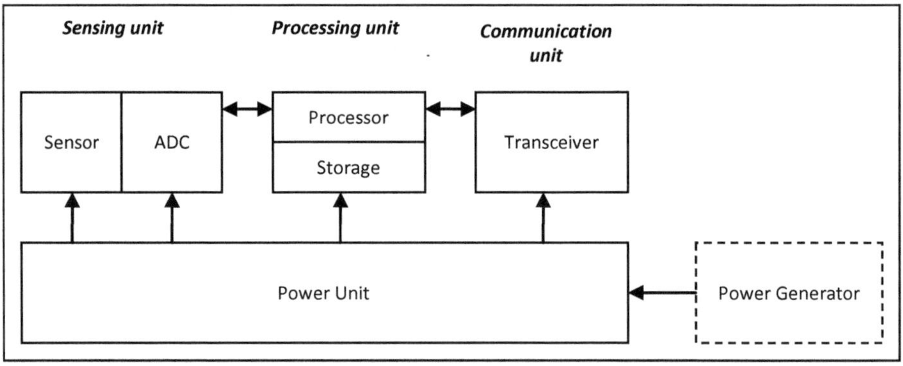

Figure 5.4 The components of the sensor node

5.4.1 Hardware Components

5.4.1.1 Analogue-to-digital Converter Unit

In the modular approach with sensing unit, the sensor modules should contain various sensor interfaces which are available through a connector that links the sensing unit and processing modules. The interface could include an 8-channel, 10-bit A/D converter, plus, a serial port. This would allow the processing module to connect to a sensor module, including modules that use analogue sensors as well as digital smart sensors. Components of ADC unit are shown in Figure 5.5.

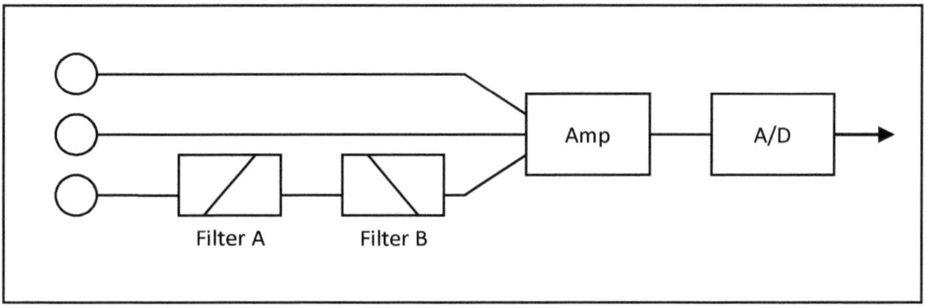

Figure 5.5 The components of the ADC unit

5.4.1.2 Sensing Unit

The sensing unit is responsible for the conversion of raw data signal from the core part of the sensor such as input parameters: temperature, pressure, or acceleration measurement to the processing unit (Table 5.6).

Table 5.6 A range of measured sensor values

Parameter	Value	Units
Temperature	-15 to +85	°C
Pressure	10	bar
Vibration	5	g

Sensing units are usually composed of two subunits: sensors and analogue-to-digital converters. The analogue signals produced by the sensors based on the observed phenomenon are converted to digital signals by the ADC and then fed into the processing unit. This is typically analogue to digital conversion, but depending on the design of the electronics of the sensor conversion might be done straight after receiving the signal or in some cases after the pre-processing of the raw signal, for example when analogue filters are used.

This part of the design is variable and depends on the measured parameters. Design specification should consider modular assembly depending on the needed sensor board configuration. The ADC unit will require A/D, anti-aliasing and single channel filter.

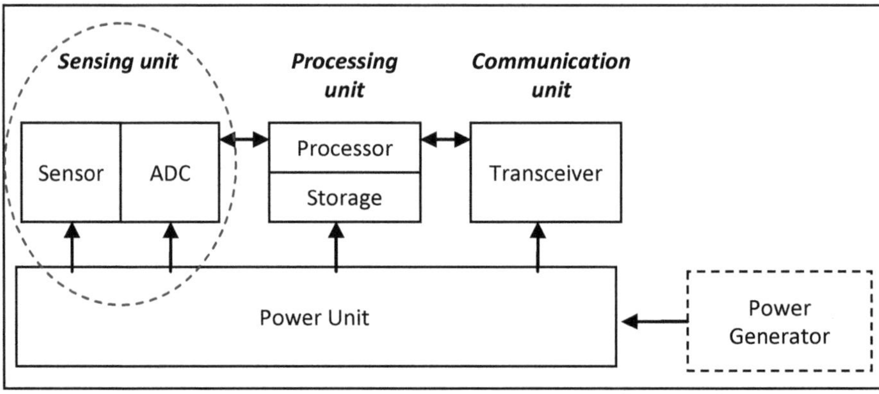

Figure 5.6 The components of the sensor node – processing unit

5.4.1.3 Power Sources

The powering of intelligent sensors is a sensitive issue influencing everything from the possible application of the sensor to the architecture of the sensor. Different sources of power brings advantages and disadvantages. A standard suggestion for the powering of the remote wireless sensor will be a battery source. Use of the battery as a source brings a tight restriction on the power consumption and the life of the device without maintenance. To reduce the power consumption the processor should have three sleep modes: idle, which just shuts the processor off, power down, which shuts everything off except the watch-dog and power save, which is similar to power-down, but leaves an asynchronous timer running. Power consumption equates to battery life. Long battery life is desired, and in some applications one to five years is required. The processors, radio and a typical sensor load consumes about 100 mW. This figure should be compared with the 30 μW draw when all components are in sleep mode. The overall system must embrace the philosophy of getting the work done as quickly as possible and then going into sleep mode. This is a third key constraint on the software design for wireless networked sensors.

Power harvesting opportunities are a natural trend to improve the lifespan and maintenance free time of the installed sensors. Some sensor applications require a high level of self-sufficiency from the device, and in these cases the harvesting power method enables them to operate without maintenance (battery changing operation).

The alternative of scavenging power is possible from vibrations, acoustic or millimetre wave energy through the use of sensor resonators or piezo-electric sensors. The options increase as the size and power consumption diminish (Table 5.7).

Table 5.7 Comparison of power sources

Power source	Advantages	Disadvantages
Battery powered	Long lasting source	Additional cost
	Cheap solution	Environmentally unfriendly
	Reduced power management	Limited/uncertain life span
	Reliable technology	Maintenance required
Online power supply	No need for energy saving power management	Cabling distance restrictions
	Reduced cost of the device	Closeness to the source
	Maintenance free	No mobility possible
	Reliable	Reliable on the continuity of external supply
Power harvesting	Maintenance free – no battery replacement	Complicated power management
	Enhance lifespan of device	Additional cost of device
	Environmentally friendly	Only applicable in right environment
	Modern approach with big prospects	Need for additional source or power storage when device is stopped

5.4.1.4 Housekeeping and Information Processing

A system must be created to oversee the measurement, communications and housekeeping functions. The system specifications will include:

- Measure and sample data from the transducers at pre-defined intervals, log the data and perform basic diagnostic functions such as comparison with programmed thresholds, with event triggering.
- Receive instructions from a programmer including changes to logging and diagnostic functions.
- Organise transmission of data packages at required intervals and events.
- Manage system start-up and power management. A hardware platform will be constructed to accommodate the specification, in the context of the miniature format.

The processing unit (Figure 5.7), which is generally associated with a small storage unit, manages the procedures that make the sensor node collaborate with the other nodes to carry out the assigned sensing tasks. There are many processors available that meet the power and cost targets as well as data processing requirements (Table 5.8). In a given network, thousands of sensors could be continuously reporting data, creating heavy data flow. Thus, the overall system is memory constrained, but this characteristic is a common design challenge in any wireless sensor network (White 1997).

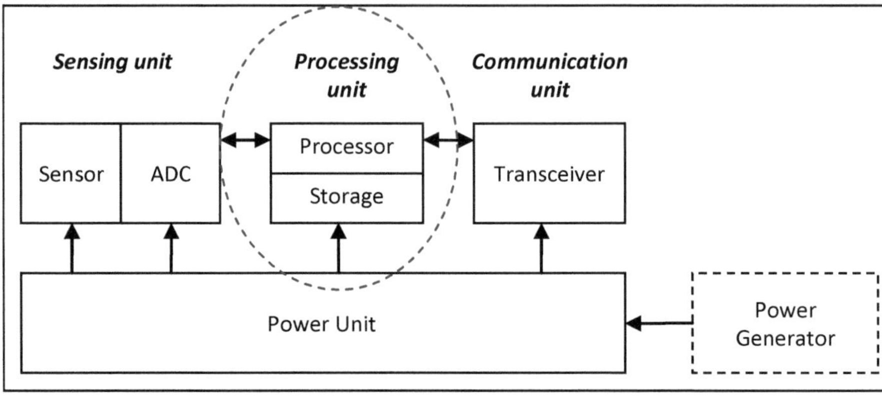

Figure 5.7 The components of the sensor node – processing unit

Table 5.8 A sample of minimum requirements

Characteristics	Description	In use
Hardware resources	CPU	55 MHz
	RAM Memory capacity	1 MB

An intelligent sensor differs from a conventional sensor as it is designed to be a standalone device performing more advanced tasks than data acquisition. A sensor should be capable performing most of the functions automatically. These functions include:

- data collection from the sensors;
- standard tests performed on the device;
- continuous monitoring of the parameters;
- first instance decision making;
- basic alarms for the control system;
- advanced cooperation with condition monitoring system;
- communication with the advanced condition monitoring system;
- TCP stack;
- self test function;
- low power mode (sleep); and
- register.

5.4.2 ZigBee as a Suggested Communication Technology

To fully understand the advantages of ZigBee it may be helpful to review the basics of the 802.15.4 standard and how ZigBee builds upon it. The IEEE standard is based directly on the "direct sequence spread spectrum" (DSSS) transmission scheme using binary phase shift keying for 868/915 MHz and offset-quadrature phase shift keying for 2.4 GHz. On top of this structure, ZigBee defines layers for network, security and application profiles. Its network layer handles network topologies of star, mesh and cluster trees. Of these, mesh and cluster trees are probably of most interest for industrial needs. Mesh or peer-to-peer networks provide more than one path through the network for a wireless link. This makes them highly reliable in environments characterised by a lot of RF interference. Cluster-tree networks are hybrids of mesh and star topologies. They provide reliability while keeping power drain to a minimum in battery powered nodes. ZigBee also has facilities for a sleep mode that conserves power. Thus most battery powered wireless sensors will probably take the form of radio frequency devices (RFDs). One of the principal attractions of ZigBee networks is that they are self-forming and self-healing. This means messages can pass from one node to another *via* multiple paths. If one path becomes unavailable, nodes have enough intelligence to reroute traffic around it. Further, there are provisions for security such as 128-bit encryption. The quality of service definitions provides a guaranteed time slot for devices that must gain access to a network quickly. Applications in this class include security alarms and medical alert devices. Finally, 802.15.4/ZigBee networks are optimised for low-duty cycle transmissions. New nodes are typically recognised and connect within 30 ms. The process of waking up a sleeping node and transmitting data takes about 15 ms, as does accessing a channel and transmitting (Lynch *et al.* 2003).

ZigBee offers at three different operational frequency ranges, each one having a corresponding data rate (Table 5.9).

Table 5.9 Frequency range for ZigBee (Mitchell *et al.* 1999)

Frequency	Data rate
2.4 GHz	250 kbps
902 MHz–928 MHz	40 kbps
868 MHz–870 MHz	20 kbps

ZigBee has been designed with very low latency as one of the initial criteria and therefore in its current production form does achieve very low latency characteristics. The main conditions to consider are shown in Table 5.10.

Table 5.10 Response time (Mitchell *et al.* 1999)

Condition	Response time
Enumeration of new nodes	30 ms
Wake-up time of node	15 ms
Time to access a channel	15 ms

5.4.2.1 ZigBee Interference

Details surrounding the interference encountered by ZigBee from other forms of wireless communications vary quite considerably, from practically interference-free to catastrophic loss of data throughput.

ZigBee devices can operate in three bands, 2.4 GHz, 915 MHz and 868 MHz, each having different issues with interference. The 2.4 GHz band could be assumed to have the highest interference tolerance if one considers that it uses 16 channels to transmit data, which allows for data to be transmitted down different channels depending on the traffic on each. However, this is the highest number of wireless communications competing for the 2.4 GHz spectrum compared to the two other bands, due to the higher potential bandwidth/data rate. By allowing a higher bandwidth, data is able to be transmitted quicker and therefore along with ZigBee's low duty cycle this should dramatically reduce the chances of interference occurring whilst any specific device is transmitting.

If any interference were to be experienced on the single channel, such as background noise or electro-magnetic interference, EMI, successful packet transmission would be greatly reduced if not completely stopping data throughput. Reducing the percentage of successful data transmission causes an increase in data retransmission, which increases the power requirement of the ZigBee devices significantly, contrary to ZigBee's objective of a low powered wireless standard.

5.4.2.2 Network Topologies Offered by ZigBee Protocol

The layout and networking of the nodes in a ZigBee network are primary factors, due to the topologies limiting the protocols used.

The main and most conventional topology is called "star". In this type of topology, all the nodes are communicating to the single central node, functioning as a receiver (Figure 5.8). This kind of communication benefits from the minimum power requirement for the set of RFDs to communicate to each other, but limits the distance because all the nodes have to stay in the range covered by the central receiver. An extension of the range can be achieved only by increased radio power, which results in increased power consumption. Some characteristics are:

- limited range;
- very long battery life;
- very simple; and
- very reliable.

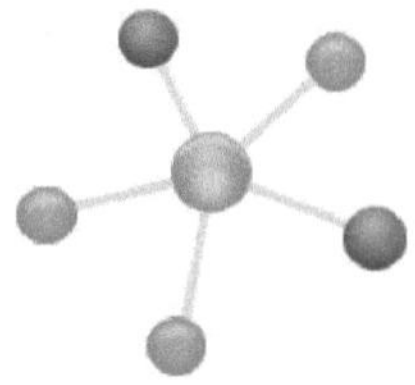

Figure 5.8 Star mesh

The second level of topology will be "tree". In this topology, there is more than one receiver in the network (Figure 5.9). Other nodes are specified to fulfil the role of the receivers passing data from the most distant nodes to the main receiver. This enables locating nodes much further away than in the star topology, extending the range x times the number of branches y in the tree. Nodes responsible for the passing data will increase the amount to power used for the communication. Now they have to transmit not only their own messages but also messages from the other nodes. Another disadvantage from this type of communication is that nodes working as a transmitters and receivers will be responsible for the communication for the whole branch. In the case of failure of the local receiver the whole branch will be disengaged. Some characteristics are:

- extended range;
- increased power consumption; and
- single point failures – device failure disables all children

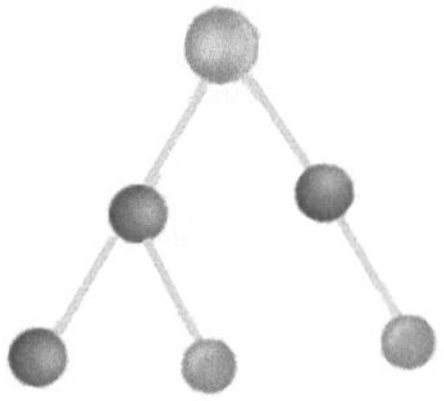

Figure 5.9 Tree mesh

The MESH topology (Figure 5.10) is the most advanced topology offered by ZigBee protocol. In MESH communication every node can be transmitter and receiver. The decision about the route is made automatically by the protocol itself. The networks built on this topology have the advantage of a much extended range and the use of self-healing to transmit data anywhere in a network. However, this requires the use of routers instead of RFDs and requires constant operation (always on), drawing a lot more power from each node. Some characteristics are:

- extended range;
- unparalleled adaptability – self healing;
- increased resolution through node interaction; and
- increased power usage

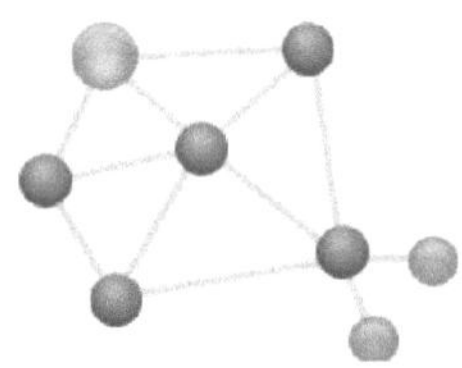

Figure 5.10 Mesh topology

Cluster tree – interconnected small networks (star or MESH). The best option however seems to be with the use of a cluster tree topology (Figure 5.11) where end devices may be low-power RFDs increasing the battery life or allowing power harvesting. It also has the benefit of self-healing and an extended range drawn from the star networks being connected by a tree style backbone of routers, which can be wired and are able to reroute or select new parent devices like in a MESH network. Some characteristics are:

- low power usage – end devices in star topology
- extended range – through mesh backbone
- set preferred transmission routes
- self-healing – except for RFDs

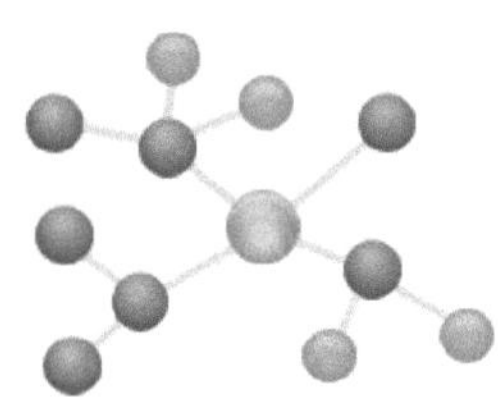

Figure 5.11 Cluster tree

5.4.2.3 Performance and Network Reliability Assessment

The overall reliability of the implemented system will rely on the integrity of each component, the effective transmission of data and also the integration of the system with sufficient forms of redundancies, to cover all foreseeable problems. This can be split into the hardware and software of the system and will be described separately below.

Component failures (hardware). This is one of the main reliability issues that cannot be controlled by the end-manufacturer and includes all components ranging from surface mount resistors to the ICs and sensors. Therefore, rigorous testing should be implemented to certify that the devices will be able to function correctly in their applications, in varying conditions and environments. This does not mean that they will have to be proven to extremes but they should be certified to work within the operating conditions considered by the end-manufacturer and regulations issued with the devices. Warranties and lifespan predictions should also be issued to allow the end-user to choose the appropriate component for the application, such as applications where the device will be operating for several years. The current Crossbow product warranties are for 1 year from shipping, which will cover most defective components and non-operational problems such as faulty connection/soldering.

Hardware redundancies. When the devices are operating in their final configurations, there should be sufficient redundancies in critical operation areas, such as the backbone routing of data in a tree topology. Depending on the number of sensors operating in a network and also the criticality of the sensing, there should not be the need to provide redundancies in the individual node hardware. This can be seen in applications (Figure 5.12) such as condition monitoring on production line machinery, where the condition of machines tends to degrade slowly and the number of sensors will be fairly high. In applications such as critical sensing on aircraft structures or other future uses, there may be the need to investigate hardware redundancies. However, the simplest and most effective way to combat this is to increase the number of sensors for each application. This can be addressed by driving down the cost of these devices with increasing sales, so that operators consider the benefits of implementing more devices outweigh the increased cost.

OTAP over the air programming (hardware). One of the conditions required for OTAP is battery power of above 3.6 V, which will limit its use to the early stages of a devices life once it has been installed. This can be prolonged by the replacement of batteries periodically; however this counters the idea of remote sensors that require little or no human interaction. OTAP should not be required once the devices have been installed other than for minor configuration changes soon after installation, which will be permissible. If reprogramming is required after this the voltage of the nodes should firstly be checked to ensure that the process will be successful.

Data transmission (software). To ensure that the data sent in a network is effectively received by the intended final destination the transmission protocols, data integrity and security will need to be considered. The first concern is guaranteeing that data will get to the specified destination. This has been addressed in the network topologies section (Section 5.4.2.2), to find the most effective method depending on the situation the devices are in. For the Dynamite project this was considered to be the cluster tree topology, which allows for several routers to act as a wired backbone transmitting data from low-powered RFDs in star topologies associated with one of the routers. The reliability of this system will depend on the backbone of routers communicating *via* a MESH network allowing for routers to conduct self-healing of routes if a device fails. Thus, in Figure 5.10, if any one of the three routers were to fail the other two would conduct a longer hop to the next appropriate router. The end-devices associated with the failed router would also become orphaned devices and would attempt to rejoin the network *via* another router.

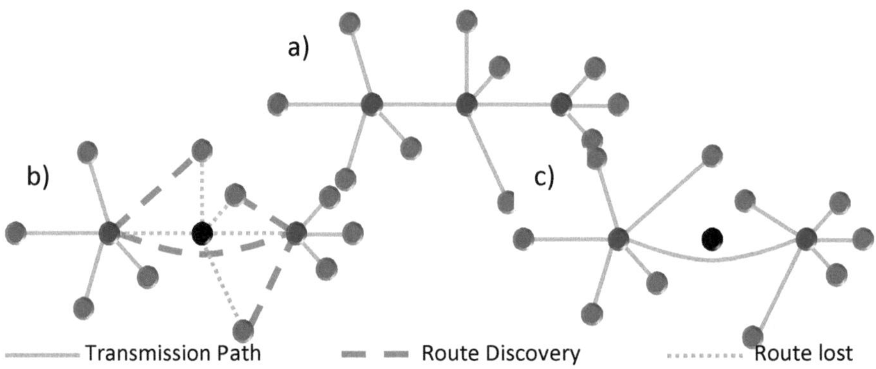

Figure 5.12 Cluster tree redundancy. (a) In normal configuration, (b) loss of second router, transmission re-routing, and (c) final configuration, with failed router

Security is another concern for data transmission, ensuring that the data has not been tampered with before it is received by the final destination. This is of little concern at the moment due to the low-security risk of the applications that the ZigBee devices will be used for, however it is advisable to get in-depth information about security.

Data integrity is one of the main concerns for network reliability; without accurate data transmission there is no use for the system. In order to reduce the amount of corrupted data in the received packets there needs to be a reliable wireless communication, which has been proven with previous and current ZigBee applications. There also needs to be ways to check the data for errors and re-issue it if any are found. This is conducted using the FCS in the data frame, with a 16-bit International Telecommunication Union – Telecommunication Standardization Sector (ITU-T) and Cyclic Redundancy Check (CRC).

A CRC considers a block of data as coefficients of a polynomial that is then divided by a pre-determined polynomial and the coefficients of the result used as the redundant data bits, the CRC. The CRC is then multiplied back with the pre-determined polynomial at the receiver and compared to the data. Alternatively, the data can be divided by the polynomial to calculate a CRC and then the two compared, if the two are the same then the data was sent without errors. The polynomial used by the IEEE 802.15.4 standard is $G_{16}(x) = x^{16} + x^{12} + x^5 + 1$.

*Data processing***.** To minimise the amount of power used by the Crossbow devices, each of the sensors and other processes may be shut down, to allow only essential applications to run at any given time. Since the excitation for the sensors will be provided by the mote, these sensors will be controlled by the ZigBee processor and will easily be shut down when not in use.

The Crossbow motes feature two processors, with the main ATMega128L processor being used for all application control and active processing such as analogue-to-digital conversions (ADC). The second is an Atmel AT45DB041 serial flash used to store data and measurements along with allowing it to write to the program memory of the main processor whilst conducting OTAP. Although it would not be able to conduct pre-transmission processing of the data it will be able to store over 100,000 measurements. This data can then be processed using the main processor, whilst the other functions sleep until the data is ready to transmit, only requiring 4 mA to read the data of the secondary processor.

Whilst conducting calculations the processor can also move to using the external oscillator, which is slower but requires lower power than that of the internal oscillators on the processor. This can also be combined with dropping the voltage supplied to the processor to the lower limit of its operating voltage, extending the battery life. As long as digital processing is occurring and not analogue, which requires a higher accuracy and therefore more power, power saving will be improved without affecting the processing.

5.5 Power Reduction Methods Available in ZigBee Protocol

The powering of intelligent sensors is a sensitive issue influencing everything from the possible application of the sensor to the sensor architecture. Different sources of power bring advantages and disadvantages. A standard suggestion for the powering of the remote wireless sensor will be a battery source. Use of the battery as a source brings a tight restriction on the power consumption and the life of the device without maintenance. To reduce power consumption the processor should have three sleep modes: idle, which just shuts the processor off; power down, which shuts everything off except the watch-dog; and power save, which is similar to power-down, but leaves an asynchronous timer running. Power consumption equates to battery life. Long battery life is desired, and in some applications, one to five years is required. The processors, radio, and a typical sensor load consumes about 100 mW. This figure should be compared with the 30 μW drawn when all components are in sleep mode. The overall system must embrace the philosophy of getting the work done as quickly as possible and then going into sleep mode. This is a third key constraint on the software design for wireless networked sensors.

In order to keep the power consumption down to a minimum the processes used at each layer of the ZigBee architecture have been considered, finding the most power efficient method of performing tasks. The most significant ways to reduce the power are:

- very low duty cycle;
- orthogonal signalling;
- warm-up power loss – DSSS;
- more power efficient when blindly transmits than blindly receive;
- recovery effect in batteries; and
- cost based routing algorithm – link quality and hop count.

The main expenditure of energy and power is when the ZigBee device is transmitting or receiving data; therefore the easiest way to reduce power consumption is to decrease the amount of time the device is transmitting or receiving. This is carried out by lowering the duty cycle of the devices transmission/reception frequency.

5.5.1 Orthogonal Signalling – Used for 2.45 GHz

In order to lower P_{avg}, as well as lowering the duty cycle the peak current also needs to be kept to a minimum. When studying the current characteristics of data processing it is found that the peak current tracks the *symbol rate* rather than data rate. Multi-level signalling can be used to lower this; however, simple application may result in a loss of sensitivity, which can be detrimental to the low-power goal. This can be resolved by using orthogonal signalling, for which ZigBee uses 4 bits/symbol. Therefore, whilst keeping the same bit (data) rate the number of bits *per* symbols increases; binary (2 bits/symbol) to 16-ary (4 bits/symbol).

It is useful to note that the bits/symbol is the modulations power of 2, *i.e.*, 16-ary is 2^4 = 4 bits/symbol.

5.5.2 Warm-up Power Loss – DSSS

Due to the use of sleep periods for the ZigBee devices and a low duty cycle, the active periods of IEEE 802.15.4 are very low. A large disadvantage would be if the device transceiver took a long time to warm-up, with the warm-up period being dominated by the settling time of the channel filters. By using DSSS a form of wideband filter the device will benefit from a shorter settling time. Also due to the wider spaced spectrum the lock-on time is deceased significantly.

5.5.3 Transmitting and Receiving

In any IEEE 802.15.4 device the power required to receive is greater than the power required to transmit due to the high number of filters and processing required in the receiver, $Rx > Tx$. This means that it is more power efficient to blindly transmit than blindly receive.

5.5.4 Recovery Effect in Batteries

All batteries exhibit an effect called the 'recovery effect' where short bursts of power, rather than the equivalent average current, can extend the battery life. This is used to great effect with low-power protocols such as IEEE 802.15.4 where the duty cycle is low.

5.5.5 Cost Based Routing Algorithm – Link Quality and Hop Count

When sending data between devices the number of retransmissions and route discovery requests should be kept to a minimum in order to reduce the power consumption of the devices. This is carried out by improving the quality of the routes used to transmit data between end devices by:

- minimising the number of hops required to reach the destination device
- avoiding low quality links

The quality of the links is determined by collecting link quality indicators (LQIs) from previous data transmissions and using this along with the number of hops, calculating the overall cost of using different routes. This can then be used in conjunction with the signal-to-noise ratio (SNR) to differentiate between a corrupt packet sent with low signal strength and a packet sent with high signal strength and interference.

The quality of the links between devices is one of the issues that are being considered for future versions of the ZigBee specification, with the following areas being addressed:

- node power remaining;
- node power source;
- transmitter P_{out};
- *etc.*

5.5.6 Power Consumption Tests

The power consumption of the communication modules has been tested (Figure 5.13), by unseeing application simulated condition monitoring environments. The data transfer and frequency of data exchange was increased to speed up the time required for the tests. The results from the tests are conclusive that the power consumption from the ZigBee protocol is relatively small compared to the power consumption from the other alternative technologies. The tests represent condition monitoring process foreseen for the sensors, accelerated for the test purposes by about 450 times.

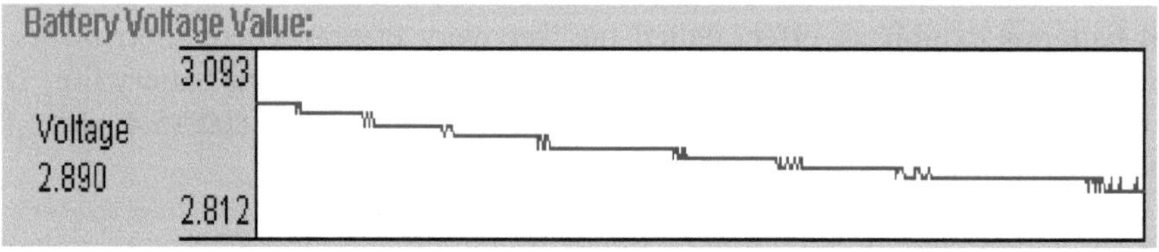

Figure 5.13 Power consumption *vs.* time for single unit reporting to base

5.6 Conclusions

Intelligent wireless sensors have become extremely important for monitoring complex plants using multiple sensors. Monitoring a machine would require fundamental knowledge of the physical measurands to be monitored as well as its characteristics. As an example, accelerometers are needed for vibration measurements in various directions, looking for amplitude of vibration and natural frequencies of the system. MEMS accelerometers usually have low bandwidth due to the finite dimensions of the internal parts. However, conventional devices in miniature form will achieve the required performance and cost.

Proper integration of various constituent elements into a sensor module is important for enhancing the performance of microsensors platform. Integration increases reliability of the sensor or allows for multiple quantities to be measured in one chip. It also allows for the integration of signal processing, wireless communication, remote powering modules and ease of field installation.

Powering the sensors is currently an issue if battery replacement is not permitted when sensors are placed in difficult to reach areas. Power harvesting may be the only solution.

The challenge is to further reduce the energy consumption by optimising energy awareness over all levels of design. Reducing start-up time improves the energy efficiency of a transmitter for short packets. Since the ADC subsystem is in the sensors front-end, it is important to implement a sleep mode operation for example.

Tests performed on the communication module included power requirements, bandwidth, range and the commercial outlook of the ZigBee standard. The device has proved its ability to receive as well as transmit data foreseen for the condition monitoring applications expected from an intelligent sensor. An additional advantage of the tested module was integrated with the programmable microprocessor. This means that the the communication module can be programmed and gives control over the sensing unit.

The key targets for intelligent sensors are to research and utilise novel technologies that can perform the required functions robustly, inexpensively and at extremely low power.

References

Abidi AA (1995) Low-power radio-frequency ICs for portable communications. Proc IEEE 83:544–69

Asada G, Dong M, Lin TS, Newberg F, Pottiea G, Kaiser WJ (1998a) Wireless integrated network sensors: low power systems on a chip. Proceedings of the 24th European Solid-State Circuits Conference, (ESSCIRC 98), Editions Frontieres, Paris, pp. 9–16

Baronti P, Pillai P, Chook V, Chessa S, Gotta A, Hu F (2007) Wireless sensor networks: a survey on the state of the art and the 802.15.4 and ZigBee standards. Computer Communications 30:1655–1695

Bojko T (2005) Wireless sensors network for vibration measurements. Pomiary Automatyka Kontrola 5:12–15

Boltryk PJ, Harris CJ, White NM (2005) Intelligent sensors – a generic software approach. School of Electronics and Computer Science, University of Southampton, SO17 1BJ, UK

Chuang J, Thomson DJ (2005) Noise reduction in RF cavity wireless strain sensors. Proc SPIE 5768:344–53

Crossbow Technology Inc (2006) http://www.xbow.com

Culler, D, Hill, J, Horton M, Pister K, Szewczyk R, Woo A (2005) http://www.sensorsmag.com/networking-communications/mica-the-commercialization-microsensor-motes-1070

Esteban J, Starr AG, Willetts R, Hannah P, Bryanston-Cross P (2005) A review of data fusion models and architectures: towards engineering guidelines. Neural Computing and Applications. Springer, London Ltd 14:273-281, ISSN 0941-0643 (paper) 1433-3058 (online)

EZURIO report (2006), www.ezurio.com

Han T, Shi W (2001) Surface acoustic wave devices based wireless measurement platform for sensors. Dept of Instrumentation, Shanghai Jiaotong University, Proc SPIE – The International Society for Optical Engineering 4601:14–19

Ihler E, Zaglauer, HW, Herold-Schmidt U, Dittrich, KW, Wiesbeck W (2000) Integrated wireless piezoelectric sensors. Proc SPIE – The International Society for Optical Engineering 3991:44–51

Kahn JM, Katz RH, Pister KSJ (1999) Next century challenges: mobile networking for "SmartDust". Proc 5th International Conference on Mobile Computing and Networking (MobiCom 1999), 271–278, Seattle WA, Aug 1999

Kiefer KF, Swanson B, Krug E, Ajupova G, Walter PL (2003) Wireless sensors applied to modal analysis. Sound and Vibration 37:10–17

Liu RC, Zhou L, Chen X, Mau ST (2001) Wireless sensors for structural monitoring. In: Strong motion instrumentation for civil engineering structures. Kluwer, Dordrecht, 253–266
Lynch JP, Arvind S, Kincho LH, Kiremidjian AS, Kenny T, Carryer E (2003) Embedment of structural monitoring algorithms in awireless sensing unit. Structural Engineering and Mechanics 15:285–297
Maser K, Egri R, Lichtenstein, A, Chase, S (2003) Development of a wireless global bridge evaluation and monitoring system (WGBEMS). Proc Specialty Conference on Infrastructure Condition Assessment: Art, Science, Practice 91–100
Mekid S (2006) Further structural intelligence for sensors cluster technology. Manufacturing. Sensors 6:557–577
Mitchell K, Sana S, Balakrishnan VS, Rao, V, Pottinger HJ (1999) Micro sensors for health monitoring of smart structures. SPIE Conference on Smart Electronics and MEMS 3673:351–358
Moschler WW, Hanson G (2004) Wireless microwave based moisture sensors for hardwood drying kilns. AIChE Annual Meeting, Conf Proc AIChE Annual Meeting 1707–1708
Moteiv Corporation, http://www.moteiv.com
Online article:
http://www.issnip.unimelb.edu.au/about/issnip_research_themes/intelligent_sensors
Patra JC, Kot A C, Panda G (2000) An intelligent pressure sensor using neural networks. IEEE Transactions on Instrumentation and Measurement 49:829–835
Pottiea GJ, Clareb L P (2004) Wireless integrated network sensors: Towards low cost and robust self-organizing security networks. Electrical Eng Dept, UCLA, Los Angeles CA. http://intranet.daiict.ac.in/~ranjan/sn/papers/spie3577-12.pdf
Powner ET, Yalcinkaya F (1995) Intelligent sensors: structure and system. Sensor Review 15:31–34
Prentza A, Nokas G, Dermatas E, Tsoukalis A, Koutsouris D (2004) Design of wireless network of sensors for continuous monitoring of vital biosignals. IEE Eurowearable 03:79–85
Shen X, Wang Z, Sun Y (2004) Wireless sensor networks for industrial applications Proc 5th World Congress on Intelligent Control and Automation. June 15-19, Hangzhou, P.R. China
Spencer B F (2003) A study on building risk monitoring using wireless sensor network MICA mote. First International Conference on Structural Health Monitoring and Intelligent Infrastructure, Japan, 353–363
Straser EG, Kiremidjian AS, Meng TH, Redlefsen L (2003) A modular, wireless network platform for monitoring structures. Proceedings – SPIE The International Society for Optical Engineering, 3243:450–456
Wang N, Zhang, N, Wang M (2006) Wireless sensors in agriculture and food industry - Recent development and future perspective. Computers and Electronics in Agriculture 50:1–14
White N (1997) Intelligent sensors. Sensor Review 17:97–98

Bibliography

Allan R (2005) Technology Report: Wireless sensors, wireless sensors land anywhere and everywhere. Electronic Design 35(16)
Andrew JF (2003) Integrity-based self-validation test scheduling. IEEE Trans on Reliability 52:162–167
Atashbar MZ, Bazuin BJ, Krishnamurthy S (2003) Design and simulation of SAW sensors for wireless sensing. Proc IEEE Sensors 2003, IEEE Cat No 03CH37498, Part 1, 1:584–9
Ibarguengoytia PH, Sucar LE, Vadera S (2001) Real time intelligent sensor validation. IEEE Transactions on Power Systems 16:770–5

Kohlstrand, KM, Danowski C, Schmadel I, Arms, SW (2003) Mind the gap: using wireless sensors to measure gaps efficiently. Sensors Magazine 20:14–25

Murphy RR (1996) Biological and cognitive foundations of intelligent sensor fusion. IEEE Transactions on Systems, Man and Cybernetics, Part A: Systems and Humans 26:42–51

Pister K, Kahn J, Boser, B (2001) SmartDust: Autonomous sensing and communication in a cubic millimeter. http://robotics.eecs.berkeley.edu/ pister/SmartDust/

Reindl L, Shrena I, Kenshil S, Peter R (2003) Wireless measurement of temperature using surface acoustic waves sensors. Source: Proc IEEE International Frequency Control Symposium and PDA Exhibition, Jointly with the 17th European Frequency and Time Forum (IEE Cat No 03CH37409) 935–41

Sachenko A, Kochan V, Turchenko V (2003) Instrumentation for data gathering. IEEE Instrumentation and Measurement Magazine 34–41

Staroswiecki M (2005) Intelligent sensors: A functional view. IEEE Transactions on Industrial Informatics 1(4): 238–249

Ye W, Heidemann J, Estrin D (2002) An energy-efficient MAC protocol for wireless sensor networks. IEEE Infocom 1567–1576

Chapter 6
MEMS Sensors

Samir Mekid and Zhenhuan Zhu

Abstract. The latest trends of MEMS sensors are summarised in this chapter. The aim of this chapter is to present an example of the working prototype of the multi-measurand, wireless, intelligent sensor incorporating intelligence that would allow controlling the way sensors take measurements and communicate with the system. This chapter presents the functionality of the internal sensor control system also called "house keeping" and its capability with:

- measurement using various strategies;
- transmission of data;
- sensing part control;
- power management; and
- self diagnostic possibilities.

This chapter introduces the way the house keeping system functions within the test bed application, presenting the experience in the application and identifying the procedures for the software and hardware.

6.1 Introduction

Microelectromechanical systems (MEMS) are normally highly integrated devices that combine electrical and mechanical components, which range in size from the sub micrometre level to the millimetre level and include component numbers from a few to millions. A sensor is defined as a device that detects the value or the change of value of a physical quantity and converts the value into a signal for an indicating or recording instrument.

MEMS devices are widely applied to inkjet-printer cartridges, accelerometers, miniature robots, micro-engines, locks, inertial-sensors, microtransmissions, micromirrors, micro-actuators, optical scanners, fluid pumps, transducers, chemical, pressure and flow sensors.

These systems can sense, control and activate mechanical processes at the microscale, and function individually or in arrays to generate effects on the macro scale. The microfabrication technology enables fabrication of large arrays of devices, which individually perform simple tasks, but in combination can accomplish complicated functions.

In industrial scenes, MEMS sensors, accelerometers and gyros, are often used to detect motion such as vibration, shock, angular rotation, linear motion and tilt, because these sensors can provide lower power, compact and robust sensing. Multi-axis sensing and more accurate data can be provided by using multiple sensors.

Vibration monitoring for machine diagnosis is an important industrial application for accelerometers, for example, in machine maintenance. An accelerometer-based vibration analyser can detect abnormal vibrations, analyse the vibration signature and help identify its cause.

Accelerometers are often used for structural testing. This is because the vibration signature of a structure changes when a structural defect occurs, such as a crack, bad weld or corrosion. The structure may be the casing of a motor or turbine, a reactor vessel or a tank. The test is performed by striking the structure with a hammer, exciting the structure with a known forcing function. This generates a vibration pattern that can be recorded, analysed and compared to a reference signature.

Mechanical accelerometers, such as the seismic mass accelerometer, velocity sensor and mechanical magnetic switch, detect the force imposed on a mass when acceleration occurs. The mass resists the force of acceleration and thereby causes a deflection or a physical displacement, which can be measured by proximity detectors or strain gages. Many of these sensors are equipped with dampening devices such as springs or magnets to prevent oscillation.

Acceleration sensors also play a role in orientation and direction-finding. In such applications, miniature tri-axial sensors detect changes in roll, pitch, and azimuth (angle of horizontal deviation), or *X*, *Y* and *Z* axes. Such sensors can be used to track drill bits in drilling operations, determine orientation for buoys and sonar systems, serve as compasses and replace gyroscopes in inertial navigation systems.

A servo accelerometer, for example, measures accelerations from 1 □g to more than 50 g. It uses a rotating mechanism that is intentionally imbalanced in its plane of rotation. When acceleration occurs, it causes an angular movement that can be sensed by a proximity detector.

Among the newer mechanical accelerometer designs is the thermal accelerometer. This sensor detects position through heat transfer. A seismic mass is positioned above a heat source. If the mass moves because of acceleration, the proximity to the heat sources changes and the temperature of the mass changes. Polysilicon thermopiles are used to detect changes in temperature.

In capacitance sensing accelerometers, micromachined capacitive plates (CMOS capacitor plates are only 60 μm deep) form a mass of about 50 μm. As acceleration deforms the plates a measurable change in capacitance results. However, piezoelectric accelerometers are perhaps the most practical devices for measuring shock and vibration. Similar to a mechanical sensor, this device includes a mass that, when accelerated, exerts an inertial force on a piezoelectric crystal.

In high temperature applications where it is difficult to install microelectronics within the sensor, high impedance devices can be used. Here, the leads from the crystal sensor are connected to a high gain amplifier. The output, which is proportional to the force of acceleration, is then read by the high gain amplifier. Where temperature is not excessive, low impedance microelectronics can be embedded in the sensor to detect the voltages generated by the crystals. Both high and low impedance designs can be mechanically connected to the structure's surface or secured to it by adhesives or magnetic means. These piezoelectric sensors are suited for the measurement of short durations of acceleration only.

Piezoresistive and strain gage sensors operate in a similar fashion, but strain gage elements are temperature sensitive and require compensation. They are preferred for low frequency vibration, long-duration shock and constant acceleration applications. Piezoresistive units are rugged and can operate at frequencies up to 2000 Hz.

MEMS based accelerometers is an implementation of combining the acceleration principle with microelectronic and mechanic techniques, which significantly extends the sensing range of traditional sensors and offers revolutionary improvements in cost, size and performance. Features of MEMS sensors are their very small size, high accuracy and reliability, robustness and self-calibration, which have wide applications in the automotive, aerospace and consumer electronics sectors.

The advantages of using multiple sensors over a single sensor to improve the accuracy of acquired information about an object have been recognised and employed by many engineering disciplines ranging from applications such as a medical decision-making aid system to a combined navigation system. For example, recently some researchers began to use heterogeneous sensor data fusion to improve the accuracy of MEMS gyroscope; they called this technology the "virtual gyroscope".

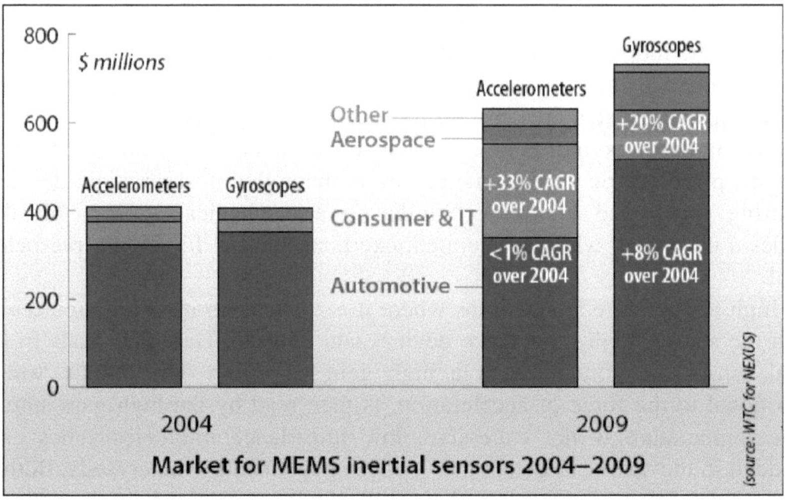

Figure 6.1 Market for MEMS inertial sensors

The market for MEMS inertial sensors (accelerometers and gyroscopes) is set to grow from $835 million in 2004 to over $1360 million in 2009 (Figures 6.1 and 6.2). Currently, the main applications are in the automotive industry. These markets are well established and growth rates range from a stagnant 1% for airbag acceleration sensors up to 8% for gyroscopes used in ESP units and GPS navigation assistance. Exciting for MEMS inertial sensors is the market opportunity for mobile applications and consumer electronics. Over the next few years, an annual growth rates exceeding 30% for accelerometers is predicted. Mobile phones in particular will provide multi-axis accelerometers with interesting opportunities in menu navigation, gaming, image rotation, pedometers, GPS navigation and the like. Gyroscopes are largely servicing markets for image stabilisation and hard disk drive (HDD) protection in camcorders.

In contrast to the automotive sector, consumer applications feature relaxed specifications. Failure rates for automotive electronic control units (ECU) that house airbag accelerometers must be less than 50 failures *per* million and down to a few failures *per* million for application-specific integrated circuits ASICs. Car manufacturers deploy reliable, high performance accelerometers that are relatively expensive (up to $5 to measure lateral acceleration).

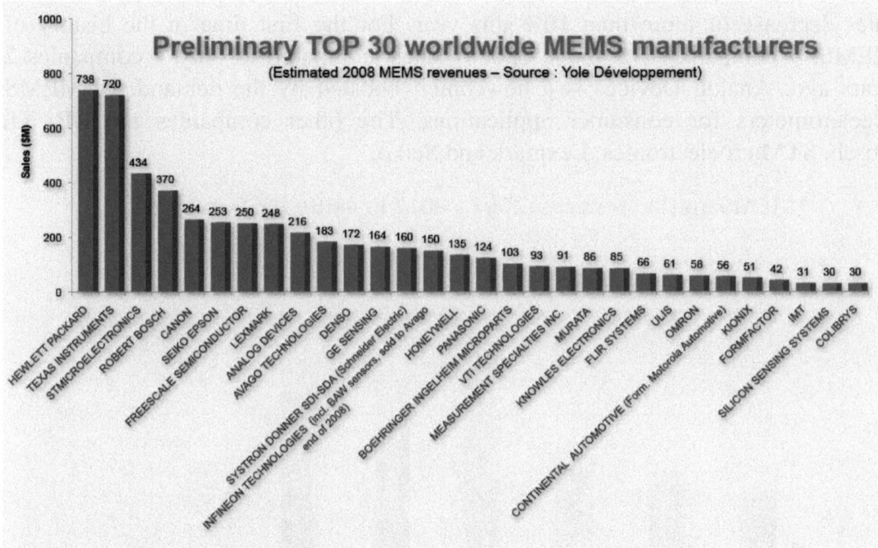

Figure 6.2 Top 30 worldwide MEMS manufacturers

Figure 6.3 provides all further information on these forecasts with a detailed evolution of the 15 main product families. The figure details the evolution of the sales *per* applications fields. In terms of size, the consumer applications are expected to remain the main application area for MEMS devices. Medical, automotive and telecommunications applications are the three other big areas. Europe is very strong in the automotive, medical and life science business, with leaders like Bosch, Infineon, VTI Technologies and Roche.

YOLE Développement has published the ranking of the 30 major MEMS companies worldwide, ranked by sales (Figure 6.3). This ranking does not take into account the profitability of the companies as this data is not available and is extremely difficult to estimate. So for the fifth year, let us understand the evolution of the 30 largest MEMS companies and the place of the European one.

Although automotive applications continue to fuel the MEMS market, the real driver for growth comes from consumer applications. STMicroelectronics (consumer MEMS business unit), Analog Devices or Avago Technologies record more than 20% of annual growth rate from 2006 to 2007; the applications driving this growth are mainly MEMS in mobile phones, gaming systems or sport applications.

Knowles acoustics with its MEMS silicon microphone and Avago Technologies with FBAR components are two companies of the top 30 that show a growth rate exceeding 35% (compared to 2007). Hewlett Packard becomes the first MEMS manufacturer with more than $850M in 2007, thanks to the HP inkjet print head based on its innovative scalable printing technology (unveiled in 2005) and good financial health of the company. Texas Instruments (TI) has a record sales decrease of more than 10% this year. For the first time in the history of MEMS, 9 companies are above $200M sales, compared to only 4 companies 2 years ago. Analog Devices is a newcomer, boosted by the demand for MEMS accelerometers for consumer applications. The other companies are HP, TI, Bosch, STMicroelectronics, Lexmark and Seiko.

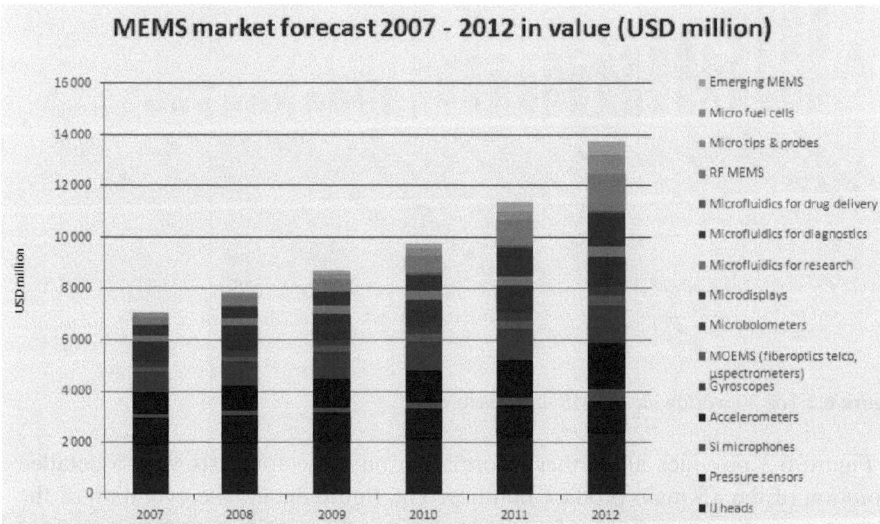

Figure 6.3 MEMS market forecasts by applications (source: YOLE development)

6.2 State-of-the-art of MEMS

MEMS can be viewed as a natural extension of the microelectronic revolution that has so markedly influenced engineering since the 1960s. In 1990, the *New York Times* picked the field of micromachines as a top prospect for influencing engineering in the waning years of the 20th century and this prediction looks to have been very much on target.

Silicon micromachining was first started at Bell Laboratories back in the 1950s where the piezoresistive effect of silicon was discovered. The gauge factor of silicon was about two orders of magnitude higher than that of a metal, which was widely used in strain gauge pressure sensors. In the 1960s, a silicon bar was adhesively bonded to the metal diaphragm, directly replacing a metal strain gauge. Later, a silicon diaphragm was micromachined out of silicon wafers, and silicon piezoresistors were directly formed on the silicon diaphragm, which dramatically improved the sensor performance, yield and cost. In the 1970s and 1980s, silicon micromachining technology became more mature and manufacturable. Silicon pressure sensors and accelerometers are the two major applications for silicon micromachining. Other sensors such as optical sensors, flow sensors and regulators also blossomed during this period of time. In the early 1990s, the first fully integrated accelerometer for air bag applications was introduced, where the mechanical accelerometer together with its readout and signal conditioning circuitry were integrated onto a single silicon chip.

Silicon micromachined sensors have numerous advantages over conventional sensors: high performance, small size, low cost and light weight. There are two major factors making silicon micromachining technology attractive: the near perfect mechanical property and the readily available fabrication technology. Silicon is almost a perfect mechanical material, which is very important for sensors. Thanks to the integrated circuit industry, silicon is one of the most widely studied and understood elements. Structurally, silicon has a crystalline orientation. The lattice structure makes its mechanical properties highly predictable and repeatable. This same crystal structure makes silicon extremely strong. It is stronger than steel and does not have any mechanical hysteresis. The strength combined with the repeatable behaviour make silicon appropriate for a number of mechanical as well as electromechanical uses.

In addition to excellent mechanical properties, silicon micromachining technology is derived from the well established semiconductor manufacturing processes. Precise geometry control through photolithographic techniques can be directly transferred from the semiconductor industry with minimal development. Silicon sensors also possess extremely small size and light weight properties, which can be critical to several applications such as medical diagnosis where size and weight are important.

The photolithographic processes developed for silicon enable the development of microscopic structures. Feature sizes of less than 1 μm, (100 times smaller than the diameter of a human hair) are readily achievable. This means devices can be made incredibly small and can fit into spaces previously impossible to attain. The low cost and small size make distributed sensing systems technically and economically viable where conventional sensors cannot perform due to their size and cost limitations. The advantage of using silicon sensor technology is not only in the increased price/performance ratio, but also in its extreme high volume production due to batch fabrication technology. The integration of the silicon sensors and microelectronics creates a new generation of "smart sensors", which establishes the basis for producing sensor based systems or subsystems entirely on a single silicon chip. This will dramatically increase the performance and the functionalities of sensing systems while reducing the system cost significantly.

As mentioned earlier silicon is not the only micromachined material for sensors. Systron Dormer Inertial Division (a BE1 Sensors & Systems Company) based in Concord, California uses quartz to produce micromachined gyroscopes that employ the Coriolis effect, where a rotational motion about the sensor's longitudinal axis produces a voltage proportional to the rate of rotation. The sensor replaces traditional spinning wheel and fibre optic gyros, which consume higher power, are heavier and have lower operating lifetimes. The sensor consists of a microminiature double-ended quartz tuning fork and supporting structure, all fabricated chemically from a single wafer of monocrystalline piezoelectric quartz, similar to quartz watch crystals.

These gyroscopes are currently in use in a wide variety of applications ranging from yaw sensing to the measurement of angular rates of the space shuttle astronauts during untethered space walks and navigation of the Maverick missile. Concurrently, the area of artificial intelligence where advances in neural networks are being complemented with genetic programming techniques and fuzzy control will formulate the basis for a new generation of intelligent electronic sensing and decision making embedded in the same chip. MEMS will also be particularly useful in optical applications due to their small volume, low energy and small force required to switch photons. Recent progress in optical MEMS technology has demonstrated the possibility of fabricating micro-optical and micromechanical elements on the same wafer with batch processing techniques, thus opening the door to a new class of integrated optics that combines free-space optics with optomechanical sensing and actuating elements.

So far MEMS sensors have been applied widely in the following fields:

- In the consumer arena, these sensors can add an intuitive man-machine interface to game controllers and to portable equipment, such as mobile phones, MP3 players and PDAs, allowing the user's wrist, arm and hand movements to interact with applications, navigate within and between pages, or move characters in a PC game. MEMS accelerometers are also essential for virtual reality games to sense movements of the players. MEMS sensors are also being used in digital cameras to compensate for unintentional movement while pictures are being taken. In the emerging market for robotic toys, accelerometers and gyroscopes sense the robot's movements so that it is "aware" of its position in space.
- In the computer segment, MEMS sensors help provide data integrity protection in laptops and other portable devices. In the case of a free fall or other abnormal movement, a MEMS sensor promptly instructs the system to stop all reading and writing operations and move the magnetic head on the hard disk drive to a safety position.
- In the automotive field, MEMS devices have many applications, including airbag sensors, anti-theft alarms and navigation systems. In the last example, they are used in "dead-reckoning" systems where monitoring of motion and distance travelled is used to maintain correct digital-compass readings in the temporary absence of the GPS signal.
- In the industrial sector, accelerometers are being used as vibration detectors in washing machines, dishwashers and other new home appliances to alert users to unbalanced loads and to detect excessive wear of mechanical parts before a failure occurs. Security systems are another important application area: antitheft alarms based on MEMS accelerometers can detect movement in any desired axis, protecting cars, briefcases, laptops and other mobile hardware from unauthorised removal and detecting movement of doors and windows.

6.3 Characteristics of MEMS Sensors

The following are some of the more important sensor characteristics.

Transfer Function

The transfer function shows the functional relationship between physical input signal and electrical output signal. Usually, this relationship is represented as a graph showing the relationship between the input and output signal, and the details of this relationship may constitute a complete description of the sensor characteristics. For expensive sensors that are individually calibrated, this might take the form of the certified calibration curve.

Sensitivity

The sensitivity is defined in terms of the relationship between input physical signal and output electrical signal. It is generally the ratio between a small change in electrical signal to a small change in physical signal. As such, it may be expressed as the derivative of the transfer function with respect to physical signal. Typical units are volts-kelvin, millivolts, kilopascal, *etc.* A thermometer would have "high sensitivity" if a small temperature change resulted in a large voltage change.

Span or Dynamic Range

The range of input physical signals that may be converted to electrical signals by the sensor is the dynamic range or span. Signals outside of this range are expected to cause unacceptably large inaccuracy. This span or dynamic range is usually specified by the sensor supplier as the range over which other performance characteristics described in the data sheets are expected to apply. Typical units are Kelvin, Pascal, Newton, *etc.*

Accuracy or Uncertainty

Uncertainty is generally defined as the largest expected error between actual and ideal output signals. Typical units are Kelvin. Sometimes this is quoted as a fraction of the full-scale output or a fraction of the reading. For example, a thermometer might be guaranteed accurate to within 5% of FSO (full scale output). *Accuracy* is generally considered by metrologists to be a qualitative term, while *uncertainty* is quantitative. For example, one sensor might have better accuracy than another if its uncertainty is 1% compared to the other with an uncertainty of 3%.

Hysteresis

Some sensors do not return to the same output value when the input stimulus is cycled up or down. The width of the expected error in terms of the measured quantity is defined as the hysteresis. Typical units are Kelvin or percent of FSO.

Nonlinearity (Often Called Linearity)

The maximum deviation from a linear transfer function over the specified dynamic range is of interest here. There are several measures of this error. The most common compares the actual transfer function with the "best straight line," which lies midway between the two parallel lines that encompass the entire transfer function over the specified dynamic range of the device. This choice of comparison method is popular because it makes most sensors look the best. Other reference lines may be used, so the user should be careful to compare using the same reference.

Noise

All sensors produce some output noise in addition to the output signal. In some cases, the noise of the sensor is less than the noise of the next element in the electronics, or less than the fluctuations in the physical signal, in which case it is not important. Many other cases exist in which the noise of the sensor limits the performance of the system based on the sensor. Noise is generally distributed across the frequency spectrum. Many common noise sources produce a white noise distribution, which is to say that the spectral noise density is the same at all frequencies. Johnson noise in a resistor is a good example of such a noise distribution. For white noise, the spectral noise density is characterised in units of volts root (Hz). A distribution of this nature adds noise to a measurement with amplitude proportional to the square root of the measurement bandwidth. Since there is an inverse relationship between the bandwidth and measurement time, it can be said that the noise decreases with the square root of the measurement time.

Resolution

The resolution of a sensor is defined as the minimum detectable signal fluctuation. Since fluctuations are temporal phenomena, there is some relationship between the timescale for the fluctuation and the minimum detectable amplitude. Therefore, the definition of resolution must include some information about the nature of the measurement being carried out. Many sensors are limited by noise with a white spectral distribution. In these cases, the resolution may be specified in units of physical signal/root (Hz). Then, the actual resolution for a particular measurement may be obtained by multiplying this quantity by the square root of the measurement bandwidth. Sensor data sheets generally quote resolution in units of signal/root (Hz) or they give a minimum detectable signal for a specific measurement. If the shape of the noise distribution is also specified, it is possible to generalise these results to any measurement.

Bandwidth

All sensors have finite response times to an instantaneous change in physical signal. In addition, many sensors have decay times, which would represent the time after a step change in physical signal for the sensor output to decay to its original value. The reciprocal of these times correspond to the upper and lower cut off frequencies, respectively. The bandwidth of a sensor is the frequency range between these two frequencies. Below is an example of sensor performance characteristics of an accelerometer.

To add substance to these definitions, the numerical values of these parameters are identified for an off-the-shelf accelerometer, Analog Devices' ADXL150.

Transfer Function

The functional relationship between voltage and acceleration is stated as

V (Acc)= 1.5V +(Acc *x* 167 mV/g).

This expression may be used to predict the behaviour of the sensor and contains information about the sensitivity and the offset at the output of the sensor.

Sensitivity

The sensitivity of the sensor is given by the derivative of the voltage with respect to acceleration at the initial operating point. For this device, the sensitivity is 167 mV/g.

Dynamic Range

The stated dynamic range for the ADXL322 is +2g. For signals outside this range, the signal will continue to rise or fall, but the sensitivity is not guaranteed to match 167 mV/g by the manufacturer. The sensor can withstand up to 3500 g.

Hysteresis

There is no fundamental source of hysteresis in this device. There is no mention of hysteresis in the data sheets.

Temperature Coefficient

The sensitivity changes with temperature in this sensor, and this change is guaranteed to be less than 0.025%/C. The offset voltage for no acceleration (nominally 1.5 V) also changes by as much as 2 mg/C. Expressed in voltage, this offset change is no larger than 0.3 mV/C.

Linearity

In this case, the linearity is the difference between the actual transfer function and the best straight line over the specified operating range. For this device, this is stated as less than 0.2% of the full-scale output. The data sheets show the expected deviation from linearity.

Noise

Noise is expressed as a noise density and is no more than 300 □g/Rt Hz. To express this in voltage, we multiply by the sensitivity (167 mV/g) to get 0.5 □V/Rt Hz. Then, in a 10 Hz low-pass-filtered application, we would have noise of about 1.5 □V RMS, and an acceleration error of about 1 mg.

Resolution

Resolution is 300 □g/Rt Hz as stated in the data sheet.

Bandwidth

The bandwidth of this sensor depends on choices of external capacitors and resistors.

6.4 Specification of Multi-MEMS Sensor Platform

6.4.1 Introduction

Conventional sensors are bulky and expensive. They require cables for power supply and extraction of the signal. Cables in particular are becoming a major problem for installation, especially when retrofitting monitoring systems. Suggestion from the industry is that ideally a sensor should be very small, low cost, wireless and if possible self-powered. Additionally, in order to increase its general applicability, it should be able to measure several common measurands.

Current technology makes it possible for the sensing elements to be made to be very small, and their power requirements can be reduced. A very important benefit of continuing advances in micro technology is the ability to construct a wide variety of MEMS including sensors and RF components. These building blocks enable the fabrication of complete systems in a low cost module, which include sensing, signal processing and wireless communications. Wireless transmission requires a significant amount of power. As the initial reports suggest, it is possible to "harvest" ambient power with resonant devices for example tapping vibration. This power can be used in a burst for intermittent transmission, potentially leaving sufficient for monitoring when combined with the right communication technology. Together with innovative and focused network design techniques this will enable simple deployment and sustained low power operation.

The application can be exploited in a network design to enable sustained low-power operation. In particular, extensive information processing at nodes (distributed processing), hierarchical decision making, and energy conserving routing and network topology management methods are all seen as being necessary for the exploration of the full potential for this technology.

6.4.2 Objectives

The main objective for this specification is to identify the state-of-art in MEMS sensor technology design and match the requirements to the need of the condition monitoring sector. In the following, a couple of suggested specified directions for the design of a multi-measuring device for the proposed sensors range will be discussed. The specification includes general requirements for the functioning of the device. Functional specification range for possible hardware solutions will be presented and alternatives analysed. Resolution of hardware design should be done prior to host design and software specification, which will take place in future work projects.

The second aspect of the specification is to investigate commercial solutions for wireless transmission mediums, *e.g.*, Bluetooth, ZigBee or WiFi. The device should receive as well as transmit so that it can be programmed. Wireless standards (international and *de facto*, *e.g.*, Bluetooth or 802.11) will be reviewed in the context of power requirements, bandwidth, range and commercial outlook.

There are four main sections that obviously influence on the design of the sensor:

- signal sources – sensing unit;
- information – information processing;
- power source – power management; and
- radio frequency communication – communication protocol.

The intelligent sensor should be able to test data and automatically decide if the system is functioning normally. If the results from the data suggest an abnormality, a decision about the severity of the fault should follow. In the case of the detection of a faulty situation, the sensor should be able to make a decision on what to do next. This decision is based on automated reasoning. Intelligence is also necessary to perform self diagnostics to ensure that the microsensor is working properly. A validation of the data acquisition unit therefore needs to be incorporated into the communicated diagnostic information. Decision making and reasoning will be the part of the advanced capabilities of the intelligent sensor designed to operate in a network system with minimum of unnecessary traffic. These functions are an essential part of the intelligent sensor and the intelligent sensors based system.

Depending on the communication model with the other elements in the system (communication strategies) there might be various requirements to the hardware and possible applications of the sensor. Considering the different possible levels of distribution possible within a condition monitoring system, three profiles of communication are being offered. These communication strategies demonstrate the use of distributed processing methods and will provide experience and results from the tests. The strategies represent the range of distributed processing, from the minimal distribution of intelligence as in a conventional data acquisition system, to the other extreme, where distribution of the data processing offers elements virtually independent from the supervisory system. The middle profile offers a balanced solution with distributed processing capability and elements cooperating and being controlled by the central supervisory stations. These three different strategies are designed to illustrate the use of different levels of distribution, and illustrate how the communication strategies might influence required hardware involved in the intelligent sensors.

Different levels of intelligence, hardware cost and data transfer have been used to influence the design of the hardware. In the early stages of the project accurate specification of the hardware needs and industrial expectations cannot be fully defined without tests. Tests should be aimed at developing a few variants of the sensor. Based on the results from the experiments and performance assessments, the final specification is expected to emerge as experience is gained.

6.4.3 Possible Profiles of Intelligent Sensors

6.4.3.1 Autonomy Intelligent Sensor – Profile 1

Profile 1, called "autonomy", has an advanced level of independence between the intelligent sensors and the supervisory stations. The high level of distributed processing provides the independence for the system elements. All the elements in the system should have the capability of performing independent data processing; additional functions should be installed to provide a degree of redundancy. The intelligent sensors designed in this way should be capable of performing the condition monitoring process without the need for the supervisory station to be involved in the process. The supervisory stations' role is separated from the activity of the intelligent sensors. The supervisory station uses its ability to communicate with the sensors to generate a general status of the plant and adjust control according to the information delivered by the sensors. When creating a standalone automated condition monitoring station, it might be expected that the requirements according to the hardware and complicity of the software will grow rapidly, but use of the network keeps this to an absolute minimum and it is less important for the system. This profile demands that the functions of the intelligent sensor are partly redundant from the centralised condition monitoring station. This means that the intelligent sensor requires more components on the side of the sensor, providing the expected capabilities (example processor capable processing data). To fulfil this expectation enough data processing power should be placed in the local intelligent sensor to allow for local data processing and reduce the need for control of the sensor by the supervisory station. The intelligent sensor is then designed as a standalone system, collecting, analysing and storing data and information. Therefore, the sensor can fully perform all monitoring functions without help from the supervisory station.

6.4.3.2 Cooperation Intelligent Sensor – Profile 2

Profile 2 is based on close cooperation between the local intelligent sensors and supervisory stations. This strategy is a direct implementation of the distributed condition monitoring processing concept. The supervisory and intelligent sensors are designed to work together under the direction of the supervisory station. The intelligent sensors must be equipped with enough intelligence to perform the main functions involved in data acquisition and local data processing. Under this strategy the intelligent sensor work is being coordinated and managed by the condition monitoring supervisory station. Data processing is performed in conjunction with low level decision making. The main function of the intelligent sensor is to pre-process data and change it into usable information, so only minimal results have to be sent to the supervisory station. The intelligent sensor can decide upon the severity of the fault and ask for permission to send details about the problem. Designed this way, the system aims to reduce the amount of traffic on the network. Decentralisation of the data processing will increase the reliability of the condition monitoring system as it does in the case of control systems. The intelligent sensor can perform the main functions continuously without support from the supervisory station.

Decision making and intelligence are divided between the intelligent sensors and the supervisory condition monitoring station. If necessary, the intelligent sensors can work temporarily without supervision, following the last work profile and storing data locally. This is required to prevent system failure when the central system experiences difficulties. In a system based on this profile, the intelligent sensors and supervisory station will share the maintenance schedules for performing local and central monitoring routines. In comparison with profile 1, the reduced data processing requirement of the sensor will reduce hardware and software complexity.

6.4.3.3 Slave Master Intelligent Sensor – Profile 3

Profile 3 utilises a "master/slave" concept where the intelligent sensor plays the role of the slave part, acting and performing tasks only on requests from the supervisory station. The intelligent sensor is used mainly as a data acquisition tool, collecting data from a device, under directions from the supervisory station. Requests for data are sent at intervals, which are based on information about the current status of a machine. The raw data samples are sent to the supervisory station were data processing takes place. The intelligent sensor in this scenario has restricted capabilities and intelligence; therefore all the functions are dependent on the supervisory station. This profile depends heavily on network communication and requires greater data transfer in comparison to the previous two profiles. The communication medium is used to transfer the large amounts of condition data instead of the pre-processed condition information. The intelligent sensor's hardware can be much reduced, as there is no need for data processing or storage capabilities. This profile option is designed to provide a contrast for profiles 1 and 2 and test a system working on the conventional basis. The system built with this communication strategy can be used to determine if there any significant benefits from the use of the distributed processing method and to assess the ability of the system not using it.

6.4.3.4 Simplest Intelligent Sensor – Profile 4

Profile 4 consists only of the simple sensor implemented with the transmitter. This profile is for contrast with the other profiles. The capabilities provided by this profile will not satisfy the main requirements for intelligent sensors. A tabular comparison (Table 6.1) shows differences between the selected features of the proposed profiles.

Table 6.1 Intelligent sensor profiles of communication

	Device proposal			
Feature	P1 "Autonomy" most sophisticated	P2 "Co-operation"	P3 "Slave-master"	P4 Simple sensor simplest
Sensing T:Temperature P: Pressure	T, P, A Bandwidth <40 kHz, enveloping	T, P, A Bandwidth <40 kHz, enveloping	T, P, A Bandwidth <40 kHz, enveloping	T, P only
Power	Advanced housekeeping, multi-level sleep modes, battery charging long life	Advanced housekeeping, multi-level sleep modes, battery charging long life	Simple housekeeping; sleep modes simply programmed	Only power-up on activation, or only battery less effective usable life
RF	Any standard bidirectional protocol occasional burst: Bluetooth? Zigbee? 802.11?	Any standard bidirectional protocol; Occasional burst	Any standard bidirectional protocol; Higher data rate	Simple one-way traffic; relatively high data rate
Intelligence/ self sufficiency	The IS works as a standalone data collector system fully capable of data processing	Data processing tasks are shared between IS and data collector	Only the data collector is responsible for data analysis, IS works as data collection device	No intelligence
Work schedule	The IS works with the schedule required by the type of the condition monitoring method and is independent from data collector schedule	The IS works within a schedule which can be modified by the data collector depending on the situation in the system	The data collector creates a working schedule starting and ending tasks of data collection for IS	*No schedule*

Table 6.1 (continued)

Feature	Device proposal P1 "autonomy" most sophisticated	P2 "co-operation"	P3 "slave-master"	P4 Simple sensor simplest
Transferring data	Data and results can be transmitted on request from data collector but it is not a priority task for IS	Data and information are transmitted from the IS to the data collector, additional information can be transmitted from the data collector to the IS	All data is transmitted to the data collector for data analysis, the data collector requests are based on the status of the system	All data is transmitted to the data collector for data analysis; unconditional
Hardware requirements	The IS must have enough CPU power and also enough possible storage space for data backup	The IS must be able to work temporarily as a standalone system, be able to modify program settings on request from the data collector	Hardware is reduced to conventional sensors for data acquisition with the communication capabilities	*Hardware is reduced to conventional sensors for data acquisition with the communication capabilities*
Storing data function	The IS must have the possibility to store large amounts of data, which should be easily accessible from remote devices	The IS can store data locally but, the data collector is also responsible for data storage	Data collector stores all the data, IS is used only for data collection	*No data storage*
Networkability and overall telemetry	Size limits range, may require some boosting or mobile collection	Data collector could connect to enterprise	No problems at this size	*Networking simple and off the shelf*

Table 6.1 (continued)

	Device proposal			
Feature	P1 "autonomy" most sophisticated	P2 "co-operation"	P3 "slave-master"	P4 Simple sensor simplest
Storage capacity	Capacity required for data collection processed as well as data backup and a database: Single spectrum 4 kB Single time series (for FFT) 8 kB 20 spectra history 80 kB Time series for wavelet 256 kB 20 time histories ~5MB	The IS should store samples and hold until data transfer is ordered from the data collector, also actual information should be temporarily stored in case of a fault in the data collector system Up to 5 MB	Required capacity is reduced to live sample collection Up to 256 kB	Required capacity is reduced to live sample collection Up to 256 kB
Dependency on the other elements in the system	The IS can perform all the tasks independently from other elements of the system (data collector)	As well as sharing the activity with the data collector, the IS should be able to continue main tasks without the data collector support (in case the data collector shuts down) or for verification of maintenance information	The data collector takes the main role in this profile and should be able to detect faults and instabilities in the IS system, decreasing the same dependency on this system	*Totally dependent*
Cooperation with other devices	Bidirectional: the IS should be able to exchange information with the data collector and other sensors	The IS and the data collector work together, the data collector takes the role of supervisory station and communicates with other stations	A possible option to use other systems, *e.g.*, the CS for information authentication process	*No cooperation: reading only*

Table 6.1 (continued)

Feature	Device proposal P1 "autonomy" most sophisticated	P2 "co-operation"	P3 "slave-master"	P4 simple sensor Simplest
Software development	Software for the IS can be more specialised for the device, software for the data collector includes data transfer and interpretation of the information from the IS	Creation of the software must be based on full compatibility between the IS and the data collector	Interpretation of the collected data would be in the data collector and that involves the possibility of: configuration, settings, storing information about differences between devices. IS software would be reduced to data acquisition	*IS software would be reduced to data acquisition*
Signal conditioning and processing	Amplifier A/D filters FFT wavelet statistical parameters programmable	Amplifier A/D filters FFT wavelet statistical parameters programmable	Amplifier A/D filters FFT	Amplifier A/D?
Possibility of modifications in the system	The IS creates independent systems, easy to modify; programmable	Correlation between the systems makes any changes more complicated as influencing both systems	Dependency in the systems, means that changes are not easy to implement	*No change possible*
Hardware complexity	This option increases the hardware cost of the IS, concentrating on capacities and reliability	Less than profile1, but still demanding from the equipment used	This is the cheapest hardware option for the IS and the most advanced part is the data collector, extracting and evaluating all the data	Low cost

Table 6.1 (continued)

Feature	Device proposal P1 "autonomy" most sophisticated	P2 "co-operation"	P3 "slave-master"	P4 simple sensor Simplest
Use of the system	Remote monitoring applications; relatively high costs can be accepted	Many devices to monitor Does not want a centralised system Benefits from distributed processing The communication medium takes a responsible role in such a system	Better for tasks that do not require continuous monitoring, and large time interval between tests	Better for tasks that do not require continuous monitoring, and large time interval between tests

Legend:

Do not meet requirements	Fully meet requirements	Might meet requirements	Not decided

Those requirements discussed here and other general requirements required for intelligent sensor design have been discussed in Chapter 5 on intelligent wireless sensors.

6.5 Simulation of a Multi-MEMS Sensor Platform

The application that will be discussed next includes a multi sensor platform; Imote2 from Crossbow for which two extra sensors have been added. A multi-MEMS sensor platform is implemented for maintenance of large scale equipment. The platform consists of sensing unit, processing unit, and energy harvesting system.

6.5.1 Sensing Unit

The sensing unit includes an accelerometer, ADXL311JE (Analog Devices), and a temperature sensor, the NTC thermistor. Functions of these sensors are described as follows:

The ADXL311, shown in Figure 6.4, is a low cost, low power, complete dual-axis accelerometer with signal conditioned voltage outputs, all on a single monolithic IC and is built using the same proven iMEMS® process used in over 100 million Analog Devices accelerometers shipped to date, with demonstrated 1 FIT reliability (1 failure *per* 1 billion device operating hours). The ADXL311 will measure acceleration with a full-scale range of ±2 g. The ADXL311 can also measure both dynamic acceleration (*e.g.* vibration) and static acceleration (*e.g.*, gravity). The outputs are analogue voltages proportional to acceleration. The typical noise floor is 300 g/√Hz allowing signals below 2 mg (0.1° of inclination) to be resolved in tilt sensing applications using narrow bandwidths (10 Hz). The user selects the bandwidth of the accelerometer using capacitors C_X and C_Y at the X_{FILT} and Y_{FILT} pins. Bandwidths of 1 Hz to 2 kHz may be selected to suit the application.

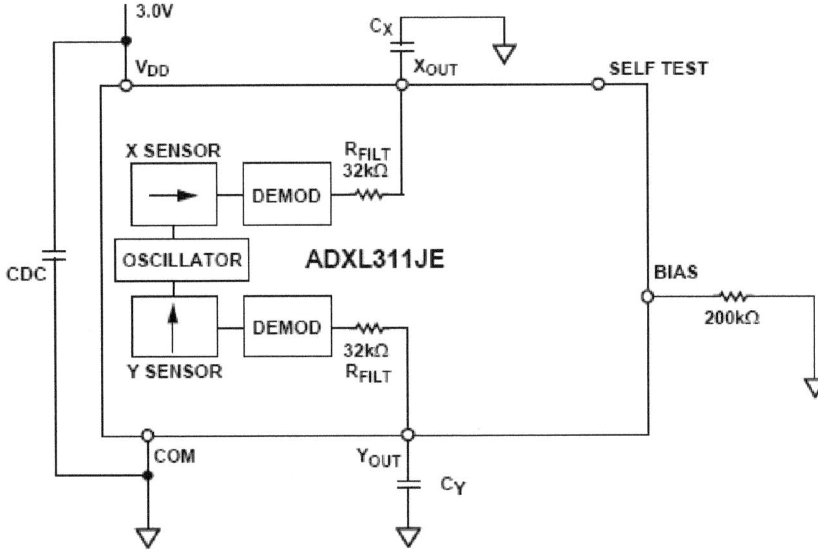

Figure 6.4 Framework of ADXL311JE

The NTC thermistors, shown in Figure 6.5, which are discussed herein, are composed of metal oxides. The most commonly used oxides are those of manganese, nickel, cobalt, iron, copper and titanium. The fabrication of commercial NTC thermistors uses basic ceramics technology and continues today much as it has for decades. In the basic process, a mixture of two or more metal oxide powders are combined with suitable binders, are formed to a desired geometry, dried, and sintered at an elevated temperature. By varying the types of oxides used, their relative proportions, the sintering atmosphere, and the sintering temperature, a wide range of resistivities and temperature coefficient characteristics can be obtained.

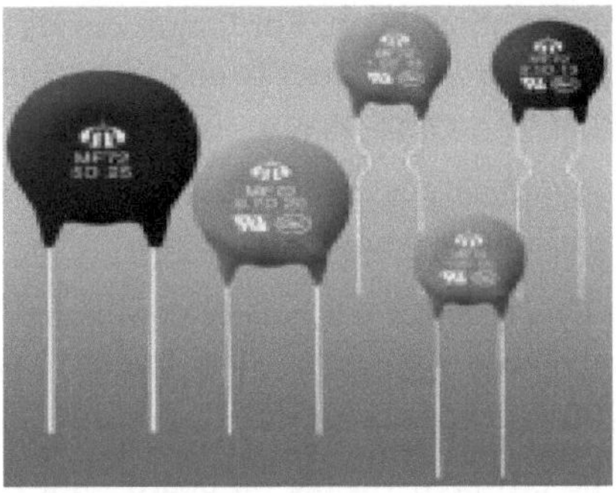

Figure 6.5 NTC thermistors

6.5.2 Processing Unit

The processing unit in this system is an Imote2, which is an advanced wireless sensor node platform. It is built around the low-power PXA271 XScale CPU and also integrates an 802.15.4 compliant radio. The design is modular and stackable with interface connectors for expansion boards on both the top and bottom sides. The top connectors provide a standard set of I/O signals for basic expansion boards. The bottom connectors provide additional high-speed interfaces for application specific I/O. A battery board supplying system power can be connected to either side.

Imote2 contains the Marvell PXA271 CPU. This processor can operate in a low voltage (0.85V), low frequency (13 MHz) mode, hence enabling very low power operation. The frequency can be scaled from 13 to 416 MHz with dynamic voltage scaling. The processor has a number of different low power modes such as sleep and deep sleep. The PXA271 is a multichip module that includes three chips in a single package, the CPU with 256kB SRAM, 32MB SDRAM and 32MB of FLASH memory. It integrates many I/O options making it extremely flexible in supporting different sensors, A/Ds, radios, *etc.* These I/O features include I2C, two synchronous serial ports (SPI), one of which is dedicated to the radio, three high speed UARTs, GPIOs, SDIO, USB client and host, AC97 and I2S audio codec interfaces, a fast infrared port, PWM, a camera interface and a high speed bus (mobile scaleable link). The processor also supports numerous timers as well as a real time clock. The PXA271 includes a wireless MMX coprocessor to accelerate multimedia operations. It adds 30 new media processor (DSP) instructions, support for alignment and video operations and compatibility with Intel MMX and SSE integer instructions. For more information on the PXA271, please refer to the Marvell datasheet.

Imote2 uses the CC2420 IEEE 802.15.4 radio transceiver from Texas Instruments. The CC2420 supports a 250 kb/s data rate with 16 channels in the 2.4 GHz band. The Imote2 platform integrates a 2.4 GHz surface mount antenna that provides a nominal range of about 30 m. For longer ranges an SMA connector can be soldered directly to the board to connect to an external antenna.

Imote2 can be powered by various means:

Primary battery. This is typically accomplished by attaching a Crossbow Imote2 battery board to either basic or advanced connectors.

Rechargeable battery. This requires a specially configured battery board attached to either the basic or advanced connectors. The Imote2 has a built-in charger for Li-Ion or Li-Poly batteries.

USB. The Imote2 can be powered *via* the on-board mini-B USB connector. This mode can also be used to charge an attached battery.

In the project, Imote2 processor radio board (IPR2400) is used. Typical power parameters of IPR2400 are shown in Table 6.2.

Table 6.2 Typical power parameters of IPR2400

Parameter	Operating Value
Supply voltage (V_{bat})	5.5 V
Charger input voltage (V_{chg})	10 V
Input voltage (V_{in})	$V_{CC\ io} \pm 0.3$ V
Storage temperature	–40 to +125°C
Operating temperature	0 to +85°C
Current in deep sleep mode	387 μA
Current in active mode (13 MHz, radio off)	31 mA
Current in active mode (13 MHz, radio *Tx/Rx*)	44 mA
Current in active mode (104 MHz, radio *Tx/Rx*)	66 mA

6.5.3 Hardware Implementation

Imote2 (IPR2400) is an advanced wireless sensor node platform. It is built around the low power PXA271 XScale processor and integrates an 802.15.4 radio (CC2420) with a built-in 2.4 GHz antenna. This processor can operate in a low voltage (0.85 V), low frequency (13 MHz) mode, hence enabling very low power operation. The frequency can be scaled from 13 to 416 MHz with dynamic voltage scaling. The processor has a number of different low power modes such as sleep and deep sleep. The PXA271 is a multi-chip module that includes three chips in a single package, the CPU with 256kB SRAM, 32MB SDRAM and 32MB of FLASH memory. It integrates many I/O options making it extremely flexible in supporting different sensors, A/Ds, radios, *etc.* These I/O features include I2C, two SPI, one of which is dedicated to the radio, three high speed UARTs, GPIOs, SDIO. Imote2 is a modular stackable platform and can be expanded with extension boards to customise the system to a specific application. Through the extension board connectors sensor boards can provide specific analogue or digital interfaces. Imote2 uses the CC2420 IEEE 802.15.4 radio transceiver from Texas Instruments. CC2420 supports a 250 kb/s data rate with 16 channels in the 2.4 GHz band. The Imote2 platform integrates a 2.4 GHz surface mount antenna which provides a nominal range of about 30 meters. For longer ranges an SMA connector can be soldered directly to the board to connect to an external antenna.

Figure 6.6 below shows multi-sensing with the Crossbow devices. In this example there are two external sensors, temperature and accelerometer. These sensors are connected to the ADC of the device. The Imote2 sensing module takes measurements and wireless module transmits to the coordinator (wireless receiver) connected to the PC based monitoring software.

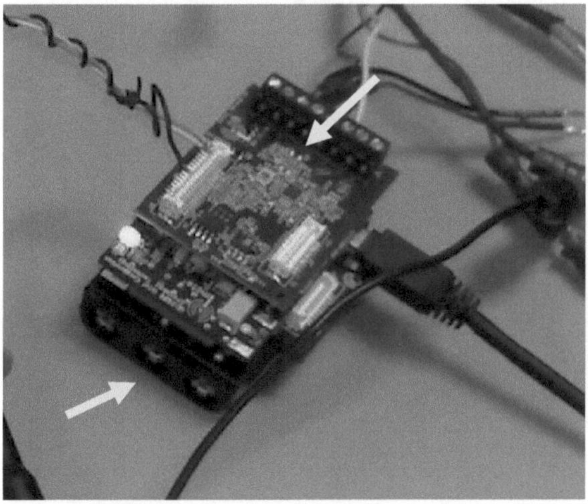

Figure 6.6 Connection of the Crossbow devices (Imote2)

Figures 6.7 and 6.8 show the combined platform for temperature and pressure sensors and the thermistor end. These are connected to the Imote2 ADC. The voltage range accepted by ADC is between 0 and 2.8 V. Sensors are set up and calibrated to send measurements in this range.

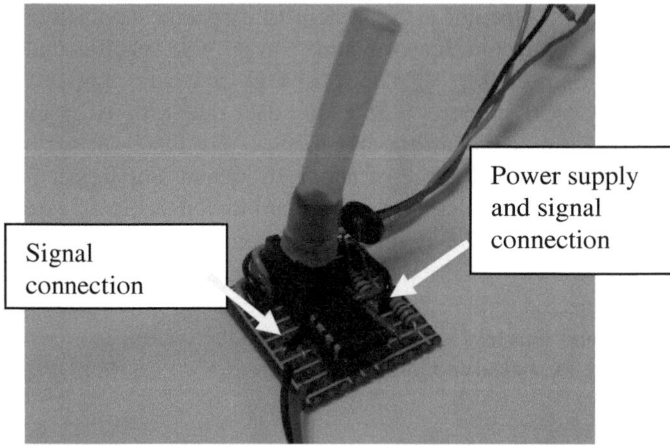

Figure 6.7 Connection of the pressure sensor

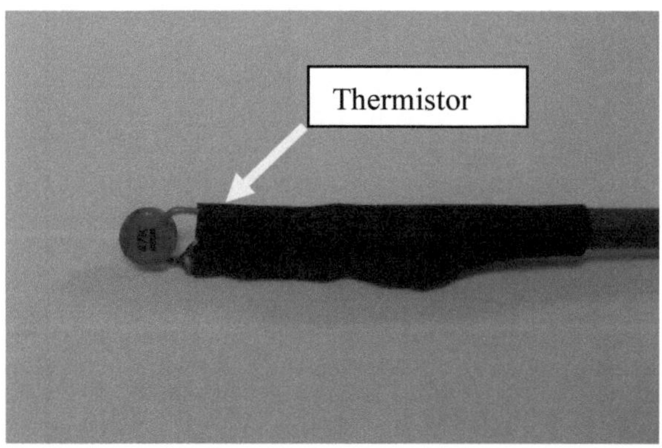

Figure 6.8 Connection of the temperature sensor

6.5.4 Data Sampling

One of the sensor functions is to record vibration data samples that might be large in size compared to the data frame available in the ZigBee transmission protocol. To enable the file to transfer the raw data sample, the recorded files (represented as a string of values in Figure 6.9) need to be divided into manageable parts. Each data frame has a 60-byte size part to be fitted into the data frame. The data sample is divided into X parts. These parts are transferred one by one in the data frame of the message as presented in Figure 6.9. The top part of the data symbolises the part of the vibration data that will be transferred in the data-packed X times.

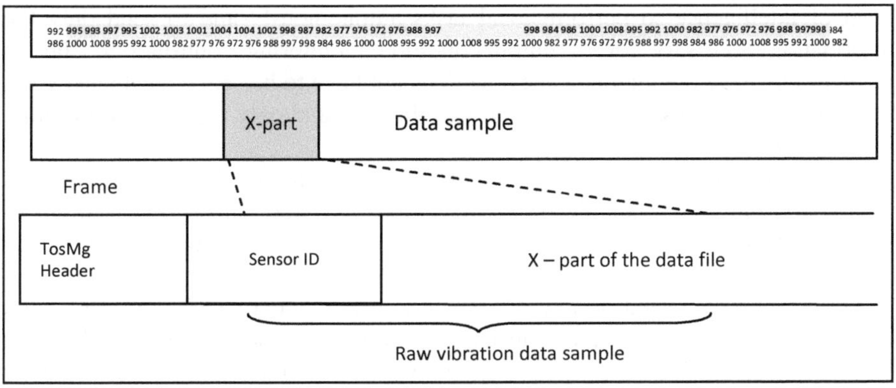

Figure 6.9 Organisation of data transfer

At the receiving end, parts of the sample have to be collected and joined together to recreate the original file that has been transmitted (Figure 6.10). The reliability of the data transfer validation after reception is guaranteed by ZigBee protocol itself.

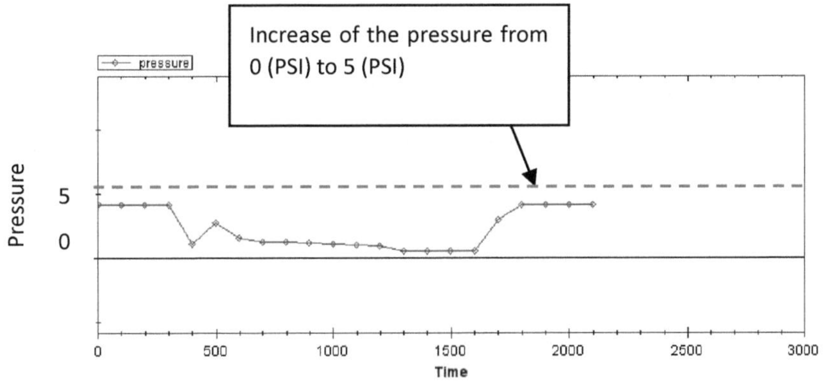

Figure 6.10 Monitoring software, pressure display (*X*: time (s), *Y*: pressure (PSI)

6.5.5 Local Decision Making Based on Condition

Intelligent communication can be restricted to a minimum by the implementation of the decision making capabilities in the local intelligent sensor. If local data processing is to take place without any need for communication with a supervisory monitoring station, sensors must possess a set of information and references enabling a decision on the condition.

For this reason, it is necessary to employ a distributed processing approach, with intelligent condition monitoring sensors that are equipped with a sufficient level of intelligence to perform local tasks autonomously and to send only the information necessary for a central condition monitoring controller to stay in control. This way the maintenance data does not have to be sent over the network. Local intelligent sensors can shift through the data and only pass on the essential information.

Depending on the communication strategy, information about the status of a device is used to initiate the communication with the supervisory station. A supervisory station will only receive information about a status if it has been changed or requested.

Outside the requested periodical transfer, an intelligent sensor will only send data that has failed the standard tests; so healthy data does not need to be sent. With this strategy unnecessary traffic is reduced to a minimum, leaving space for urgent actions and for increasing the number of devices communicating on a single segment of the network. This provides a very economical solution because the use of an extremely fast network is not required.

Figures 6.10 and 6.11 show the detailed data from the multi-sensor device. In this case a sudden change in the temperature and pressure is indicated, which triggers an alarm.

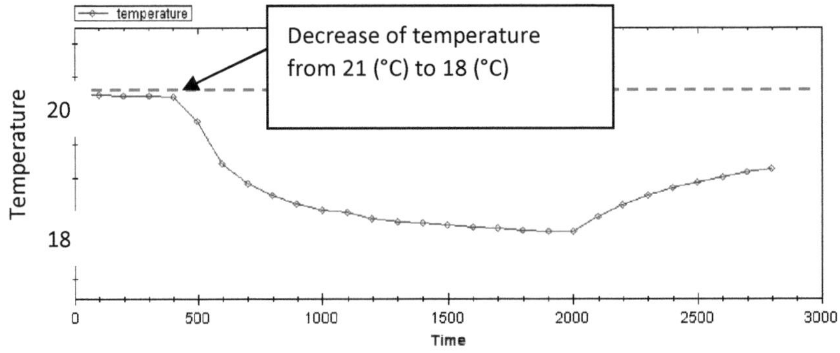

Figure 6.11 Monitoring software, temperature display (*X*: time (s), *Y*: temperature (°C))

6.5.6 Threshold with Event Triggering

Raw condition monitoring data is not typically useful for the rest of the subsystems. Often the raw data can only be used by the condition monitoring system chosen for the specific data processing task. It is the information about the condition produced from the data during the processing that is valuable for the rest of the system. The requirement to reduce the amount of data is achieved by analysing data locally. Data is converted into information, reducing size, standardising the format and making it directly usable for the other stations. To provide this kind of communication, the conversion of data into information must be performed directly at the source of the data. The intelligent sensor is able to pre-process and assess the data. Then based on the thresholds automatically decide whether the data is within the limits thereby allowing normal functioning to continue. If the results from the data suggest abnormality, a decision about the severity of the fault should follow. In the case of the detection of a faulty situation, the system should make a decision on what to do next. This decision is based on automated reasoning.

Decision making and reasoning are part of the advanced embedded software designed to operate the prototypes of the intelligent sensors. These functions are an essential part of the intelligent sensor and the Dynamite system.

In the following example, the multiple sensors continuously sense the selected types of data and simultaneously assess the value of the measurement. The values are analysed by the wireless sensor. The sensor decides on the condition of the parameter, comparing it to the value of the thresholds set for it. If the parameters exceed the limit, it will prompt an alarm for the monitoring station. The monitoring station can have a different set of limits set for the internal alarms, so an alarm from the sensor can be triggered, and either they are activated or ignored. Figure 6.12 shows the screen from the monitoring software where three sensors simultaneously monitor three parameters.

The data transfer from multiple devices is controlled automatically by the ZigBee protocol layer. Thus there are no collisions or data losses.

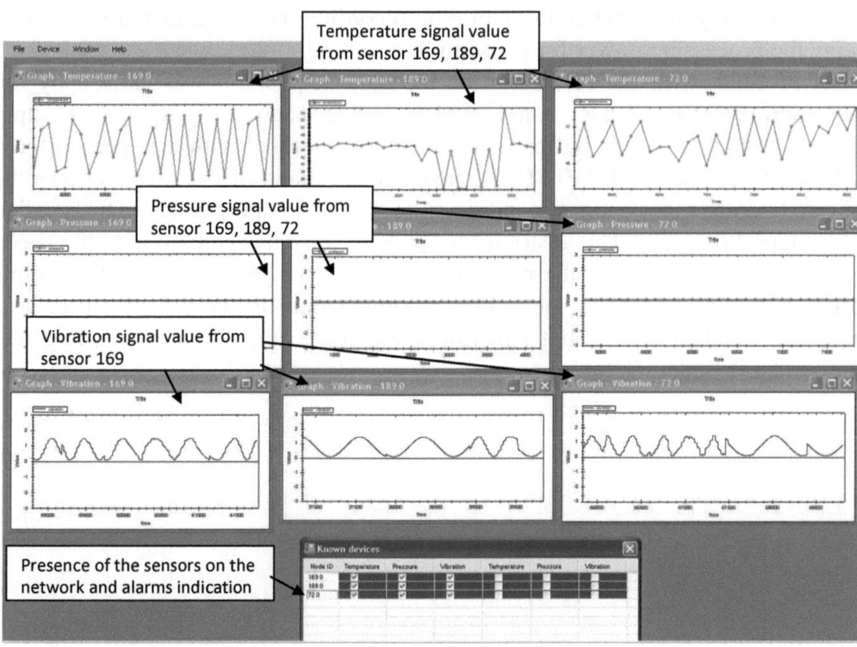

Figure 6.12 Monitoring software, selection of the sensors displayed

6.5.7 Data Pre-processing

The data received from the sensor is pre-processed in the sensor before transmission to the monitoring station. This data pre-processing can be used for first instance condition diagnostics. As an example of data pre-processing, the FFT function is performed on the vibration data to determine the dominating frequencies. The following figures show graphs plotted from the raw data of an MEMS accelerometer (sensor networks) acquired through Imote2. The vibration raw data values are transmitted to the monitoring system in unchanged format, but due to the internal pre-processing, with the FFT function sensor that can determine the monitored situation and use it for decision making. The FFT processing system uses an especially modified FFT algorithm created in cooperation with diagnostic solutions. The example shows the resolution of the vibration data sample plotted on the graph (the *X* axis is time and *Y* is the amplitude). For the test purposes, data samples were recorded, transmitted and reconstructed on the graph over a range of 800 Hz. Figure 6.13 shows two frequencies at 500 Hz and 600 Hz.

The initial test with data transfer with the use of Imote2 indicated that data collection process interferes with the transmission process. This is especially so when requested vibration data is recorded. Initial tests showed that the sampling was interrupted, and the sampling frequency was restricted by the transmission interval. The only solution to eliminate the problem was to reorganise the data acquisition process to run separately from the transmission, so that the transmission only starts when the vibration recording has finished and *vice versa*, freeing the processing power for the sampling routine.

The problem with the restriction of the data acquisition was caused by the driver routine that pools data from the ADC device. There is no automated output from the ADC to a shared buffer, so data cannot be stored with the speed offered by the ADC device clock (10 kHz). The current version of the driver allows the software to collect just one set of values from the ADC inputs when ADC is called in the internal program. This restricts the speed of the data acquisition to the speed of the main program. After the modifications of the original embedded code this routine was improved from the 1 kHz recommended by the manufacturer to a maximum speed to 2 kHz, but still low compared to the 10 kHz sampling rate promised by the manufacturer.

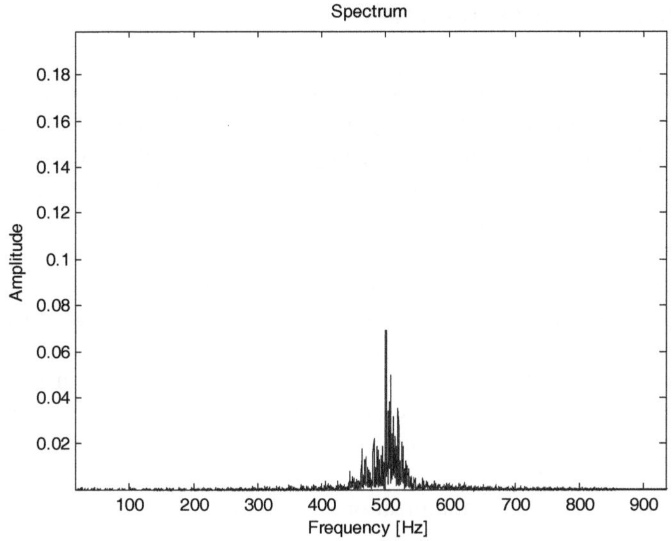

(a)

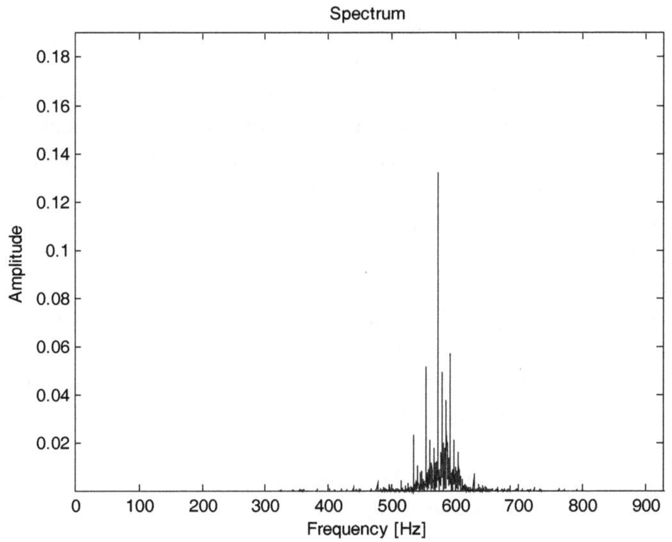

(b)

Figure 6.13 Data recorded (a) at 500 Hz and (b) at 600 Hz

6.5.8 Transmission on Intervals

To prevent the functioning of the system from being dependent on the transmission medium it is necessary for the flow of data to be reduced. To function effectively and avoid overloading the system with data, the amount of data exchange required for the condition monitoring function must be minimised.

Data from the sensor is transmitted periodically; with a time interval between sensors performing internal processing and local decision making. This reduces the amount of data transferred and recorded. Processing of data before sending it on to the network is necessary. This is an essential part of the program. This function has to be included from an early stage of the design ensuring compatibility of the embedded software. The calculation function represents the future control algorithms and data computation process. In this example it will be data reformatting to the measuring value and thresh-holding or in the case of vibration it will be FFT analysis.

Figure 6.14 shows the communication process in three possible modes:

- internal data transmission schedule
- schedule setup in the local monitoring station
- remote agent request

The first mode shows scheduled data transfer initiated by the sensor itself based on a programmed time interval. The sensor waked up, takes measurements and sends data to receiver. The data is stored and available for remote access.

The second mode to trigger the data transfer can be initiated by the monitoring software. The local acquisition software sends a request to the sensor, then wakes up, takes measurements and sends data to receiver. The data is be stored and available for remote access.

The third mode to trigger the data transfer can be initiated by the remote monitoring software or agent. The remote agent sets the request (or change to the monitored interval) on the MIMOSA database. The change in the database is picked up by the acquisition software, then the request is sent to the sensor, which wakes up, takes measurements and sends data to receiver. The data is stored and available for remote access.

The more detailed algorithm of the embedded program is shown in Figure 6.15. The UML diagram explains the logical procedure and shows all the functional steps in the program.

MODE1: Scheduled data transfer initiated by the by the sensor it self based on programmed time interval. The sensor wakes up, takes measurements, sends data to receiver. The data is stored and available for remote access. *MODE2: The s*econd option to trigger the data transfer can be initiated by the monitoring software. The local acquisition software sends request to the sensor, then wakes up, takes measurements, sends data to receiver. The data is stored and available for remote access. *MODE3:* The third option to trigger the data transfer can be initiated by the remote monitoring software or agent. The remote agent sets the request (or change to the monitored interval) on the MIMOSA database. The change in database is picked up by the acquisition software, then it sends the request to the sensor, which will wake up, take measurements, send data to receiver. The data is stored and available for remote access.

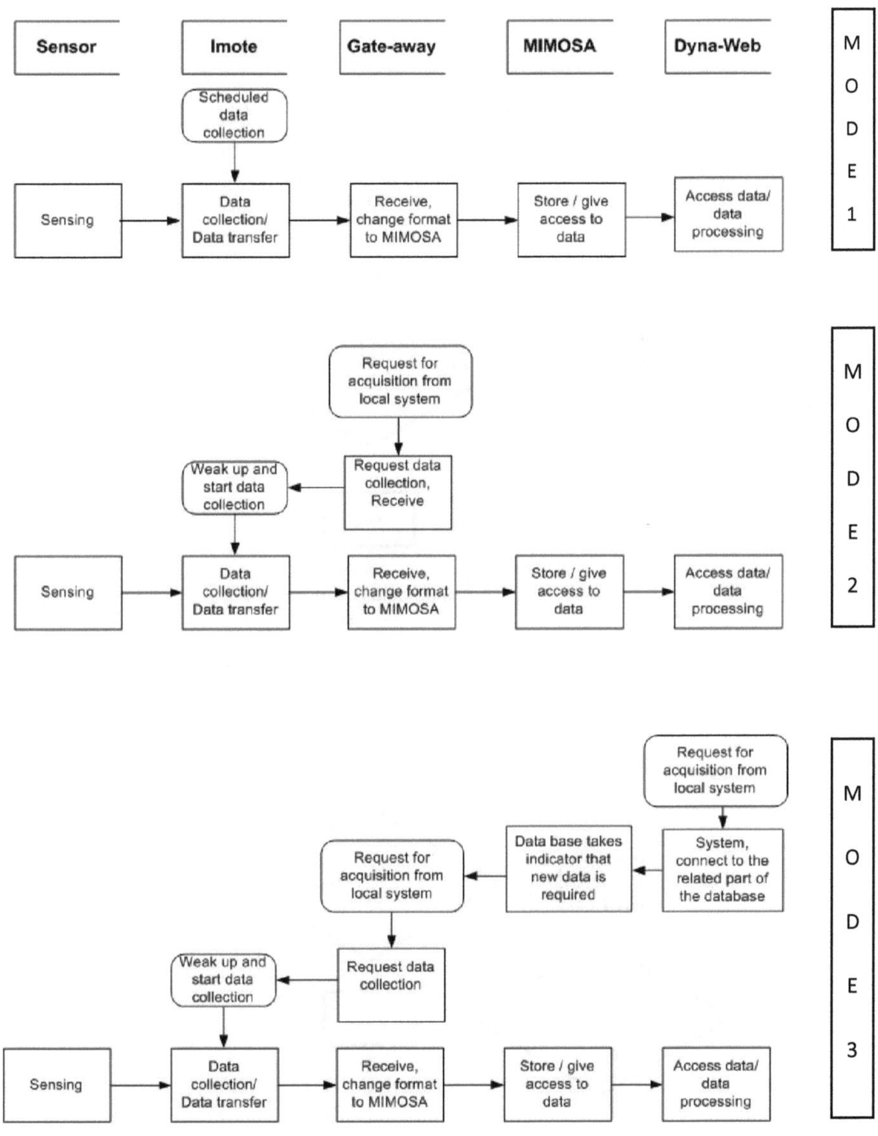

Figure 6.14 Function iteration diagram

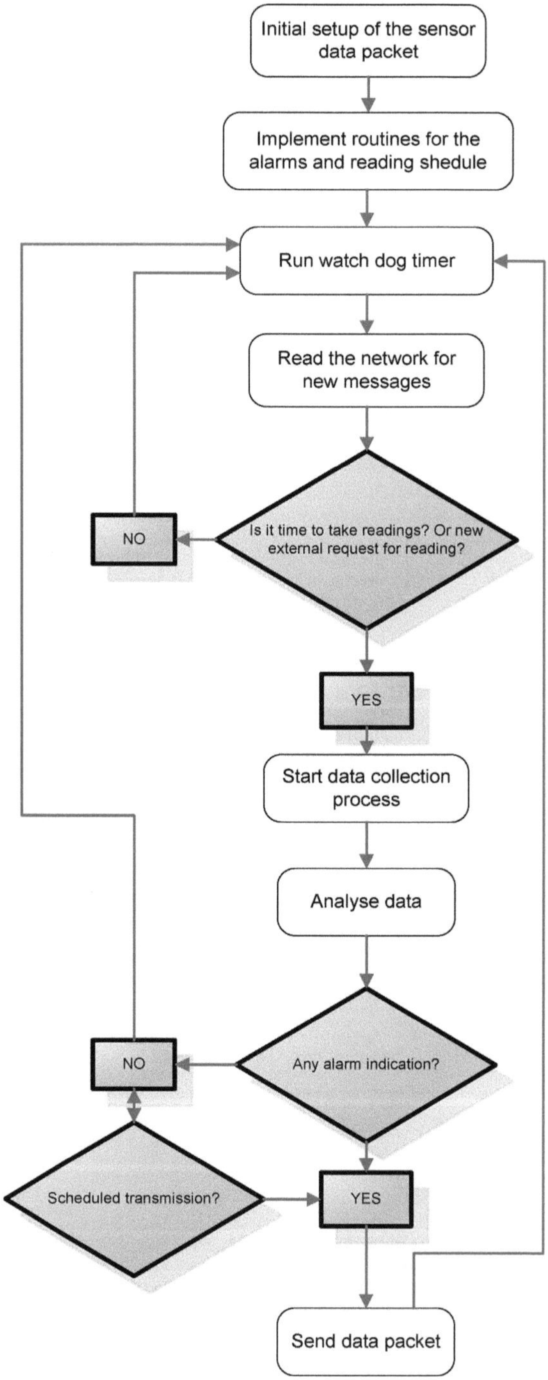

Figure 6.15 Function iteration diagram

6.6 Power Management

One of the most important constraints on sensor nodes is the low power consumption requirement. Power restrictions influencing design are:

- Very little electrical power available from the batteries, although plenty of other sources (like vibration, temperature or pressure).
- Requires efficient processing and careful selection of processor power and duty cycle to enhance the life spam of the batteries.
- Power consumption of transducers and interfaces must be considered.

The main task of a sensor node in a sensor field is to detect events, perform quick local data processing and then transmit the data. Power consumption can hence be divided into three domains: sensing, communication and data processing.

The sensing power varies with the nature of applications. Sporadic sensing might consume less power than constant event monitoring. The complexity of event detection also plays a crucial role in determining energy expenditure. Higher ambient noise levels might cause significant corruption and increase detection complexity.

A transceiver unit connects the node to the network. The radio module plays a very important role in sensor design. Wireless transmission can be responsible for the draining of batteries, so the control algorithm in the sensor must configure the sensor to operate in either transmit, receive or power-off mode. The transmission *via* radio signals contains no buffering, so each bit must be serviced by the processor at the time of receiving it.

Because sensor nodes carry limited, generally irreplaceable, power sources power consumption has to be minimised. High quality of service provision in the sensor network protocols has to be traded to focus primarily on power conservation. The presented profiles of communication with built-in intelligence have been designed to show that the communication based on intelligent decision mechanisms can prolong network lifetime at the cost of lower throughput or higher transmission delay. The basic hardware will use a fraction of a watt of power and consist of commercial components of a square inch in size. During the hardware development it was expected that many different types of hardware would be created to meet the needs for a variety of applications.

6.6.1 Sleep Mode

Using the power management function of the available API the mote can be put into a deep sleep (drawing 525 uA) for a specified period of time.

To put the mote into deep sleep mode use the following method:

- PowerManagement pm = new PowerManagement;
- pm.DeepSleep(30).

The mote will go into a deep sleep state for 30 s.

When Imote2 recovers from the deep sleep state, it is rebooted. Information that needs to be preserved over reboots must be written to the flash memory. The sample application XSensorITS400Sleep demonstrates how to accomplish this by using ExtendedWeakReference.

6.6.2 Performance versus Power Consumption

Considering the power consumption of Imote2 (Crossbow 2007), the system requires an average of 4.5 V and 180 mA. The minimum battery requirements that can power the Imote2 platform are 3.5 V and 174 mA. Assume the typical data collection cycle battery discarding time (about 12849 s = 3.57 h) and each cycle including data collection, data processing and transmission will last 95 s. A standard battery will last 180 cycles. The total working time depends on the frequency of the data collection. Assume that the application requires 10 measurements *per* day for 18 days. The time available must be long enough to recharge the second battery pack *via* the energy harvesting system. This battery may be used as a replacement for the main battery. Figure 6.16 shows the Imote2 discharge cycle lasting about 3.57 h. After the voltage goes below the 3.5 V the system cannot be powered sufficiently. Imote2 stops working and the battery regenerates.

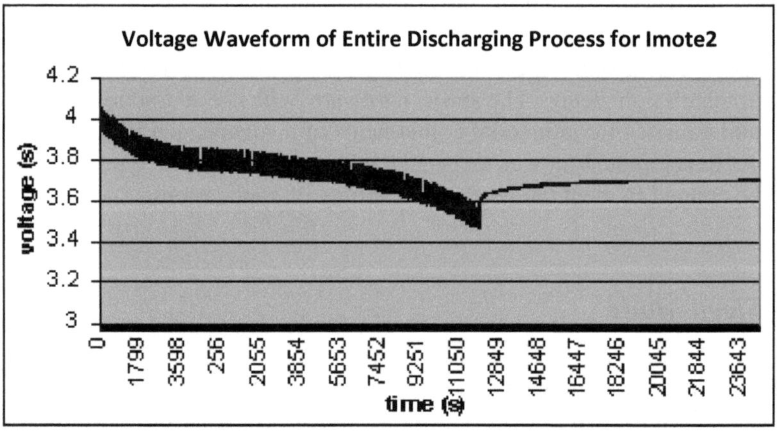

Figure 6.16 The battery discharge process

6.6.3 Energy Harvesting System

An energy harvesting system normally consists of energy transducers, an energy converting subsystem and energy storing components. Energy transducers are mechanical energy resulting from vibration, stress and strain; thermal energy from furnaces and other heating sources even biological ones; solar energy from all forms of light sources, ranging from lighting, light emissions and the sun; electromagnetic energy that captured *via* inductors, coils and transformers; wind and fluid energy resulting from air and liquid flow; and chemical energy from naturally recurring or biological processes. An energy converting subsystem converts all kinds of external energies into electrical power. Energy storing components are divided into two: one using a super capacitor storing converted electrical energy and the other ones uses rechargeable batteries storing energy.

6.6.4 Energy Transducers

6.6.4.1 Piezo Film

In the experiment, FS-2513P shown in Figure 6.17, a piezo film provided by PROWAVE, was used in evaluation. The polymer film is composed of polyvinylidene fluoride (PVF2). The strain constant is 10–20 times larger than normal piezo ceramic and it is therefore ideal for converting mechanical energy into electrical energy. Its features are the following:

- large mechanical-electrical coefficient;
- low mechanical and acoustic impedance;
- high resistance to moisture;
- pliant, flexible, tough and lightweight; and
- self generated voltage, noncontact, rustless and free of sparking.

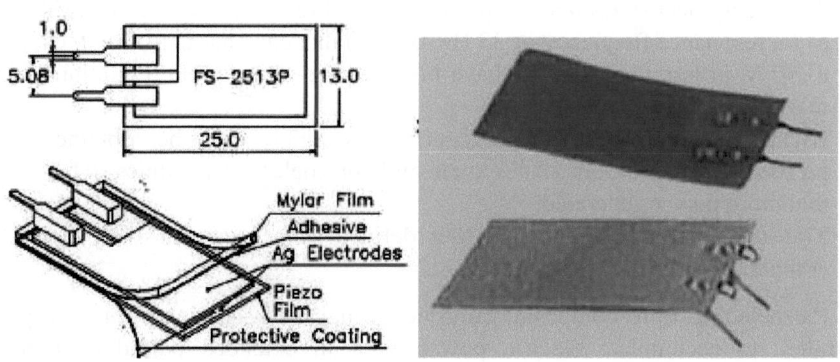

Figure 6.17 FS-2513P piezo film transducer

6.6.4.2 Piezo Buzzer

In order to explore the possibility of scavenging more energy by using piezo elements, a piezo buzzer is reversely used as an energy harvesting source. Here three piezo transducers are selected for experiments (Figure 6.18).

TFM43 41.PT55L030H ABT-422-RC

Figure 6.18 Three piezo transducers selected for experiments

ABT-422-RC, a piezo transducer from PRO SIGNAL, is with resonance frequency 5.5 kHz. 41.PT55L030H and TFM43, provided by UNBRANDED, are with 4.4 kHz and 4.5 kHz, respectively. It is expected to further explore better results by selecting low resonance frequency.

6.6.4.3 Piezoelectric Fibre Composites

Mechanical vibration energy also can be converted into electrical energy by piezoelectric materials such as piezoelectric crystalloid, ceramics, polymer and piezo fibre composites. Piezoelectric fibre composite transducers are made by Advanced Cerametrics Incorporated (ACI). There are a number of advantages of piezoelectric fibre composite transducers over bulk ceramics:

- Due to their composite nature, piezoelectric fibre composites are lighter, more flexible and robust than bulk piezoelectric ceramics.
- ACI's datasheet states that the piezoelectric fibre composites produce 500 V(p-p) (at resonance frequency of 35 Hz, 0.9 lbf) that can charge a 400 μF capacitor to 50 V in less than 4 s and have been configured to yield a continuous 145 mW of power.
- Higher piezoelectric voltage coefficients (g33) can be obtained from piezoelectric fibre composites than bulk piezoelectric ceramics. This means that more power generated.
- Transducers can be created to user defined shapes inexpensively (including complex shapes).

Piezoelectric fibre composite transducers may be used singly, or multiply in parallel, to accumulate electric power over an extended period of time for energising low power microelectronics (Perpetuum website).

6.6.4.4 Electromagnetic Generators

Vibration energy harvesting is receiving a considerable amount of interest as a means for powering wireless sensor nodes. An electromagnetic generator utilises discrete components and captures energy from a low ambient vibration level based upon real application. In general, the generator uses magnets arranged on an etched cantilever with a wound coil located within the moving magnetic field. Magnet size and coil properties have been optimised. The amount of energy generated by this approach fundamentally depends upon the quantity and form of the kinetic energy available in the application environment and the efficiency of the generator and the power conversion electronics.

Perpetuum has refined the principle of gathering and converting unused mechanical vibration into useable electrical energy to power industrial wireless sensors. PMG serials of products, vibration-based generators, have been developed. For example, PMG17-120 is applied in industrial field and can output power 45 mW when it works at 15 Hz and 1.0 g. Its part parameters and shape are shown in Figure 6.19.

POWER OUTPUT					
OUTPUT WITH V_{OUT} = 4.5V					
CHARACTERISTIC		MIN	TYP	MAX	UNIT
0.025g (RMS)	CENTRE	0.9	1.0	1.1	mW
	50% POWER BANDWIDTH	2.0	2.5	3.0	Hz
0.1g (RMS)	CENTRE	4.0	4.5	5.0	mW
	50% POWER BANDWIDTH	6.0	7.0	8.0	Hz
1.0g (RMS)	CENTRE	40	45	50	mW
	50% POWER BANDWIDTH	13	15	17	Hz
MAXIMUM OUTPUT VOLTAGE		8.6	9.0	11.06	V

Figure 6.19 A PMG17 vibration generator and its parameters

6.6.4.5 Solar Panels

A solar panel can convert sunlight or fluorescent light indoor into DC power supply. SA-0640 is a product of solar panel and is shown in Figure 6.20. Its parameters are the following:

- product: SA-0640
- manufacturer: SOLAREX
- SOLAR PLATE, 0.3 W
- current, output max: 0.04 A
- depth, external:2 mm
- length/height, external: 152 mm
- length, lead: 150 mm
- power, DC: 0.3 W
- spectral distribution, STC: 1.5
- temperature, STC cell: 25°C
- voltage, nom: 6 V
- voltage, output max: 7.5 V
- width, external: 55 mm.

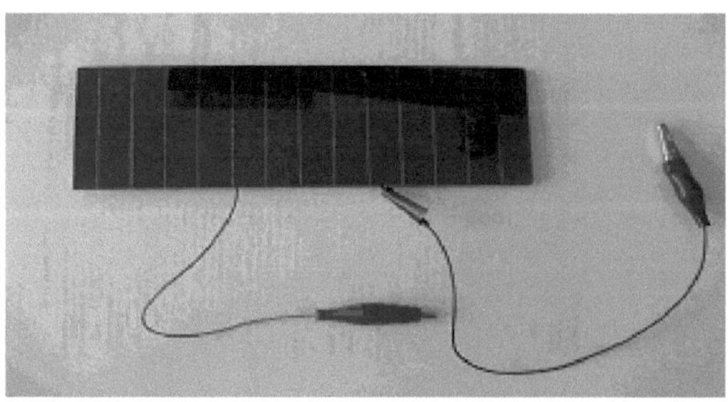

Figure 6.20 Solar panel SA-0640

6.6.4.6 Heating Transducer

Whenever a temperature difference exists in any environment, it is possible to convert heat energy directly into electrical energy. As shown in Figure 6.21, the thermoelectric converter Thermo Life is a very small and compact energy source. When the temperature difference is 5°C, at matched load it can generate 3 V of voltage, 10 μA of current and 30 μW of power; 10°C temperature difference generates 5.5 V of voltage, 25 μA of current and 135 μW of power at matched load. The maximum temperature difference is up to 100°C, according to the datasheet of the Thermo Life transducer.

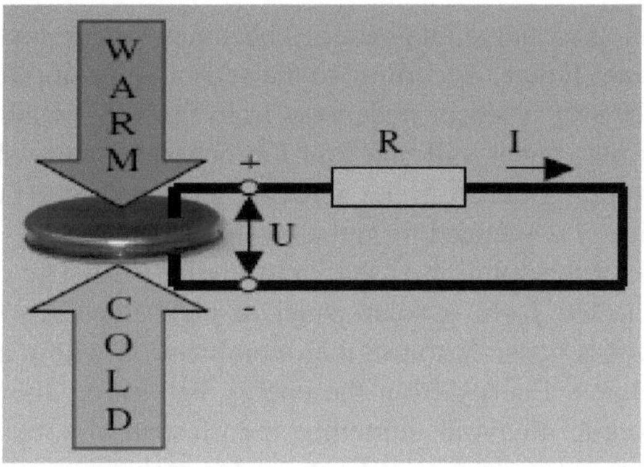

Electrical Parameter	@ ΔT=5 K	
Open circuit voltage	V	5.2
Short-circuit current	μA	22
Resistance	kΩ	236
Voltage @ matched load	V	2.6
Current @ matched load	μA	11
Power @ matched load	μW	28

Figure 6.21 Thermo Life application and its parameters

6.6.5 Energy Converting and Storing Subsystems

According to the energy consumption analysis of an Imote2 node, a power supply subsystem should provide voltage supply of 3.3 V or more and more than 44 mA current supply for the entire sensor node. Because there exist some unexpected factors, a certain redundancy quantity should be considered for these parameters.

From the results of experiments in the simulating process, a single energy harvesting device can only produce less than 5 V voltage and less than 5 mA current under indoor conditions, so it is impossible for a single energy harvesting source to provide enough power to directly drive a sensor node to work. Also it is not a good design to increase greatly the number of energy harvesting devices to capture enough energy to support an entire system because the volume of a node inflates sharply. However, it is possible to use captured energy to recharge the batteries.

A possible solution, by combining rechargeable batteries and a scavenging energy mechanism, is proposed in terms of the reasons listed as below. The power consumption of a wireless sensor node will become smaller as a development of SoC (system on chip) technology and research reducing power consumption of IC chips. The lifespan of rechargeable batteries becomes longer, and their capacity to store energy becomes larger. According to the self-organisation feature of wireless sensor networks (WSN), a sensor node exits from the WSN where it once worked if its energy runs out, but it still can join the network again when its energy is restored. While dropping out from the network, the sensor node can hoard energy little by little, which is scavenged from the outside environment.

The framework of the solution is shown in Figure 6.22. The energy harvesting source can be selected from a solar panel, a piezoelectricity component or a thermo-generator. It is possible to use their combination within allowing range of a sensor node volume. Energy from the energy harvesting source is stored in a supercapacitor. In order to avoid shortening the lifespan of a rechargeable battery, its manufacturer provides strict specifications for recharging voltage and current supplies. For example, the specification of Panasonic batteries indicates 3.3 V rechargeable voltage and less than 4 mA rechargeable current for VL serials. To obtain a recharging voltage with a fixed value, it is unavoidable to use a DC-DC converter in the circuit, which normally has a larger range of input voltage and appointed or adjustable output voltage. A ZENER diode is used to protect the supercapacitor and DC-DC device. A rechargeable battery with small volume and larger capacity is recommended since it is easy to form a cordwood structure for different applications. A microprocessor can monitor the voltage values of two batteries so that the corresponding switch signal is sent according to their states.

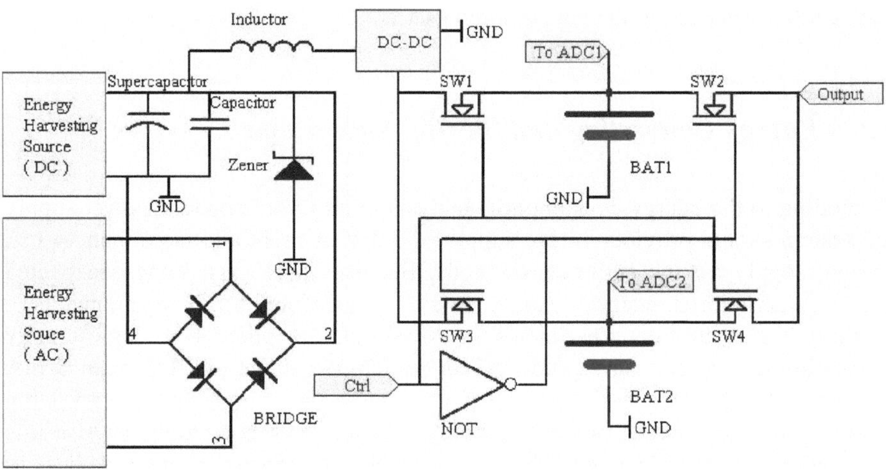

Figure 6.22 A solution of energy harvesting system architecture

There are two issues that may affect the design of an energy harvesting system. One is how to capture maximum energy from environment and the other is how to extend the lifespan of rechargeable battery for as long as possible.

It is feasible to design a combiner of energy harvesting using several sources from environment for example to maximise the energy captured depending on the application. For example, a combiner consisting of a solar panel and a piezoelectric generator is used to capture energy from fluorescent light and machine vibration in a workshop with larger motors. A combiner of multi-buzzers with different frequency resonance can capture more energy from vibration.

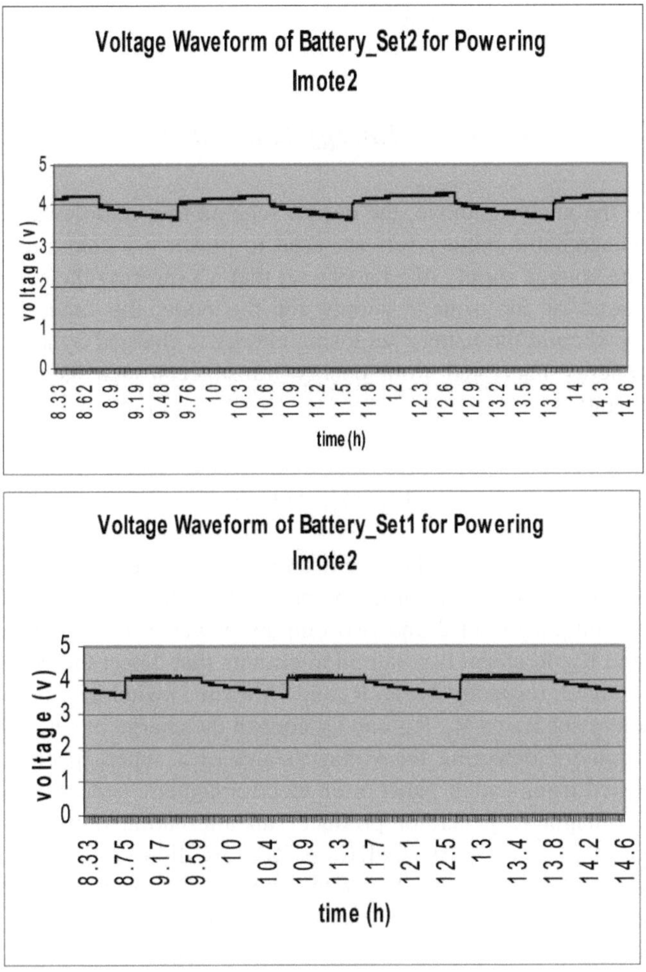

Figure 6.23 Battery charging cycles for powering Imote2

In the scheme the lifespan of a wireless sensor node is as long as that of recharging batteries. So it should be firstly considered to select rechargeable batteries with a long lifespan. Rechargeable batteries are normally divided into four types: nickel cadmium (NiCd), nickel metal hydride (NiMH), lithium based (Li+) and sealed lead acid (SLA). Li+ batteries have a longer cycle lifetime and involve a lower rate of self-discharge. Secondly, effective methods for extending the lifespan of rechargeable batteries are to design the recharging circuit and switch scheduler reasonably and to reduce battery aging caused by frequent charge–discharge cycles.

Middleware has been used to alternately charge the batteries. Hence, while three batteries are used the other three are charging. Figure 6.23 shows the alternating charge applied to batteries.

6.6.6 Implementation of an Energy Harvester

According to the analysis above, the design scheme of a middleware is proposed here. Two rechargeable battery sets are used to power a wireless sensor node in turn. When the voltage supply of a battery set that is powering the node declines to the lower bound of the voltage supply for the node, the other battery set is switched to work, and the battery set losing energy is charged by a supercapacitor storing the energy harvested from energy transducers such as solar panels or vibration generators.

6.6.6.1 Hardware Structure and Implementation

The schematic diagram of middleware is shown in Figure 6.24. Here, R_7 R_8 and Q_2 form a switch circuit to control the output of the battery set 1, and R_{13}, R_{14} and Q_4 control the output of set 2 and two outputs power a wireless sensor node in turn. R_5 R_6 and R_{11} R_{12} form two sampling circuits that detect the voltage states of battery sets 1 and 2, respectively. R_3 R_4 and Q_1 form a switch circuit to control the charge of battery set 1, and R_9, R_{10} and Q_3 control the charge of set 2. R_1 R_2 form a sampling circuit for detecting the voltage states of a supercapacitor storing the energy harvested from a solar panel or an electromagnetic generator, respectively. As the electromagnetic generator produces an alternating output, a full bridge rectifier is used to convert AC to DC. A ZENER diode is used to protect the supercapacitor and rechargeable batteries. The whole system is controlled by a PIC microprocessor and there are four LEDs to indicate the states of the four switches. The circuit of R_{16} R_{19}, S_1 and C_3 starts the system and controls the PIC processor to turn LEDs on or off in order to save power consumption on LEDs. In other words, a double-function button S_1 is used to start the system after the system is powered and it is used to control LEDs on or off when the system is working. The schematic circuit diagram of the middleware is shown in Figure 6.24. The prototype of the middleware is shown in Figure 6.25.

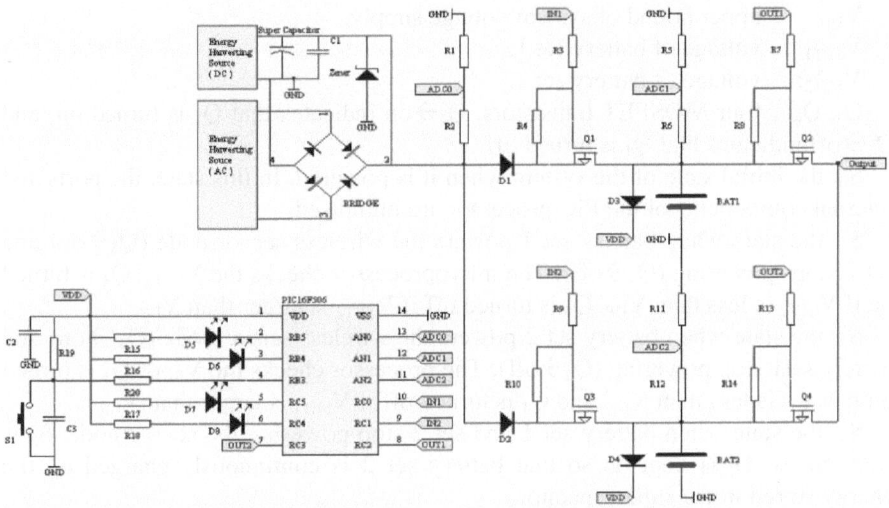

Figure 6.24 Hardware structure of middleware

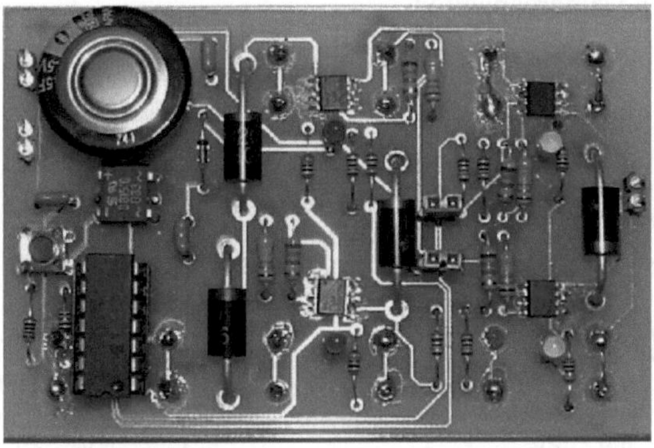

Figure 6.25 Prototype of the middleware

6.6.6.2 The Work Process of the System

In order to clearly express the work process of the system, the dynamic behaviour of the system is divided into five states, and the conversion between states is described by a finite state machine in Figure 6.26. The parameters and states are defined as follows:

V_{LOW}: lower bound of system voltage supply, which include V_{LOWL} and V_{LOWH}. V_{LOWL}: a threshold value, and the wireless sensor node stops working if its voltage supply is less than it. V_{LOWH} also is a special threshold value between V_{LOWL} and V_{UP}, the value is set in order to avoid frequently switching between states.

V_{UP}: upper bound of system voltage supply.
V_{SET1}: voltage of battery set 1.
V_{SET2}: voltage of battery set 2.
Q_1–Q_4: four MOSFET transistors. Q_1→on indicates that Q_1 is turned on, and Q_1→off indicates that Q_1 is turned off.

S_0: the initial state of the system when it is powered. In this state, the ports and internal components of the PIC processor are initialised.

S_1: the state when battery set 1 powers the wireless sensor node (Q_2→on) and set 2 stops powering (Q_4→off). The microprocessor checks the V_{SET2}. Q_3 is turned on if V_{SET2} is less than V_{UP}, Q_3 is turned off if V_{SET2} is larger than V_{UP}.

S_2: the state when battery set 2 powers the wireless sensor node (Q_4 →on) and battery set1 stop powering (Q_2→off). The processor checks the V_{SET1}. Q_1 is turned on if V_{SET1} is less than V_{UP}, and Q_1 is turned off if V_{SET1} is larger than V_{UP}.

S_3: the state when battery set 1 and set 2 stop powering the sensor node. S_3 is from S_1, so Q_3 is kept on so that battery set 2 is continuously charged by the energy stored in the supercapacitor.

S_4: the state when battery set 1 and set 2 stop powering the sensor node. S_4 is from S_2, so Q_1 is kept on, so that battery set 1 is continuously charged by the energy stored in the supercapacitor.

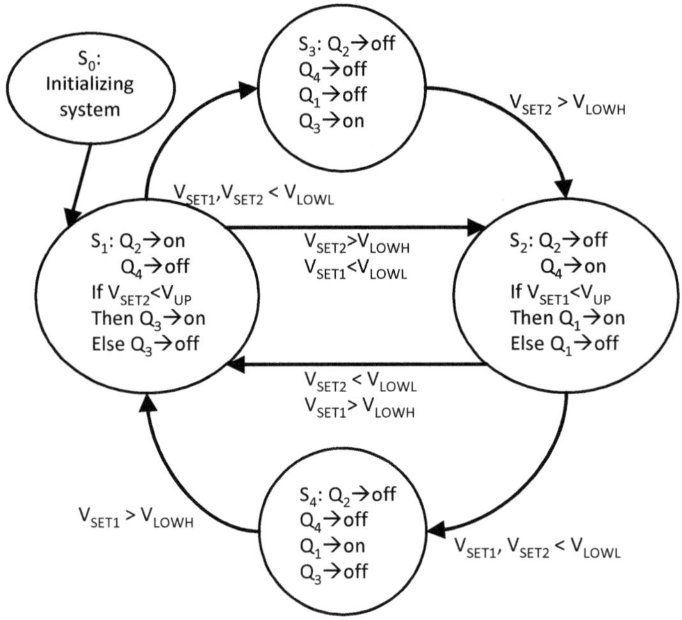

Figure 6.26 State transfers

6.7 Conclusions

The thorough identification of the state-of-art in MEMS sensor technology has resulted with a range of suggestions for the hardware design: modularity of design, multiuse of the processor (for data processing and housekeeping) and the possibility of power scavenging.

One of the main conclusions is that the design has to match the requirements to the need of the condition monitoring sector. More information about specific requirements suggested by the industrial partners should resolve this issue.

The multi-measuring sensors platform could host various types of sensors and constitute a network of sensors within a plant. The general requirements for the functioning of the device have been defined and described in as much detail as possible at this stage.

The functional specification range for a possible new hardware has been presented and potential solutions have been analysed.

References

ADXL311JE datasheet from Analog Devices, Norwood, MA

Albarbar A, Mekid S, Starr A, Pietruszkiewicz R (2008) Suitability of MEMS accelerometers for condition monitoring: an experimental study. Sensors 8:784–799, ISSN 1424-8220

Atashbar MZ, Bazuin BJ, Krishnamurthy S (2003) Design and simulation of SAW sensors for wireless sensing. Proc IEEE Sensors 2003, IEEE Cat No 03CH37498, part 1, 1:584–9

Crossbow (2007) Imote2 hardware reference manual. Berkeley, CA, Sept 2007

Dixon R, Bouchaud J (2006) "MEMS Inertial. Sensors go consumer," think small! vol 1, Issue 2

http://www.aldinc.com/pdf/EH300Brochure.pdf

http://www.perpetuum.co.uk/resource/PMG17%20%20Technical%20Specification%20Rev%20 2%200.pdf

http://www.perpetuum.co.uk/resource/PMG17-120_dsheet.pdf /, accessed on 02/2009

http://www.poweredbythermolife.com/thermolife.htm

Madni AM, Wan LA (1998) Microelectromechanical systems (MEMS): an overview of current state-of-the art. In: Proceedings of the IEEE Aerospace Conference, vol 1, IEEE, Snowmass, 1998, 421-427.

Malan R, Teschler L (2004) Wireless sensors. Machine Design, May 6, 76:(9), ISSN 0024-9114

Raghunathan V, Kansal A, Hsu J, Friedman JK, and Srivastava MB, (2005) Design considerations for solar energy harvesting wireless embedded systems. IEEE Int Conf Information Processing in Sensor Networks (IPSN), April, 2005

Staroswiecki M (2005) Intelligent sensors: a functional view. IEEE Trans Industrial Informatics 1(4)

Web report 2005 http://www.sensornetworks.net.au/intsens.html

Web report 2005 http://www.sensorsmag.com/articles/0402/40/

Wilson J S (2004) Sensor technology handbook. Newnes Publishers, Oxford

Chapter 7
Lubricating Oil Sensors

Jari Halme, Eneko Gorritxategi and Jim Bellew

Abstract. The necessity of a proper sensor to analyse the properties of fluids (lubricating oil in particular) is a reality. The use of intelligent sensors for online lubricating oil analysis will permit in a short period of time the optimisation of its life, cost reduction and the detection of problems in lubricated machinery. The various sensor techniques described here are important in analysing and detecting different lubrication features. Four types of lube sensors are focused in the lubrication system: fibre optic laser absorption and scatter sensors for solid contaminants, optical particle detectors for wear particles, water sensors for water detection in oil and oxidation sensors for oil degradation detection.

7.1 Introduction

Industrial machinery such as compressors, gas turbines, windmills, power and propulsion engines, large shaft lubrication and hydraulic systems suffer from stops and failures because of undetected degradation and sudden contamination of the lubricating oil. This is one of the main root causes of several dramatic failures. Information pertaining to the condition of the fluid is still a rudimentary, unsafe and inaccurate process in many industries. Traditional off-line lubricating oil analysis carried out in the laboratory is not able to detect and solve the early stages of degradation, and occasionally the reaction time is too long.

The necessity of a proper sensor to analyse the properties of the lubricating oil is nowadays a reality. For example, the Tribology Action Campaign, carried out in 1992 by the Institution of Mechanical Engineers and DTI (Department of Trade and Industry – UK) suggests that British industry could earn about 1.5 billion pounds *per* year, by means of the application of tribological notions in all production, maintenance and design processes.

Taking all this into account, the use of intelligent sensors for online lubricating oil analysis will permit in a short period of time the optimisation of its life, cost reduction and detecting problems in lubricated machinery. Critical machinery can take advantage of increased reliability, the operational staff could avoid the dangerous tasks of sampling and the industry that supplies lubricants could reduce inadequate and excessive maintenance tasks. Many times, lubricating oil is changed when the oil is still in good condition or in the worst case when the degradated oil cannot ensure proper functionality of the component. Finally, it is very important to highlight that lubricant monitoring helps in increasing system or machinery life and to perform proactive maintenance. Environmental problems associated with lubricant replacement are also reduced, minimising the waste of inappropriate oil changes.

One of the challenges of the industry is to carry out real-time oil analysis through a new sensor generation. Sensors can be the solution of many unsolved problems such as:

- Accessing lubricated points in machines working in extreme conditions or in inaccessible engines.
- Detecting early stages of degradation and wear and abnormal performance in lubricants, hydraulic oils and greases, which are the main root cause of many severe machine failures due to improper maintenance tasks.
- Real knowledge of what is happening in the mechanical components in a system, for example, one can precisely predict where the problem is if one type of wear particles is detected.
- Establishment of a good predictive and proactive maintenance with the corresponding saving in costs for the European Industry.

7.2 State-of-the-art

7.2.1 Oxidation

Agoston *et al.* (2008) used online sensor assembly consisting of an IR transmitter, IR detector and a narrowband IR filters by evaluating the oil oxidation number from the absorption of the selected IR band and a reference band. Kasberger and Jakoby (2007) showed a new design for integrated, online lubrication oil sensor system on a mono-mode IR waveguide, which can be fabricated in thin film technology. The sensor can be used for oil deterioration (*i.e.*, oxidation) measurements based on IR absorption in oil. The ongoing development of an mono-mode IR waveguide in the mid infrared region fabricated by thermal components was further reported by Kasberger *et al.* (2008).

Kumar and *et al.* (2005) reported about an oil colour based online engine oil monitoring device, which is based on a blue LED below the glass slide that spreads a uniform oil film on the glass surface. An LRD (light receiving device) was installed on the other side of the glass to detect penetrating light for monitoring oil property changes. A respective LED-LDR based online system, which uses blue, red and green light transmittance for a low cost oil condition monitoring analyses was reported by Parikka *et al.* (2004). Also an optical, digital camera based online system for monitoring colour by calculating different transmitted wave lengths from pictures has been reported by Halme (2002a).

Smiechowski and Lvovich (2003) have created online, iridium oxide potentiometric sensors for acidity determination of industrial lubricant. One version of the sensor was constructed on a macro scale and the other on a MEMS. The tests with diesel oil showed a good correlation between TAN/TBN (total acid number/total base number) and sensor output, where the macro scale sensor had a better sensitivity. However, despite the good correlation, the sensors were not considered as ready for industrial usage due to their problems with stability

7.2.2 Viscosity

Agoston *et al.* (2005a) used a thickness shear mode microacoustic resonator for an engine oil viscosity sensor. According to their experiments the sensor was well suited for oxidation induced viscosity changes caused by thermal deterioration. However, the sensor was not able to detect changes induced from interactions such as viscosity modifier additives outside the limited penetration depth of the acoustic wave. It is also reported by Agoston *et al.* (2006) that a microacoustic sensor indicates online lubrication base oil viscosity caused by oil deterioration products. The reported sensor was also sensitive to sedimentation of deterioration products. As viscosity is strongly dependent on temperature, temperature measurement is needed for scaling the measurements.

Turner and Austin (2003) tested different electrical motor oil monitoring techniques for viscosity measurements. According to their test the magnetic permeability measurement did not correlate with viscosity changes. However, dielectric properties of measured oil correlated with viscosity changes and was seen as a a promising technique for motor oil monitoring. Dochowski (2006) reported about a Hydac sensor that is able to detect viscosity changes based on changes in dielectric value.

7.2.3 Corrosion

Agoston *et al.* (2005b) studied corrosiveness of lubrication oil by using multiple resistive copper thin films of different thickness. The sensor indicates both the acid and corrosive sulphur compound contents in oil. The surface treatment of thin films is crucial and was seen a potential source for measurement errors. The thin film corrosive sensor was reported to be sensitive for corrosion by Agoston *et al.* (2006). Peratec (1994) also reported that an electrical resistive sensor can be used to determine the level of corrosivity of the fluid.

7.2.4 Water

Liu *et al.* (2000) used upper and lower grid capacitance sensors through marine diesel engine oils to detect from the relative variation of the dielectric constant, such as water; it is also sensitive to particles. Agoston *et al.* (2006) used an open, two-electrode permittivity sensor structure for indicating water contamination changes in lubrication oil of biogas engines. Dochowski (2006) also reported a capacitive, dielectric value-based sensor for detecting changes in the relative water content in lubrication oil.

7.2.5 Particles

Hunt (1993) and Roylance and Hunt (1999) have listed wear particle detection methods suitable for online analysis. Among those methods are magnetic flux, magnetic inductance, magnetic plugs, inductive coil, atomic absorption and emission spectroscopy, mass spectroscopy, X-ray fluorescence spectroscopy, IR, ultra violet, visible light, narrow band laser light and colour imaging. Different methods have different properties and limitations. For example, light obscuration-based methods calculate water bubbles as particles and only some methods can deviate air bubbles from particles (Toms and Toms 2006). In general, magnetic and light obscuration-based methods can be considered among the most inexpensive methods.

Peratec (1994) reported about an online particle monitoring system that measures the total abrasiveness index of a fluid by monitoring particles. Liu *et al.* (2000) connected in series an online ferrography detector and online grid capacitance sensors. They reported that they could detect particles in engine oil and classify non-ferrous particles from ferrous particles with the connected detectors. Electrostatic sensing has been reported to be an effective online method for the detection of wear rate due to metal particles in oil lubricated rolling bearings (Harvey *et al.* 2007, Morris *et al.* 2003). An optical, digital camera-based online lubrication particle detection and classifying system was reported by Halme (2002b).

7.2.6 Others

Peratec (1994) has reported about an electrochemical probe that correlated with the microbiological activity of cutting fluid. In the same report, a microwave sensor was used to measure the thickness of "cream" film that floats on the surface of the cutting fluid.

7.3 New Sensor Developments

7.3.1 Detection of Solid Contaminants

7.3.1.1 Fibre Optic Solid Contaminant Sensors

The VTT Technical Research Centre of Finland has developed two fibre optical sensor prototypes. One prototype is for laser scattering and the other for laser absorption. The scatter sensor end consists of a probe like rod the diameter of which is about 5 mm. One probe sensing side is free while the other end is cabled with optical fibres; one optical fibre end for low power laser light output and the other one for collecting inputting scatter response. The sensor heads for the absorption measurements are identical to the scatter probes, except for outputting laser beams that are transmitted from the one side of the cavity and through the lubrication to the other side of the cavity to a collecting fibre. Thus there are two probes for the absorption measurement and both probes contain only one fibre. Figure 7.1 shows the measurement cavity with installed scatter and absorption probes. The cavity is designed for low pressure applications (preferable below 2–3 bar). The defined temperature specification is below 60–70°C.

Figure 7.1 Fibre optic contaminant sensor measurement cavity on a Petri dish with installed probes for scatter (probe above) and absorption (probes on the right and left) measurements

While the oil flows through the cavity, the light responses due to the solid contaminations as well as other light affecting properties in the oil are collected with the sensor probes installed, and transmitted to the measurement unit. In the upper segments (1.1 and 1.2) of Figure 7.2 are shown a schematic operational principle of light absorption measurement and in the lower segments (2.1 and 2.2) light scatter measurement, respectively. While relatively clean oil flows through the absorption measurement cavity, small discontinuations (non-uniformities, eddies, *etc.*) in the oil and flow introduce small deviation into passing light and on the other side of the cavity the response is slightly reduced and deviated (segment 1.1). In the contaminated case, instead, the light is absorbed due to the solid particles. Also the deviation in the response increases while the particles are flowing through the absorption measurement cavity (segment 1.2). In the case of light scatter measurements, only small portion of introduced light is scattered back to the measurement probe (segment 2.1). However, while the solid contamination increase, both the deviation and the mean light power increases in the scattered response (segment 2.2). In Figure 7.3 is a schematic general view of the light scatter and absorption measurement system shown and Figure 7.4 displays the real installed measurement cavity at the field shown. The length of the cables is 30 m; thus while the probes and the measurement cavity are out in the field and in direct contact with the oil, the measurement unit with sensitive photo optic cell can be located in an office like environment that is more suitable for the electronics. Actually, the length of the optical cable could be several kilometres without practically any noticeable attenuation.

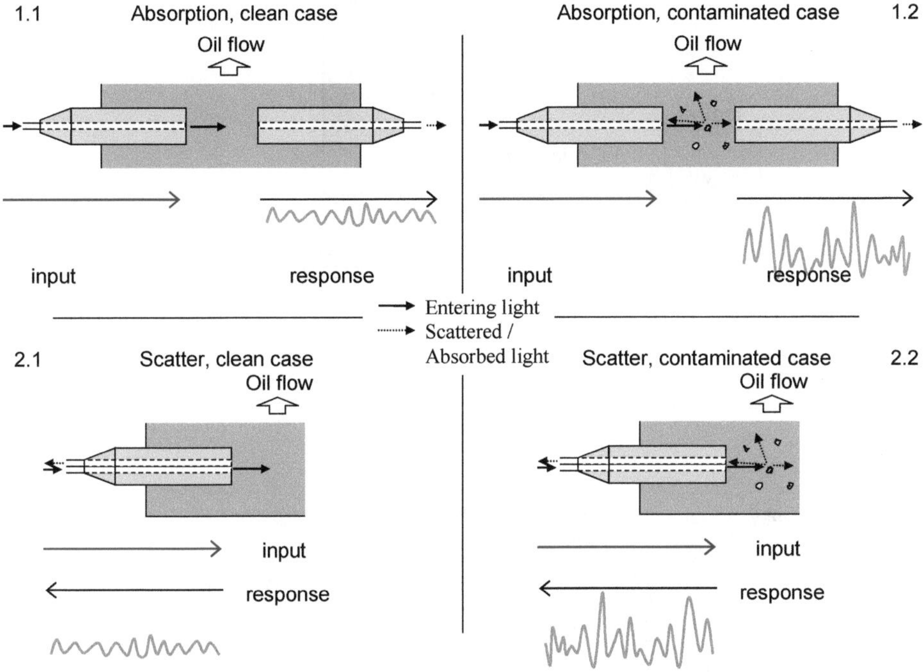

Figure 7.2 Light scatter and absorption measurement in lubrication. The upper segments (1.1 and 1.2) are schematic picture of the light absorption sensor and absorption of light due to the discontinuities and absorptive agents in the lubrication. The lower segments (2.1 and 2.2) are a schematic picture of the scatter sensor and scattering of light due to the discontinuities in the lubrication sensor

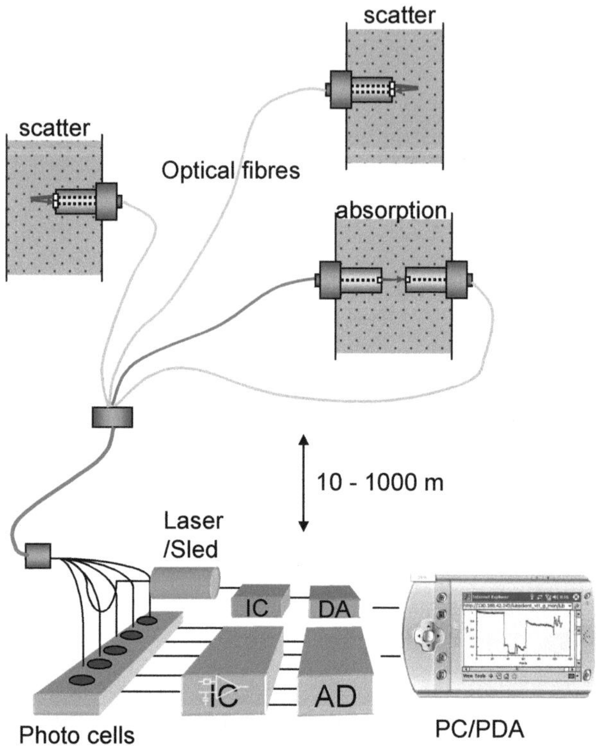

Figure 7.3 A general view of a low cost fibre optic local area detection principle for contaminants

Figure 7.4 Fibre optic contaminant sensor measurement cavity installed in the foundry at the Volvo site in the Skövde factory in Sweden. Both the scatter and absorption are measured while the hydraulic oil flows through the cavity. (Photo taken by Ronny Karlsson)

The measurement unit (Figure 7.5) provides the laser light for the outputting optical fibres and probes as well as read the responses from the inputting probes and fibres. For each channel (scatter and absorption) there is a laser. The lasers are commercial low cost fibre coupled laser diodes. The type of laser is LPM-660-SMA, which includes a laser diode of 660 nm (red) and a coupled fibre for high temperature and harsh environments. The mode of the laser beam is multimode, which delivers higher power from the diode with the cost of coherence. The maximum power of this laser type is 25.5 mW. The photo detectors installed in to the measurement unit are commercial low cost detectors with installed signal amplification with a low noise transimpedance and specified power coupling. The detectors have BNC output and removable treaded couplers allowing convenient mounting of external fibres, instead of even lower cost components but varying laborious (expensive) installation. The type of the photo detector is PDA36A-EC.

Figure 7.5 Measurement unit (top) during testing. The measurement cavity is in the front of the measurement unit

The measurement unit needs a power supply of 220 V. While the power cable is connected, the system is on. In addition to two analogue fibre optic outputs and two analogue fibre optic inputs, the unit consists of two analogue outputs. One is for scatter and the other for absorption. The output ranges for both channels are 0–10 V and the connectors are of BNC type. The size of the box is 40 cm × 30 cm × 10 cm. Two coaxial cables are connected to the analogue outputs. One is for scatter and another for absorption (surprise, surprise). The other ends of the coaxial cables are screwed to an AD-card. The type of the AD-card is USB-1608fs from Measurement Computing Corporation as reported in D4.2.2. The AD-card is operated by measurement software installed to a PC.

Test Results

The scatter and absorption sensors and measurement unit were tested with the hydraulic fluid provided by Volvo. Altogether five samples were used for the testing. First the samples were analysed with a commercial, laboratory particle counter. The results of these analyses were used for the reference for the analysis of the responses of the scatter and absorption sensors. Figure 7.6 shows the reference particle counting results. Figure 7.7 shows the measured responses of developed lube sensors. It can be concluded from the results that the particle contamination has an effect on the developed lube sensors. It seems that the measured absorption (light through the oil) has the best correlation with smaller particle classes and that the measured scatter correlates with bigger particles. However, the samples for the scatter and absorption measurements were taken to the measurement cavity in the small and stable glass tubes (Figure 7.8). Thus the situation is not the same compared to the real flow of fluid at the field. Also the sedimentation of the particles has an effect on the measurements.

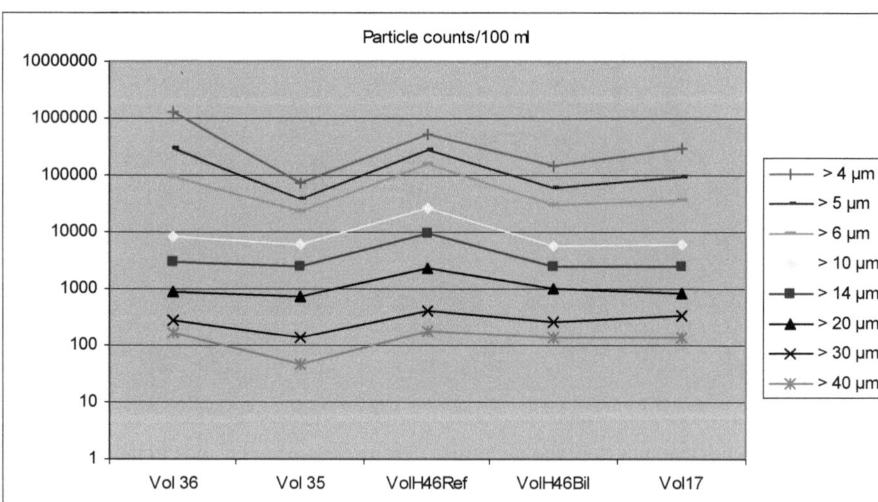

Figure 7.6 Calibrated particle counting responses with five Volvo oil samples

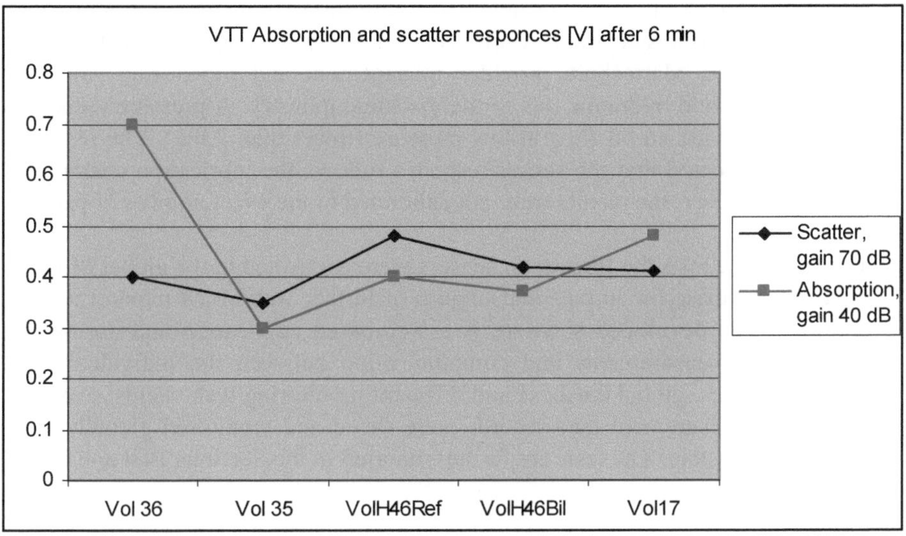

Figure 7.7 Scatter and absorption sensor responses with five Volvo oil samples

Figure 7.8 An oil sample in a small glass tube (top) used for the testing of the developed lube sensors. The measurement cavity is shown at the bottom

The fiber optic sensor was installed for online measurements at Volvo's and Goratu's hydraulic systems and Martechnic's test rig. Goratu is a machine tool manufacturer and Martechnic provides management and monitoring solutions fuel, lubrication and hydraulic oils to the maritime industry. A pressure reduction was built to provide an oil flow at low pressure (lower than 2 bar). The result of online tests indicated that the sensor response follows the lubrication contamination trend. However, the results were not calibrated to the exact number of particle size classes.

For the online tests the fiber optic sensors were connected to the global Mimosa SQL server database for storage and support of further web-based monitoring and diagnostics. The developed software is a web-based application and during the testing of the measurements and communication between the individual lube measurement sites, global database and different monitoring web clients, over one million lube measurement data records were saved and monitored globally with the developed system. The tests are further reported in the Sections 14.4 and 14.5.

7.3.1.2 Particle Sensors

Over the years the lubrication machine's wear has been analysed, so a lot information has been obtained, mainly regarding the type of wear, the degree of wear and one indication about the support that the machines need. Tekniker developed an optical particle detector (OPD) based on image analysis, which is able to detect the size and shape of the particles in the lubricating oil (Figure 7.9). The optical detection sensor uses light blockage particle detecting technology.

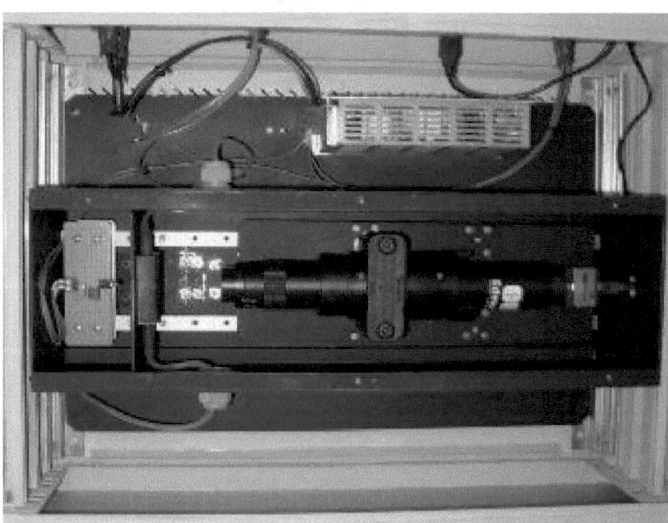

Figure 7.9 Global design of optical particle detector

The OPD sensor is an automated oil debris analysis instrument. It is a particle counter and shape classifier that identifies sizes and trends of wear debris in all types of lubricants and hydraulic fluids. The operation of the sensor is based on technology that combines laser imaging (image processing) and artificial intelligence to characterise wear debris. A representative oil sample is drawn from a lubricating system and taken to the instrument for analysis. The sensor draws the sample fluid through a fluidic cell whose back is illuminated by a lamp. The light transmitted through the fluid is then imaged onto (captured by) a video camera with macro focusing optics (zoom lens). The collected images are analysed by software to determinate the type of particles present and to sort the particles by their size following the corresponding ISO standard (Figure 7.10).

In order to classify the particles by their size as set up by the ISO standard, a calculation of the amount of oil that is measured in the cell is done and the number of particles is extrapolated to the amount of oil that is determined by the ISO.

As the oil sample is back-illuminated the system does not see a direct image of the particle but its shadow instead, so the contrast between the particle and the background (clean oil) is quite high, which improves the classification method by shape. Otherwise, because of this kind of illumination the colour of the particles is lost, and it is actually known that their colour can provide some information about their origin, *e.g.*, which part of the engine is failing and thus producing this kind of particles; this is something to into account for a possible improvement of the system.

On the other hand, to have a more accurate extrapolation of the amount of analysed oil to the ISO standard the objective lens chosen for the video camera has a lateral amplification, as the correlation between the image pixel size and the real sample size is 1 to 1 (1 pixel on the image is 1 μm on the cell). So that particles as small as 1 μm are detected, getting the biggest image of the sample possible. This set up has the disadvantage that the illumination of the sample has to be controlled to be as uniform as possible in the whole part of the sample to be analysed; it is not easy to illuminate the system by just one LED. To avoid a gradient illumination of the sample, that is, a gradient illuminated image, a correction of the illumination is done by software.

The image's treatment was been carried out by Matrox Imaging Library (MIL). This software is a high-level programming library with an extensive set of optimised functions for image capture, image processing (*e.g.* point-to-point, statistics, filtering, morphology, geometric transformations, FFT and segmentation), pattern recognition, registration, blob analysis, edge extraction and analysis, measurement, metrology, character recognition, 1D and 2D code reading, calibration, graphics, image compression, display and archiving.

Designed to facilitate development and increase productivity, MIL offers a common C API (application programming interface) that supports Matrox Imaging's entire hardware line and an intuitive and easy-to-use function set. MIL also includes ready-made interactive dialogues for handling file I/O, adjusting function parameters, manipulating image data (*e.g.*, for pattern recognition model and character recognition font definition) and managing results, all geared towards simplifying application development.

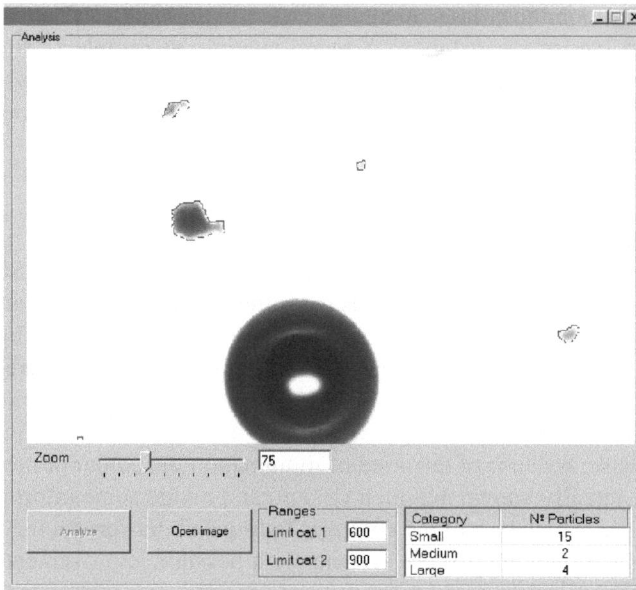

Analysis information

Original image:
C:\Analysed_Samples\Camara nueva\441770.jpg

Analysed image:
C:\Analysed_Samples\Camara nueva\441770_cont_6.jpg

Ranges
Small_threshold: 4
Large_threshold: 14

Categories
Small: 73242
Medium: 36621
Large: 36621

Figure 7.10 Results window and report from the OPD sensor

7.3.2 Water Detection

7.3.2.1 Water Sensor Development

Near infrared spectroscopy (NIRS) is a spectroscopic method using the near infrared region of the electromagnetic spectrum (from about 800 to 2500 nm). NIRS is based on molecular overtone and combination vibrations.

Water is perhaps the most common measurement made in the near infrared (NIR) range. This is due to its strong effect on product properties and chemical reactivity of the starting materials. From an analytical perspective, water is easy to analyse due to it is relatively strong signal compared to the hydrocarbon background.

Water is one of the most destructive contaminants in almost all lubricants and hydraulic oils. It attacks additives, induces base oil oxidation and interferes with oil film production. Low levels of water contamination are normal in engine oils. High levels of water ingression merit attention and are rarely correctable by an oil change. Some problems related to water contamination are:

- Long idling in wintertime causes water condensation in crankcases, which leads to loss of base number and corrosive attack on surfaces, oxidation of the oil, *etc.*
- Emulsified water can mop up dead additives, soot, oxidation products and sludge. When mobilised by flowing oil, these globular pools of sludge can knock out filters and restrict oil flow to bearings, pistons and the valve deck.
- Water sharply increases the corrosive potential of common acids found in motor oil.

There are three different ways in which form the water may be present in the oil:

- Saturated: the oil can absorb a certain amount of humidity in a dissolved way.
- Emulsion: emulsion is the spreading and presence of very small drops of water that are held highly stable in suspension in the oil.
- Free: larger "parts" of water mixed with oil without any stable connections in independent phases.

The option to measure water in oil by means of infrared allows us to reliably recognise all three types of water in oil statuses.

This is a great advantage compared to the technologies existing on the market. Any other sensors, *e.g.*, capacitive measuring methods, can generally only reliably detect water that is present in a saturated form in oil. Free water often leads to faulty measurements that no longer offer a reliable statement regarding the real status of the oil.

There are two water sensors being developed, both are based on infrared spectroscopy. The water sensor developed by Tekniker uses a NIR range between

1250–1300 nm whereas the water sensor developed by Martechnic GmbH of Hamburg operates in the mid-in (MIR) for red range of 3500 nm. The latter sensor is called AHHOI (automatic H_2O inline) and its technique evolved from a European Commission development programme and has recently been successfully launched into the commercial maritime industry. The MIR range is suitable for a wide range of different type of oils, including synthetics and marine engine oils, and is most ideal for measurement of water in oil due to the minimum effect of other parameters in the oil. A water sensor operating in the NIR range is shown in Figure 7.11 and the AHHOI sensor operating in the MIR range in Figure 7.12.

Figure 7.11 Sensor assembly of water sensor operating in the NIR range

Figure 7.12 AHHOI water sensor operating in the MIR range

Test Results of a Water Sensor Operating in the NIR Range

Validation of the sensor was carried out with real samples at the Tekniker Industrial Laboratory. Hydraulic lubricating oil samples were selected and a group of artificially contaminated samples was obtained. The correlation between these oil samples and the spectra in the 900–2500 nm region was studied. According to the data found in the bibliography the range of 1380–1480 nm was chosen. The following table and graphics show the results obtained for the wavelength range of 1380–1480 nm: Results for all samples are shown in Table 7.1 and Figure 7.13.

Table 7.1 Water sensor (NIR) test results for the wave-length range1380–1480 nm

Parameter	Treatment interval	Calibration		Validation	
		Slope	Correlation	Slope	Correlation
Water	1380–1480 cm^{-1}	0.946231	0.972744	0.864510	0.948788

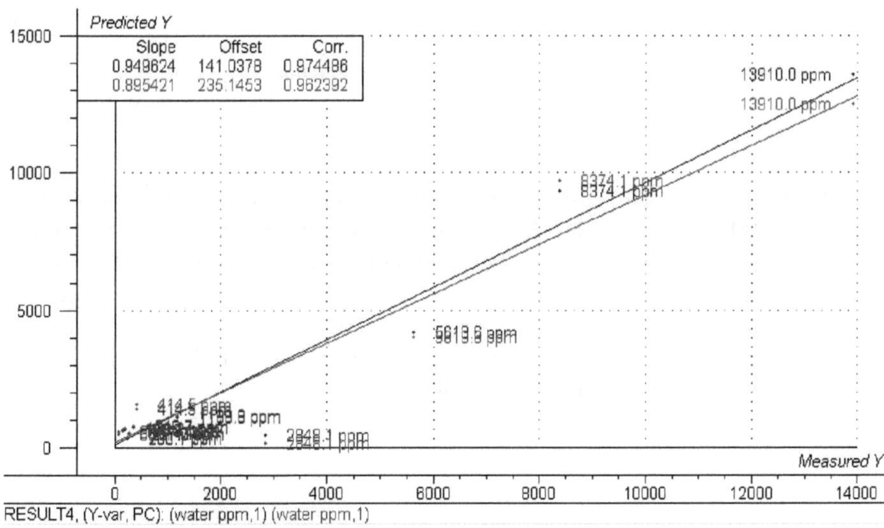

Figure 7.13 Water prediction results in the range of 1380–1480 nm

Validation of Water Sensor (NIR) in a Real Machine

Some validation activities have been carried out to check the water content prediction for the sensor. These tests have shown a good correlation, but the sensor signal was influenced by external factors like air bubbles, particles or temperature and pressure changes (Figure 7.14).

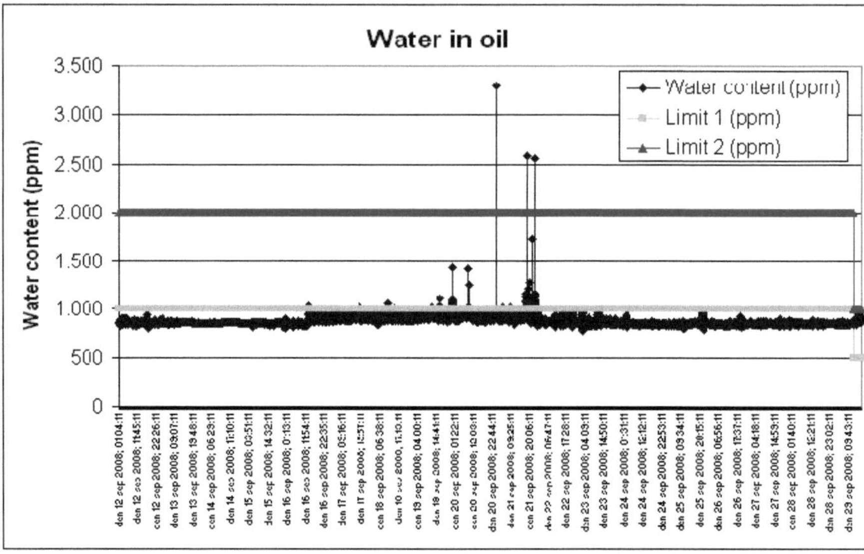

Figure 7.14 Water prediction results in the range of 1380–1480 nm

Properties of the AHHIO Water Sensor Operating in the MIR Range

The AHHOI water-in-oil sensor is able to detect reliable water contents of up to 1.0% vol (10000 ppm). The development of this sensor was strongly related to the demand and accuracy of a Karl Fischer laboratory device. It is designed for maximum oil operating pressures from 3–10 bar. The sensor measuring cell of the system is protected by a built-in oil filter and set to a constant pressure of about 1 bar by means of a pressure reducing valve. A comparison between the AHHOI sensor and a laboratory test is shown in Figure 7.15.

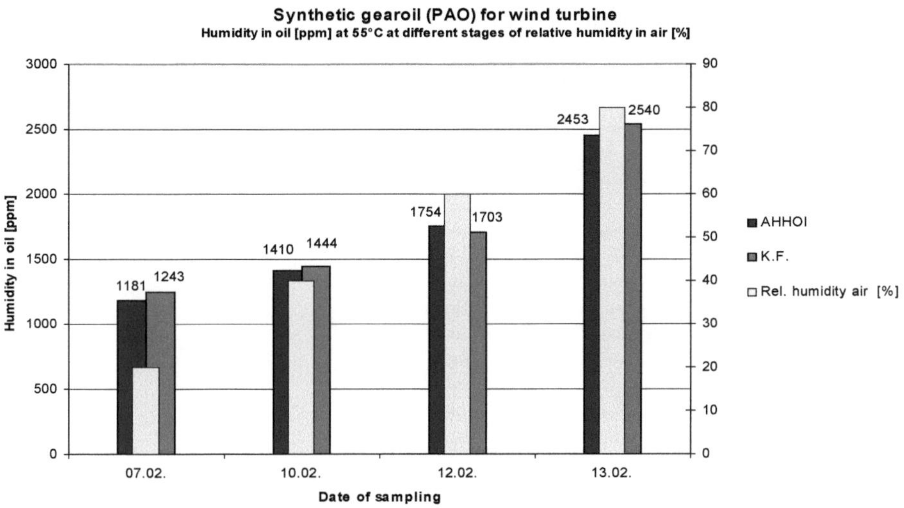

Figure 7.15 Water-in-oil: comparison AHHOI and laboratory test (Karl Fischer) – gear oil

Test Results of the AHHOI Sensor

The following test results were generated during a demonstration of a simulated maritime application of a stern tube/tail end shaft assembly where salt water ingress is a significant problem. A test rig replicated the application and circulated lubricating oil was progressively contaminated with known quantities of water. The subsequent deterioration in the lube oil quality was monitored with the results below (Figure 7.16):

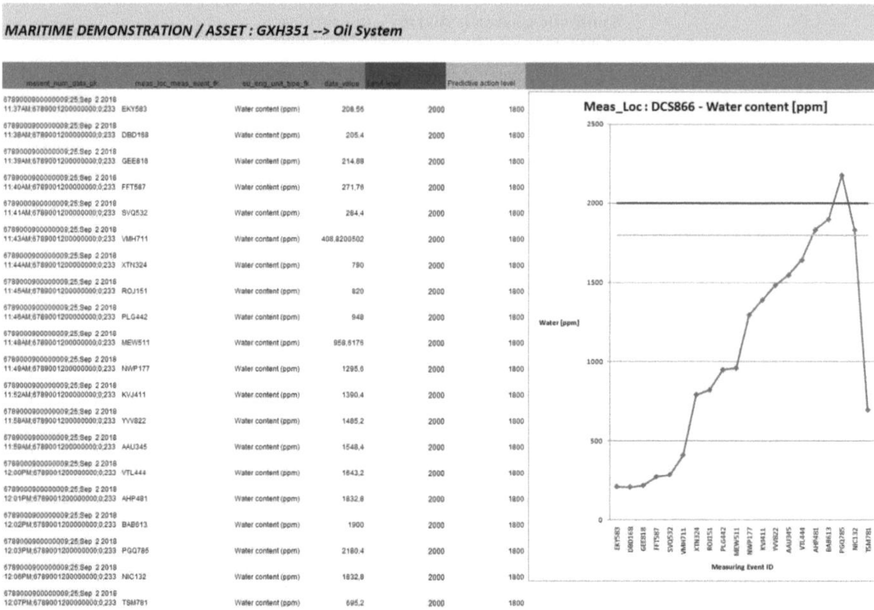

Figure 7.16 Demonstration of the AHHOI sensor

These results accurately reflect the actual conditions that prevail in a number of maritime installations where the AHHOI has delivered consistent and reliable performance.

7.3.3 Lubrication Deterioration by Ageing

Tekniker developed a sensor to monitor and to predict lubricating oil degradation, which uses visible spectroscopy (Figure 7.17). The sensor is very useful for early stages of degradation and the measurements are direct and fast, and the lubricating oil status can be continuously monitored.

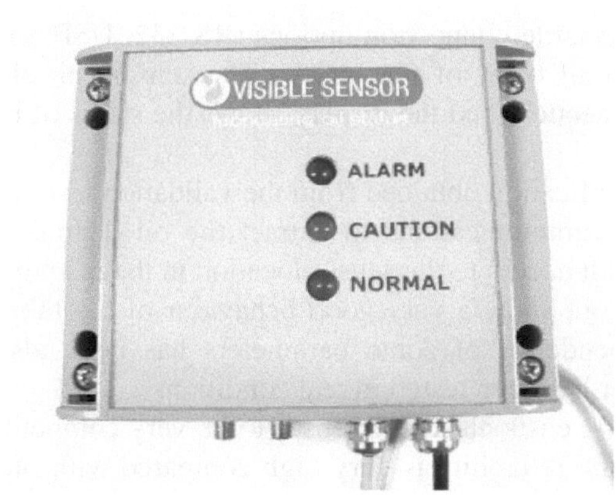

Figure 7.17 Oxidation/degradation sensor

The visible sensor uses light absorption in the visible spectra to predict the degradation status of the lubricating oil. Micro technologies allow a very robust and smart design with different characteristics depending on the final application that the sensor can be installed.

Air bubbles are one of the most important issues that affect signal precision and stability. The sensor is also sensitive to particles and dust.

Validation tests have been developed to check the precision of the sensor. The validation tests show a clear trend that identifies a change in lubricating oil degradation (Figure 7.18).

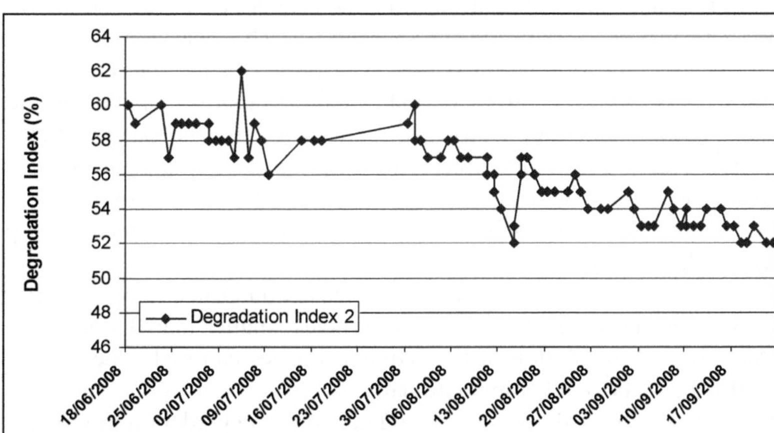

Figure 7.18 A three month trend of the visible sensor (% degradation index)

The output of the sensor is the remaining useful life of the monitored lubricating oil, which is calculated by means of an algorithm implemented into the electronics of the sensor.

The sensor has versatile connection options (RS 232, USB and 4–20 mA) and can be connected to all types of machinery. The sensor can also work autonomously, without connection, and the display shows the status of the lubricating oil all the time.

An important specification obtained from the validation test is the possibility to implement wireless communications to extract the oil degradation information, because of the difficult access to the sensor location in the machine.

The tests carried out show a very good behaviour of the lubricating oil status prediction. The dependency of some parameters has been also identified and solved and the sensor has been tested in real conditions.

Low manufacturing costs allow the sensor to be very competitive with sensors on the market, and the reliability is very high compared with other types of sensors. Windmills have been identified as the most promising market for the sensor, where lubricating oil monitoring is still very poor and the interest in sensors is quickly growing.

7.4 Conclusions

Historically the most challenging area for sensors in machine monitoring has been the quality of fluids such as lubricating and hydraulic oils. Four different types of oil sensors were developed: fibre optic laser absorption and scatter sensors for solid contaminants, CCD camera-based particle senor, oxidation sensor and two water sensors. Fibre optic laser absorption and scatter sensors for solid contaminants use optical fibres as data and energy channels. The measurement principle in the cavity is light absorption and scatter. The results of the measurements indicated that the first prototype gives a good cleanliness index for solid particle content in the measured lubrication. However, it does not give the exact particle count number. The measurement principle is immune to electromagnetic noise, has a low signal/power attenuation, thus long cables (<1000m) are possible. Possible further developments would be to use a more sensitive photo detector and power feedback for the light source.

The particle sensor uses a CCD camera as detection principle with an illumination system. The sensor is able to detect particles smaller than 1 micron in size. The extremely high resolution of the acquired images can make the image treatment too difficult, above all computationally.

The oxidation sensor_allows the early detection of degradation in the oil status using visible spectroscopy combined with a proper warning management and data transmission. The oxidation sensor measures the transmittance of light in the visible range (380–780 nm) of the light spectra, correlating the absorption of the light with the degradation status of the lubricating oil. Optical measuring principles have some limitations and problems, which have been detected and measured. Wear debris and air bubbles in the lubricating oil are external factors that can affect the quality of the sensor signal. Avoiding this type of external influence is critical with lubricating oil.

The main purpose of the online water sensor is to measure the water content of lubricating oil. The water sensors measure the transmittance of the light in the both the NIR range and MIR of the light spectra, correlating the absorption of the light with the water content of the lubricating oil. As with the oxidation sensor, the optical measuring principles have some limitations and problems, which have been detected and measured.

All the lube sensors developed were successfully connected to the Mimosa database for storage and support further web-based monitoring, diagnostics and prognostics. These web-based services will be a good solution for engineers to read the up-to-date data anywhere by means of their PC, PDA and mobile phone *via* accessing WiFi and the mobile internet.

With the developments described above critical parameters such as water, oxidation and contaminants can be identified in real time and assessed with a high degree of accuracy and reliability. The newly developed sensors finally provide the means for applying condition-based monitoring and e-maintenance techniques to a broad range of machines and applications.

References

Agoston A, Dörr N, Jakoby B (2006) On-line application of sensors monitoring lubrication oil properties in biogas engines. IEEE Sensors 2006, 1099–1102

Agoston A, Ötsch C, Jakoby B (2005a) Viscosity sensors for engine oil condition monitoring – application and interpretation of results. Sensors and actuators A, 121:327–332

Agoston A, Svasek E, Jakoby B (2005b) A novel sensor for monitoring corrosion effects of lubrication oil in an integrated manner. Proc IEEE Sensors, 2005, 1120–1123

Agoston A, Schneidhofer C, Dörr N, Jakoby B (2008) A concept of an infrared sensor system for oil condition monitoring. Elektrotechnik & Informationstechnik 125:71–75

Dochowski J (2006) Sensors eye fluids and lubricants. Machine design 78:88–96

Halme J (2002a) Condition monitoring of oil lubricated ball bearing using wear debris and vibration analysis. Proc Int Tribology Conf AUSTRIB'02, Vol II Frontiers in tribology, Perth, Australia, 2-5.12.2002, 549–553

Halme J (2002b) Wear particle recognition based on neural networks, fuzzy logic and pattern recognition methods. TUKEVA, Academy of Finland Research programme, Tampere, Finland, 3.1.2002, 4 p

Harvey TJ, Wood RJL, Powrie HEG (2007) Electrostatic wear monitoring of rolling element bearings. Wear 263:1429–501

Hunt T (1993) Handbook of wear debris analysis and particle detection in liquids. Elsevier Applied Science, 488 p, ISBN 1-85166-962-0

Kasberger J, Jakoby B (2007) Design of a novel fully integrated IR – absorption sensor system. IEEE Sensors 2007, 515–518

Kasberger J, Saeed A, Hilber W, Hingerl K, Jakoby B (2008) Towards an integrated IR-absorption microsensor for online monitoring of fluids. Elektrotechnik & Informationstechnik 125:65–70

Kumar S, Mukherjee P, Mishra N (2005) Online condition monitoring of engine oil. Industrial lubrication and tribology, 57:260–267

Liu Y, Liu Z, Xie Y, Yao Z (2000) Research on an online wear condition monitoring system for marine diesel engine. Tribology International 33:829–835

Morris S, Wood R, Harvey T, Powrie H (2003) Electrostatic charge monitoring of unlubricated sliding wear of a bearing steel. Wear 255: 430–443

Parikka R, Vidqvist V, Vaajoensuu, E (2004) Low-cost analysers based on light transmittance for oil condition monitoring. AIMETA Int Tribology Conf, 14–17.9.2004, Rome, Italy, AITC, 725–734

Peratec (1994) Automatic online fluid monitoring. Industrial Lubrication and Tribology 46:8–9

Roylance B, Hunt T (1999) Wear debris analysis. Coxmoor Publishing Company's Machine and Systems Condition Monitoring Series, 128 p, ISBN 1-90189-202-6

Smiechowski M, Lvovich V (2003) Iridium oxide sensors for acidity and basicity detection in industrial lubricants. Sensors and actuators B, 96: 261–267

Toms L, Toms A (2006) Lubricant properties and test methods. In: Totten G (ed) Handbook of lubrication and tribology. Vol 1, Application and maintenance. Taylor and Francis Group, New York, 30.1–30.33, ISBN 0-8493-2095-X

Turner J, Austin L (2003) Electrical techniques for monitoring the condition of lubrication oil. Measurement science and technology 14:1794–1800

Chapter 8
Smart Tags

Adam Adgar, Alan Yau and Dale Addison

Abstract. This chapter introduces the use of 'smart tags' or radio frequency identification (RFID) technology in the dynamic maintenance framework. Although the technology has already been well developed and also implemented in a range of commercial systems, the benefit of implementing the technology into the e-maintenance framework has been seen as key to finding new and innovative solutions. The RFID technology permits the collection of electronic data on many of the actors within the maintenance system, thus directly supporting the main e-maintenance concepts – extensive and up-to-date data on which to base maintenance decisions.

The structure of the chapter is given as follows: Firstly, the main features of the technology, including both passive and active tag types together with tag readers and frequency bands are outlined. Then an overview of the background to the use of RFID in commercial systems is given followed by a review of existing application areas of RFID specifically in maintenance systems. Next, the potential applications/scenarios in which RFID technology is envisioned to be useful in the maintenance domain are presented. The technical work behind the demonstrators developed as part of the project is then described. Finally the contents of this chapter are summarised.

8.1 Introduction

RFID has actually been utilised by companies for many years to improve the efficiency of supply chains. Profound benefits have been accrued from the adoption of barcode technologies, and RFID represents another enabling technology that can streamline operations even more.

The main reason that this technology is receiving such attention currently is that several large organisations have mandated the use of this technology by their suppliers. This has occurred on both sides of the Atlantic, for example, Tesco (UK), Metro (Germany), Marks & Spencer (UK) in Europe and the Department of Defense and Wal-Mart in the U.S.A. have all announced their intention to adopt this technology.

Radio frequency identification, commonly shortened to RFID or 'smart tag' technology, is a technology primarily used for automatic identification and data capture and allows for non-contact reading of information in order to track and monitor physical objects. There is huge interest in the application of RFID in business operations, especially in manufacturing and supply chain activities. The main drivers for use of the technology are primarily cost and efficiency. Recent technological advances have brought down the costs of both tags and readers, and there has been considerable effort made to establish industry standards.

A key benefit of RFID technologies is the automation of the process of identification of individual objects together with automatic capture of this information. This can contribute significantly to operational efficiency and control. Many applications for RFID technology have been found and many are still yet emerging. Physical tracking of goods and inventory management are common areas of interest.

Falling prices of tags and readers and the rapid pace of the standards development process is making RFID technology an increasingly viable option for many companies, however this is still confined largely to the operations involved in the supply chain.

8.2 Overview of the Technology

8.2.1 Technical Basics

RFID is an automatic identification technology whereby digital data encoded in an RFID tag is captured by a reader using radio waves. In simple terms, RFID is very similar to barcode technology but it uses radio waves to capture data from tags, rather than optically scanning the barcodes on a label. RFID does not require the tag or label to be seen to read its stored data. This is one of the key characteristics of an RFID system.

Information is sent to and read from RFID tags by a reader using radio waves. In passive systems, which are the most common, an RFID reader transmits an energy field that "wakes up" the tag and provides the power for the tag to operate. In active systems, a battery in the tag is used to boost the effective operating range of the tag and to offer additional features over passive tags, such as temperature sensing. Data collected from tags is then passed through familiar communication interfaces (cable or wireless) to host computer systems in the same manner that data scanned from barcode labels is captured and passed to computer systems for interpretation, storage, and action.

There are several versions of RFID that operate at different radio frequencies. The choice of frequency is dependent on the operational requirements and the read environment – unfortunately this is not a technology with 'one size fits all' applications.

Three primary frequency bands are being used for passive RFID systems and a further two for active (powered) systems. These frequency bands are detailed in Table 8.1.

Table 8.1 Primary frequency bands used for RFID

Band	Frequency	Transfer rate	Read range	Notes
LF low frequency	125–134 kHz	Low	Short < 0.5 m	Most commonly used for access control, animal tracking and asset tracking Signals can pass through liquids but not metal objects
HF high frequency	13.56 MHz	Moderate	Medium < 1.5 m	Used where medium data rate and read ranges up to about 1.5 m are acceptable Signals can pass through liquids but not metal objects
UHF* ultra high frequency	865–956 MHz	Moderate/ High	Medium < 5 m	Offers the longest read ranges, up to approx. 3 m and high reading speeds Signals cannot pass through liquids or metal objects
UHF ultra high frequency	433 MHz	High	Very long < 100 m	Active tags Signals cannot penetrate liquids/metals
Microwave	2.45 GHz	Very high	Long < 10 m	Active tags Signals cannot penetrate liquids/metals

* Due to the national difference of air frequency usage regulation, RFID in 868 MHz is generally used in European countries and 915 MHz is used for non-European countries.

RFID is a wireless identification technology that consists of three key components: RFID smart tags, RFID readers and RFID middleware. An RFID smart tag, also called transponder and RFID tag, is a tiny compact silicon-chip containing memory, modulator and antenna (Finkenzeller 2003, Glover and Bhatt 2006). The smallest and thinnest one is just 0.05×0.05 mm (Uldrich 2007). The memory is an electrically erasable programmable read-only Memory (EEPROM), which is a type of non-volatile memory, containing a serial number and user defined area. The modulator is for modulating and demodulating a radio-frequency (RF) signal to carry and transfer messages. The antenna is for receiving and transmitting the RF signal. In order to detect and receive the RF signal emitted from RFID tags, a standard compatible RFID reader is needed. The communication between hardware and software application is the role of RFID middleware.

8.2.1.1 RFID Tags

RFID tags consist of an integrated circuit (IC) attached to an antenna, typically a small coil of wires, plus some protective packaging (like a plastic card) as determined by the application requirements. RFID tags can come in many forms and sizes. Data is stored in the IC and transmitted through the antenna to a reader. RFID tags are either *passive* (no battery) or *active* (self-powered by a battery). Tags also can be read-only (stored data can be read but not changed), read/write (stored data can be altered or re-written) or a combination, in which some data is permanently stored while other memory is left accessible for later encoding and updates.

8.2.1.2 RFID Smart Labels

So called smart labels are a particularly innovative form of RFID tag and operate in much the same way. However, a smart label consists of an adhesive label that is embedded with an ultra-thin RFID tag inlay (the tag IC plus printed antenna). Smart labels combine the read range and unattended processing capability of RFID with the convenience and flexibility of on-demand label printing. Smart labels also can be pre-printed and pre-coded for use. In on-demand applications, the tag inlay can be encoded with fixed or variable data and tested before the label is printed, while the label can contain all the barcodes, text, and graphics used in established applications. Smart labels are called smart because of the flexible capabilities provided by the silicon chip embedded in the tag inlay. A read/write smart label also can be programmed and reprogrammed in use, following initial coding during the label production process.

8.2.1.3 Tagging Mode (Active *versus* Passive)

There are three common types of RFID technologies, active, passive and semi–passive, in the market (Gerst *et al.* 2005, Anon 2006, Hodges and Harrison 2004, Ward *et al.* 2006).

The active RFID smart tag is fully battery powered so it supports longer range communication. Many commercial systems now claim ranges of up to 100 m. Such systems can continuously emit a signal for asset/personnel tracking and security purposes. Also, it can support the use of different condition based techniques such as a temperature sensor for monitoring different temperature at different locations and a vibration sensor to detect movement. Typically the tags are expensive, costing 8–30 € or more and are used to track high value goods like vehicles and large containers of goods. Shipboard containers are a good example of an active RFID tag application.

Passive RFID smart tags do not contain a battery. Instead, they draw their power from the radio wave transmitted by the reader. The reader transmits a low power radio signal through its antenna to the tag, which in turn receives it through its own antenna to power the integrated circuit (chip). The tag will briefly converse with the reader for verification and the exchange of data. As a result, passive tags can transmit information over shorter distances (typically 3 m or less) than active tags. Typically the read range is dependent on the reader antenna type – portable readers have shorter ranges than some fixed readers with large antennas. The passive tags have a smaller memory capacity (limited up to 2K) and are considerably lower in cost (less than 1 €) making them ideal for tracking lower cost items. The technology also supports embedding temperature sensors in passive RFID smart tags but they are not suitable for continuous logging tasks. Passive tags are well standardised and durable, and are therefore suitable for inventory control and logistics.

Addressing to the read range limitation of passive RFID, *semi-passive RFID* has also been developed, embedding a battery power source in a passive smart tag in order to boost up the read range.

8.2.1.4 Read-only *versus* Read-write

There are two basic types of chips available on RFID tags, namely read-only and read-write. The read-only chips are programmed with unique information stored on them during the manufacturing process – often referred to as a number plate application. It is not possible to modify the information on read-only chips, which limits their flexibility for certain uses.

Read-write chips, however, allow the user to add information to the tag or write over existing information when the tag is within range of the reader. Read-write chips are more expensive than read-only chips as would be expected, but they may be used in much more sophisticated applications. Applications for these chips may include field service maintenance or item attendant data – where a maintenance record associated with a mechanical component is stored and updated on a tag attached to the component. Another method used is something called a WORM chip (write-once, read-many). It can be written once and then becomes read-only afterwards.

8.2.1.5 RFID Readers

A reader is basically a radio frequency transmitter and receiver, controlled by a microprocessor or digital signal processor. The reader, using an attached antenna, captures data from tags then passes the data to a computer for processing. As with tags, readers come in a wide range of sizes and offer different features. Readers can be affixed in a stationary position (for example, beside a conveyor belt in a factory or dock doors in a warehouse), portable (integrated into a mobile computer that also might be used for scanning barcodes), or even embedded in electronic equipment such as print-on-demand label printers.

8.2.1.6 Key Attributes

Passive smart label RFID systems offer unique capabilities (Anon 2004b) as an automatic data capture system in that they:

- provide error–free, wireless data transmission that is battery-free and maintenance-free;
- do not require line-of-site scanners for operation;
- allow stored data to be altered during sorting or to capture workflow process information; and
- work effectively, even in harsh environments, with excessive dirt, dust, moisture, and temperature extremes.

A summary of some important RFID Tag characteristics is shown in Table 8.2.

Table 8.2 Summary of important RFID tag characteristics

Band	Notes
Power	Active tags have a battery on the tag. The battery may be used to boost read/write range, allow for larger memories, or add sensory and data logging capabilities, such as temperature sensing. Passive tags receive all of their energy from the read/write device that "powers" the tag to allow it to transmit data
Memory	Passive tags (non-battery) typically have anywhere from 64 bits to 1 kilobyte of non-volatile memory. Active tags, such as those used in military tags, have memories as high as 128 kilobytes.
Storage	The majority of passive tags use EEPROM memory. Some are laser programmed at the silicon level. Many active tags utilise battery-backed SRAM
Frequency	Passive high-frequency systems typically operate at 13 MHz) and low-frequency at around 125 kHz)

Table 8.2 (continued)

Band	Notes
Range	Passive high-frequency (HF, typically 13 MHz) and low-frequency (LF, around 125 kHz) systems typically exhibit a read range of less than 1 m. The size of a tag (antenna area) will have a significant impact on the read range. Some applications limit the read range to around 150 to 200 mm. Some newer technologies (UHF systems) do have a longer read range that can be 6 to 8 m, but these systems are intended for pallets and shipping crates. The read range depends on many factors, most importantly the size of the antenna (on the transponder and the reader) and the reader's output power. If transponders are not powered there is a trade off between read range and size
Size	The smallest smart labels are approximately 25 mm square. RFID tags in general have been shrunk as small as 2 mm^2
Resilience	Durability depends on the tag manufacturing process. Much work is being done by tag vendors to ensure high durability in harsh environments. The typical operating temperature for an RFID tag is between -25°C and 70°C, storage temperature typically is between -40°C and 85°C. Both ranges vary from manufacturer to manufacturer depending on the tag's components. Some industrial tags on the market will withstand temperatures as high as 250°C. Most tags would withstand fairly high pressures, especially if moulded. X-rays do not affect tags but this does depend on the intensity of radiation
Mounting surface	In general tags cannot be mounted directly on metal in near proximity due to interference. They require a non-metallic spacer (*e.g.*, air, cardboard, plastic, *etc.*) to ensure proper function
Multiple tags	All RFID technology being used in logistics applications has the ability to read multiple tags at the same time. This is important for tagging requirements of machines components, *etc.*
Antenna	UHF tags are highly affected by the physical properties of the items to which they are affixed. It may be necessary to use a range of tag types to accommodate a variety of items

8.2.2 RFID Software Considerations

Many software providers, including some of the main ERP vendors, have developed packages to deal with RFID reader and printer/encoder management, plus "tag data capture event" management. As with most business applications software, packages are typically customised to meet customers' requirements rather than being ready to go 'off the shelf'.

8.2.3 RFID Standards

The development of standards for smart tags is still very much a topic of interest to industrial and research organisations. However there are several groups working to define smart label standards that will deal with the physical characteristics of the smart label, what information it carries and how the data is represented. Barcode standards were developed under technical subcommittee SC31 and it is generally expected that RFID standards comprising data structures and the methods of communicating between tags and readers would be submitted into this committee for ratification as ISO standards (Anon 2004b).

At the present time, the only two operating frequencies that are of major interest are the HF 13.56 MHz and the UHF 865 to 956 MHz range. There is little interest in developing standards for the lower frequencies range.

There are currently no global public bodies that govern RFID frequency bands. In principle, every country can set their own rules and already variances between allowable frequencies are beginning to appear between continents and countries. Organisations that govern frequency allocation for RFID in Europe include ERO, CEPT, ETSI and national administrations. In the USA, the regulatory body is the Federal Communications Commission (FCC). The development of global systems is unsurprisingly hindered by the lack of standardisation. Some standards have been formulated regarding RFID technology, with examples including ISO 11784 and 11785, ISO 14223/1, ISO 10536, ISO 14443, ISO 15693 and ISO 18000. Complications of utilising the technology also occur because of national regulations such as power restrictions in certain frequency bands, interference with military bands, and health and environmental issues, *e.g.* the European Waste Electrical and Electronic Equipment Directive does not permit disposal of RFID tags.

The two most well accepted standards are ISO international standards and EPC global industrial standards (Magellan Technology 2006). ISO standards define RFID in a more generic way and it is driven by manufacturers. It does not concern itself with data, rather focusing on data access. Reflecting global requirements is the aim. Thus, it covers the entire range of air interface frequencies. In contrast, EPC standards are industry specific, where lots of retailers and suppliers are included. It does not cover the entire range of frequencies, but rather the frequencies of interest only (860–960 MHz). The aim at first addresses what RFID is needed for in each specific industry (Anon 2006). In 2006 the International Standards Organization approved the EPC Gen 2 Class 1 UHF standard and published it as an amendment to its 18000-6 standard (Type C). The standard relates to RFID air interface for item management using devices operating in the 860 to 960 MHz range (O'Connor 2006).

The following tables list the details of the EPC Global RFID class structure and ISO air-frequency standard.

Table 8.3 RPC global RFID class

EPC Global RFID class structure	
Class 1: Passive identity tag	Simple, passive and read only non-volatile memory
Class 2: Passive functional tag	Extension of class 1 tag with up to 65 kB of read-write memory
Class 3: Semi-passive tag	Extension of class 2 tag with a built-in battery power to increase the read range
Class 4: Active *ad-hoc* tag	A built-in battery power to activate the circuit and power the transmitter for broadcasts a signal to reader
Class 5: Reader tag	Built-in power, able to communicate with other class 5 tags and devices

Table 8.4 ISO Air-frequency standard

ISO 18000 Air frequency standards	
18000-1	Generic parameters for the air interface for globally accepted frequencies
18000-2	Parameters for air interface communications below 135 kHz
18000-3	Parameters for air interface communications at 13.56 MHz
18000-4	Parameters for air interface Communications at 2.45 GHz
18000-6	Parameters for air interface communications at 860 to 960 MHz
18000-7	Parameters for air interface Communications at 433 MHz

8.2.4 Costs Involved

The implementation of RFID systems incurs costs over and above those invested in a barcode infrastructure. Additional investment is required for tags, printer/encoders, readers, software and professional technical services to integrate these components. One of the largest expenditures is expected to be the costs of the actual tags, as typically many thousands will be required. The high cost of RFID tags has been one of the biggest drawbacks of wide-scale adoption of the technology. Tag costs are decreasing but still have some way to go before they are more widely acceptable. A smart label costs approximately 0.40 € *per* label (in large quantities).

8.2.5 Advantages and Disadvantages

It is clear that automated reporting of real-time, accurate data can provide tremendous advantages in all kinds of industries. In addition, the ability of RFID to support asset tracking and the creation of a life history or maintenance record of an asset is of great interest to a number of industries.

RFID technology enables reading of tag information from a greater distance, even in harsh environments, avoiding problems related to dirty tags/readers, non-line of sight situations or hazardous situations. In addition, the information on RFID tags can be dynamically updated. As items pass by the reader they may be read automatically, providing potentially large savings in labour costs or a substantial increase in throughput of scanned items.

A summary of the advantages of RFID technology (with particular emphasis on the benefits over existing barcode systems that are traditionally used in maintenance scenarios) are shown in the list below:

- no line of sight requirement;
- tags can tolerate a harsh environment;
- long read range;
- data may be updated;
- multiple tag read/write at same time and location; and
- real-time tracking of people, items, and equipment.

The main disadvantages of the technology have to be the implementation costs and technical risk. However, more basic problems can occur – product packaging may need to be redesigned, business processes may need to be changed, *etc.* Also there are well known problems utilising the systems in the vicinity of liquids and metals, which can inhibit the transmission of radio signals significantly.

8.2.6 Privacy Issues

Many privacy issues have been discussed, notably in the media but these have more to do with basic trust between organisations and customers. There is a great deal of legislation that enforces good practice and the question associated with RFID is more related to 'unauthorised' use of data that has been transmitted from an RFID tag.

In general there is no personal identification information on the tag. However, as the technology advances and increased data storage capabilities emerge this may become more of an issue. Solutions to such problems may involve the use of encryption to protect any data that is required to be prevented from unauthorised access.

8.2.7 Applications for RFID

Applications fall into two principal categories: firstly, short range applications where the reader and tag must be in close proximity (such as in access control) and secondly, medium to long application, where the distance may be greater (such as reading across a distribution centre dock door). The Dynamite project demonstrated applications of smart tags that fit into both of these categories.

A list of typical applications of smart tags is shown below:

- access control for people: to work place, dangerous/secure equipment, computer or vehicle;
- access control for vehicles: secure access on site, road tolling, instant fuel payment;
- manufacturing automation: control of flexible manufacturing processes, labelling key components for later rework/recycling;
- logistics and distribution: tracking parcels from shipment to customer, tracking goods from manufacture through to retail;
- retail: supply chain management, stock taking, reducing loss through shrinkage, reverse logistics, product availability;
- maintenance: plant and equipment, fixed assets, patients; and
- product security: tamper evidence, product authentication, anti-counterfeiting.

The typical industrial sectors that have applied smart tags are shown in the list below. This list is by no means comprehensive, however, it is notable that the maintenance field has been much slower than others to adopt such technology. The typical sectors are:

- retail and consumer packed goods;
- clothing and apparel;
- food and drink manufacturing;
- leisure industry and services;
- logistics and transport;
- healthcare and pharmaceuticals;
- building and construction;
- IT, electrical and electronics;
- defence;
- automotive; and
- farming and livestock.

A typical use of RFID technology in the Dynamite concept is where RFID tags are placed on different machines or certain key assets such as replacement machinery parts and specific tools. While the RFID smart tag is scanned by the reader, the smart tag identification number and memory can be retrieved wirelessly. Based on that, users can identify the attached asset either by matching the smart tag identity number with asset record in database or reading the asset code and other relevant information directly from memory.

8.3 Real-time Locating Systems Using Active RFID

Real-time locating systems (RTLS) using active RFID is an important application area for indoor objects tracking and positioning. Three core components in an RTLS system including multiple sensing devices, position engine/algorithms and user interfaces. Four fundamental positioning techniques are time of arrival (ToA), angle of arrival (AoA), time difference of arrival (TDoA) and received signal strength indication (RSSI). Recently, some hybrid approaches such as LANDMARC were suggested specific for increasing the reliability and accuracy in operating in indoor and noisy environments.

8.3.1 Time of Arrival

ToA is a technique to measure the travel time or propagation delay of a radio signal from a transmitter to one or more receivers. In this case, transmitters are active RFID smart tags and receivers are readers. By multiplying the propagation speed of the signal by the propagation time measured (t_i-t_0), the distance between transmitter and receiver can be found. In order to determine the position in a 2D plane, at least three reference points (readers) are needed. For a 3D plane, four readers are needed. Since the locations of the readers are known, the position of the tag can be calculated as described in Figure 8.1.

This method is suitable for accurate indoor positioning and the calculation is simple; however, the clocks of the tag and the readers must be synchronised and the precision of the time measurement has to achieve in nanoseconds. So, special hardware is needed. Additionally, the hostile conditions in the industrial environment will cause variations in the propagation speed of the radio signal and this will affect the distance calculation.

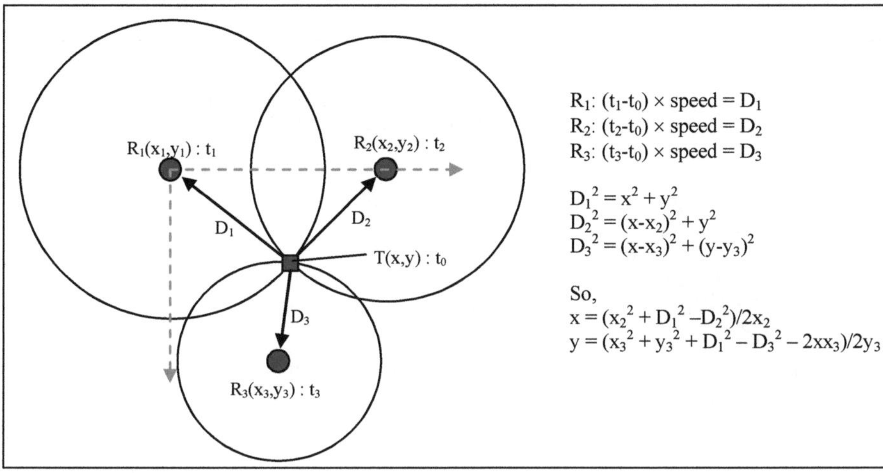

Figure 8.1 Illustration of the time of arrival method of location

8.3.2 Time Difference of Arrival

The time difference of arrival (TDoA) method can be considered as the intersection of hyperbolas in 3D. The method is difference to that of ToA, where only the time that a transmitter (tag) sends a signal to the readers is recorded. In the case of TDoA it requires that the receivers (readers) record the time at which the signals have been received. This means that TDoA is operated based on the measured time difference between departure from one and arrival at the other. Generally, three or four readers are fixed at some known positions. They synchronously receive a signal from the tag and record the received signal's time. Then, a location engine will calculate the received signal's time difference between each of the readers and transform them into an estimated tag position like ToA does (Figure 8.2).

Since the principles of TDoA and ToA are very similar, their problems are also similar. TDoA also requires clock synchronisation of all readers and tags, which means that special hardware is needed.

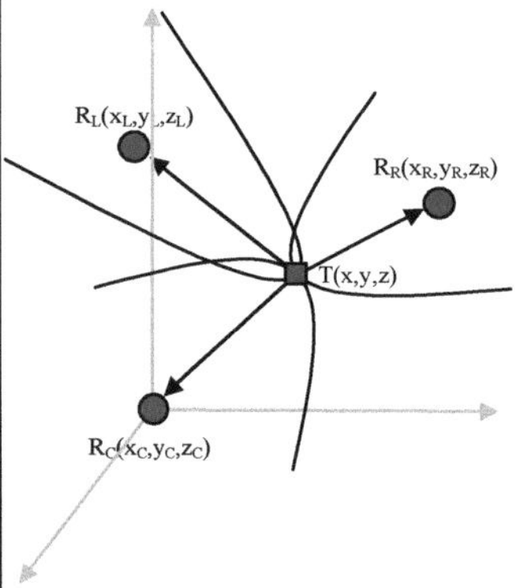

T --- The travel time (T) from the emitter at T(x,y,z)
c --- The pulse propagation rate (c)

$$T_L = \frac{1}{c}\left(\sqrt{(x-x_L)^2+(y-y_L)^2+(z-z_L)^2}\right)$$
$$T_R = \frac{1}{c}\left(\sqrt{(x-x_R)^2+(y-y_R)^2+(z-z_R)^2}\right)$$
$$T_Q = \frac{1}{c}\left(\sqrt{(x-x_Q)^2+(y-y_Q)^2+(z-z_Q)^2}\right)$$
$$T_C = \frac{1}{c}\left(\sqrt{(x-x_C)^2+(y-y_C)^2+(z-z_C)^2}\right)$$

Assume R_C is centre (0,0,0),

$$T_C = \frac{1}{c}\left(\sqrt{x^2+y^2+z^2}\right)$$

Then, the time differences of arrival are:

$$\tau_L = T_L - T_C = \frac{1}{c}\left(\sqrt{(x-x_L)^2+(y-y_L)^2+(z-z_L)^2} - \sqrt{x^2+y^2+z^2}\right)$$
$$\tau_R = T_R - T_C = \frac{1}{c}\left(\sqrt{(x-x_R)^2+(y-y_R)^2+(z-z_R)^2} - \sqrt{x^2+y^2+z^2}\right)$$
$$\tau_Q = T_Q - T_C = \frac{1}{c}\left(\sqrt{(x-x_Q)^2+(y-y_Q)^2+(z-z_Q)^2} - \sqrt{x^2+y^2+z^2}\right)$$

Finally, (x,y,z) can be calculated by the algorithms described as ToA

Figure 8.2 Illustration of the time difference of arrival method of location

8.3.3 Angle of Arrival

The angle of arrival (AoA) method is a technique to determine the direction of the propagation of the radio signal of a tag received at a reader. Using direction sensitive antennas on a receiver (reader), the direction to the transmitter (tag) can be obtained. Two or more readers are fixed in positions and antenna directions, the position of a tag transmitted to readers can be determined by using simple triangulation. For each reader, the AoA of the signal received from the same tag is calculated by the location engine to determine the position of the tag.

As illustrated in Figure 8.3, the accuracy is based on the precision of the angle measurements. Even just a degree error can cause a huge difference in the estimation of position for a long range application. Although this can be improved by increasing the number of antenna arrays (readers) used, no matter how many readers are used, with sensitive angle measurements like reflection of radio signals the problem cannot be solved, especially in tight indoor industrial environments. Therefore, it is more suitable for direct line of sight measurements in a clear area between tags and readers.

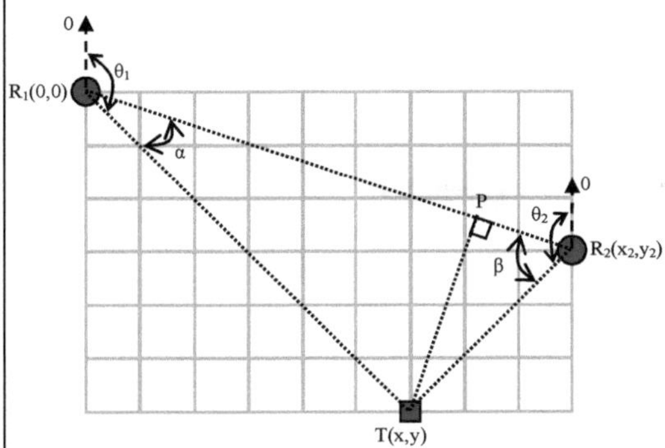

θ_1 θ_2 R_1 R_2 are known

$\alpha = \theta_1 - 90 - \tan^{-1}(y_2/x_2)$
$\beta = \theta_2 - \tan^{-1}(x_2/y_2)$

$(R_1R_2)^2 = x_2^2 + y_2^2$

Triangulation:
$R_1R_2 = PT/\tan(\alpha) + PT/\tan(\beta)$

Then,
$PT = R_1R_2 \times \{[\tan(\beta) \times \tan(\alpha)] / [\tan(\alpha) + \tan(\beta)]\}$
$R_1P = R_1R_2 \times \{\tan(\beta) / [\tan(\alpha) + \tan(\beta)]\}$

$(R_1T)^2 = (R_1P)^2 + (PT)^2$

$x = R_1T \times \sin(180 - \theta_1)$
$y = R_1T \times \cos(180 - \theta_1)$

Figure 8.3 Illustration of the angle of arrival method of location

8.3.4 Received Signal Strength Induction

The received signal strength induction (RSSI) is the simplest approach for locating a mobile node (tags) by measuring the signal strength and not the time and angle. It uses several readers simultaneously to track the location of a tag. At least three readers are needed for determining the location of the assets or person. The calibration of the signal strength at various points in a predefined area is necessary. Because the signal strength will continuously decrease with distance during propagation, a particular path-loss model can be generated. Then, the distance between a tag and a reader can be reversely found by converting the value of the signal strength at the reader into a distance when the signal output power of the tag is known. Finally, three particular distances between a tag and readers are found. The exact position of the tag is determined through a triangulation calculation (Figure 8.4).

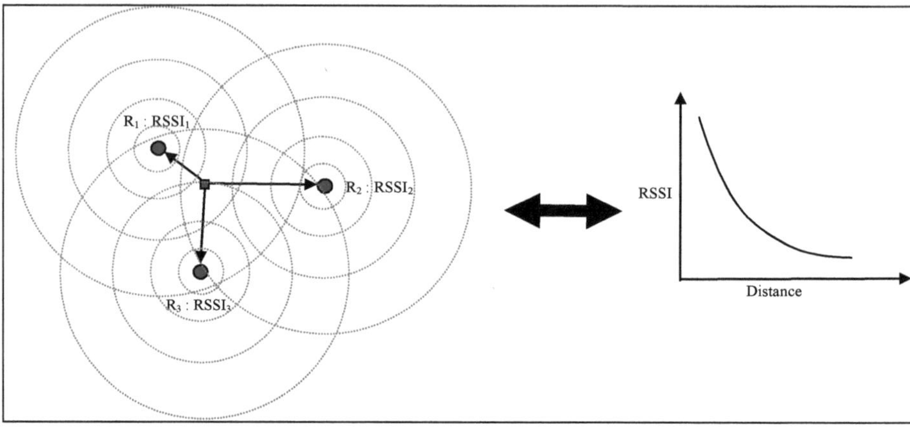

Figure 8.4 Illustration of the received signal strength induction method of location

8.3.5 LANDMARC

LANDMARC is a new positioning approach specific for indoor and noisy environment positioning of smart tags. The theory is based on comparing and calculating the RF signal distance between numerous static reference smart tags and movable tags in order to approximate the location of movable tags. The method is in many ways similar to the RSSI approach described earlier, however in this case different readers receive RF signals emitted from every active smart tag, including the reference tags and the movable tags. The system is expected to deliver a high level of accuracy, however at the cost of many more reference tags. However, if the trends of reducing tag costs continue, then this type of system may become more attractive for developers in the future.

8.4 Background to Applications of RFID

Today's commercial enterprises and industrial production systems are constantly under pressure to manage their maintenance resources as efficiently and cost effectively as possible. The opportunity to coordinate different resources like personnel, spare parts, equipment, tools, *etc.*, is a key factor to success. Web-based on-line computerised maintenance management systems (CMMS) provide good options to allow engineers access to critical data; however the systems in place today are far from perfect. Particular weak points include the collection of sufficiently accurate, up-to-date information and the use of that data in the most sensible way to make good maintenance decisions. This chapter focuses on the first problem – namely how can data be collected more rapidly, efficiently and more accurately.

Data entry or collection is often time consuming in existing CMMS; errors or missing data are commonly found in such databases. Automatic data capture systems are becoming more commonly used. For example, the barcode is often used as one option for accurately capturing information, *e.g.* it is often used in spare part inventory management. Barcodes have their own problems however, in that they are easily damaged, become covered with oil/dirt, *etc.*, especially in an industrial environment.

RFID is rapidly emerging as the replacement for the barcode. It can be combined with CMMS to establish high speed, accurate and reliable data collection in industrial and commercial environments. Additionally, RFID contains embedded memory and supports updating of storage information. Based on that, field engineers can immediately and accurately identify items, manipulate storage information, such as machinery data, sensor identification, audit trails of maintenance activities, spare part information and use of maintenance tools. Under the new wireless maintenance environment, engineers can remotely perform identification of problems and standardisation of decision making in a highly efficient and convenient way.

The Dynamite research has focused on integrating RFID and CMMS. The main strategy involves the extensive use of stored and transmitted electronic information in order to ensure instant access to up-to-date, accurate and detailed information. The aim of this chapter is to report the new maintenance information flow and a design of RFID tags template in order to effectively store maintenance information in passive RFID tags and embed them into a web service based CMMS infrastructure. Ultimately, by taking advantage of current mobile wireless internet technology, engineers can access the internet at any time and any location *via* RFID and internet equipped mobile computers and PDAs to retrieve information from RFID tags and query CMMS.

8.5 Review of RFID Applications in Maintenance

Despite the widespread use of RFID technology in supply chain management, the number of applications specifically in maintenance is still relatively low. Some of the world's largest companies and organisations have invested heavily in the use of smart tags to support their enterprise management systems. Widely known examples are those of Wal-Mart and the US Department of Defence (McGinty 2004, Weinstein 2005). There are, however, some limited examples of the use of RFID technology in maintenance systems and these are briefly summarised below.

Several aircraft manufacturers and airlines, for example Boeing and Delta, employ RFID in their predictive maintenance programme by using smart tags to track airplane engine parts. The huge amounts of money involved in such assets give excellent incentives to the companies to reduce costs by better tracking their assets. It has been reported that Boeing's new 787 Dreamliner will have around 2,000 high-memory passive tags (Brewin 2004). In addition the importance of such systems has given rise to planned collaboration between manufacturers (Boeing, Airbus) in the setting up of standards for RFID in the aerospace industry.

Examples of RFID being used in the maintenance of high value assets such as airport terminals have been reported (Legner and Thiesse 2006). The study described outlines the many benefits of the application of such technology, including better planning, control and documentation of the work of the maintenance technicians.

Early reports have been noted in the academic literature of the incursion of RFID into facilities management (Ko 2009). Here the information technology applications used in the maintenance of buildings have been enhanced by using data collected from RFID systems. Data management modules have been developed to collect both building usage and maintenance data.

RFID technology can help to address many maintenance problems in vehicle fleet management. RFID tags placed on critical assets requiring regular maintenance can give accurate maintenance and usage history as well as reduce parts inventory and eliminate obsolete parts. Service data can be recorded directly onto RFID tags. That way service history is readily available anywhere the fleet vehicle goes. Movements of tagged vehicles can be recorded in RFID enabled locations or depots (Anon 2009).

Computerised maintenance management systems have seen limited uptake of the RFID technology; one notable exception is that of the MicroMAIN software (http://www.micromain.com). This was reported as early as 2004 as incorporating an RFID middleware solution, so that assets can be managed with radio frequency identification in addition to, or instead of barcodes (Anon 2004a).

8.6 Applications and Scenarios

RFID technology is often considered as a wireless barcode technology, able to store large amounts of data. However, the technology has improved exponentially and large amounts of data can be stored within the internal memory of smart tags no larger than a small coin. Engineers often have limited or no access to a network, wired or wireless, and therefore, the internal memory of the smart tag removes the need for a network connection. Maintenance tasks are stored on the smart tag for retrieval at a later date. The current memory size of passive RFID smart tags is limited (from 1 Kbit to 32 Kbytes), however, they are capable of storing asset codes and status and progress of previous maintenance works. Within two years it can be anticipated that RFID smart tags technology will increase sufficiently to allow the tags to be smaller and contain more text, diagrams and maintenance schedules.

Currently, passive tags are most widely used because they are low tag cost, do not incur battery replacement cost, have long tag life, and are small and thin compared with active tags. Toll applications and inventory tracking applications are two common examples of their usage. However, for some applications specifically requiring high-speed identification, long distance area monitoring or continuous sensor monitoring, active tags are more recommended. Container and manifest tracking applications are examples of active tags usage.

In a small company it would not pose a problem for engineers to identify asset and replacement parts, however, within a large company with many assets in different locations it may prove more difficult to locate a particular asset. Therefore, the asset identification system is designed to fully utilise both the RFID tag identity code and the content stored in the smart tag internal memory for identification and query purposes. It is considered as a middleware system able to interlink users, smart tags and web-services. Based on the system, users can primarily identify the asset based on the information stored inside the smart tag.

The mobile assets tracking system is a real-time locating system (RTLS) utilising active RFID technology. The tracking system can be used to track and monitor the position of valuable movable assets. Similar to passive smart tags, active smart tags will be attached to movable assets and personnel. Because the active RFID smart tag operates in active mode, it will continuously send the new data to the monitoring system. To ensure data reliability the position of the smart tag is identified using multiple readers; in addition this ensures the signal strength is kept at a maximum. Using multiple readers allows the position of the smart tag to be calculated by a positioning engine, and the result is used to update the record in a database. The PDA can query the updated asset positions *via* the corresponding web-service and display them on the screen.

The proposed application areas of the RFID technology have been organised into four main areas, namely *users*, *tools*, *machines* and *spare parts*. These categories represent many of the important assets involved in maintenance scenarios. The categories are described in the following sections, together with the typical information that is likely to be collected and/or processed in each case. For reference a schematic diagram representing each of the categories is shown in Figure 8.5.

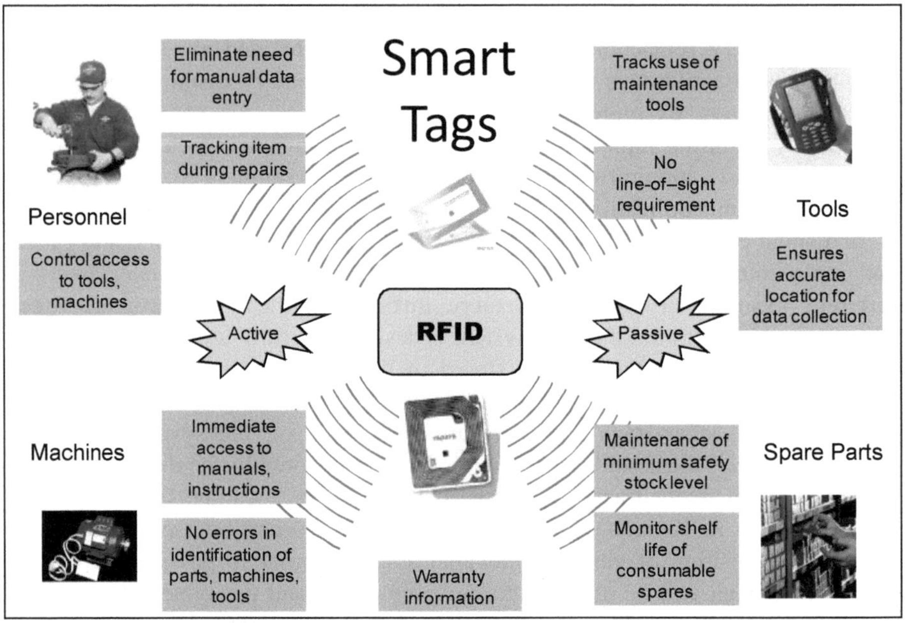

Figure 8.5 Selected RFID application areas in maintenance

8.6.1 Tools

Smart tags attached to tools have several purposes. Firstly, tracking of tools provides information on the location or the current user of a particular tool. Given that the types of tools used in maintenance are generally high value, relatively scarce items, it is important that the company keeps track of these assets. Examples of the sorts of tools considered are vibration data collectors, infrared cameras, acoustic emission kits, *etc.* In addition, tags can be used to store records of calibration of the tools and also to keep a check on whether the users of the tools have been suitably trained.

8.6.2 Spare Parts

The spare parts inventory is critical for the effective maintenance of assets in all companies. The value of spare parts stock can be a considerable; therefore it is critical that it is managed effectively. Smart tags may be used to track periodic use of inventory stock and subsequent replenishment. Data collected from these activities can soon be utilised to perform more sophisticated activities such as ensuring minimum stock level is not surpassed and also predictive parts ordering.

Many examples of this type of inventory control have been implemented in companies all over the world. It is one of the common features of the many supply chain management systems that have been reported in the literature for many years. However, it is still a critical part of any maintenance system. To ensure that the most effective decisions are made, due note of the relevant spare parts inventory must be taken. Hence this is one of the areas of RFID application that was considered in the project.

8.6.3 Machines

Production machines are a prime candidate for the application of smart tags. These are the main production assets and therefore it is critical that they are correctly identified and recorded. Tags can be used to uniquely identify a machine and allow the accurate collection of relevant maintenance data. Since the tags may be used to store information, a lot of useful information may be included on the tag to assist any operator who interrogates the tag. The machine will be uniquely identified, and maintenance actions carried out on the machine (by whom, at what time, *etc.*) may be recorded. Other useful information such as the equipment supplier, details of the installed sensors, measurements made from the machine *etc.*, may all be useful in typical maintenance activity. Taking the concept to a further level it is feasible to also include items such as details of existing work orders, standard maintenance procedures and bill of materials on the tag.

8.6.4 Personnel

There are many advantages in allocating smart tags to individual personnel. For example, any maintenance action or manual data input may be attributed to that staff member. Spare parts and tools can be signed out by individual staff members and the data can be recorded. For safety reasons, specific personnel can be excluded from performing specific tasks for which they have not been trained.

8.7 Smart Tag Demonstrators

In the Dynamite demonstration system, the database server is located at a remote site and connected *via* the internet. The web services server is located at a local seever to host all necessary web-services and web pages to support different maintenance activities and database records manipulation. A lot of sample assets information was pre-created in the database. Those sample data were collected from the use of real machines, equipment and systems at a university and an industrial plant. The following section will present the implementation on two RFIDs, the PDA-based asset identification system and the mobile assets tracking system (Figure 8.6).

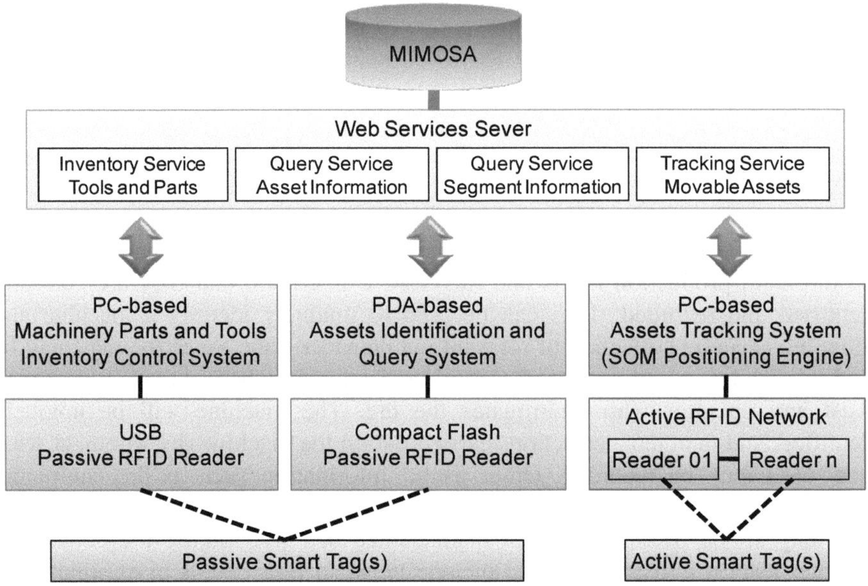

Figure 8.6 System architecture

8.7.1 Inventory Tracking (Passive)

The electronic inventory control system was developed and installed on a personal computer as this is the most likely hardware format to be found in a typical store location. This does limit the portability of the platform. especially when compared to the PDA applications to be discussed later; however it was possible to install multiple systems at different self-service checking points within the plant, *i.e.*, store rooms and the entrance of working areas. This allows maintenance engineers to check-in and check-out their required spare parts and tools for maintenance. If an asset is misplaced the system can query the most recent movement and identify the last person to use it and also the present or last location (Figures 8.7 and 8.8).

Figure 8.7 Self-service checking machine with virtual image

Figure 8.8 Common format USB smart tag reader

8.7.2 Asset Identification and Query System for PDAs (Passive)

The demonstration system for assets identification and query was successfully created to use passive RFID to identify different types of maintenance information. In order to better organise and manage the maintenance information, six types of fundamental information templates have been designed for machines, machinery parts, tools, facilities, locations and personnel. Their roles are to provide a more convenient and easy way to categorise and standardise the format for saving on an RFID smart tag and displaying the data on a PDA screen. One example machine template for machine includes the template identity code, the template format, the template record and the database query statement. All this information is stored in the MIMOSA database, so the user can easily specify their information template for a particular purpose.

For passive RFID, 13.56MHz I.CODE SLI is selected to be used in the demonstration system. A compact flash RFID reader is connected to the PDA and a USB reader is connected to the PC for reading and writing information to passive smart tags. The memory of I.CODE SLI passive smart tag is 1024 bits organised in 32 blocks of 4 bytes each. Some metal-on-mount smart tags are considered since they are suitable for mounting on the metallic surface of a machine. Figure 8.9 shows the equipment currently utilised in the development of a passive RFID system.

Figure 8.9 PDA-based passive RFID system

Two platforms, web-based and PDA-based, have been developed as shown in Figure 8.10. The web-based platform is a group of multiple ASP.Net web pages mainly for dealing with the task of asset identification, the query asset component and sub-components. Each ASP page are designed for different purposes including displaying basic information of assets, displaying images, schematics and diagrams, querying related asset information including spare parts and maintenance instructions. The PDA-based platform is a window mobile application. It is designed for bridging an RFID device and web-services to manage the content stored inside smart tags. It also supports export of the smart tag carried information to other maintenance software used. Through predefined information templates, the software can download all necessary information from the database in order to auto-generate content to store inside the memory of the smart tag (Figure 8.11).

The main role of both platforms is to allow users to effectively manage the information in order to display it on the screen and store it in the smart tags; and to utilise the information for auto filling of different online forms to query information of details, report failure and record the latest maintenance status.

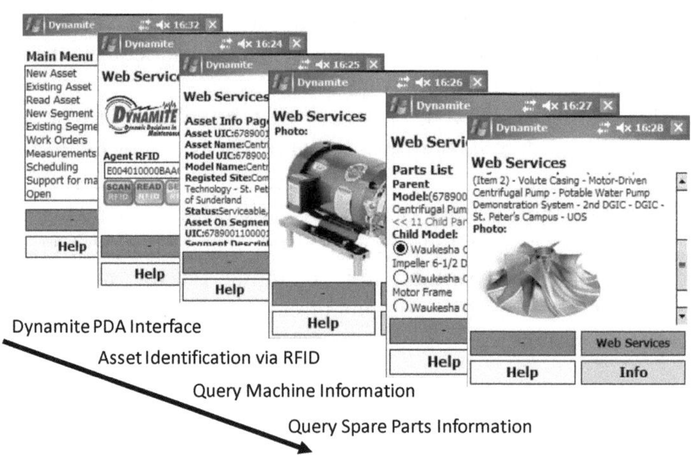

Figure 8.10 Multiple ASP.net web pages

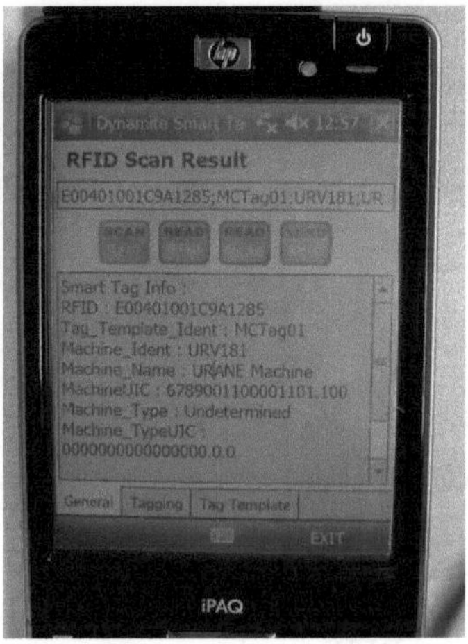

Figure 8.11 PDA-based platform

8.7.3 Mobile Assets Positioning System (Active)

Typically, there are three deployments for using active RFID for tracking: (1) mobile smart tags with fixed location of readers, (2) fixed location of smart tags with mobile readers and both are mobile and (3) hybrid. The first one is for the centralised processing scenario; one positioning system is used to track and mange all movable smart tags. The second one is much simpler because each reader contains a basic calculating power to determine the nearest position based on identifying fixed active smart tags nearby. The final one is a hybrid scenario; each movable reader must have full positioning power to continuously calculate the distance between it and at least three detected movable active tags in order to estimate their positions (Figure 8.12).

For active RFID, 433 MHz Wave trend products are used. Up to 255 L-RX201 active RFID readers can be connected by network cable to form a Wave trend reader-network. The first reader is connected to a PC *via* the serial RS232 protocol to transfer the reader data to the PC. Typically, the life of smart tag is estimated at 5 years and the transmission time interval is approximately 1.5 s. If the reader uses a L-AN100 (Whip) antenna, the read range can achieve up to 35 m.

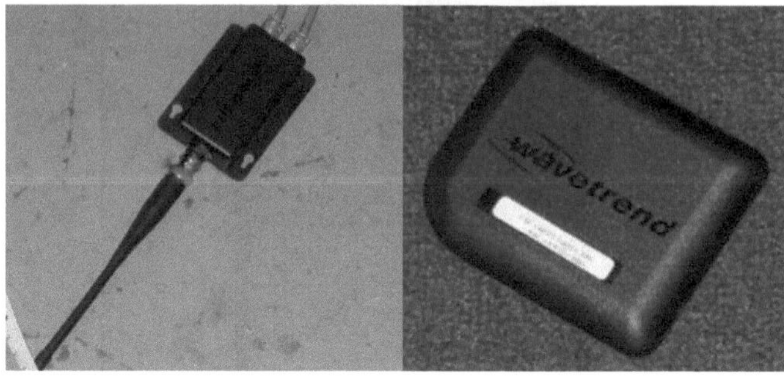

Figure 8.12 Active RFID system components

In the experiment setup, we only consider the first deployment because our current active RFID equipment does not support PDA to receive the active RFID signal. There are four active RFID readers connected as shown on the map (see Figure 8.13). Those readers will continuously receive RF messages transmitted from active smart tags. All information such as the RSSI signal, sensor value and stored information are bound as a single message. All messages from different smart tags will be collected by the tracking application as shown in Table 8.5. Each row in the table represents a set of RSSI values of smart tags from four readers.

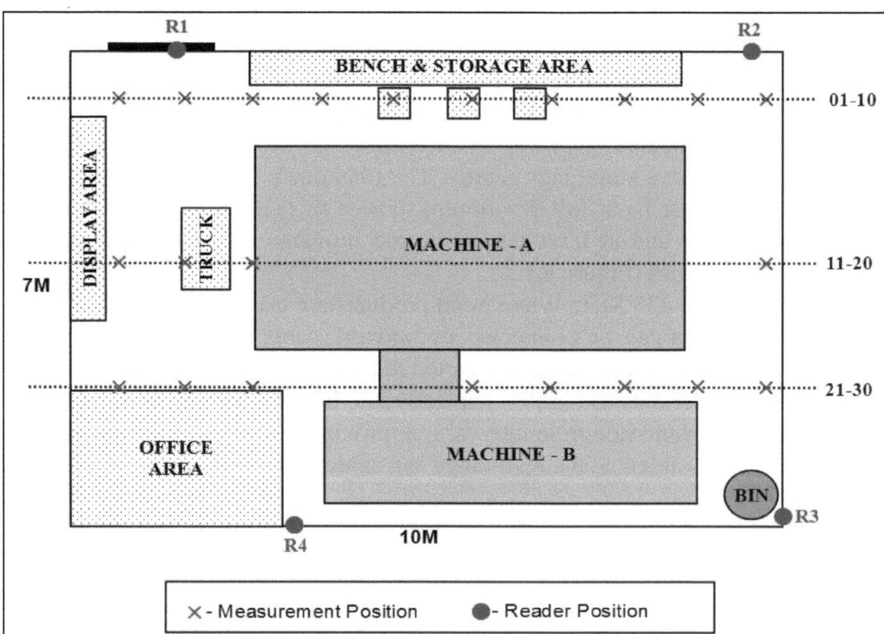

Figure 8.13 Map of the active RFID tracking application demonstration

Table 8.5 Example RSSI values from the application

Reader1 RSSI	Reader2 RSSI	Reader3 RSSI	Reader4 RSSI
78	69	75	83
70	80	75	79
69	85	78	76
69	70	84	75
68	78	70	74
71	85	85	79
79	78	80	79

Then the positioning engine will read the table row-by-row to calculate the position and estimate the position of the smart tag. Currently, a non-linear artificial neural network called a self-organising map (SOM) is used as the positioning algorithm (Figure 8.14). When four RSSI values for a smart tag feed into the SOM position engine, one corresponding colour output will be activated to represent a corresponding position on the map. Finally, the results will be displayed on the map (see Figure 8.15) and the corresponding record in the database will also be updated. PDA users can call specific asset tracking web-services reversely to retrieve the asset tracking records and displaying them on the screen.

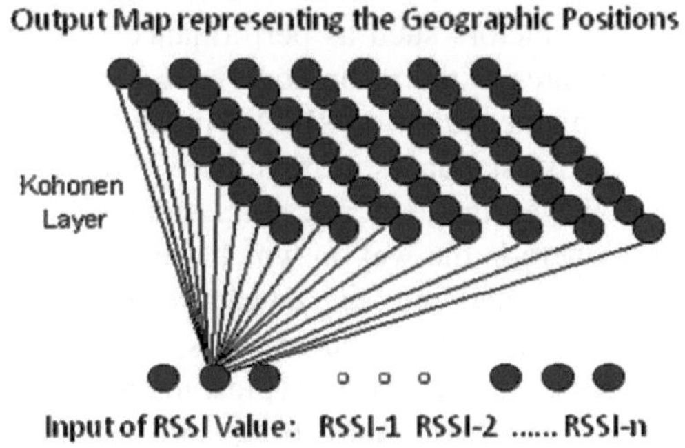

Figure 8.14 Representation of the Kohonen self-organising map

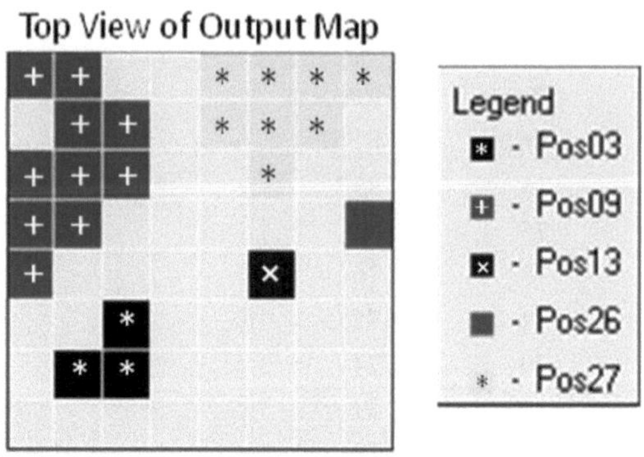

Figure 8.15 Schematic representation of the positioning matrix

8.8 Conclusions

This chapter has given an overview of RFID or smart tag technology and the importance it is likely to play in the concept of dynamic decisions in maintenance. An overview of the technology in its different forms has been presented to give an understanding of the key features and benefits. The current uses of the technology in engineering and commercial systems have been outlined together with its proposed uses within the Dynamite concept. The current state-of-the-art has been described in terms of key factors such as performance and cost. In addition the advantages and disadvantages of the technology have been clearly stated. Several key challenges in the implementation of RFID schemes have been identified, the most notable ones being cost, standards, performance, reliability and privacy.

Most importantly, however, the demonstrator systems developed are described. The key important issue here is the way in which the information gleaned by the RFID systems can be integrated into the Dynamite data framework. It is only through such clearly identified linking mechanisms that the data can be of use in the decision making process. In each case a typical maintenance scenario has been outlined and also the way in which data can be collected, processed and utilised for decision making has been illustrated.

The use of RFID technology to provide information has been designed within the context of the Dynamite concept and has been validated by means of the demonstrators based on the use case scenarios. The integration of RFID technologies is thought to add considerable benefit to the decision making capabilities of the maintenance system developed.

References

Anon (2004a) Frequently asked questions about RFID technology. Zebra Technologies 2004, available at: http://www.scansource.eu/en/education.htm

Anon (2004b) Maintenance software is RFID enabled. ThomasNet Industrial Newsroom, 10.8.2004, available at: http://news.thomasnet.com/fullstory/454373

Anon (2006) A summary of RFID standards. RFID Journal, online, available at: http://www.rfidjournal.com/article/articleview/1335/1/129, accessed: 15.4.2009

Anon (2009) RFID Solutions for fleet maintenance. GAO RFID Inc, Whitepaper, available at: http://fleetmaintenance.gaorfid.com/

Brewin B (2004) Delta, Boeing to test RFID on engine parts. Computerworld, June 2004, p 1, 61

Finkenzeller K (2003) RFID Handbook: fundamentals and applications in contactless smart cards and identification, 2nd Edition, Wiley, New York

Gerst M, Bunduchi R, Graham I (2005) Current issues in RFID standardisation. Proc Workshop on Interoperability Standards, Interop-ESA Conference, 22-25.2.2005, Geneva, Switzerland, Hermes Science Publishing, London, p 13

Glover B and Bhatt H (2006) RFID essentials – theory in practice. O'Reilly Media, Cambridge MA

Hodges S and Harrison M (2004) Demystifying RFID: Principles and practicalities. Whitepaper, Auto-ID Centre, Institute for Manufacturing, University of Cambridge

Ko C-H (2009) RFID-based building maintenance system. Automation in Construction 18:275–284

Legner C and Thiesse F (2006) RFID-based maintenance at Frankfurt airport. Pervasive Computing, IEEE, 5:34–39

Magellan Technology (2006) A comparison of RFID frequencies and protocols. Whitepaper, Magellan Technology, Sydney, March 2006

O'Connor MC (2006) Gen 2 EPC protocol approved as ISO 18000-6C. RFID Journal, online, available at: http://www.rfidjournal.com/article/view/2481, accessed: 26.3.2007

Uldrich J (2007) Hitachi's RFID takes a powder (HIT), online, available at: http://www.fool.com/investing/value/2007/02/20/hitachis-rfid-takes-a-powder.aspx, accessed: 15.6. 2009

Ward M, van Kranenburg R, Backhouse G (2006) RFID: Frequency, standards, adoption and innovation. JISC Technology and Standards Watch, May 2006

Weinstein R (2005) RFID: A technical overview and its application to the enterprise. IT Professional 7:27–33

Bibliography

Adgar A, Addison JFD, Yau C-Y (2007) Applications of RFID technology in maintenance systems. Proc 2nd World Congress on Engineering Asset Management, Harrogate, UK, 11-14.6.2007

Business Wire (2009) Tego, Inc. demonstrates world's first high-memory, passive RFID. RFID Solutions Online, online, available at: http://www.rfidsolutionsonline.com/article.mvc/Tego-Inc-High-Memory-Passive-RFID-Tagging-0001, accessed: 5.6.2009

McGinity M (2004) RFID: is this game of tag fair play? Communications of the ACM, 47:15–18

Chapter 9
Mobile Devices and Services

Erkki Jantunen, Christos Giordamlis, Adam Adgar and Christos Emmanouilidis

Abstract. E-maintenance and the use of mobile devices offer the flexibility to initiate maintenance-related applications at flexible locations while networked in unstructured environments. They are becoming a key enabling factor in achieving ubiquitous data and services availability, by retrieving information from heterogeneous data sources. Personal digital assistant (PDA) devices play a key role in bringing mobile maintenance management closer to daily practice on the shop floor.

The use of PDAs enables maintenance personnel to directly gain access to information stored at different locations, whether this be in the back office, on the monitored machinery or even outside of the plant. PDAs are becoming a ubiquitous information and services mediator, bringing the right maintenance-related information and e-applications to the right place at the right time to anyone authorised to gain access. Mobile users can again consult and act upon information, such as data relevant to monitored machinery, e.g., condition monitoring readings, the current machine state, maintenance actions history and scheduling, spare parts availability, maintenance activities instructions. In this way, mobile devices and services constitute a mobile assistant, enabling information and applications accessibility by mobile personnel, greatly expanding the available toolset with enhanced data and knowledge processing capabilities.

9.1 Introduction

Mobile devices, typically in the form of a PDA, are central to the implementation of mobile e-maintenance solutions. In the DynWeb architecture a PDA is a multi-purpose device, communicating with smart tags, intelligent wireless sensors, web-services and semantic web information. The rapid growth in the mobile devices market indicates that a device-type choice is not straightforward. It further indicates that as technology and costs are changing rapidly in this area and new products, devices and software enter the market, ensuring a high level of interoperability is more important than tying a solution to a specific type of device.

This need can be addressed by specifying a minimum set of hardware requirements for the mobile device, a set of web-service applications to support the e-maintenance functionality, and a common interoperable format for data exchanges, tailored to the maintenance function needs. Undoubtedly this trend will continue and many more devices will become available in the future.

It can be reasonably predicted that costs will reduce whilst power and functionality will increase. In this setting, the primary factors on which the hardware decision was made concerned the key technical features of the device, with special attention on connectivity issues, as this is very important to enable rapid and reliable communication with other devices.

Secondary factors such as unit robustness, aesthetics, *etc.*, were not considered due to the rapidly changing nature of the market. Applications were defined as typical web-services with published interfaces, while database support observed MIMOSA (www.mimosa.org) specifications to ensure services and data interoperability.

A PDA is a tool that enables the maintenance technician or engineer to interface and communicate with the surrounding world (Jantunen *et al.* 2008). It is a mobile user interface that provides access to the computerised maintenance management system (CMMS) employed for the management of maintenance personnel, materials and activities. By enabling fast and flexible access to remote web-services, the PDA can operate on a semantic web based platform, such as DynaWeb developed within the context of the Dynamite architecture, offering many support tools to the maintenance engineer.

The PDA also provides access to the identification of the machine and to condition monitoring data, as well as diagnosis of the machine condition. Usually the PDA is employed in thin client architecture, *i.e.*, most of the data is located and processing takes place in the central computers providing e-maintenance services. Since it is not always easy to automatically diagnose the condition of a machine, it is natural to include the condition monitoring and signal analysis capabilities also on the PDA. This dual functionality both takes into advantage the powerful capabilities of a remote server system, as well as the flexibility offered by the mobile device to bring information and services accessibility at mobile locations, such as on the shop floor or even outside the plant.

9.2 Mobile Devices in Maintenance Management

Since their inception (Chess *et al.* 1995), mobile agents have been used in a wide variety of applications (Jantunen *et al.* 2007). There are several advantages in employing mobile computing compared to conventional desktop applications. Among other things, mobile computing offers the flexibility to initiate applications at flexible locations in unstructured networked environments, to quickly and efficiently search for and retrieve relevant information from heterogeneous data sources, to perform tasks while utilising limited or intermittent connectivity and to provide asynchronous services to client requests (Samaras 2004).

Adding the ease and flexibility of carrying a handheld wireless device, mobile computing has the potential to transform the way in which a range of industrial management, monitoring and control tasks are performed (Buse and Wu 2004, Arnaiz *et al.* 2006). This potential is still largely unexplored in maintenance management. Although the usage of wireless devices within an e-maintenance framework has been suggested in the past (Lee 2001), integrated maintenance management solutions based on combined usage of wireless sensing, RFID tags, handheld devices and central or remote server-side computing and data-offices (Lampe *et al.* 2004, Legner and Thiesse 2006, Wittenberg 2003) have only recently started to emerge.

Part of the difficulty is attributed to the challenge of integrating equipment, devices, computing resources and codes from heterogeneous sources (Bartelt *et al.* 2005, Trossen and Pavel 2005), this being further complicated by the considerable complexity of optimising the management of maintenance in modern industry (Arnaiz *et al.* 2006). In DynaWeb the usage of PDA devices plays a key role in bringing mobile maintenance management closer to the daily practice on the shop floor. PDAs are used in synergy with intelligent sensing devices and smart tags on the lower-end of the data processing architecture, but also with server-side databases, data processing and remote access applications at the higher-end of the architecture.

Here the hand-held device is employed within thin-client server architecture (Jantunen *et al.* 2007). The "mobile worker" (technician), equipped with the PDA, approaches the monitored machinery. The PDA is equipped with an RFID tag-reader enabling the automatic identification of the equipment/component and thus it becomes possible to automatically retrieve relevant data from the central system (historic and reference data database) and quickly present it to the user. Furthermore, the PDA can access measurements logged at the intelligent sensing device (sensing agent) and combine/compare those with the automatically retrieved related historical and reference data from the central database, but also with domain knowledge from the central KBS (server KBS).

Thus the PDA becomes a ubiquitous expert advisor and, at the same time, a flexible data collector. Within this architecture there can be several intelligent sensing devices, distributed across the plant, which can wirelessly transmit data *via* short range RF either directly or indirectly *via* data logger gateways from the shop floor. In this manner, instead of an inflexible, costly and rather inaccessible wired monitoring structure, we have a flexible, easy to deploy and operate wireless e-maintenance architecture, which can become a powerful, efficient and easy to use tool for the maintenance engineer, while, at the same time, can be integrated with the organisation ERP.

9.3 Role of PDA Within DynaWeb

The PDA is a key element in the DynaWeb structure and it is arguably the most essential element in making mobile e-maintenance possible in practice (Jantunen *et al.* 2007). Here the e-maintenance concept is defined as the system or approach that enables maintenance information and application services availability at the exact location where they are needed, *i.e.*, in most cases near the machinery. Following this definition, the role of PDA is defined as a mobile user interface to the maintenance web services platform, DynaWeb. In the Dynamite project the key features are: (1) smart tags, (2) intelligent MEMS and oil sensors, (3) wireless data acquisition, (4) semantic web and (5) cost effective maintenance.

The information flowing from each of these sources is required to be collected, assimilated, analysed, processed and presented to the user at the PDA. Therefore, all the above information sources should be made accessible to the mobile user. This is achieved through the choice of an appropriate PDA and associated hardware, the development of suitable software and the customisation and integration of adequate wireless networking solutions. Thus the PDA supports:

- The use of smart tags for equipment component identification, storage and handling of component data, details of maintenance actions and machine diagnosis results.
- Wireless communication with intelligent sensors and local signal analysis and diagnosis.
- Communication with the CMMS system for handling asset and spare parts information, work orders, asset identification and localisation, *etc.*
- Communication with the semantic web platform, DynaWeb.
- Cost-effectiveness analysis.
- Logical and efficient display of raw data, processed data and summary information based on the above features.

Table 9.1 summarises the main PDA functions developed. Many of the inputs are from sensors that can be connected to the PDA or from Mimosa-compliant database tables. Naturally most of the output goes to the Mimosa database, either located in the PDA or at the server, depending on the availability of a wireless connection.

Table 9.1 Summary of the PDA functions supported in DynaWeb

PDA actor/role	Optional hardware	Input from	Output to
Definition of production machinery - Segments - Assets - *etc.*	None	The maintenance engineer defines it on site	Mimosa - Segments - Assets - Tables
Definition of measuring systems - Sensors - Locations - *etc.*	None	The maintenance engineer defines it on site	Mimosa - Measuring locations - Sensors - Tables
Identification of machinery components at plant floor	RFID reader	RFID tag Mimosa database	
Condition monitoring measurements - Vibration	DynaWeb vibration sensor USB host port	Sensor	Mimosa - Spectrum - Parameters
Condition monitoring measurements - Temperature - Pressure - Vibration	DynaWeb wireless sensor package DynaWeb gateway	Gateway	Mimosa - Parameters
Condition monitoring measurements - Oil analysis, water content, particles, *etc.*	DynaWeb oil sensors	Mimosa (oil sensors send data to server and that can be studied with the PDA)	The settings of oil sensors can be modified through WEB connection
Manual monitoring of production machinery	None/unconnected manual sensors	Manual input	Mimosa
Diagnosis of machinery condition		Mimosa	
Prognosis of machinery condition		Mimosa	
Work order creation - Manual creation of work orders	None	Manual input of work orders and work order steps	Mimosa - Work orders - Work order steps
Work order scheduling, user interface to scheduling service	None	Mimosa - Work orders - Work order steps	

Table 9.1 (continued)

PDA actor/role	Optional hardware	Input from	Output to
Handling of work orders and work order steps that have been assigned to the maintenance engineers	None		Mimosa - Work orders - Work order steps
Guidance to carry out the maintenance work	None	Maintenance guide database	

The specifications of the handheld PDA device, including hardware and software requirements have been previously defined (Jantunen *et al.* 2007). The primary hardware elements include the PDA itself, the wireless hardware and the RFID or smart tag hardware. Among the choices, the following may be outlined: the hardware device must be a commonly available 'off the shelf' product to allow ease of use by any potential system user. The processor speed should be sufficient to satisfy processing and display requirements. A processor at least as efficient as Intel XScale PXA270 should be suitable for the tasks in mind. The device connectivity needs should be met by sufficient networking and interfacing support, although additional hardware (an RFID reader) is needed for RFID communication.

The screen resolution of the device is to be 640 × 480 pixels, *i.e.*, standard VGA. This is sufficient for the typical information that needs to be displayed and is supported by a growing number of devices. The chosen operating system is Windows Mobile or Windows CE, as it is a very widely used system on mobile devices (Jantunen *et al.* 2007). The operating system should be .NET compatible, for consistent development of the software modules. One consequence of this is that many popular types of PDA hardware such as Nokia Smart Phones or Blackberry (very popular in USA) cannot be used to run DynaWeb during the development phase of the project.

The device must have long battery autonomy for a typical operator shift length. Although RF data transmission is power consuming, the typical use of the device does not require continuous RF activity. The PDA device should be of a size and weight that allows easy portability around an industrial type site. It should be typically much smaller than a light laptop-computer (< 1 kg). Manual data entry should be supported by a range of standard methods. Typical PDA devices will allow several of the following methods:

- touch screen, pen browsing, usually more comfortable than a keyboard in small devices.
- Alphanumeric keyboard, numeric input to connected device is easier.
- Qwerty keyboard, some letter input is somewhat possible.
- Easier letter input. Many of the keyboard options are actually supported by the on-screen touch keyboard.

The PDA device should have a USB host to be able to collect data from a connected vibration sensor. This gives the advantage provided by active sync programmes and also allows cable communication for backup purposes.

Bluetooth communication is an interesting possibility for the PDA device as it allows connections between other PDAs to sensors and to phones, *etc.* This may be a useful feature to include in the PDA specification, *e.g.*, for device calibration or setup purposes. The RFID reader should be compatible with the PDA device. A convenient solution is *via* a compact flash card RFID reader, which can be added or removed from the PDA as required.

Additional identification capability could be included by the use of a laser barcode reader. This would provide a backup for machine identification. A final backup strategy would be to enter the Asset ID manually *via* the PDA on-screen keyboard. Other possible solutions require a free SD, CF or PCMCIA card slot.

9.4 Description of Typical PDA Usage Scenario in Maintenance Operations

In order to illustrate the typical use of the PDA within a maintenance management system, an end-user scenario has been created. The example considered in this scenario is a typical sequence of events that would occur when a maintenance technician user is performing a routine inspection tour. Prior to the defined scenario, the maintenance engineer would have defined the segments (machinery) and assets (components) he would be taking care of (see Figures 9.1 and 9.2).

The shown inspection routine is a very common set of activities that can be applied to a wide range of applications. The user actions together with the actions triggered in the software are described below. For clarity the software elements are shown in `fixed-width` font.

Step 1

The user will typically be walking around a production facility and would like to make inspections of various pieces of equipment or machinery. When approaching the machine the user can use the PDA together with the onboard RFID reader to scan the tag of that particular machine. This will return the unique RFID code of that tag, ensuring that the machine is correctly identified. The software uses the unique ID of the tag (unique to that machine) to perform a query on the MIMOSA database in DYNAMITE. The query will return the model ID of that asset, so that the user is clear as to what the piece of equipment is and what its function is.

`Query_Asset(RFID)`

retrieves asset info including "Model Code"

`(= model_db_site, model_db_id, model_id)`

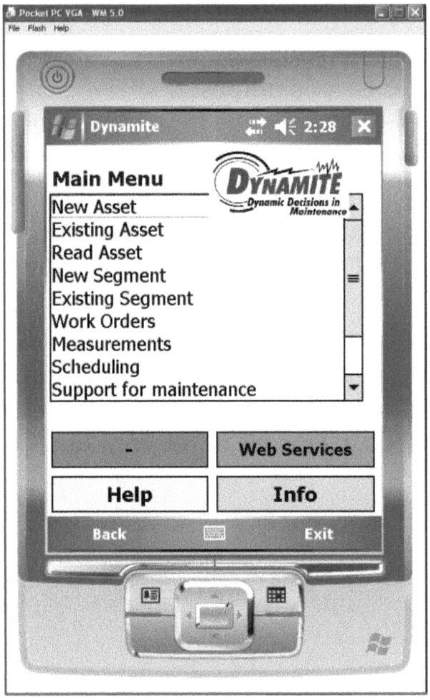

Figure 9.1 The main menu in the PDA user interface

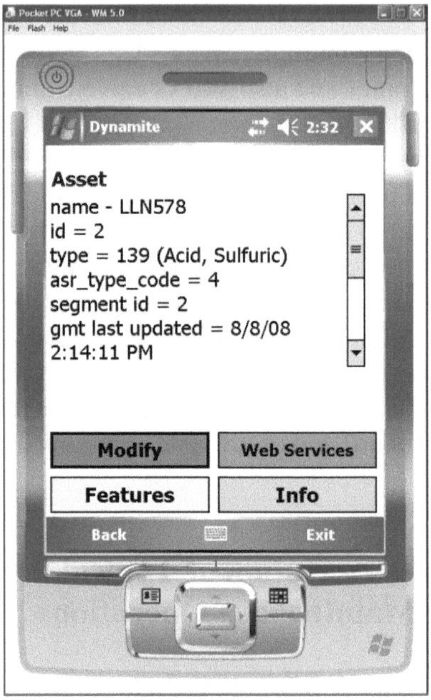

Figure 9.2 Detailed information about the asset

Step 2

Once the asset and model ID have been identified in this way, the next step is to determine the model code. Another query is used to return the model code of that asset. The model code and other related information may then be presented to the user on the screen of the PDA. The user can then confirm that the machine is the correct one and that the inspection process can begin.

```
Query_Model (Model Code)
```

to obtain the model information.

Step 3

All of the possible events that could occur (*e.g.*, machine failures) are pre-defined in the MIMOSA database and hence it is not necessary for any information to be stored on the PDA. The events need to be determined. This is achieved by again performing a query on the asset_event table in the MIMOSA database. This table holds the different types of possible events for assets. These include both operating events and failure events such as time-stamped facts, logged operating events, failure "modes", failure "effects", failure "mechanisms", defect events, abnormal situations, alarm events, *etc.* An event is *not* work performed by an entity, but a log of the actual events that occurred at an asset over time.

```
Query_ModelEvent (Model Code)
```
for a list of "Event type codes"
```
(= ev_db_site, ev_db_id, event_type_code)
```

Step 4

From the list of asset events determined earlier, the type of events applicable to the particular machine must be determined. For the purpose of the demonstration of this scenario, the event to be flagged will be that the machine is dirty (and therefore needs cleaning).

At this point the user will be presented with a checklist of observations for this machine. All the observations that have been made can be recorded. For instance, if the possible faults for a machine were as shown in the list below, then the correct action would be to check/select the "cleaning LTA" option to indicate that the machine was dirty.

Wrong component installed	
Balancing LTA	
Alignment LTA	
Broken seal	
Cleaning LTA	✓
Belt resonance	
Shaft eccentricity	
Emission, recorded	

Figure 9.3 Sub-set of possible machine observations specific to the machine being observed, *e.g.*, "belt resonance" would not appear on equipment with no belt

The entry in the MIMOSA database will be "cleaning LTA"

Note: LTA = less than adequate
```
Query_AssetEventType(Event Type Code)
```
returns all different possible event definitions for that particular machine.

Step 5

The next step is to determine the agent identity (usually a maintenance technician). This can be approached in two ways. The PDA can be determined to be the agent, in which case a code can be assigned to uniquely identify the PDA. Alternatively the PDA user can be found by scanning an RFID badge on the user, or by password login at the start of the user session.
```
Query_Agent(RFID)
```
this retrieves the agent (user) information from the database.

Step 6

The next stage is that an asset recommendation is created.

`Create_AsssetRecommendation(asset code, agent code, time, recommendation comment)`

this creates a new recommendation record.

From the previous examples it is clear that the software can guide the user through typical maintenance procedures extremely efficiently. The communications with the database are quite complex; however, this is almost completely hidden from the user.

To continue the example, the following steps may be completed. Details of the data base communication are omitted for brevity.

Step 7

An agent (in this case a piece of software residing on the server) checks in the database for any automatically generated or user inputted data that describe occurrence of significant new events.

Step 8

An agent (again on the server) translates the new event to a recommendation. This is achieved by a set of web services (state detection, diagnosis) for rapid and consistent decision making. Alternatively, the agent could be triggered from the PDA instead of running continuously or at regular intervals. For efficiency of operations with large databases there would most likely be a combination of the above approaches. A service would be requested (condition monitoring or state detection, and diagnosis) from the Mimosa event data.

Step 9

Recommendation is translated to a work order by a software agent routine on the server.

Step 10

The PDA may then be used to display the current list of work orders, see Figures 9.4 and 9.5. This can be achieved *via* a simple query on the database. These results can be filtered by user to present an individual work list for each individual system user (maintenance technician). Alternatively an agent can be programmed to examine the work order table in Mimosa and then send these work orders to the maintenance personnel, *i.e.*, to the correct PDA. This would be done by a web service (`IssueWorkOrdertoPDA`) to identify the most suitable PDA. The PDA can also be used to help carry out the maintenance work (see Figure 9.6).

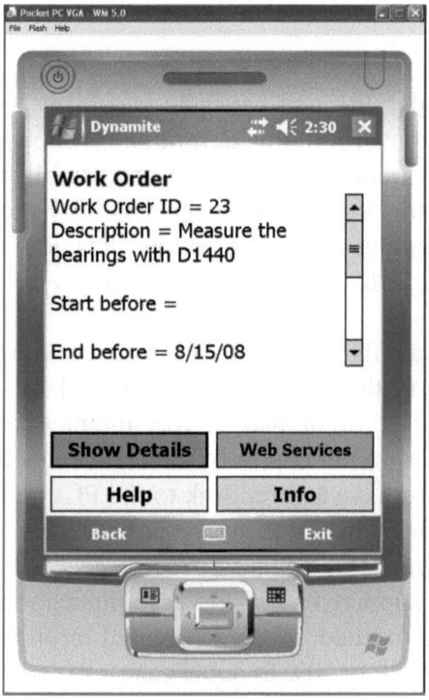

Figure 9.4 Managing work orders

Figure 9.5 Managing work order steps

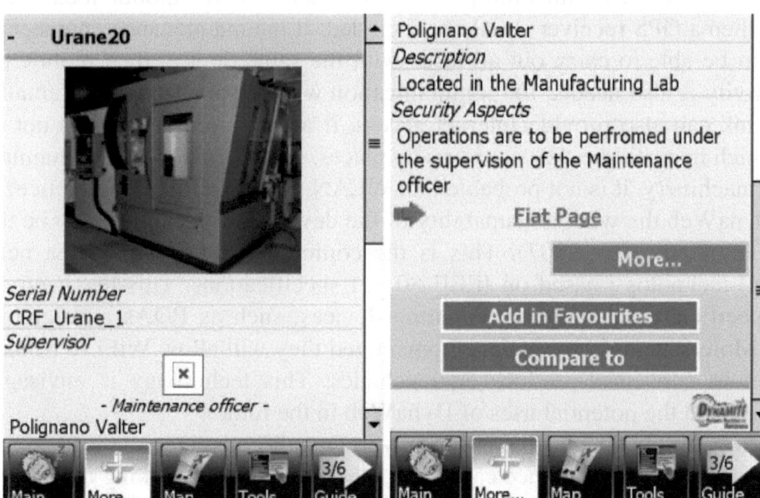

Figure 9.6 Presenting an object's information

9.5 Wireless Communication

The PDA is expected to provide anytime availability and have the capability to perform all the tasks listed in the previous sections. From this definition it follows that the PDA must be equipped with wireless connectivity, so as to establish communication between the maintenance centre, the machine, the tags and the sensors. Different technologies are used for wireless communication, depending on the requirements for the connectivity range, throughput and operating environment.

A wireless local area network is commonly available in many built spaces providing Internet access and through that access to the needed web services. However, WLAN might not always be available and it might also be very impractical for regular heavy duty data transmission, such as partial copies of the maintenance related database and therefore it is natural to also have a wired link to the PCs that support the maintenance personnel. Usually the wired link today is based on a USB connection. The same USB (host port) connection can also provide access to sensors and data acquisition components that support it. For the communication with smart tags an RFID communication link is needed, usually provided through an RFID tag reader that is plugged into a PDA, CF or SD. In some cases, the sensors and data acquisition equipment may support a Bluetooth connection, which is commonly available in other devices, such as printers and phones.

For complex sites with multiple building facilities, if outdoor localisation is needed then a GPS receiver should be included. If the maintenance engineer is expected to be able to carry out all tasks with the same device, then mobile phone connectivity is also needed for communication with the phone, fax and email. The phone link can also provide Internet access, if a local area network is not available, which may still be the case in many places. In fact, in the case of maintaining mobile machinery, it is not probable that WLAN would be found in practice.

In DynaWeb the wireless capability of the device has been chosen to be that of WiFi (Jantunen *et al.* 2007). This is the common wireless local area network (WLAN) technology based on IEEE 802.11 specifications. This has been developed specifically for mobile computing devices, such as PDAs and laptops, in LANs. More standards are in development and they will allow WiFi to be used by many more consumer products and vehicles. This technology is envisaged to greatly extend the potential uses of DynaWeb in the future.

The only disadvantages that may apply to DynaWeb include the following. Wi-Fi can suffer from interference caused by other wireless networking devices, operating at the 2.4 GHz band, such as cordless phones and microwave ovens. Power consumption is fairly high compared to some other standards, making battery life and heat a concern. Wi-Fi networks have limited range, typical router/antenna combinations might have a range of 45 m indoors and 90 m outdoors. Wi-Fi interference can prevent access, caused by overlapping channels, which can be a problem in industrial or commercial environments.

9.6 Technical Requirements

From the maintenance engineer's point of view, an optimal PDA would have all the communication features described above and it would also be small and easy to carry around (Jantunen *et al.* 2008). Unfortunately, the demand of small size and light device is somewhat contradictory to the technical capability of the device. Especially the screen size and resolution are of great importance when the device is being used for, *e.g.*, showing a video of how to dismantle a device or when a report has to be filled in.

When looking at the market it would seem that a small size is appreciated more than the screen size, *i.e.*, it seems that devices that are capable of VGA resolution are typically used for making condition monitoring measurements and communication with the CMMS. A full-size keyboard would be the nicest way of writing information in reports and messages. Unfortunately a keyboard does not fit nicely into a small device.

Today it seems that small keyboards are becoming more popular in devices used in communication. However, it should also be remembered that the main task of a maintenance engineer is to keep the machines running and not to communicate with other people or systems. In fact, it seems to be a common experience that maintenance engineers do not like writing reports, *i.e.*, it is something that is considered a necessary evil and is also avoided if there comes an opportunity to do so.

Another issue is that the so called 'prose'-type maintenance reports, which in a verbal way define what has been done, are not a good way of collecting valuable historical data that could support the maintenance planning work. Instead all reports should be made in a very formal way so that the CMMS can easily use the data. This issue has usually been solved using check lists where the user ticks the options that correspond to what has been done. Luckily this kind of methodology fits the PDA well and is also easier from the maintenance engineer point of view.

9.7 Practical Limitations Today

There is rapid trend in the development of both PDA hardware and software to manufacture devices that are faster, equipped with larger memory and enhanced features, while at the same time the software development tools are increasingly enriched in order to take better advantage of the hardware development (Jantunen *et al.* 2008). Currently, there seems to be quite a number challenges in practice. In many places the supporting environment does not provide wireless communication networks or the intermittent nature of the connection hampers the completion of the PDA tasks.

Furthermore, if all the features of the PDA are simultaneously used, the battery pack does not last for one working day, *i.e.*, the device has to be turned off for some periods or it has to be connected to electrical power source. The rapid evolution of software development platforms brings in new features on a daily basis but at the same time seems to be rather vulnerable and suffers from infantile problems, which in many cases are easy to handle but nonetheless remain annoying and need to be addressed.

9.8 Mobile User Interface Issues

The small form factor of mobile computing devices and their usage pattern by mobile users give rise to special mobile user interfaces requirements, both functional and non-functional. Functional requirements are directly related to user needs for the intended application. Non-functional requirements are introduced not by user needs but as constraints posed by the employed technologies and standards, as well as by special requirements related to the application field, *e.g.*, security, interference, *etc.*

When designing user interfaces for small factor devices a common pitfall is to attempt to fit within a single small screen all the features that are expected to be available on larger screens of desktop applications. Instead, one should carefully consider the typical user interface design patterns that are applicable to the smaller screen of PDAs and produce a dedicated mobile user interface design. Indeed, it is not practical to seek to make available all the functions intended for the mobile user within a single high-level overall user interface. Smart interface design can make them indirectly available, while working within the constraints of the small screen. Table 9.2 summarises typical screen design characteristics, whilst Table 9.3 summarises application navigation options for the same design patterns (Ballard 2007).

Screen design for mobile devices seeks to overcome the small form factor limitations by embedding enhanced browsing and navigation options in smart user interface designs. From simple list and table-based selection to menu and tab-based selection, with a limited number of clicks, touches or stylus integration these designs offers much greater functionality than would have otherwise been available on such small screens.

Table 9.2 Mobile user interface design patterns

UI design patterns	Design	Applicability	Use	Rationale
List	Vertical, text wrapping, simple	Scroll and select devices with small screen; up to two columns	Web-based applications	Intuitive; avoids side bars and tables; works on most devices

Table 9.2 (continued)

UI design patterns	Design	Applicability	Use	Rationale
Table	Includes all table features	Stylus-based devices	Not suitable for web-apps; avoided for large tables; combined with application launch buttons	Simple and without soft-keys; good when icons are used as table cells
Location selection	Set usage context (*e.g.*, shop floor; office, home, *etc.*)	All devices	Useful when different applications are needed for different usage context	Can support LBS (location-based services) selection
Returned results	BIG design issue; return results as web pages with or without scripting or within application windows	All devices but check if scripting is supported	Good for flexible list displays	If negligible fetch display, then improved navigation; otherwise scrolling may be preferable
Menus	List of commands broken down in compact categories	Scroll select devices with button input	Good for 'expert' navigation in more complex applications	Reduces numbers of keypresses
Tab navigation	As in desktop tabbed navigation	Stylus devices; also for scroll and select with four-way button navigation in some cases	Complex navigation tasks	Employ tabs to embed rich navigation options in applications
Breadcrumbs	Can be combined with drop-down lists; similar to desktop breadcrumbs	Not for scroll and select devices	Complex web-based applications	More content fits on screen

Application accessibility is also addressed from a different perspective in mobile devices, as not all application launcher options need to be readily available at the highest interface level. Instead, a contextual adaption of application availability implies that appropriate application accessibility has to be offered, depending on the context of the mobile device usage.

Table 9.3 Application navigation in mobile devices

UI design patterns	Design	Applicability	Use	Rationale
List	Back-next-done options for list	All devices but implementation varies	Good for lists within categories	Intuitive
Alphabetic	Handles complex application lists	All devices but implementation varies	Good for applications that lend themselves to alphabetical listing	Reduces need for direct text entry
Softkey and buttons	Users gain experience in using the device for complex applications	All devices but implementation varies	Good for all cases except web browsers and text messages	Fast execution by expert users

9.9 Trends

When considered from the hardware development point of view, it can be said that the modern PDA devices are getting closer to smart phones, *i.e.*, a device that was used for calendar and office applications in its latest version might also have mobile communication capability and *vice versa*, *i.e.*, smart phones are getting closer to PDAs and mobile computing devices are also equipped with a platform for office solutions customised for the PDA (Jantunen *et al.* 2008).

Luckily, from the developer's point of view, the same is partially true with the development of programming tools, *i.e.*, the same development tools can be used for programming hardware based on different technology, which especially holds for web programming. However, neither the hardware nor the software can be considered mature enough, and there is a strong technology push to enhance their technical capabilities and overcome the many technical challenges to ensure increased system reliability.

Mobile devices have been demonstrated to play a key role in e-maintenance as they offer many important benefits (Emmanouilidis *et al.* 2009b), such as:

- multi-connectivity features, *e.g.*, connection with smart sensors, RFIDs, WiFi/ Ethernet, mobile phone networks , peripheral devices;
- the possibility of use as information mediators to connect to devices deployed at the operations level, such as RFIDs and wireless sensors;
- the possibility of use as information mediators to provide access and exchange of information with the decision layer of enterprise operations, *e.g.*, CMMS, ERP;
- the possibility of use as mobile actors to initiate applications at multiple locations, 24 h a day;

- the possibility of use as mobile user interfaces, enabling data entry at multiple locations, 24 h a day; and
- the possibility of acting as mobile data loggers and data banks, offering local data logging and processing capabilities right at the point of need.

In this setting, mobile devices in e-maintenance are expected to follow and benefit from the general trends that can be identified in further advancing the software and hardware of such devices. The clear pattern here is that PDA hardware and software are rapidly growing in terms of maturity and capabilities and as ongoing developments promise to mitigate, to some extent, the limitations posed by the small form factor of such devices. This does not imply, and it would be wrong to assume, that mobile computing devices will be equipped with all the capabilities of their power desktop counterparts.

Nonetheless, many of the concerns about the limited resources that are currently available to PDAs are not considered to be critically linked to hampering the deeper penetration of such devices in an industrial usage context. On the contrary, it is expected that apart from an enhancement of their performance capabilities, mobile computing devices will increasingly feature enhanced connectivity options, including conformance to the most popular wireless networking standards. Furthermore, as the available tools for application development grow to maturity, we can expect an increasing number of vendors and application providers to offer tools that enable such devices to operate with increased interoperability features, thus better fitting the needs of a modern enterprise for improved maintenance products, applications and services.

While the above remarks hold with respect to the likely future capabilities and trends of the usage of mobile computing devices in industrial settings, it is of particular interest to focus also on the likely impact of other enabling technologies, which have been made more accessible to the daily e-maintenance practice, through the usage of PDA devices, namely technologies such as RFIDs, wireless sensors, web-based computing and smart optimisation and decision making tools.

The combined usage of mobile computing and RFID technology for asset and maintenance management is expected to benefit not only from advances in mobile computing technology but critically also from the increasing maturity of RFID technology. Such technology offers a strong business case, particularly in supply-chain management (Angeles 2005).

There are currently three significant limiting factors for the future development of RFID enabled mobile computing solutions in e-maintenance (Roussos and Kostakos 2009). First, RFID technology needs to lower associated costs. In industrial maintenance management this concern is less prevalent compared to logistics and supply chain applications, as the costs associated with deploying RFID applications and tags are a significantly lower fraction of the equipment cost, compared to the case of consumer products. A second issue is related to environmental reasons. Decommissioning and disposal of RFID devices is not an issue that has been resolved. Again, this is of little or moderate concern, when RFIDs are integrated with equipment, as in such cases the disposal issue has to be resolved for whole machinery components or equipment. Thirdly, application development, deployment and maintenance costs can still be considerable, as still more advances are needed in the direction of ensuring end-to-end interoperability of the developed modular solutions (Jun *et al.* 2009).

With respect to using PDAs as mobile and wireless data logging and analysis devices, wireless sensing and wireless sensor networks (WSN) combined with mobile computing devices need to take into account current power consumption and connectivity issues and limitations (Emmanouilidis *et al.* 2008, 2009a). When considering practical application development and deployment, WSN solutions and platforms are still in need of further advancements in WSN operating systems and middleware to achieve improved node coordination, support smart sensor behaviour, reduced power consumption and improved and connectivity with mobile devices (Emmanouilidis and Pistofidis, 2009).

It is of particular interest to identify techniques to balance data processing activities between those that need to be performed at the sensor level (*e.g.*, data collection, novelty detection), PDA (*e.g.*, data transfer, diagnosis, work order rescheduling) and those that may be performed at the PDA and definitely also in the back office (historical data processing, prognosis, spare part management, purchases, resource assignment, work order planning, link with CMMS/ERP, *etc.*).

Integration issues are of great importance and are likely to be critical success factors for the adoption of mobile computing technology in e-maintenance. When discussing integration, one should distinguish between data integration and services integration, while taking into account interoperability requirements. It was argued earlier in this chapter that data integration and interoperability in the maintenance field can be pursued by seeking to conform to acceptable formal and *de-facto* standards, such as XML and MIMOSA; services interoperability should continue to be pursued by working within a service-oriented architecture, with clearly defined web services, actors and user roles (Jantunen *et al.* 2008). More effort is also needed to close the loop with existing CMMS and ERP systems with maintenance-related decision support tools and mobile computing to make the PDA mobile device a truly ubiquitous tool for everyday maintenance engineering practice.

It is expected that the push for focused R&D in the direction of integrating emerging technologies within e-maintenance will be further strengthened. Important issues that should not be neglected in this direction are related to how to handle interference and noise, as well as security and privacy considerations. Such issues have not received sufficient attention so far and more effort must be devoted in these directions.

9.10 Conclusions

We can easily see that the future of effective maintenance lies on the broad shoulders of mobile devices and associated technologies, which give the maintenance engineer the necessary information he requires to efficiently carry out the maintenance work. PDA devices offer great potential for the development of overall maintenance management systems. One of the weaknesses of current systems is that typically not enough information is available to allow rapid, consistent and cost-effective decisions to be made. Maintenance activity is a complex business, requiring a wide range of knowledge – not only about the assets being maintained. For instance, to make sensible maintenance decisions, one may typically require information on production schedules, spare part inventories, equipment suppliers, staff availability, machine measurements from sensors or from the laboratory, *etc.*

The DynaWeb concept includes many of these issues. An extensive database is used to store all relevant information. This information is fed from many of the components of the software that have been developed. However, the PDA is the key component that enables the user to bring together all of this information without being bound by location and time constraints. Decisions may be made automatically by the software or with some manual intervention. This ensures that the maintenance recommendations are flexible to fit with the end-user's needs.

As the cost of the PDA devices decreases and the power increases, the benefit to even smaller companies of such technology can be appreciated. The numerous features of such devices, often supplied as conforming to standards, such as WiFi, Bluetooth, I/R, USB interface, *etc.*, allow users to select a system according to their exact requirements. In addition, comprehensive software applications, including basic office tools, *etc.*, allow the user to take advantage of features like document preparation and reporting, numerical analysis and data trending.

Thus, mobile devices and services are powerful tools for the future of maintenance, providing the maintenance engineer the necessary information and application launching capacity to carry out the maintenance work in an effective way.

References

Angeles R (2005) RFID Technologies: supply-chain applications and implementation issues. Information Systems Management 22:51–65

Arnaiz A, Emmanouilidis C, Iung, B, Jantunen E (2006) Mobile maintenance management. Journal of International Technology and Information Management – JITIM 15:11–22

Ballard B (2007) Designing the mobile user experience. Wiley, New York

Bartelt C, Fischer T, Niebuhr D, Rausch A, Seidl F, Trapp M (2005) Dynamic Integration of heterogeneous mobile devices. Proc DEAS, First Workshop on Designing and Evolution of Autonomic Application Software, 21.5.2005, St. Louis MO. ACM, New York,

Buse DP, Wu, QH (2004) Mobile agents for remote control of distributed systems. IEEE Trnsactions on Industrial Electronics 51(6) Dec.

Chess D, Grosof B, Harrison C, Levine D, Parris C, Tsudik G (1995) Itinerant Agents for Mobile Computing. Journal IEEE Personal Communications 2(5) Oct.

Emmanouilidis C, Pistofidis, P (2009) Design requirements for wireless sensor-based novelty detection in machinery condition monitoring, Proc 4th World Congress on Engineering Asset Management (WCEAM), 28-30.9.2009, Athens, Greece. Springer, Berlin

Emmanouilidis C, Katsikas S, Giordamlis C (2008) Wireless condition monitoring and maintenance management: a review and a novel application development platform. Proc 3rd World Congress on Engineering Asset Management and Intelligent Maintenance Systems Conference (WCEAM-IMS) 27-30.10.2008, Beijing, China, 2030–2041, Springer, London

Emmanouilidis C, Katsikas S, Pistofidis P, Giordamlis C (2009a) A wireless sensing development platform for ubiquitous condition monitoring. Proc 22nd International Congresss on Condition Monitoring and Diagnostic Engineering Management (COMADEM), 9-11.6.2009, San Sebastian, Spain. Fundación Tekniker, ISBN 978-84-932064-6-8

Emmanouilidis C, Liyanage JP, Jantunen, E (2009b) Mobile solutions for engineering asset and maintenance management. Journal of Quality in Maintenance Engineering 15:92–105. Emerald Group, ISSN 1355-2511

Jantunen E, Adgar A, Arnaiz A (2008) Actors and roles in e-maintenance. Proc 5th Conference on Condition Monitoring and Machinery Failure Prevention Technologies. CM & MFPT. 15-18.7.2008, Edinburgh, UK. Coxmoor, Oxford

Jantunen E, Arnaiz A, Adgar A, Iung B (2007) Mobile technologies for dynamic maintenance. Maintenance Management, Proc 3rd International Conference on Maintenance and Facility Management, Rome, Italy, 27-28.9.2007. CNIM Italian National Committee for Maintenace, ISBN 978-88-95405-02-5

Jun H-B, Shin J-H, Kim Y-S, Kiritsis D, Xirouchakis P (2009) A framework for RFID applications in product lifecycle management. International Journal of Computer Integrated Manufacturing 22:595-615

Lampe M, Strassner M, Fleisch E (2004) A ubiquitous computing environment for aircraft maintenance. Proc.SAC'04, 14-17.3.2004, Nicosia, Cyprus. ACM, New York

Lee J (2001) A framework for web-enabled e-maintenance systems. Proc 2nd Int Symp on Environmentally Conscious Design and Inverse Manufacturing, EcoDesign'01. IEEE, New York

Legner C and Thiesse F (2006) RFID based maintenance in Frankfurt airport. IEEE Pervasive Computing, IEEE

Roussos G and Kostakos V (2009) RFID in pervasive computing: State-of-the-art and outlook. Pervasive and Mobile Computing 5:110–131

Samaras G (2004) Mobile agents: what about them? Did they deliver what they promised? Are they here to stay? Proc 2004 IEEE International Conference on Mobile Data Management (MDM'04). Springer, Berlin

Trossen D and Pavel D (2005) Building a ubiquitous platform for remote sensing using smartphones. Proc 2nd Annual Int Conf Mobile and Ubiquitous Systems: Networking and Services, Mobiquitous'05. IEEE, New York

Wittenberg C (2003) A requirement analysis for the use of mobile devices in service and maintenance. Proc 2003 IEEE International Conference on Systems Man and Cybernetics 4:4033–4038

Chapter 10
Wireless Communication

Nicolas Krommenacker, Vincent Lecuire, Nicolas Salles, Serafim Katsikas, Christos Giordamlis and Christos Emmanouilidis

Abstract. Although wireless technologies suffer from link problems due to error-prone channels, they are increasingly employed in industrial environments and provide noticeable advantages in terms of mobility, reduction of costs, *etc.*

Current wireless technologies and their assessment for e-maintenance applications are reviewed in this chapter. However, no technology is efficient enough to fulfil the entire functional and technological requirements of such applications. The concepts of wireless gateway and collector are needed to ensure interoperability of several wireless technologies, and these are presented herein. Finally, hardware and software prototypes are presented.

10.1 Introduction

Communication architectures for industrial systems have evolved through three main paradigms. The first one is the paradigm of parallel wiring where devices are connected with a point-to-point wiring approach. This one became obsolete with the introduction of field bus technology, which allowed the use of only one line to provide power, control and configuration functions to devices. This paradigm created a multitude of proprietary solutions, which are on the market today. However, new requirements such as bandwidth, flexibility, scalability, *etc.*, and Internet technologies with its associated applications enabled recognition of the Industrial Ethernet as the third paradigm.

Different organisations such as the Industrial Automation Networking Alliance (IAONA), the Industrial Ethernet Association (IEA) promote Ethernet as "the standard in the industrial environment". Wired networks that behave in a deterministic way and meet hard real-time constraints are used each time.

This is changing with the exponential growth of the wireless market. There are countless situations in the industry where the benefits of WLANs come across very well. At the field level particularly, the large number of devices together with the increasing need for mobility make the use of WLANs very promising. In addition, wireless technology provides several advantages such as cable cost reduction, the installation facility of equipment within hazardous areas (in presence of aggressive chemicals, capable of damaging cables), and the flexibility to perform rapid plant reconfigurations.

Wireless technologies will probably become a new paradigm in industrial networks (Egea-Lopez *et al.* 2005). The foundation of the Wireless Industrial Networking Alliance in 2003, a coalition of industrial end-user companies, technology suppliers, and others interested in the advancement of wireless solutions for industry, is a good example of this trend. Note that wireless communications have been long used in industry but their use has been limited to solve very specific problems, rarely as the main communication infrastructure.

Wireless technologies have the potential of providing significant benefits in factory and industrial automation systems. However, they suffer from link problems, which are inherently less reliable than wired links. Wireless channels are error-prone and the packet losses are inevitable. First, radio signals decrease with the distance between the transmitter and the receiver. In addition, waveforms propagate from the transmitter in multiple directions and may undergo reflection, diffraction or scattering. Moreover, new challenges emerge such as:

- *Power supply:* cable replacement highlights the problem of supplying devices with energy. Alternatives to cables must be found for energy supply such as wireless energy transmission and energy scavenging methods. In addition, mobile devices are dependent on battery power. The energy consumption of the network interface can be significant and sometimes not directly productive (passive listening, routing, *etc.*). In this case, energy becomes a scare resource and it is desirable that devices consume as little energy as possible.
- *Interference immunity:* two environmental effects can impact wireless communication performance: noise and radio wave propagation.
  - Noise is interference coming from other RF and electromagnetic emitting sources found in the environment such as wireless devices (other wireless networks or cordless phones), electric motors, microwave ovens or welding equipment. Originally, ISM radio bands were used for non-commercial applications in industrial, scientific and medical applications. Today, this radio spectrum can be internationally used by anybody without a license. So, interferences with other devices or technologies can occur. For example, 802.11b/g and Bluetooth RF technologies coexist in the same 2.4 GHz ISM band.

  - During propagation, radio waves travel and interact (this includes reflection, diffraction and scattering) with the environment. So, multiple copies of a source signal arrive at a receiver from various directions over multiple paths. This phenomenon is called multi-path distortion and affects wireless communication performance.

- *Data security:* security considerations depend on the sensitivity of the transmitted data. The propagation of radio waves can easily go outside the controlled area. If a high degree of data security is required, data encryption coupled with intrusion detection systems can be used. In industrial communication, the degree of security often depends on the security policy of the company, which prescribes clear rules. Even if there exist strong protection mechanisms preventing unauthorised access, malicious attacks are possible and still represent a severe threat that requires special attention.
- *Reliability and fault tolerance*: the most common use of ISM bands concern fault-tolerant communication applications such as Internet access in office environments. In industry, machine downtimes involve high costs and time failures become unacceptable. So, the reliability of the wireless channel becomes an important issue.
- *Throughput and strict delay requirements:* many industrial applications require low throughput. However, the use of multi-media traffic (voice, video) can change the trend. Moreover, delay requirements differ with the industrial application and the type of data. For instance, minimising delay jitter is important for periodic control communications.
- *Operational range:* the effective range is influenced by physical obstructions (walls and other structures or furniture) and electrical interference (other wireless devices or electrical noise) present in the environment. If the transmission range is insufficient, an application may simply not work or may require repeaters or additional access points. Range is the most difficult criteria to assure because many factors (interferences, power emission) and application constraints (mobility, bandwidth, number of devices) affect it.
- *Cost:* the cost of the wired networks has remained constant or has increased slightly over the years. By contrast, the cost of wireless communications has dropped dramatically due to their growing popularity in office communication applications. So, the use of wireless solutions for industrial communication such as factory automation, environmental monitoring systems and others will reduce the cost of wiring and maintenance.
- Support for large and varying number of devices: industrial applications are likely to serve a large number (in the order of hundreds) of field devices and sensors. Wireless technology must be consistent with this point. With e-maintenance applications, the number of devices could generally be much smaller.

Many types of wireless communication systems exist and to classify them, several parameters should be considered such as cost, transmission range, *etc.* Wireless networks can be also divided according to the physical media used for transmission. Figure 10.1 summarises the main categories of wireless communication technologies according to the media used.

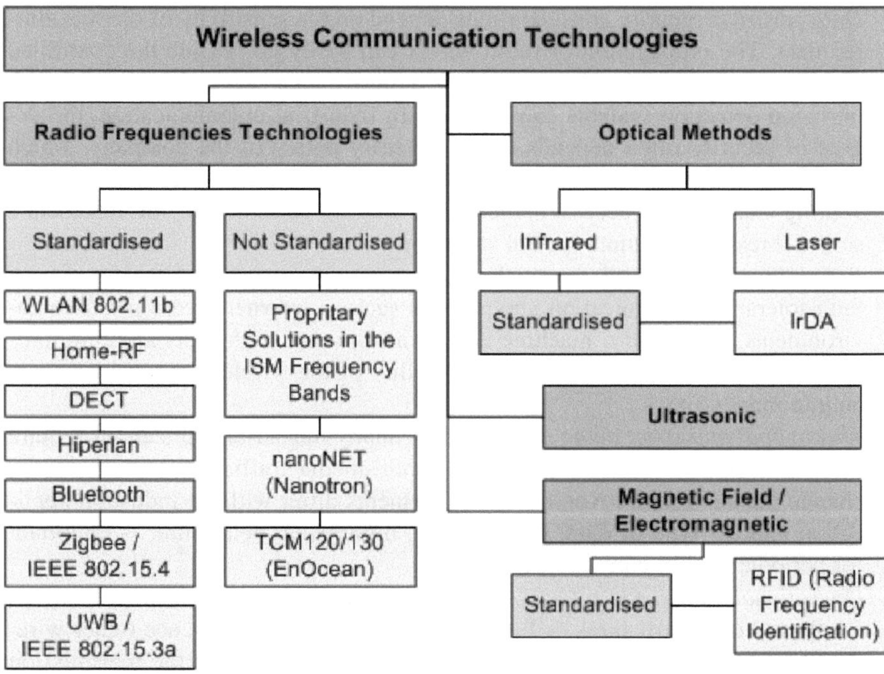

Figure 10.1 Wireless communication technologies

In the following section, a review of the most relevant wireless technologies for e-maintenance applications is presented. Note that this chapter only focuses on wireless LAN and PAN technology; cellular technologies (GSM, GPRS, EDGE, UMTS etc) are not considered here.

10.2 State-of-the-art

10.2.1 WLANs (IEEE 802.11)

IEEE 802.11 standards are already broadly used and are commercially available under the references 802.11a, 802.11b and 802.11g. Today, the standard continues to evolve through several improvements such as quality of service (802.11e), power control (802.11h), mesh networking (802.11s), roaming (802.11r), external interworking (802.11u), vehicular (802.11p) and throughput (802.11n).

Network Topology

802.11 networks can be deployed in infrastructure mode (where nodes communicate *via* a coordinator, usually the access point) or in *ad hoc* mode (direct communication between the nodes). With the infrastructure mode (Figure 10.2), access points (AP) are used to interconnect wireless stations (STA). Each AP manages communications within its transmission range, the so called basic service set (BSS), by using management frames (authentication, association, beacon, etc). For instance, a beacon's frame body contains the service set identifier (SSID), timestamp, and other pertinent information regarding the access point. The operational range of a BSS can be extended by interconnecting several AP through a distribution system (DS) which is often based on an existing wired network. The extensive coverage *via* an extended service set (ESS) allows mobile stations to move from one BSS to another.

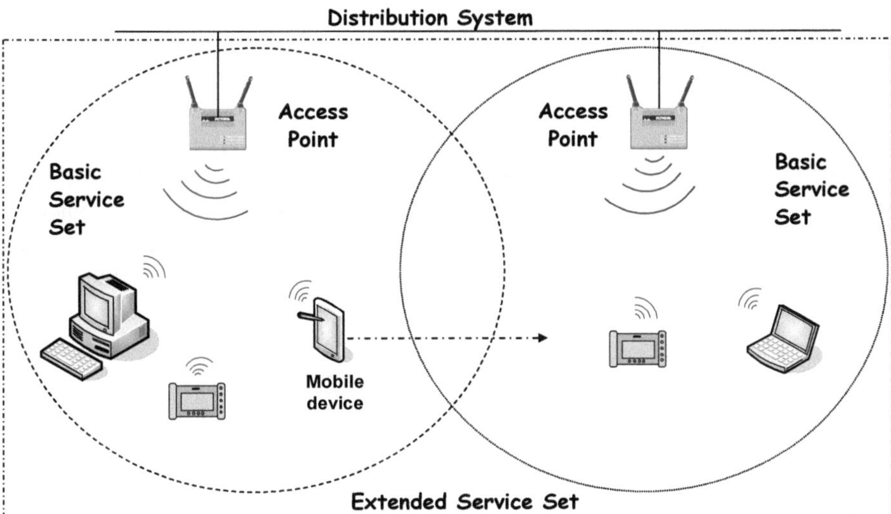

Figure 10.2 Infrastructure mode

The *ad hoc* mode is the most basic wireless topology. It does not utilise access points and the set of stations that have recognised each other are connected *via* wireless media in a peer-to-peer fashion (Figure 10.3). All stations need to be within range of each other to communicate directly. This form of network topology is referred to as an independent basic service set (IBSS). When communicating nodes belong to different IBSS, communications are possible through multiple nodes (multi-hop *ad hoc* network) and imply routing protocols. Consequently, it is more difficult to control network end-to-end delay with this mode.

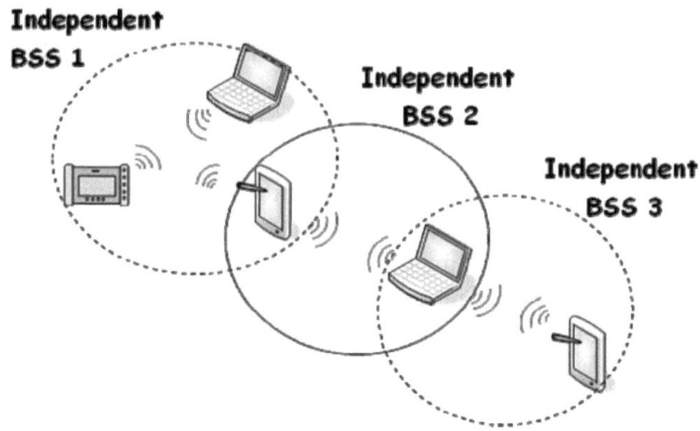

Figure 10.3 *ad hoc* mode

PHY Layer

The 802.11 MAC layer can be implemented on a variety of underlying physical layers. The transmission delay is dependent on the PHY specification. Two PHY layers are described below. The first one is the high rate/direct sequence spread spectrum (HR/DSSS) PHY. It is the enhanced physical layer defined by 802.11b, which supports data transfer at up to 11 Mbps. The HR/DSSS PHY specified in IEEE 802.11b standard (IEEE802.11b 1999) uses the 2.4 GHz frequency band as the RF transmission media. The physical protocol data unit (PPDU) format is depicted in Figure 10.4. PPDU consists of a PHY preamble (used for bit synchronisation purposes), a PHY header and a data part, which contains the MAC protocol data unit (MPDU), including the MAC header, the data payload and the frame checksum.

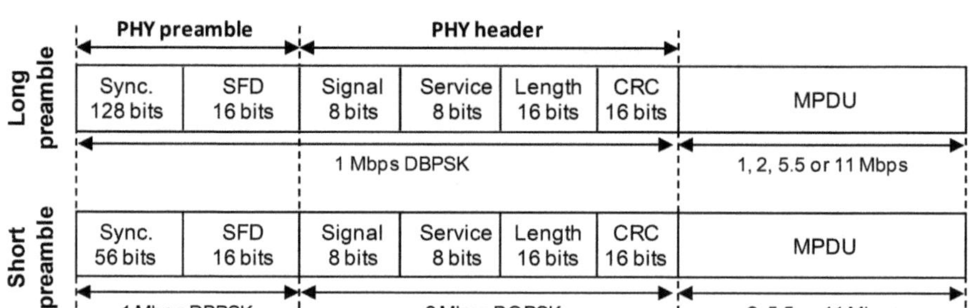

Figure 10.4 PPDU format with HR/DSSS PHY

The orthogonal frequency division multiplexing (OFDM) PHY supports data rates at up 54 Mbps. The OFDM modulation is used with 802.11g (IEEE802.11g 2003) and 802.11a (IEEE802.11a 1999) technologies. It also supports complementary code keying (CCK) modulation for backward compatibility with 802.11b.

OFDM PHY first divides a high-speed binary signal to be transmitted into a number of lower data rate sub carriers. There are 52 subcarriers, of which 48 subcarriers carry actual data and 4 subcarriers are pilots that facilitate phase tracking for coherent demodulation. Each lower data rate bit stream is used to modulate a separate subcarrier from one of the channels in the 2.4 GHz band. OFDM PHY supports eight different data rates, 6, 9, 12, 18, 24, 36, 48 and 54 Mbps. Forward error correction is performed by bit interleaving and convolutional coding; the coding rate depends on the transmission data rate selected. PPDU consists of a PHY preamble, a signal field and a PHY data part, as shown in Figure 10.5.

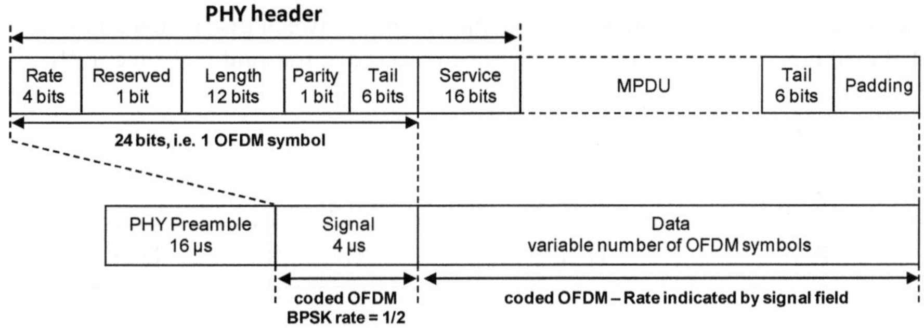

Figure 10.5 PPDU format with OFDM PHY

From the modulation and the data rate of the PHY layer, the transmission range can be calculated. Table 10.1 shows some typical indoor and outdoor ranges according to the used mode.

Table 10.1 Transmission distance for 802.11 PHY mode

Mode	Outdoor range		Indoor range	
	Metres	Feet	Metres	Feet
1 Mbps DSSS	550	1804	50	164
2 Mbps DSSS	388	1275	40	133
5.5 Mbps CCK	235	769	30	98
11 Mbps CCK	166	544	24	79
6 Mbps OFDM	300	984	35	114
12 Mbps OFDM	211	693	28	92
18 Mbps OFDM	155	508	23	76
24 Mbps OFDM	103	339	18	60
36 Mbps OFDM	72	237	15	48
48 Mbps OFDM	45	146	11	36
54 Mbps OFDM	36	119	10	32

802.11 MAC Layer

The medium access control (MAC) layer provides and maintains communications between 802.11 nodes (stations and access points) by coordinating access to a shared radio channel, which is the wireless medium. Two mechanisms to access the medium have been defined: the distributed coordination function mode (DCF) and the point coordination function mode (PCF). Both PCF and DCF alternate within a contention free period repetition interval named Superframe, as represented in Figure 10.6. The duration of the Superframe and the PCF period are manageable parameters, CFPrate and CFPMaxDuration, respectively, which are maintained by the AP. Both determine the balance between distributed and scheduled medium access and influence the time reserved to transmit real-time traffic.

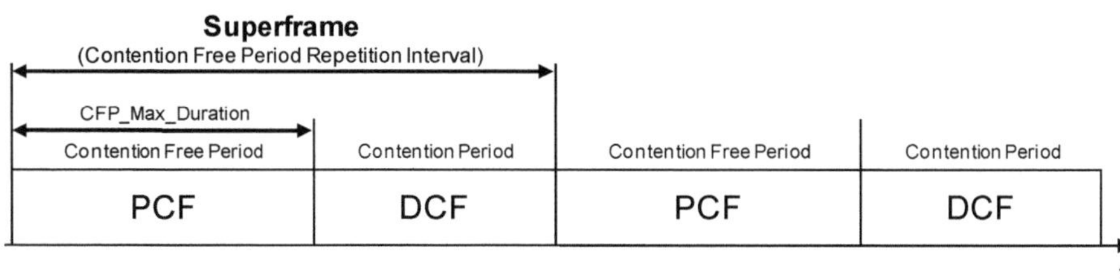

Figure 10.6 PCF and DCF modes alternation

The DCF mode is the mandatory mode. In this mode, a node must first gain access to the medium before transmitting frames. To achieve that, a basic access mechanism is employed, the so-called CSMA/CA protocol. It uses the 802.11 physical (PHY) layer to perform the task of carrier sensing. So, each node must check that the channel is idle for some minimum amount of time, before initiating a transmission. This amount of time is called the distributed inter-frame spacing (DIFS) time. Since multiple nodes may be accessing the medium at the same time, collisions can occur. A back off timer is used to reduce the collision probability. A node computes a random back off time, which is an additional interval beyond the DIFS time. The node that wishes to transmit must verify that the medium is still idle after the elapsed time. To provide reliable data services the 802.11 standard defines an explicit acknowledgment to inform the source node of the outcome of the previous transmission. If the receiving node detects no errors in the received frame, a positive acknowledgment (ACK) must be send to the source after a short inter-frame spacing (SIFS) time. The SIFS time not as long as the DIFS time so that the receiving node is given priority over other stations that are attempting to get transmission opportunities. Otherwise, the source node will assume that a collision (or loss) has occurred and will retransmit the frame. To reduce collisions, the standard also encompasses an optional RTS/CTS reservation mechanism, which implies short control frames exchanges prior to data transmission.

The second mode, called PCF, is an optional mechanism that provides asynchronous, time-bounded and contention free access control. The PCF mode relies on a centralised access control that requires the presence of a node that acts as point coordinator. Thus, PCF is used only in infrastructure mode where the AP operates as point coordinator. PCF is built over the DCF mode and provides a contention free period in which nodes may have contention-free access to the channel. A polling-based access mechanism is used with a round-robin scheduler. The point coordinator polls individual stations to transmit without contending for the channel. At the nominal beginning of each contention free period, the AP senses the medium. If the medium is idle after a priority inter-frame spacing (PIFS) time, a beacon frame including a delivery traffic indication message is sent. The PIFS time is shorter than DIFS to ensure that PCF is given priority over DCF frame transmission. However, the beacon frame that signals the beginning of the contention free period may be delayed due to a busy medium. This problem is well known under foreshortened CFP delay (Krommenacker and Lecuire 2005).

A typical sequence of frames during PCF is shown in Figure 10.7. The AP polls each station, according to a polling list, by sending a CFPoll frame. The list of all pollable stations is built on the CF-pollable subfield using by stations at the time of the association and reassociation request frames. A polled station may only transmit a data frame without contending for the channel. Stations always respond to a poll. If there is no pending transmission, the response is a null frame containing no payload. In the case of an unsuccessful transmission, the station retransmits the frame after being repolled or during the next contention period. A CFACK frame is used to acknowledge receipt of the previous data frame. Note that for efficient medium use it is possible to piggyback both the acknowledgement and the CFPoll onto data frames. All stations will hear the message due to the shared medium, so the acknowledgment is not necessarily for the polled station. This continues for CFPMaxDuration. The AP explicitly terminates the contention free period by transmitting a contention free (CFEnd) frame. A new DCF period starts. If the polling list is empty before the CFPMaxDuration, the AP may shorten the period in order to provide the remainder of the repetition interval for the DCF mode.

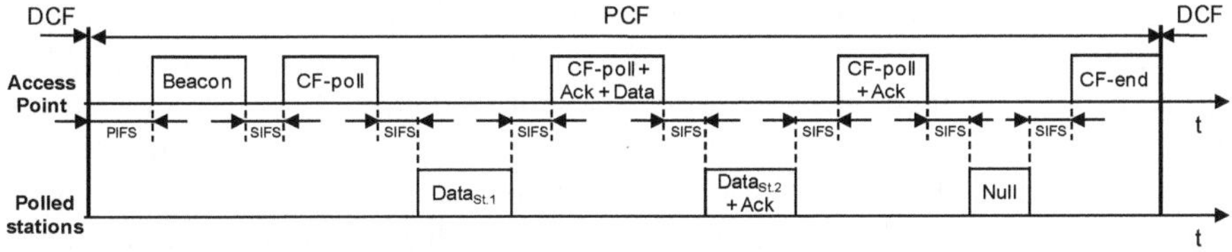

Figure 10.7 PCF frame transfer

Others communications may be supported during the contention free period. A polled station can send data either to a pollable station or a non-pollable station that does not appear in the polling list. If the data message received by AP is destined to a pollable station, it sends a piggybacked Data+CFPoll frame to the destination station at the next round. Otherwise, it will send a data message to the non-pollable station before continue its normal operations.

10.2.2 Bluetooth (IEEE 802.15.1)

The Bluetooth standard is an open, global standard defining the complete system from the radio to the application. The description of the different layers in Bluetooth stack can be found in the standard (IEEE 802.15.1 2002).

Network Topology

All devices can be easily interconnected to coordinate and exchange information using an infrastructure-less short-range wireless connection. A set of Bluetooth devices sharing a common channel is called a piconet. As shown on the left side of Figure 10.8, a piconet is a star-shaped configuration in which one of the devices performs the role of master and all other devices operate as slaves. Up to seven slaves can be active and served simultaneously by the master. Additionally, slaves can stay synchronised with the master in a parked state. They are not active on the channel but can become active without using the connection establishment procedure. The master controls the channel access, for all slaves, active and parked. Since most envisioned Bluetooth applications involve local communication among small groups of devices, a piconet configuration is ideally suited to meet the communication needs of such applications (Bhagwat 2001).

When many groups of devices need to be active simultaneously, each group can form a separate piconet. The slave nodes in each piconet stay synchronised with the master clock and hop according to a channel-hopping sequence that is a function of the master's node address. Since channel-hopping sequences are pseudo-random, the probability of collision among piconets is small. Piconets with overlapping coverage can coexist and operate independently. Nonetheless, when the degree of overlap is high, the performance of each piconet starts to degrade.

In some usage scenarios, devices in different piconets may need to communicate with each other. Bluetooth defines a structure called scatternet to facilitate inter-piconet communication. A scatternet is formed by interconnecting multiple piconets. As shown on the right side of Figure 10.8, the connections are formed by bridge nodes, which are members of two or more piconets. A bridge node participates in each member piconet on a time-sharing basis. After staying in a piconet for some time, the bridge can turn to another piconet by switching to its hopping sequence. By cycling through all member piconets, the bridge node can send and receive packets in each piconet and also forward packets from one piconet to another. A bridge node can be a slave in both piconets or be a slave in one and a master in another.

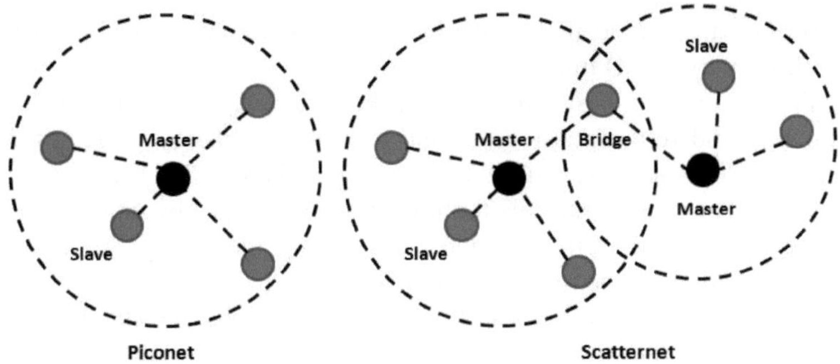

Figure 10.8 Piconet and scatternet topologies

802.15.1 PHY Layer

The Bluetooth transceiver operates in the 2.4-GHz ISM band, which is globally available for license-free use. According to country regulations, 79 channels spaced 1 MHz apart are defined for Europe and the U.S. and 23 RF channels spaced 1 MHz apart are defined for Spain, France, and Japan. Bluetooth is a frequency-hopping spread-spectrum system to combat interference and fading. This means that the radio hops through the full spectrum of 79 or 23 RF channels using a pseudo-random hopping sequence. The hopping rate of 1,600 hops *per* second provides good immunity against other sources of interference in the 2.4-GHz band. The basic data rate is 1 Mbps, and a theoretical 720 kbps payload, which is achieved using a simple modulation technique (Gaussian frequency shift keying, or GFSK). A more complex modulation technique (differential phase shift keying, or DPSK) can be optionally supported to achieve higher data rates, 2 and 3 Mbps, but GFSK keeps the radio design simple and low cost.

802.15.1 MAC Layer

The piconet channel is divided into 625 μs intervals, called slots, where a different hop frequency is used for each slot. The channel is shared between the master and the slave nodes using a frequency-hop/time-division-duplex (FH/TDD) scheme whereby master–slave and slave–master communications take turns. Slave-to-slave communication is not supported at the piconet layer. If two slaves need to communicate peer-to-peer, they can either form a separate piconet or use a higher layer protocol to relay the messages *via* the master. There are two types of connections that can be established between a master and a slave: the synchronous connection-oriented (SCO) link and the asynchronous connectionless (ACL) link.

SCO links provide a circuit-oriented service with constant bandwidth based on a fixed and periodic allocation of slots. SCO links have been designed to support time-bounded information like voice. They do not necessary provide reliable data transfer since data are unacknowledged. In order to ensure that SCO links due not reserve too much of the available bandwidth, a Bluetooth master is restricted to a total of three active SCO links.

ACL links provide a best-effort and packet-oriented service using a fast acknowledgment and retransmission scheme to ensure reliable transfer of data. In addition, ACL links support multi-cast messages whereas SCO links are strictly point-to-point. The master controls the traffic on ACL links by employing a polling scheme to divide the piconet bandwidth among the slaves. A slave is only allowed to transmit after the master has polled it.

In order to provide flexibility for Bluetooth applications, the standard specifies a variety of packet types with different combinations of payload length, slots occupied, FEC codes and ARQ options. An adaptive packet selection can be chosen to be optimised for the traffic and channels presented.

Bluetooth provides four types of packets that can be sent over an SCO link: HV1, HV2, HV3 and DV packets (HV stands for high-quality voice). The following Table 10.2 summarises the different configurations for these packets.

Table 10.2 Bluetooth SCO packet types

Packet types	HV1	HV2	HV3	DV
Packet header (bytes)	None	None	None	1D
Payload length (bytes)	10	20	30	20
Channel utilisation (%)	100%	50%	33%	100%
FEC code	1/3	2/3	None	2/3D

Note: items followed by D relate to data field only.

- *The HV1 packet* carries 10 information bytes, which are protected with a rate 1/3 FEC. This packet has to be sent every two time slots, *i.e.* every 1.25 ms. The HV1 SCO connection enables a master–slave communication at 64 kbps rate using all the Bluetooth capacity. As a result, only one slave can be active in the piconet.
- *The HV2 packet* carries 20 information bytes, which are protected with a rate 2/3 FEC. This packet has to be sent every four time slots, *i.e.* every 2.5 ms. The HV2 SCO connection enables a master–slave communication at 64 kbps rate using 50% of the Bluetooth capacity. The bandwidth available can be used to enable a second HV2 SCO connection or to enable ACL traffic.
- *The HV3 packet* carries 30 information bytes, which are not protected. This packet has to be sent every six time slots, *i.e.* every 3.75 ms. The HV3 SCO connection enables a master–slave communication at 64 kbps rate using 33% of the Bluetooth capacity. The bandwidth available can be used to enable up to two HV3 SCO connections or to enable ACL traffic.
- *The DV packet* is a combined data-voice packet. The payload is divided into a voice field of 80 bits and a data field of up to 150 bits.

The ACL link, in addition to the use of different FEC protections, adds the choice of multi-slot packets. Table 10.3 summarises the packets for the ACL link.

Table 10.3 Bluetooth ACL packet types

Packet types	DH1	DM1	DH3	DM3	DH5	DM5
Slot time	1	1	3	3	5	5
Packet header (bytes)	1	1	2	2	2	2
Payload length (bytes)	0–27	0–17	0–183	0–121	0–224	0–339
FEC code	None	2/3	None	2/3	None	2/3

A slave can send an ACL packet if it has been addressed by the master in the previous slot. To ensure data integrity, ACL packets are retransmitted. Only a single ACL link can exist between a master and a slave. The master schedules ACL packets in the slots are not reserved for the SCO links. The specifications define seven kinds of ACL packets, three DM (data-medium rate) packets, three DH (data-high rate) packets and one AUX packet.

- *DM packets* are coded with a rate 2/3 FEC; they contain a 16-bit CRC code and are retransmitted if no acknowledgement is received. Three DM packets have been defined, DM1, DM3 and DM5, which cover 1, 3 and 5 time-slots, respectively.
- *DH packets* are similar to the DM packets, except that the information in the payload is not FEC encoded. Similar to the DM packets, three DH packets (DH1, DH3 and DH5) have also been defined.
- *The AUX packet* is like a DH1 packet, but has no CRC code and is not retransmitted.

10.2.3 ZigBee (IEEE 802.15.4)

ZigBee is the name of a specification for a suite of high level communication protocols using small, low-power digital radios based on the IEEE 802.15.4 standard for wireless personal area networks (WPANs). While use of the name "ZigBee", relating to the ZigBee alliance (http://www.zigbee.org/), and IEEE 802.15.4 occasionally appear to be almost synonymous, a clear distinction should be made. The IEEE 802.15.4 is strictly a specification of a new low-power air interface and the accompanying MAC protocol (IEEE802.15.4 2006), while the scope of the ZigBee consortium reaches higher layers of the stack as well.

Network Topology

IEEE 802.15.4 standard defines two device classes named as FFD (fully functional devices) and RFD (reduced functional devices). An FFD is supposed to have very powerful processing ability. FFDs can act as PAN (personal area network) coordinator and can talk to any other device in the network in theirs vicinity. An RFD is supposed to be a device with basic processing power. RFD's cannot act as coordinators and can talk to (associate) only one FFD at a time.

PAN based on 802.15.4 supports peer-to-peer as well as star network topology (Figure 10.9). In a star topology, one of the FFD-type devices assumes the role of network coordinator and is responsible for initiating and maintaining the devices on the network. All other devices, known as end devices, directly communicate with the coordinator. Star topology is the simplest form of a WPAN, this is similar to Infrastructure BSS in 802.11.

Another topology is a peer-to-peer PAN. This is similar to *ad hoc* BSS in 802.11. Similarly piconet and scatternet exist in Bluetooth. In a peer-to-peer topology, P2P cluster trees can be formed using coordinators as cluster heads.

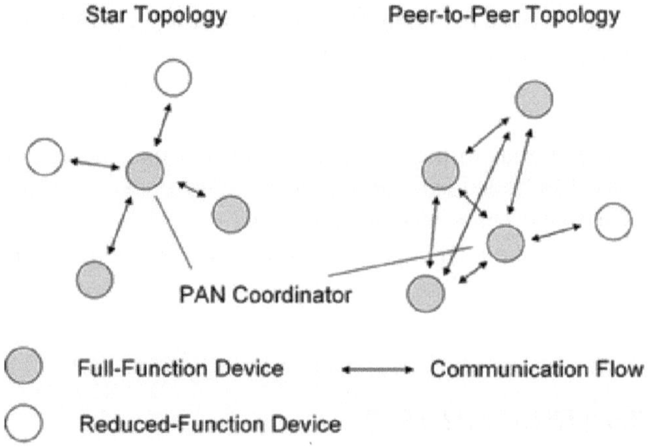

Figure 10.9 Topologies supported by IEEE 802.15.4

802.15.4 PHY Layer

The physical layer (PHY) specifies operation in the unlicensed 2.4 GHz, 915 MHz and 868 MHz ISM bands. In the 2.4 GHz band there are 16 ZigBee channels, with each channel requiring 3 MHz of bandwidth. The centre frequency for each channel can be calculated as FC = (2400 + 5*k) MHz, where k = 1, 2, ..., 16.

The radios use direct-sequence spread spectrum coding, which is managed by the digital stream into the modulator. BPSK is used in the 868 and 915 MHz bands, and orthogonal QPSK that transmits 2 bits *per* symbol is used in the 2.4 GHz band. The raw, over-the-air data rate is 250 kbit/s *per* channel in the 2.4 GHz band, 40 kbit/s *per* channel in the 915 MHz band, and 20 kbit/s in the 868 MHz band. Transmission range is between 10 and 75 m (33 ~ 246 ft), although it is heavily dependent on the particular environment. The maximum output power of the radios is generally 0 dBm (1 mW). Physical layer specifications are summarised in Table 10.4.

Table 10.4 Physical layer specification

	868 MHz	915 MHz	2450 MHz
Data rate	20 kbps	40 kbps	250 kbps
Modulation	BPSK	BPSK	O-QPSK
Number of channels	1	10	16

802.15.4 MAC Layer

Two kinds of networks (PAN), namely beacon-enabled and non-beacon-enabled, are specified by the standard. The PAN coordinator decides the network type. In non-beacon mode, the basic channel access mode is "carrier sense, multiple access/collision avoidance", also called unslotted CSMA/CA.

In the case of the beacon-enabled network, the PAN coordinator sends periodic beacons. This mode introduces the superframe structure (Figure 10.10) to divide time into different transmission periods: beacon (for synchronisation purposes), contention access period (CAP), contention-free period (CFP) and an inactive period, where the PAN coordinator may go to sleep and save energy.

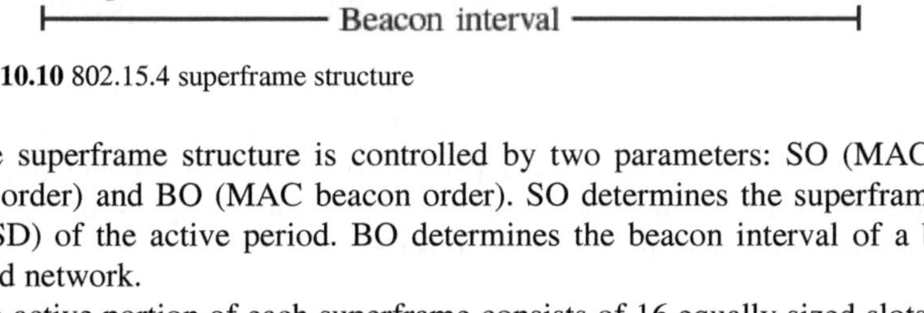

Figure 10.10 802.15.4 superframe structure

The superframe structure is controlled by two parameters: SO (MAC superframe order) and BO (MAC beacon order). SO determines the superframe duration (SD) of the active period. BO determines the beacon interval of a beacon-enabled network.

The active portion of each superframe consists of 16 equally sized slots, which may be used by CAP or CFP periods. CAP starts immediately after the beacon. During CAP, any device wishing to communicate competes with other devices using a slotted CSMA/CA mechanism. CFP, if present, is divided into guaranteed time slots (GTS) assigned and administered by the PAN coordinator to allow communication between a device and the coordinator within a dedicated portion of time. So, only registered devices are allowed to transmit packets into the CFP period. No more than 7 GTSs can be simultaneously allocated in the superframe. The GTS direction is specified as either transmit or receive. Each device may request one transmit GTS and/or one receive GTS.

10.2.4 Assessment of Previous Technologies to Support E-maintenance Applications

In this section, we present some comparison elements of previous wireless technologies, based on new challenges emerging from the use of wireless solutions in industrial applications and presented at the beginning of this chapter (Section 10.1).

Power Supply

A critical factor in the 802.11 Wireless LAN technologies is power consumption. To sustain network connectivity, 802.11-enabled devices must have power to listen for traffic, including beacons, periodically. Also, power is necessary in an active mode to transmit and receive data. In theory, lower throughput (802.11b *vs.* 802.11a/g) would result in lower battery power consumption. However, the length of time needed to transmit/receive a meaningful amount of application data can be five times longer on an 802.11b network than on an 802.11a/g WLAN (Texas Instruments 2003, Atheros 2003).

Solutions exist for better power efficiency, such as power saving mode. Moreover, this mechanism is used to reduce energy waste at the 802.15.4 MAC layer. So, ZigBee is fundamentally efficient in terms of battery performance. Battery lifetimes from a few months to many years are feasible using power-saving modes and battery-optimised network parameters. However, the penalty for saving power *via* sleep mode is greater latency on the delivery of new incoming packets.

Bluetooth was originally designed for cable replacement purposes in communication between computer equipment, mobile telephones and peripherals. The typical operational range is approximately 1 m. Under these conditions of use, the energy consumption is very low in comparison to 802.11 technologies (typically, 45 mA *vs.* 300 mA). An increase in the operational range implies an increase in the output power and, thus, more energy consumption.

Interference Immunity

Interference due to the use of multiple devices, from one technology, inside the same area are mostly reduced by the implemented mechanisms at the PHY and MAC layers. For example, frequency hopping (using Bluetooth) or CSMA/CA protocol reduce the probability of interference with other users.

In addition, interference within the same frequency band is always a critical factor of wireless solutions. Because IEEE Std 802.11, IEEE Std 802.15.1 and IEEE 802.15.4 specify operations in the same 2.4 GHz unlicensed frequency band, there is mutual interference between wireless systems, which may result in severe performance degradation. However, it means that both technologies are used in the same area. This fact is dependent on the policy of the company prescribing clear rules in the industrial area. Moreover, some coexistence mechanisms (adaptive frequency, packet scheduling, *etc.*) and recommended practice exist (802.15.2, 2003).

This mostly reduces the question of interference to that of unwanted radiation. Industrial environments are radio-hostile with typical areas (dust, extreme temperature, vibration, steel constructions, *etc.*) and strong electromagnetic fields near machinery. Very limited information is available on the electromagnetic ambient levels in industrial environments but some measurements of radio wave propagation have been undertaken (Kjesbu 2001, Catrysse *et al.* 2005). Table 10.5 illustrates some of the results obtained and the fact that that there was no measurable effect in the 2.4 GHz band.

Table 10.5 Measurements on interference of nearby machinery

Type of machinery	Freq. band (MHz)	Max. emission level (dBμV/m)
Computer	1–150	138
Frequency converter	1–200	170
Electrical motor	1–200	167
Welding machine	1–50	140
CNC center	1–400	169
Weaving machine	1–2000	156
Press brake	1–1600	158
Punch press	1–1600	169
Laser cutting machine	1–1700	162

Data Security

IEEE 802.11 initially included an optional cryptographic confidentiality algorithm, the wired equivalent policy (WEP). However, several serious weaknesses were identified in 2001 (Borisov *et al.* 2001). The IEEE 802.11i amendment (also known as WPA2) was released in 2004 and should definitely be considered for each application of 802.11.

Bluetooth can operate in two security models: at the service level (after the channel has been established) or at the link level (before the channel has been established). The latter is maintained using a unique 48-bit Bluetooth device address, two secret keys and a pseudo-random number regenerated for each transaction. The two secret keys are used for authentication (128 bits) and encryption (8–128 bits). The authentication key, often called link key, is shared between two or more devices and is the basis for all security transactions between them. It is used in the authentication routine and as one of the parameters when the encryption key is derived. A third mode exists but does not provide any security mechanism.

ZigBee provides security based on three main principles (Peirera 2004). First is simplicity, where every layer originating a frame is responsible for securing it, rather than having multiple layers do so. Second is directness. Keys are exchanged directly between each source and destination device. Third is end-to-end security, where data proceeds without having to be decrypted and re-encrypted at each hop.

Throughput and Strict Delay Requirements

Although IEEE 802.11, 802.15.1 and 802.15.4 did not originally target industrial applications, all have mechanisms to support applications with guaranteed response time requirements. The polling mechanism (PCF mode) specified in 802.11 standards could achieve the time-critical requirements of industrial communication systems. This mechanism is similar to those used by many current field buses and could serve time-triggered applications. Bluetooth SCO links, which are the native mechanism for carrying two-way symmetric voice transmissions at 64 kbps rate, enable typically periodic-oriented data communication. Time-constrained traffic can be transmitted when the beacon-enabled mode is used in Zigbee. This mode divides the channel into 16 time slots included slot reservation mechanism (GTS) to guarantee the use of the transmission channel. However, inadequacies have been identified and mentioned in several studies.

In practice PCF functionality is not implemented in commercially available IEEE 802.11a/b/g stations, and some shortcomings of this mode exist: beacon delay and shortened contention free period, unpredicted time for polled stations (Krommenacker and Lecuire 2005). It is necessary to develop this mode to achieve real-time services. One example of such a product is the Siemens SCLANCE-W industrial device (Siemens 2005), which is an IEEE 802.11-compliant device coupled with a dedicated transfer mode for transferring time-critical data (Figure 10.11).

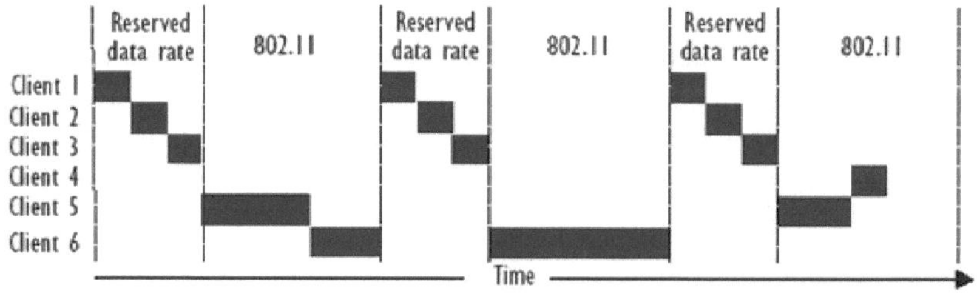

Figure 10.11 Dedicating data transfer rate with Scalance-W product

Note that the new IEEE 802.11e amendment introduced a new hybrid contention free mode that seems really beneficial when considering a wireless 802.11 field bus (Karanam *et al.* 2006).

The protocol efficiency, and consequently the throughput, is also limited in Bluetooth (Lobello *et al*, 2005) and Zigbee (Salles *et al.* 2008), mainly due to number of communications during each cycle (3 and 7, respectively for 802.15.1 and 802.15.4) and the problem of the minimum cycle time (dependent on the access protocol and configuration parameters).

Operational Range

802.11 technologies provide a typical indoor range up to 100 m with low data rate (1 Mbit/s). With high-gain external antennas and line of sight conditions, a highest operational range can be achieved and used in fixed point-to-point arrangements. In 2004, the IEEE 802.11n task group was formed to develop a new amendment to the 802.11 standard to increase data throughput and range. Today, the standard has not yet been approved, but solutions with multiple transmitter and receiver antennas (MIMO or multiple input–multiple output) exist and exploit spatial multiplexing and spatial diversity to do this.

Unlike 802.11, Bluetooth and Zigbee have been specified for WPANs, which are used to convey information over short distances among a private, intimate group of participant devices. The typical range for a low power radio transmitter is 10 meters. Under specific conditions (maximum output power, free space propagation *etc.*), a higher distance can be reached, up to 70 meters.

Cost

IEEE 802.11b/g standards are today the most widely used standards in both home and enterprise markets. As the Ethernet *de-facto* standard, the exponential increase in unit shipments support the exponential decrease in unit price. This advantage leads to the low-cost and stable supply of WLAN components.

The chip cost for Bluetooth or Zigbee was above $5 at the beginning. Today solutions down to $3 exist, and should go down to $1 *per* unit.

Support for Large and Varying Numbers of Devices

If the addressing scheme allows supporting a large number of devices in the networks, in practice, some limitations exist. For instance, 802.11 b/g networks support up to 32 users (64 for 802.11a) *per* access point. This limitation is due to strong performance degradation when the number of users is higher.

With Bluetooth and Zigbee technologies, the number of devices *per* piconet is limited to 8 and 256, respectively. To support more stations, inter-piconet communications must be established, accompanied by greater latency on the delivery of messages.

10.2.5 Conclusions

Industrial applications have different wireless requirements (power consumption, transparent data, time stamping, reliability *etc.*). The previous state-of-the-art and assessment demonstrates that:

- No technology is efficient enough to support the entire functional and technological requirements due to the diversity of applications and environments.
- For a particular application, the most efficient technology at a specific level (*i.e.*, field component) is perhaps not adapted to support at another level. For instance, 802.15.4 could be adapted for data collection at the field level, but is not suitable for file transfer at the factory level.

Furthermore, devices to integrate inside the communication system are often heterogeneous and it seems hazardous to choose one technology. All of these previous observations led to the investigation and development of the concept of wireless gateway connecting different communication areas, in which most efficient wired and wireless technologies are implemented. A recent study demonstrates technical and economical viability of wireless gateways in industrial networks (Resende *et al.* 2008). Development with respect to e-maintenance applications in particular are presented in the next section.

10.3 New Developments

E-maintenance applications involve many actors such as human actors (maintenance crew, maintenance expert), information systems (ERP, MES, *etc.*), sensors (lubrication, vibration, *etc.*) or MEMS, smart tags *etc.*, with various communication requirements, and situated at different levels of the enterprise (Figure 10.12).

The industrial ethernet paradigm mentioned in Section 10.1, changed the automation pyramid to facilitate the integration of the upper levels of the factory with process control. As shown Figure 10.12, classical e-maintenance architecture can be divided into two zones. At the decisional level, wired ethernet and WiFi communications are often used. In the operational zone, short-range communications are needed to ensure communication between mobile maintenance operators and field devices. The integration of a wireless PAN solution at this level implies:

- the development of wireless gateway to allow the interoperability of network technologies and the communication through areas; and
- the development of wireless collector providing wireless features for specific field level components (sensors, MEMS, *etc.*).

This is illustrated in the following through the development of two prototypes by the electronic communication systems manufacturer Prisma Electronics.

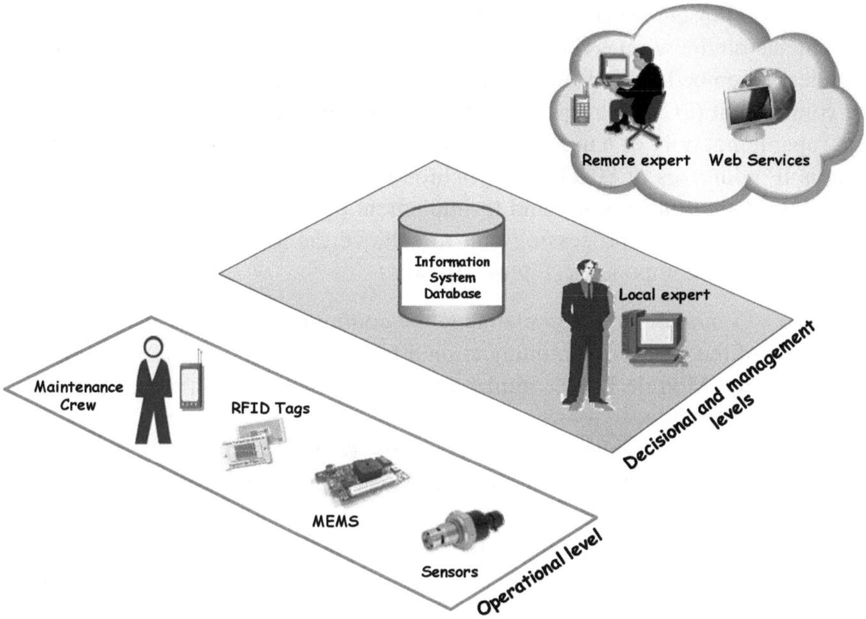

Figure 10.12 E-maintenance application architecture

10.3.1 Wireless Gateway

The Prisma Sense Gateway (Figure 10.13) is a wireless gateway developed by Prisma Electronics. The gateway acts as a hardware/software interface network, converting the ZigBee protocol to WiFi and thus enabling data transfer from operational level to decisional level and *vice versa*. This product also enables Zigbee to wired ethernet 10/100 or Zigbee to RS232/485 communications.

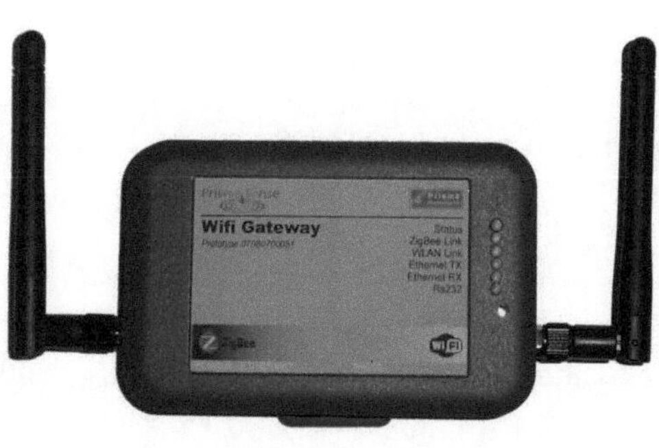

Figure 10.13 Prisma sense wireless gateway

Wireless chipsets are off-the-shelf components in order to follow the latest developments on the wireless communications. Lantronix and MAXStream XbeePro modules are embedded for WiFi and Zigbee transceivers, respectively. A processing control unit (PCU) is responsible for the implementation of the gateway logic. It is capable not only of handling packets, but also of controlling packet routing to various IP addresses of 802.11, to synchronise the sensors clock, to guarantee the reliability of the transmission and to implement power management mechanisms. The PCU comprises a processing unit to achieve the above service, with the following software components (Figure 10.14):

- The Wi-Fi module driver implements the communication of processor with the module of the wireless communication 802.11.
- The Zigbee Module Driver implements the communication of processor with the Zigbee module.
- Package processing manages all the packets received from both networks.
- Real time clock synchronisation synchronises the real time clocks of sensors that belong to Zigbee network.
- The frequency agility implements the frequency agility that was described earlier.
- The package buffer temporarily saves all the packets that will be transmitted, until their acknowledgement.
- The routing table is the address table used for the routing of packets.
- Parameters and options include all the parameters that determine the operation of the device, as well as, the programming routines and their updating.

The PCU module has been implemented using a TI processor. The software development was done using C++. The subroutine of power supply, the interfaces and the various protections EMC and EMI are designed in order to fulfil all specifications for the electronic industrial products of automatisms.

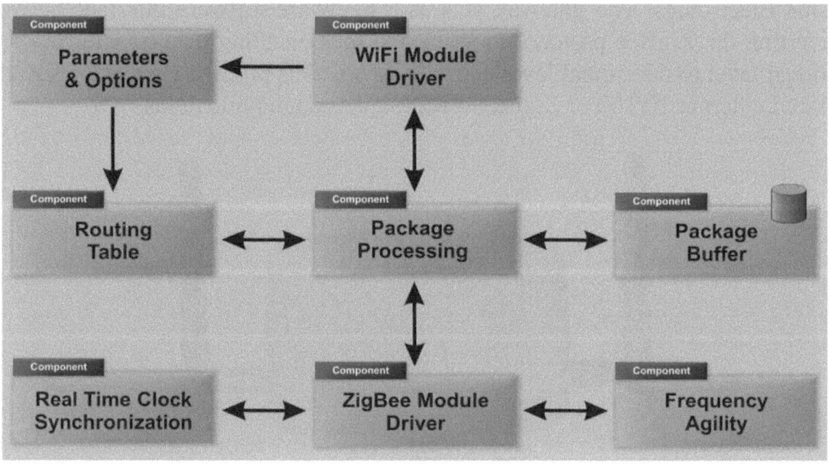

Figure 10.14 Components implemented inside PrismaSense Gateway

The gateway includes a novel operating system called the Intelligent Sensor Operating System (ISOS©). The architecture is the following (Figure 10.15):

- The task scheduler assigning task priorities and controlling their execution timing.
- The event handler receiving events and issues responses.
- The frequency agility enabling automated frequency band and channel change when "sniffing" the radio waves indicates that certain frequency bands and channels are occupied.
- The package buffer taking care of intermediate buffering of packages prior to execution.
- The real time clock enabling synchronisation of nodes and tasks.
- Measurement simulation allowing for the simulation of sensing nodes operation.
- Safe power down modes that control power operation.

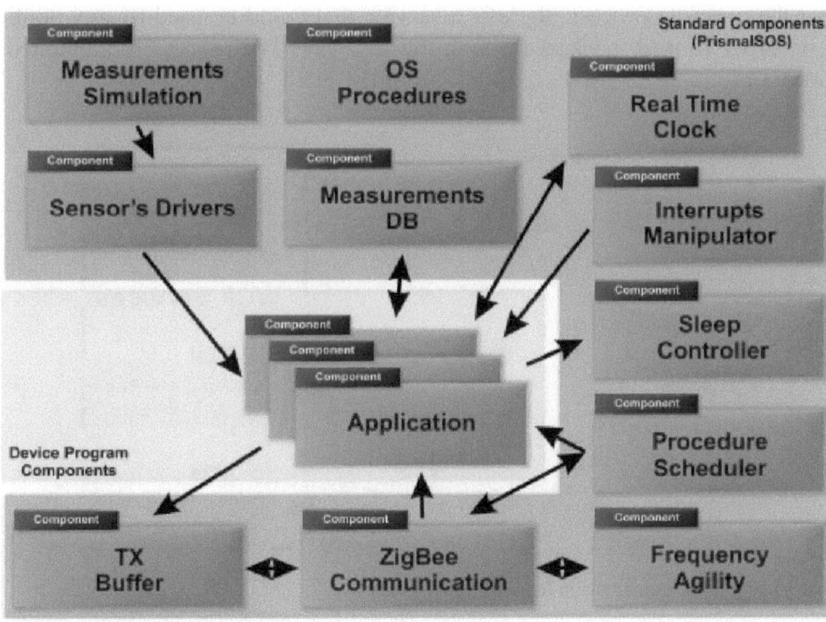

Figure 10.15 ISOS© architecture

For the gateway communication services with service side applications, a "listener" is implemented that operates on a specific gateway. Once the listener receives packets, it generates an event. Any application can use the particular service to do event handling and process the incoming packets (Figure 10.16).

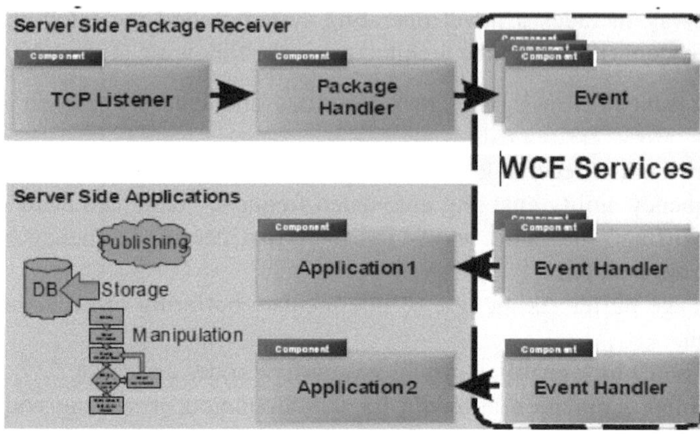

Figure 10.16 Packet reception from gateway

For the opposite process, there is a service that could be used by any application, in order to push packets through the gateway. The process is illustrated in Figure 10.17.

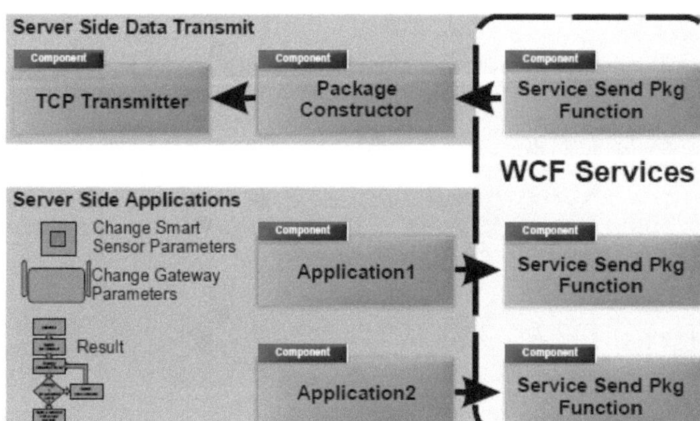

Figure 10.17 Packet transmission from server to gateway

10.3.2 Wireless Collector

Collectors can collect various physical values (temperature, lubrication, vibration *etc.*) from their close environment, depending on the number and type of sensors that they host. Figure 10.18 illustrates the collector with various hardware interfaces to connect different sensor types.

Figure 10.18 Collector hardware interfaces

Measuring values are collected by the collector and transmitted to the wireless gateway, which can retransmit all the received data to a TCP/IP network. The main characteristics of the wireless collector based on real time ISOS© are:

- minimum power consumption;
- maximum network reliability;
- minimum time and effort during collector programming.

10.4 Conclusions and Recommendations

As well as industrial applications using wireless networks, e-maintenance requires specific features: simple wireless connectivity, relaxed throughput, very low power consumption, short and long-range communication and inexpensive solutions. However, no wireless solutions were originally designed for this, and several technologies must be used to support such applications. To ensure interoperability and communications, wireless gateway and collector concepts can be used. Gateway offers data interfaces for wired/wireless and wireless/wireless communications. A collector plays the role of simple wireless interface to retrieve measuring values from sensors, before transmitting them to the network. These new developments can take advantage of wireless technology for e-maintenance applications where mobility is crucial.

References

Atheros (2003) Power consumption and energy efficiency comparisons of WLAN Products. White paper. Available at: http://atheros.com/pt/whitepapers/atheros_power_whitepaper.pdf

Bhagwat P (2001) Bluetooth: Technology for short-range wireless applications. IEEE Internet Computing 5:96–103

Borisov N, Goldberg I, Wagner D (2001) Intercepting mobile communications: the insecurity of 802.11. ACM Proceddings of the International Conference o Mobile Computing and Networking, Rome, Italy, 180–189

Catrysse J, Rayee J, Degrendele D (2005) Simulation of an industrial environment and overview of test results, COTS286 technical reports

Egea-Lopez E, Martinez-Sala A, Vales-Alonso J, Garcia-Haro J and Malgosa-Sanahuja J (2005) Wireless communications deployment in industry: a review of issues, options and technologies. Computers in Industry 56(1): 29–53.

IEEE Std 802.11a (1999) Wireless LAN medium access control (MAC) and physical layer (PHY) specifications: high-speed physical layer in the 5 GHz band. Institute of Electrical and Electronics Engineers. IEEE 802.11 WG

IEEE Std 802.11b (1999) Wireless LAN medium access control (MAC) and physical layer (PHY) specifications: higher-speed physical layer extension in the 2.4 GHz band. Institute of Electrical and Electronics Engineers. IEEE 802.11 WG

IEEE Std 802.11g (2003) Wireless LAN medium access control (MAC) and physical layer (PHY) specifications amendment 4: further higher data rate extension in the 2.4 GHz band. Institute of Electrical and Electronics Engineers. IEEE 802.11 WG

IEEE Std 802.15.1 (2002) Wireless medium access control (MAC) and physical layer (PHY) specifications for wireless personal area networks (WPANS). Institute of Electrical and Electronics Engineers. IEEE 802.15 WG

IEEE Std 802.15.2 (2003) Coexistence of wireless personal area networks with other wireless devices operating in unlicensed frequency bands. Institute of Electrical and Electronics Engineers. IEEE 802.15 WG

IEEE Std 802.15.4 (2006) Wireless medium access control (MAC) and physical layer (PHY) specifications for low-rate wireless personal area networks (WPANs). Institute of Electrical and Electronics Engineers. IEEE 802.15 WG

Karanam S P, Trsek H, Jasperneite J (2006) Potential of the HCCA scheme defined in IEEE802.11e for QoS enabled industrial wireless networks. 6th IEEE International Workshop on Factory Communication Systems, 227–230

Kjesbu S (2001) Radio wave propagation in industrial environments. Technical Reports, ABB Corporate Research

Krommenacker N, Lecuire V (2005) Building industrial communication systems based on IEEE 802.11g wireless technology. 10th IEEE International Conference on Emerging Technologies and Factory Automation, Catania, Italy

Lobello L, Collotta M, Mirabella O, Nolte T (2005) Approaches to support real-time traffic over Bluetooth networks. Proceedings of the 4th International Workshop on Real-Time Networks, Palma de Mallorca, Spain, 47–50

Pereira R (2004) ZigBee and ECC secure wireless networks. Electronic Design, http://electronicdesign.com/article/embedded/zigbee-and-ecc-secure-wireless-networks8369.aspxwww.elecdesign.com

Resende DS, Santos MM, Garzedin OS, Vasques F (2008) Technical and economical viability of wireless gateways in industrial networks. IEEE International Conference on Industrial Technology, Chengdu, China

Salles N, Krommenacker N, Lecuire V (2008) Performance study of IEEE 802.15.4 for industrial maintenance applications. IEEE International Conference on Industrial Technology, Chengdu, China

Siemens (2005) Industrial mobile communication. Scalance W, Siemens, http://www.automation.siemens.com/net/html_76/ftp/presales/SCALANCE_W_en-0406.pdf

Texas Instruments (2003) Low power advantage of 802.11a/g vs. 802.11b. White paper

Chapter 11
Semantic Web Services for Distributed Intelligence

Eduardo Gilabert and Alexandre Voisin

Abstract. This chapter presents a semantic web services platform ready to integrate intelligent processing capabilities according to open systems architecture for condition based maintenance (OSA-CBM) architecture. This platform is part of a flexible communication infrastructure, nicknamed DynaWeb, where a generic wireless device is also being developed between novel sensors, smart PDAs and existing maintenance systems of companies.

11.1 Introduction

The semantic web (SW) is a project that intends to create a universal medium for information exchange by giving meaning (semantics), in a manner understandable by machines, to the content of documents on the web. Under the direction of the web's creator, Tim Berners-Lee of the World Wide Web Consortium, the semantic web extends the World Wide Web through the use of standards, markup languages and related processing tools (Berners-Lee *et al.* 2001).

Currently web pages are designed to be read by humans, not machines. SW aims to make web pages understandable by computers, so that they can search websites and perform actions in a standardised way. The potential benefits are that computers can harness the enormous network of information and services on the web.

11.2 State-of-art in Application of the Semantic Web to Industrial Automation

A review of current knowledge automation status is given with respect to the semantic web, including current efforts in providing common reasoning mechanisms and current models on web semantics applied to the industrial world.

11.2.1 What Is Ontology?

In the context of knowledge sharing (Gruber 1992), the term ontology is used to mean a *specification of a conceptualisation*. That is, ontology is a description, like a formal specification of a program, of the concepts and relationships that can exist for an agent or a community of agents. This definition is consistent with the usage of ontology as set-of-concept-definitions, but more general. It is certainly a different sense of the word than its use in philosophy.

What is important is what ontology is *for*. Ontologies have been designed for the purpose of enabling knowledge sharing and reuse. In that context, ontology is a specification used for making ontological commitments.

For pragmatic reasons, people choose to write ontology as a set of definitions of formal vocabulary. Although this is not the only way to specify a conceptualisation, it has some nice properties for knowledge sharing among artificial intelligence (AI) software (*e.g.* semantics independent of reader and context). Practically, an ontological commitment is an agreement to use vocabulary (*i.e.*, ask queries and make assertions) in a way that is consistent (but not completely) with respect to the theory specified by an ontology. We build agents that commit to ontologies. We design ontologies so we can share knowledge with and among these agents.

11.2.2 Advantages of Semantic Web Techniques

SW offers many advantages over previous web technologies that make web applications more useful and independent from humans, but also, its use could increase companies' profits, since SW techniques improve the efficiency of these applications. The main advantages are described below (Carro-Martínez *et al.* 2005).

11.2.2.1 Improved Web Search

The first clear advantage of SW techniques is the potential of improving web search, since SW mechanisms allow for a better analysis of the items to be found. Nowadays, the keyword or word spotting search is based on true-false results. Ontology-based search uses the relationships and axioms of concepts; thus it can filter some seemingly appropriate but not desired results and add some seemingly different but actually same results. For instance, if you look for unleaded petrol, your search engine can expand the search to 95 or 98 octane unleaded petrol if both of them are derived from the concept 'unleaded petrol' in the associated ontology. Moreover, this SW-enabled search engines would not only be able to search for the object in the language of the query, but also in other languages because the system understands what the user wants.

In this way, SW technologies enable a search engine to better understand what you are looking for and increase the precision of the returned results. This is especially important in companies that have to deal with a large amount of data. SW-enabled search engines are also especially advantageous in intranets of large companies, because the business of the companies is focused on certain domains where it can be possible to define suitable ontologies.

11.2.2.2 Better Integration

The second main advantage of SW techniques is a higher potential to integrate different components. Business partners are able to better understand the syntax and semantics of their documents, to exchange them and to transfer them into the appropriate application for further automatic processing without human intervention. This also includes appropriate mapping or translation mechanisms. However, even humans can take advantage of a better integration, since ontologies can model relationships between the participants of such a process so they can uncover the hidden structure. Also on the software developer side, a higher degree of integration may also promote software component reuse in addition to a higher degree of automation.

11.2.2.3 Lexicon Flexibility and Standardisation

Theoretically, ontology mappings and translation allows users to flexibly choose the words they like. A generic ontology would facilitate the mapping and translations of ontologies. All ontologies would be mapped to this generic one and this generic one would be mapped to all of them. There would be a lack of precision but fewer ontology translators would be necessary. It could be also used for standardising the concepts, and improving the communications between different partners.

11.2.2.4 Composition of Complex Systems

In SW enabled systems, it is possible to compose numerous web services and web contents to produce one more complex system. Several SW technologies could be combined to develop a complex system with further functionalities.

11.2.3 Semantic Web Languages

SW is comprised of the standards and tools of XML, XML Schema, RDF, RDF Schema and OWL. The OWL Web Ontology Language Overview describes the function and relationship of each of these components of SW:

- XML provides a surface syntax for structured documents, but imposes no semantic constraints on the meaning of these documents.
- XML Schema is a language for restricting the structure of XML documents.
- RDF is a simple data model for referring to objects ("resources") and how they are related. An RDF-based model can be represented in XML syntax.
- RDF Schema is a vocabulary for describing properties and classes of RDF resources, with a semantics for generalisation-hierarchies of such properties and classes.
- DAML is a language created by DARPA (Hendler and McGuiness 2000) as an ontology and inference language based upon RDF. DAML takes RDF Schema a step further, by giving us more in depth properties and classes. DAML allows one to be even more expressive than with RDF Schema, and brings us back on track with our SW discussion by providing some simple terms for creating inferences.
- DAML+OIL: a successor language to DAML and OIL that combines features of both.
- OWL adds more vocabulary for describing properties and classes: among others, relations between classes (*e.g.*, disjointness), cardinality (*e.g.*, "exactly one"), equality, richer typing of properties, characteristics of properties (*e.g.*, symmetry), and enumerated classes.

11.2.4 Semantic Web Platforms

The most popular SW platforms are described in the following paragraphs.

11.2.4.1 Protégé 2000

Protégé 2000 (Noy *et al.* 2000) is probably the best-known ontology development and knowledge acquisition environment. It was developed by Stanford Medical Informatics group of Stanford University. Protégé has mainly three functions:

- construction of domain ontologies;
- customisation of knowledge acquisition forms, together with extensions for graphical widgets for tables, diagrams and other components; and
- a library which other applications can use to access and display knowledge bases.

The main idea behind the functions is to make the knowledge representation format adaptable for various ontology languages, whereas other ontology modelling tools tend to choose some specific languages to concentrate on.

The Protégé-OWL editor is an extension of Protégé that supports OWL. The Protégé-OWL editor enables users to:

- load and save OWL and RDF ontologies;
- edit and visualise classes, properties and SWRL rules;
- define logical class characteristics as OWL expressions;
- execute reasoners such as description logic classifiers; and
- edit OWL individuals for SW markup.

Protégé-OWL has an open-source Java API for the development of custom-tailored user interface components or arbitrary SW services.

11.2.4.2 Altova Semantic Works 2008

Altova Semantic Works 2008 is the visual RDF/OWL editor from the creators of XMLSpy. It visually designs SW instance documents, vocabularies and ontologies, exporting them in either RDF/XML or N-triples formats. This software works with tabs for instances, properties, classes, *etc.*, context-sensitive entry helpers, and automatic format checking.

Semantic Works 2008 allows the user to graphically create and edit RDF instance documents, RDFS vocabularies, and OWL ontologies with full syntax checking. Context-sensitive entry helpers are presented with a list of permitted choices based on the RDF or OWL dialect (Figure 11.1).

Semantic Works 2008 provides functionality for:

- visual creation and editing of RDF, RDF Schema (RDFS), OWL Lite, OWL DL, and OWL Full documents using an intuitive, visual interface and drag-and-drop functionality;
- syntax checking to ensure conformance with the RDF/XML specifications;
- auto-generation and editing of RDF/XML or N-triples formats based on visual RDF/OWL design; and
- printing the graphical RDF and OWL representations to create documentation for SW implementations.

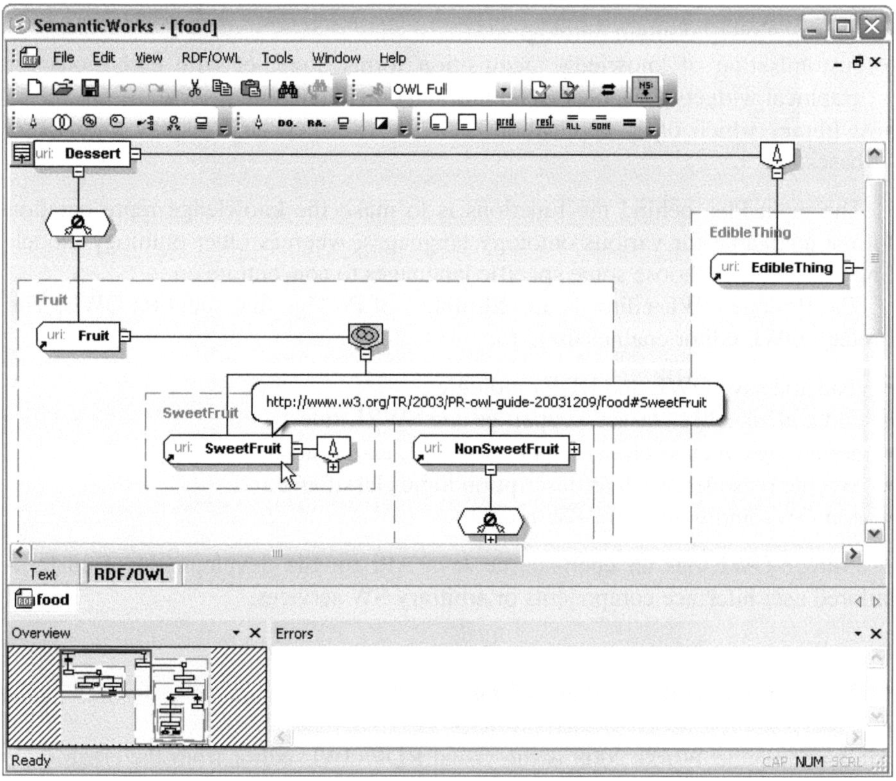

Figure 11.1 Defining an ontology with Altova Semantic Works

11.2.4.3 SMORE

SMORE is designed to enable users to markup HTML documents in OWL using Web Ontologies (Figure 11.2). Primary goals include:

- to allow the user to markup web documents with limited knowledge of OWL terms and syntax;
- to provide a way to use classes, properties and individuals from existing ontologies, do limited ontology editing or even create a new ontology from scratch using terms from web documents; and
- to provide a flexible environment to create simple web page simultaneously with markup.

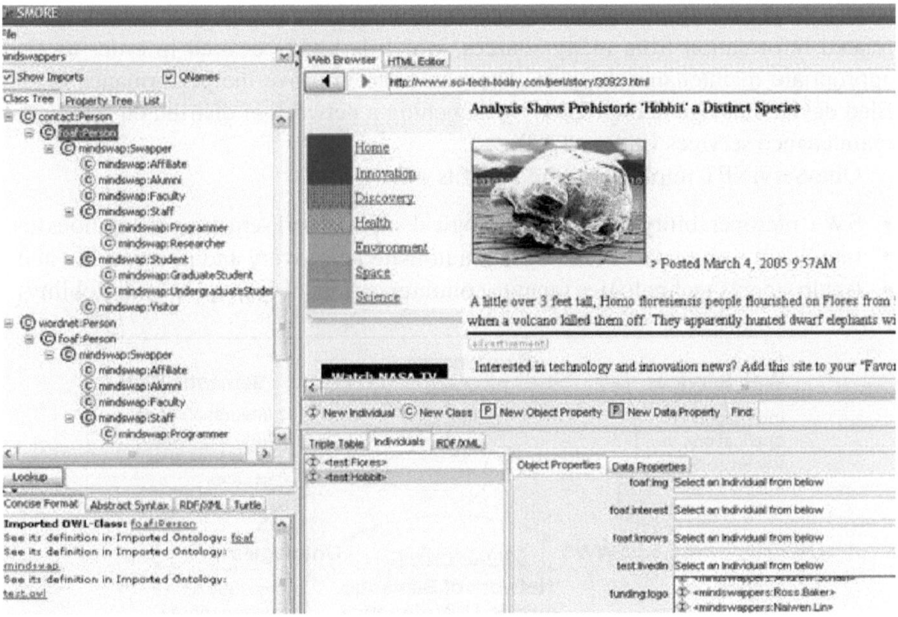

Figure 11.2 SMORE user interface

11.2.5 Semantic Web Development in Industrial Automation

In industrial environments SW developments have been performed, as explained in next paragraphs.

11.2.5.1 OntoServ.NET

OntoServ.NET (Terziyan and Kononenko 2003) is a framework for industrial semantics-enabled maintenance services organised in peer-to-peer network of services platform. It is based on web services and semantic web technologies and is meant to provide solutions for building large-scale industrial maintenance networks. It is an automated maintenance system that can integrate maintenance-related information from many sources, which is highly desired in order to give appropriate maintenance support. The goal is to improve the performance of the filed device management process by launching a network of distributed intelligent maintenance services (Figure 11.3).

OntoServ.NET implements the benefits of:

- SW (interoperability based on ontological support and semantic annotations);
- intelligent web services (modelling, automated discovery and integration); and
- (multi)agents technologies (agents communication, coordination and mobility).

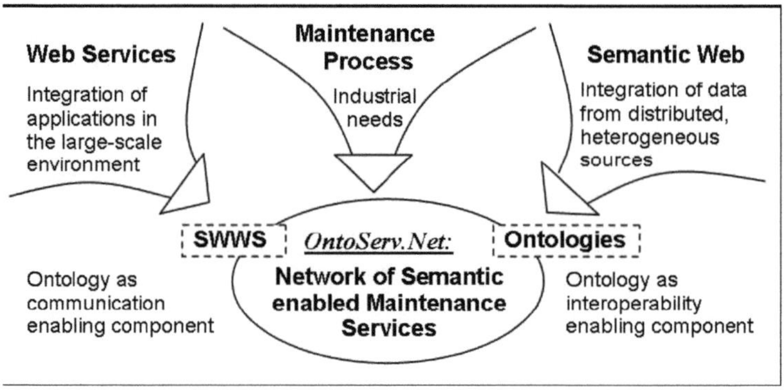

Figure 11.3 Ontoserv.Net structure

11.2.5.2 Obelix

OBELIX (ontology-based electronic integration of complex products and value chains) has developed an e-business ontology tool suite and library to support smart collaborative e-business and the realisation of innovative applications (Akkermans *et al.* 2004). The Obelix tool suite consists of an ontology server providing facilities for editing, component brokering, ontology management, and web language import and export, plus a number of ontology-based tools including an e-business scenario analysis and simulation tool for DVC models and strategies, an automatic product classifier to speed up application of content management standards and a multi-product configuration tool for online collaborative design scenarios. In addition, Obelix has delivered a modular e-business ontology library. The tool has been validated through three e-business applications: e-markets for energy trading and servicing, new digital music value chains and online design of events.

11.2.5.3 Rewerse

Rewerse is a research Network of Excellence on "Reasoning on the Web" (Bry and Schwertel 2004) that is funded by the European Commission and Switzerland within the 6th Framework Programme. The main objective of Rewerse is to provide tangible technological bases for an industrial software development of advanced web systems and applications.

For 4 years (2004–2008) Rewerse involved 27 European research and industry organisations from 14 European countries and about 100 computer science researchers and professionals playing key roles in applied reasoning. One goal of Rewerse was to spread and support the use of reasoning techniques in advanced web applications and systems. It focused on creating a competitive advantage for the European industry.

11.2.5.4 Knowledge Web

Knowledge Web (Eisenstadt and Vincent 2000) is a four-year Network of Excellence project (2004–2008) funded by the European Commission 6th Framework Programme. Supporting the transition process of ontology technology from academia to industry is the main and major goal of Knowledge Web.

The mission of Knowledge Web was to strengthen the European industry and service providers in one of the most important areas of current computer technology: semantic web enabled e-work and e-commerce. The project concentrates its efforts around the outreach of this technology to industry. Naturally, this includes education and research efforts to ensure the durability of impact and support of industry.

In comparison with the Network of Excellence Rewerse, Knowledge Web focuses on different application scenarios:

- Knowledge Web: semantic web services
- Rewerse: reasoning on geographical/time-related data, bioinformatics, personalised information systems

However, more use cases have been collected in Knowledge Web.

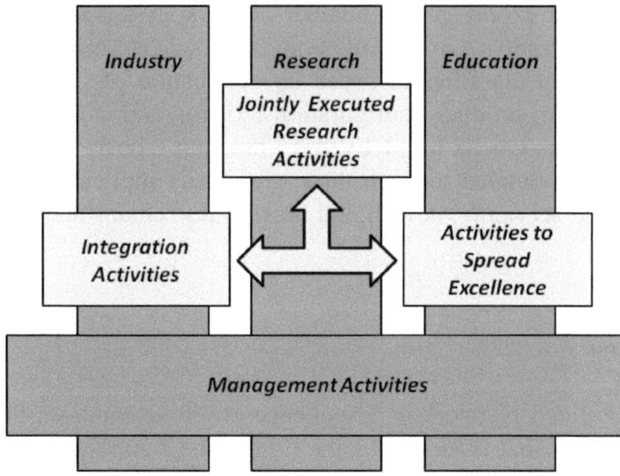

Figure 11.4 Main areas of Knowledge Web

11.2.5.5 Other Related Works

Posada *et al.* (2005) used an ontology as an improvement for building an industrial plant through a graphic design software. This ontology associates the semantics to the geometric parts from CAD (computer-aided design) reconstruction. It uses semantic compression added to simplification techniques of the geometrical data to increase the efficiency. The ontology was developed using Protégé 2000.

In airspace systems, Valente *et al.* (2005) present an architecture for operating unnamed aerial vehicles that leverages a group of military information ontologies to semantically specify and compose information services. The ontologies were developed with ontology web language (OWL) and cover a wide range of content including definitions of military information, organisations and communications.

11.3 Web Services for Dynamic Condition Based Maintenance

Web services are a well known technology used in industrial environments. They offer interoperability between independent software applications over the Internet by means of simple object access protocol (SOAP), which enables the communication.

The advantages of web services are the central issue in DynaWeb. Different software modules are able to communicate among themselves in order to perform a specific task. In this context, to provide the most convenient analysis flow, information processing is understood as a distributed and collaborative system, where there are different levels of entities that can undertake intelligence tasks. Given this, a system's architecture has been defined to identify the interactions between actors and the required functions, including four layers that correspond to the central information processing layers of OSA-CBM standard (Arnaiz *et al.* 2009, Bengtsson 2003, Lebold *et al.* 2002):

1. *Condition monitoring.* This is related to state detection. This layer receives measurements from sensors and their signal processing software. These measurements have to be compared to expected values, and an alert should be generated in the case of anomaly detection, due to values outside preset limits or changes in the usual trend.
2. *Health assessment.* The main function of this layer is to set the current health of the asset in the case of an anomaly detected by condition monitoring modules. It generates health records proposing possible faults, based on health history, operational status and maintenance tasks history. One of the existing diagnosis processes is based on previously developed systems (Gilabert and Arnaiz 2006) using Bayesian networks to facilitate a model that can work with uncertainty and can also be adapted with feedback information.
3. *Prognostics.* This module takes into account data from all the prior layers. The primary focus of the prognostic module is to calculate the future health of an asset, with account taken of the future usage profiles. The module reports the failure health status of a specified time or the remaining useful life.
4. *Decision support.* In this context this related to scheduling. A computerised maintenance management system schedules work orders based on component predictions. After that it distributes work orders to different operators (PDAs). The PDAs need to read the smart tags in order to learn about the components (Adgar *et al.* 2007).

One of the challenges is to match the semantic web concept to the maintenance function. In this way, information used over the Internet must be specified in ontologies. The ontology represents the knowledge in Internet (Fensel 2001), defining in a formal way the concepts of the different domains and relationships, with the ability to perform reasoning over this knowledge. In this case, the ontologies have been defined starting from the standard CRIS defined by MIMOSA. CRIS represents a static view of the data produced by a CBM system, where every OSA-CBM layer has been associated to ontology (Lebold *et al.* 2002).

OSA-CBM was developed around MIMOSA CRIS, which provides coverage of the information (data) that will be managed within a condition based maintenance system. It defines a relational database schema with about 400 tables for machinery maintenance information. In short, CRIS is the core of MIMOSA, which aims for the development and publication of open conventions for information exchange between plant and machinery maintenance information systems; Dynamite web services are built using CRIS standard for interoperability and internal information processing.

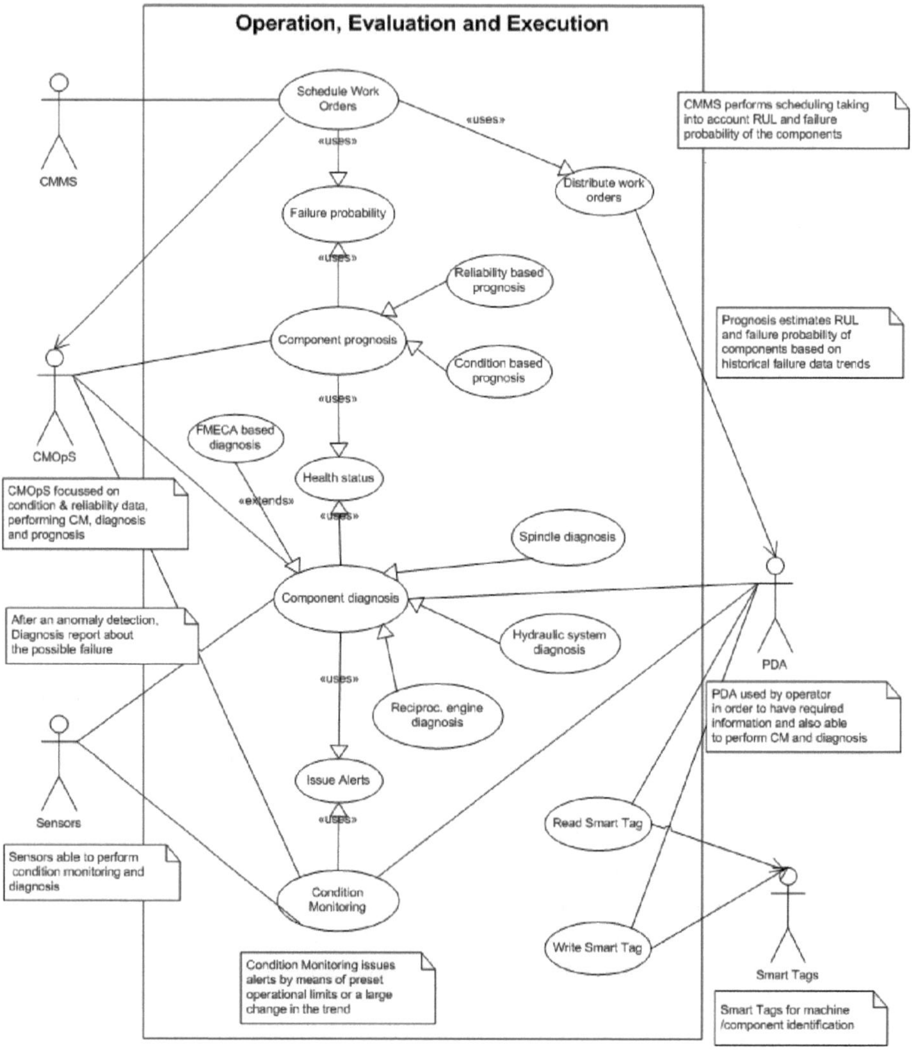

Figure 11.5 Use case diagram for operation, evaluation and execution

Another great challenge is to approach this concept for everyday computing resources (*e.g.*, SMEs), as well as for the forthcoming mobile services. In this sense, with regard to the usage of these web services, three main elements that take part in the communication are defined:

- Human machine interfaces (HMI) actor, that is, a software interface for the operator sitting at the desk or walking with the PDA and interacting with local or central database systems asking web services to process specific information.
- Agent for communicating with DynaWeb web services. The agent is able to obtain the needed data from other sources, translating it into the ontology lan-

guage. In this way, the agent acts as an interface between HMI and web service.

- Web service, performing the requested service, supported by ontologies.

With this configuration in mind, flexible data interaction architecture has been developed to provide the best access depending on agents and data repositories available. This means that, once the operator has requested a specific result, the agent can use mainly two different communication options, as shown in Figure 11.6.

- Direct communication with database can be performed if the database fulfils the MIMOSA specification. Then the agent only transmits XML commands to the web service, which in turn accesses the database for data.
- If local database is not MIMOSA compliant, the agent may choose to send the data in XML format.

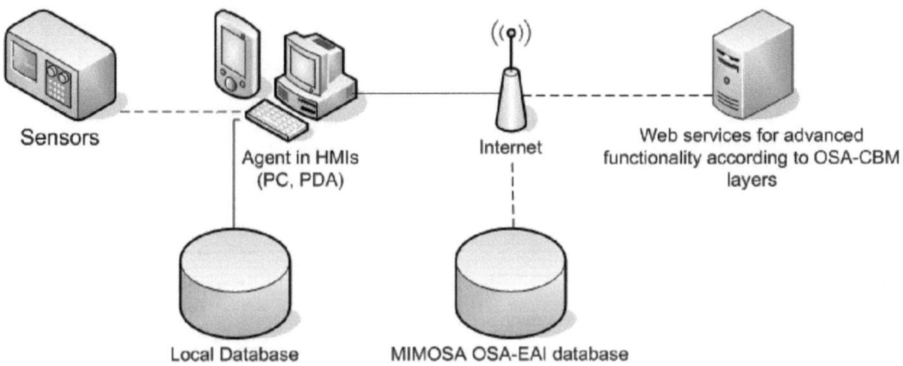

Figure 11.6 Communication architecture among HMI and web services, through the usage of an agent storing data in the MIMOSA database

As an example of this interaction, Figure 11.7 depicts a sequence diagram to request a diagnosis service, with direct communication sending data in XML format.

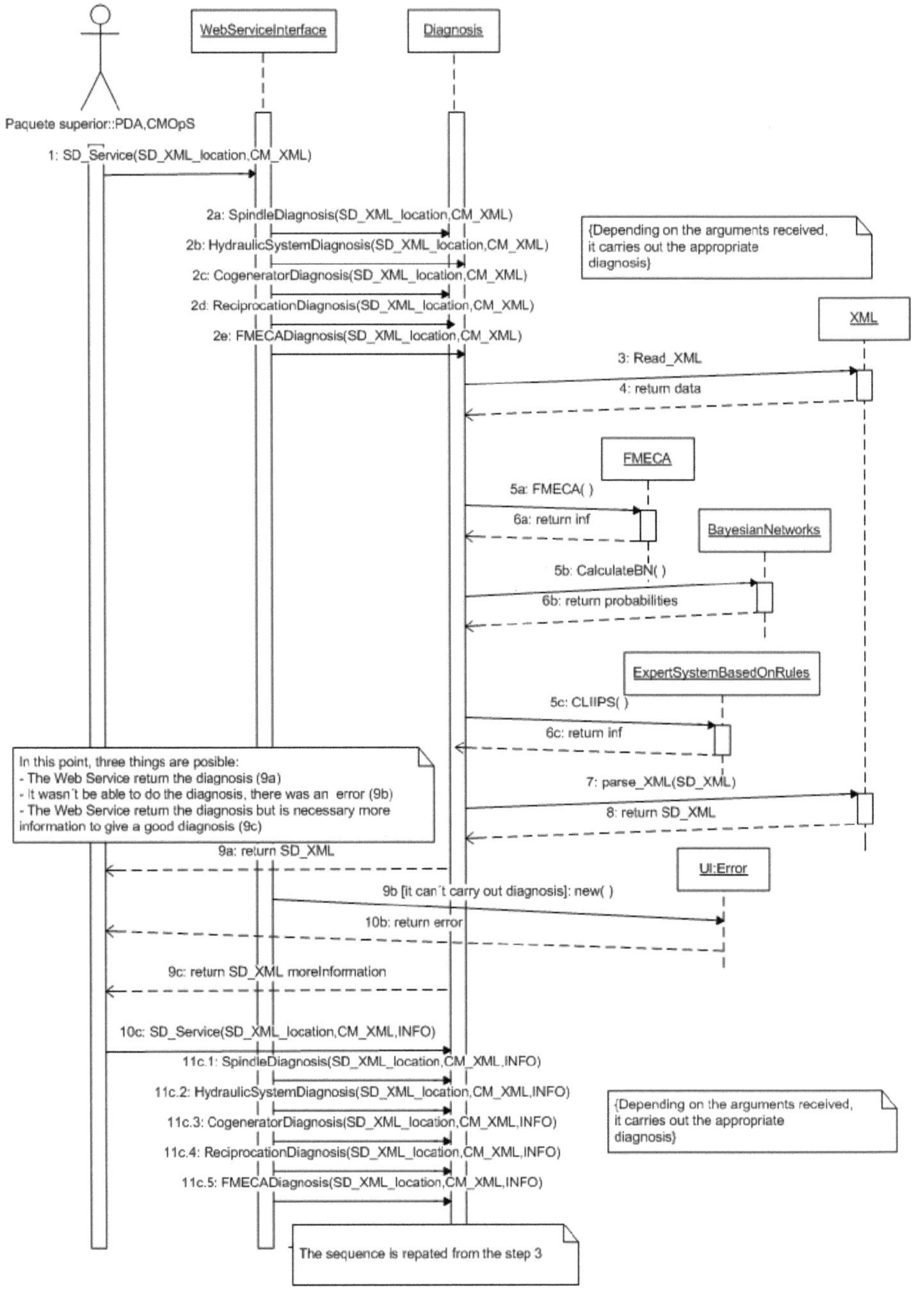

Figure 11.7 Sequence diagram for diagnosis web service

The performance of every diagnosis function follows these steps:

- Acceptance and validation of XML file.
- Extraction of data from XML file.
- Generation of diagnosis alerts with associated probabilities and severities. Depending on component, the service can manage one or more object to issue a heath assessment:
  - a Bayesian network,
  - a CLIPS expert system, based on rules, and
  - FMECA definition.
- Parse alerts to XML file.
- Return of XML file to agent.
- If it is necessary to perform a visual inspection or other test, the agent can request in a second step the diagnosis service with added information.

In this example, it is interesting to point out how both mobile PDA and CMOpS actors share same access to web services. Last, this example shows that it is possible to frame existing process systems within CRIS formatted web services to take advantage of already existing intelligence. For instance, one of the existing diagnosis processes is based on previously developed systems (Gilabert and Arnaiz 2006) using Bayesian networks to facilitate a model that can work with uncertainty and can also be adapted with feedback information. In this sense, most web services have been developed using .NET platform as a web application, using Microsoft Visual Studio .NET programming environment and the database implemented with Microsoft SQL server 2005.

11.3.1 Web Service for Condition Monitoring

One of the web services used for this e-maintenance platform will be devoted to perform condition monitoring tasks. The condition monitor receives data from the sensor modules, the signal processing modules and other condition monitors. Its primary focus is to compare data with expected values. The condition monitor should also be able to generate alerts based on preset operational limits.

The CM web service can analyse scalar data in four different forms:

- scalar values directly surpassing a direct static alert, alerts are defined by fixed limits;
- scalar values that surpass a static alert 'relative' to the original component status; depending in deviation, an alert is issued associated to a specific severity;
- a scalar value that surpasses dynamic alerts, based on the evolution of last sample values; and
- a scalar value that surpasses a given alert when combined with another parameter.

The CM service will request alert values. If they are not provided, it can also request information concerning the specifics of the component, such as component type, and then look up for information concerning these specifics. In the end, the web service will return the type of alert surpassed and the severity.

11.3.2 Web Service for Diagnosis Based on Vibration and Oil Data

The diagnosis or health assessment receives data from different condition monitors or from health assessment modules. The primary focus of the health assessment module is to prescribe whether the health in the monitored component, subsystem or system has degraded. The health assessment layer should be able to generate diagnosis records and propose fault possibilities. The diagnosis should be based upon trends in the health history, operational status and loading and maintenance history.

The aim of this web service is to identify the type of problem related to an electro mechanical component. It can cope with many different symptoms and faults, which include unbalance, misalignment, gear and bearing related problems, *etc.* It also includes processing of information specifics of multi-speed machines, such as in manufacturing systems. The web service can receive the a set of alarm types as part of the input configuration, according to measurements of vibration and oil.

The core of this service has been modelled using a probabilistic model called a Bayesian network. A Bayesian network is a model (Díez 2000). It reflects the states of some parts of a world that is being modelled and it describes how those states are related though conditional probabilities. All the possible states of the model represent all the possible worlds that can exist, that is, all the possible ways that the parts or states can be configured.

The representation is a directed acyclic graph (DAG) consisting of nodes that correspond to random variables and arcs, which in turn correspond to probabilistic dependencies between the variables. A conditional probability distribution is associated with each node and describes the dependency between the node and its parents.

In the network prototype shown above, the qualitative relationship indicated by the direction of the link arrows corresponds to dependence and independence between events. That is the, nodes higher up in the diagram tend to influence those below rather than, or, at least, more so than the other way around. On the other hand, the quantitative relationships between nodes are defined by conditional probability tables, in the case of continuous variables, by conditional probability distributions.

Many practical tasks can be reduced to the problem of classification, including Fault diagnosis. A Bayesian network helps tackle the problem of classification in a way that helps to overcome problems that other methods partially address.

A Bayesian network has the following characteristics:

- the ability to mix a-priory knowledge together with data/experimental knowledge;
- explanatory abilities;
- uncertainty management–causality management; and
- learning capabilities of both parametric and structural issues.

One important characteristic of this inference model is the adaptation ability. In this way, introducing risk values of environmental variables and expected parcel risk value, the Bayesian network adapts the weights of conditional probability tables, approximating to desired solution.

The Bayesian network was developed using the Hugin Research tool (Andersen *et al.* 1989). This software has a graphic interface on windows operating system so that Bayesian networks can be designed and it is possible to see probabilities propagation when node instances are set. An important feature is that Bayesian networks can be embedded into custom applications through an API.

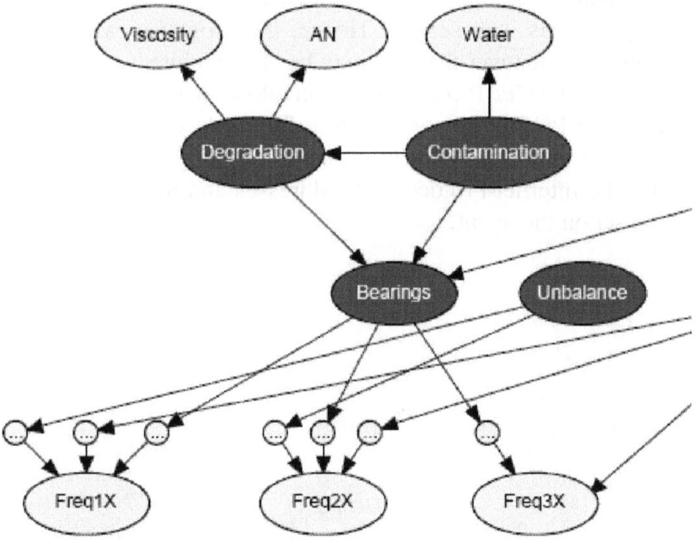

Figure 11.8 Excerpt of a Bayesian network

11.3.3 Web Service for Prognosis

A prognosis web service implements functionality to compute the remaining useful life of the component submitted to a degradation mode. The design of the prognosis web service was focused on two specific cases of prognosis: reliability based and condition monitoring based.

The development of this prognostic web service is necessary in order to compute the remaining useful life of the asset or segment. Two types of prognosis have been implemented:

- reliability based prognostic is performed taking into account influence variables through the use of the proportional hazard model (Cox 1972); and
- condition based prognosis is performed on an indicator of the health of the asset/segment.

These two types of prognosis have been chosen because they correspond to the ones required on the TELMA platform (see Chapter 14.3) that will be used for testing purposes. In the end, a unique web service was implemented and the choice between the approaches was made through the choice of the model used for prognosis. This choice was made in the call to the web service.

Since the communication architecture was performed by means of XML data, an agent was in charge of collecting data from the MIMOSA database and the specific database. The principle of operation was to deliver a client the result of the prognosis. To reach this aim, an agent was used in order to collect data and send them to the prognosis web service. Hence, the prognosis web service performs only the prognosis whatever the data may be. The architecture is described in Figure 11.9. The use of TCP/IP communication allows one to locate the client and the agent on the same PC/PDA or on different PC/PDA. In the case of a PDA client, locating the agent on a PC makes information exchange lighter between PDA and the network. The Internet Protocol (IP) addresses and the port used for communication can be set on the agent.

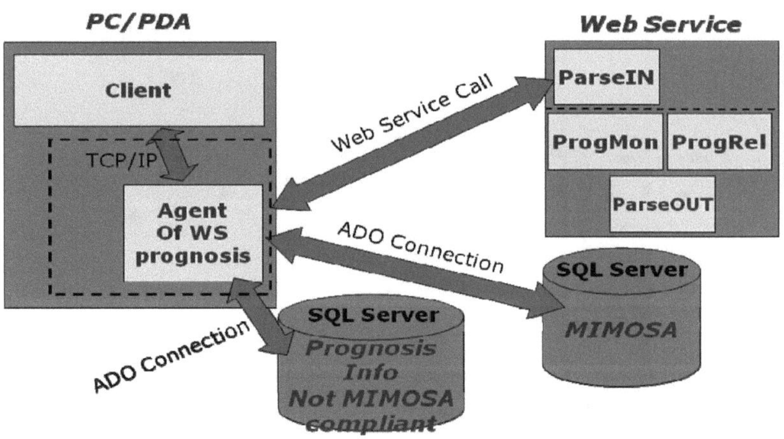

Figure 11.9 Principle of operation of the prognosis web service

The principle of prognosis is to forecast the future health of the asset/segment using a particular model. The RUL is obtained when the computed health reaches an upper or lower limit using data that begin from the last start after a maintenance action that brings back the component/function into an "as good as new" state. At the present time, the two available models for prognosis are the two described below.

11.3.3.1 Proportional Hazard Model

The proportional hazard model (PHM) (Cox 1972) assumes that the failure rate (hazard rate) of a segment/asset is the product of:

- an arbitrary and unspecified baseline failure rate, $\lambda_0(t)$, which is a function of time only and
- a positive function $g(X)$, independent of time, which incorporates the effects of a number of covariates such as humidity, temperature, pressure, voltage, *etc.*, included in the X vector.

The failure rate of a unit is then given by:

$$\lambda_0(t) = \lambda_0(t).\ g(X)$$

We implement the exponential PHM, $\lambda_0(t)$ is a constant. $g(X)$ is of the form:

$$g(X) = \exp(\beta.X)$$

where β is a vector of parameter for the covariates.

The use of PHM for an asset/segment and a hypothetical event is done by defining a model whose type field equals 'PHMexp'. The parameters of the model that have to be defined in the specific database are:

- λ_0: the value has to be specified; and
- β: for every β_i associated to a covariate, the value has to be specified and the measurement location corresponding to the covariate.

11.3.3.2 Exponential Curve Fitting

The exponential curve fitting algorithm is based on the identification of the parameter of an exponential curve to the past values of a degradation/performance indicator. The equation of the exponential curve is:

$$I = K.\exp(\lambda t - t_0)\ .$$

The identification used for the parameter identification is the least squares method. The only parameter of this model is the number of points used for the identification. Indeed, it could change since, when the degradation has just been detected only few points are representative and when it has occurred a "long" time ago, several point are be used.

11.3.4 Web Service for Scheduling

The *advanced scheduling module* has been realised as a web-service, residing at the decision support layer as defined by the OSA-CBM. The module is expected to utilise data from the prognostics layer (*e.g.*, reliability-based, empirical feature-based, model-based information) and allow for the scheduling of maintenance work orders based upon predicted condition of the equipment, thus helping to streamline the process and reduce maintenance costs.

The key implementation technology for information transport will be XML. In relation to this, the OSA-CBM standard requires the definition of an XML schema, which will define the acceptable structure for the OSA/CBM XML messages.

According to Figure 11.10, the scheduling software module provides a scheduling web services interface, which accepts input data organised in XML format. The *XML handler/parser* is responsible for processing the XML input to provide the required scheduling data to the core element of the scheduling software module. In addition, the *scheduling core* subcomponent takes responsibility for invoking the required scheduling algorithm and, along with the available data (as retrieved by the *scheduling database*), provide the produced schedule back to the scheduling web service and its invoker.

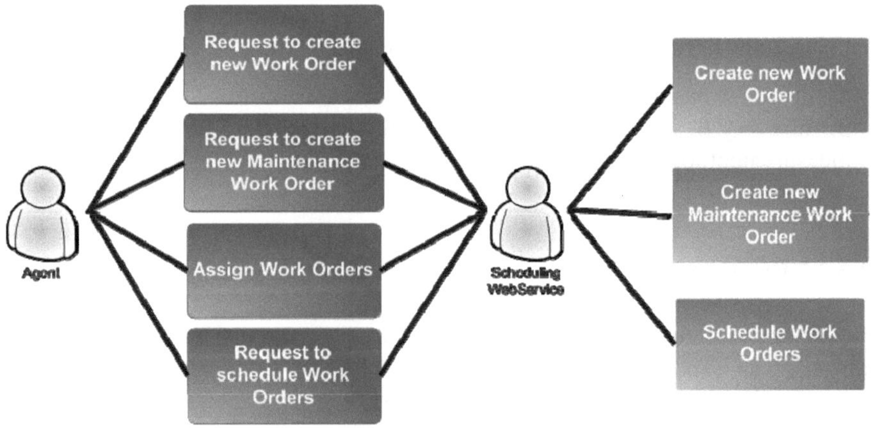

Figure 11.10 Modules of scheduling work orders

11.3.5 Testing Web Services

Different platforms have been used to test the web service and its advanced functionality:

- TESSnet: tested the web services for data acquisition, data manipulation, condition monitoring and diagnosis; and
- TELMA: tested the web services for prognosis and scheduling.

The TESSnet software tool (Gilabert *et al.* 2008) has as main target to detect mechanical problems in lubricated rotating machines in order to save time and money. It is predictive maintenance software based mainly in oil analysis, which performs an automated monitoring and diagnosis. TESSnet is the tool used in this framework to perform different tasks by means of DynaWeb's web services. It is predictive maintenance software based on oil, vibration and temperature analysis, which performs an automated condition monitoring, diagnosis, prognosis and decision support.

The platform stores measurements both from on-line and off-line sensors as well as laboratory analysis results. They are stored using a hierarchy of components: company, plant, machine, assembly, sensor and measurement.

This tool has been chosen for validation purposes, that is, to validate some web services through this application. Apart from that, some prognosis and scheduling has been added to the platform.

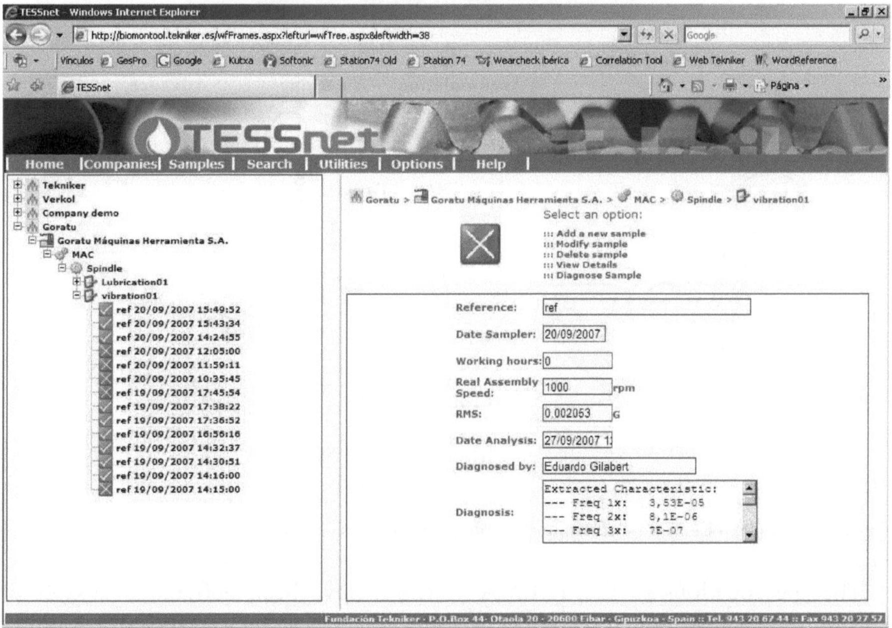

Figure 11.11 The TESSnet screen for vibration measurement

On the other hand, the TELMA platform (Levrat 2007) has been developed mainly to support in relation to e-maintenance:

- the engineering and deployment of CBM and proactive maintenance strategies; and
- the assessment of the impacts of these strategies on the performances of a manufacturing system:
  - productivity
  - quality
  - costs.

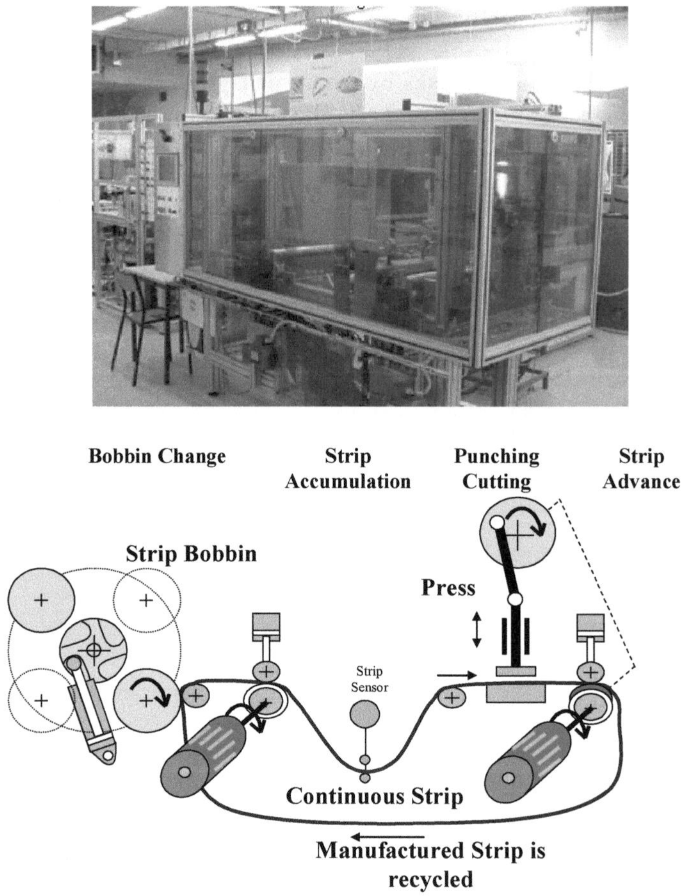

Figure 11.12 The TELMA platform and its physical process: unwinding metal strip

The specifications of the TELMA platform have been defined to answer to a group of teachers and researchers wishing to have at their disposal a training platform in the areas of maintenance, tele-maintenance and e-maintenance.

In this way, the platform is designed for:

- local use in the framework of conventional training activities;
- remote use *via* the Internet for operation on industrial e-services (*i.e.*, tele-monitoring) and for accessing to production data, performance data; and
- e-teaching and e-learning as application support of courses in the e-maintenance domains.

11.4 Conclusions

A set of semantic web services has been designed and developed in order to provide advanced functionality related to maintenance tasks. The main concept is to use them as plug-in components in a distributed environment in a company, depending on their needs. For this purpose, the MIMOSA database has been used as standard to support all communication data. In the end, the set of developed web services constitute a backbone depicting the OSA-CBM layers.

The web services integrated are very different in purpose, as cover from the data manipulation layer to decision support. They are also very different in nature. From basic data manipulation concerning vibration signals, to a sophisticated diagnostic module for diagnosis on lubricated machinery based on the use of Bayesian network. It is also interesting to point out that different algorithms are present in each layer, so they can be used depending on the needs, or can even collaborate for a global solution for the end user. At the end, the web services developed are just an example of what should be found at the end in the final product: a big number of methods and web services, at the choice of the user. This will give the real strength to this web services framework.

The end-users' needs have been studied and according to this, the web services' functionality has been developed. Moreover, previously developed applications such as TESSnet and TELMA have been adapted with the same purpose. The web services have been tested and their functionality has been validated, proving that they cope with user requirements. Next, activities related to this new approach should be addressed to provide more advanced functionality and better integration between the different components of the architecture.

References

Adgar A, Addison JFD, Yau C-Y (2007) Applications of RFID technology in maintenance systems. Proc 2nd World Congress on Engineering Asset Management, Harrogate, UK. Coxmoor

Akkermans H, Baida Z, Gordijn J, Pena N, Altuna A, Laresgoiti I (2004) Value webs: using ontologies to bundle real-world services. Free Univ, Amsterdam, Netherlands

Andersen SK, Olesen KG, Jensen FV, Jensen F (1989) Hugin – a shell for building Bayesian belief universes for expert systems. Proc 11th International Joint Conference on Artificial Intelligence 1080–1085

Arnaiz A, Jantunen E, Adgar A, Gilabert E (2009) Ubiquitous computing for dynamic condition based maintenance. Journal of Quality in Maintenance Engineering (JQME). Special issue on Condition Monitoring and ICT application 15(2): 151–166

Bengtsson M (2003) Standardisation issues in condition based maintenance. Department of Innovation, Design and Product Development, Mälardalen University, Sweden

Berners-Lee T, Hendler J, Lassila O (2001) The semantic web, a new form of web content that is meaningful to computers will unleash a revolution of new possibilities. Scientific American, May 2001.

Bry F, Schwertel U (2004) REWERSE-reasoning on the web. AgentLink News

Carro-Martínez S, Kulas A, Geisler T, Diederich J (2005) Report on existing specific Rewerse courses on the Semantic Web suitable for the industry. Deliverable T-D3 Rewerse

Cox DR (1972) Regression models and life tables. Journal of the Royal Statistical Society B 34:187–220

Díez FJ (2000) Probability and decision theory in medicine. UNED, Madrid

Eisenstadt M, Vincent T (2000) The knowledge web: learning and collaborating on the net. Proc International Distance Learning Conference (IDLCON-97), Washington, USA, 1–18

Fensel D (2001) Ontologies: A silver bullet for knowledge management and electronic commerce. Springer, New York

Gilabert E, Arnaiz A (2006) Intelligent automation systems for predictive maintenance. A case study. Robotics and Computer Integrated Manufacturing (RCIM) 22:543–54

Gilabert E, Arnaiz A, Ferreiro F (2008) TESSNET: A predictive maintenance management system. 21st Int Congress Condition Monitoring and Diagnostic Engineering Management. COMADEM2008. 11-13.6.2008, Prague, Czech Republic, 227–235 ISBN 978-80-254-2276-2. http://tessnet.tekniker.es

Gruber T (1992) What is an ontology? http://www-ksl.stanford.edu/kst/what-is-an-ontology.html

Hendler J, McGuinness DL (2000) The DARPA agent markup language. IEEE Intelligent systems, 15: 67–73

Lebold M, Reichard K, Byington C, Orsagh R (2002) OSA-CBM architecture development with emphasis on XML implementations. Proc Maintenance and Reliability Confererence (MARCON 2002)

Levrat E, Iung B (2007) TELMA: A full e-maintenance platform. Proc 2nd World Congress on Engineering Asset Management, Harrogate, UK. Coxmoor

Noy NF, Fergerson RW, Musen MA (2000) The knowledge model of Protege-2000: combining interoperability and flexibility. 2nd International Conference on Knowledge Engineering and Knowledge Management (EKAW'2000), Juan-les-Pins, France, vol 1937, pp 17–32

Posada J, Toro C, Wundrak S, Stork A (2005) Ontology modelling in industry standards for large model visualization and design review using Protégé. Proc 8th Protégé conference, Madrid

Terziyan V, Kononenko O (2003) Semantic web enabled web services: state-of-art and industrial challenges. Proc. International Conference on Web Services, Springer, ISSN 0302-9743

Valente A, Holmes D, Alvidrez FC (2005) Using ontologies to build web serviced-based architecture for airspace systems. Proc 8th Protégé conference, Madrid

Chapter 12
Strategies for Maintenance Cost-effectiveness

Basim Al-Najjar

Abstract. To survive hard international competition, it is necessary for many producing companies to enhance their competition positions by, for example, reducing production costs and maintaining and improving product quality to increase the profit margin. In order to achieve these strategic goals, strategies for making maintenance more profitable are developed, introduced and discussed in this chapter. These strategies are all developed as software modules included by an innovatively new maintenance decision support system (MDSS). Also, theoretical backgrounds are introduced, and software prototypes for making the application of any of the tools on a daily basis easier and more cost-effective are developed and discussed, and their functionalities are tested. MDSS consists of three toolsets, where each toolset consists of one to three tools that can be used for different objectives and applications. It is designed to support maintenance and production engineers to achieve dynamic and cost-effective maintenance decisions. The latter will be discussed in Chapter 13. MDSS software prototype can be utilised for several objectives, such as, to:

- enhance the accuracy of maintenance decisions;
- simulate and select the most cost-effective maintenance solution, *i.e.*, the best investment in maintenance;
- identify and prioritise problem areas; and
- map, monitor, analysis, follow up and assess maintenance contribution in company profit.

12.1 Introduction

Economic losses due to lost availability are, in general, one of the major sources of losses in producing companies. Also, the inability to fulfil the delivery time schedule is one of the causes behind losses in the customer and market shares, and consequently company profit. These types of losses are strongly influenced by the performance of maintenance and its effectiveness. The accuracy of maintenance decisions, which concerns when, where and why to stop a producing machine is essential to reduce economic losses generated due to unnecessary stoppages of a production process. This issue acquires an even higher importance in companies of intensive capital investment where stoppage time is expensive, such as the case of paper and pulp mills, refineries, power stations and engineering manufacturing industry following process production methodology (Al-Najjar 2007a).

The IT systems available for maintenance purposes give low support for the assessment and follow up the total technical and economic impact of maintenance on production, quality, working environment, company business, *etc.* (Kans 2008). The impact is restricted to the maintenance department alone, not allowing for a holistic view of maintenance role in a producing company. Consequently, the financial aspect of maintenance includes only cost analysis and control, but it does not allow follow up of the economic investment impact of maintenance on company profitability and competitiveness. Maintenance investments are usually treated as costs alone and not considered as risk capital investments that should be justified by a special maintenance payoff. Further, planning and follow up of the technical impact of maintenance usually means emphasis on some variables, such as failure rates and execution times of maintenance actions required to optimise the maintenance policy. It is hard to use only these variables to show the economic impact of invested capital in maintenance. The connection between maintenance investments and technical and economic benefits is, in other words, lacking. Thus, it is not possible to translate the technical benefits into financial measures using current maintenance IT systems and software programs.

12.2 Development of Strategies for Cost-effectiveness

Maintenance influences and is influenced by a wide range of working areas in a plant. Therefore, it participates effectively in company business (Al-Najjar 1997, Al-Najjar *et al.* 2001). Maintenance should be treated as a profit-generating centre since it is closely related to a company's internal efficiency (Al-Najjar 2007a). However, to be able to map, analyse, control and assess maintenance cost-effectiveness, relevant data and performance measures should be used.

12.2.1 Theoretical Background

In this study we consider maintenance as a sub-system in the whole plant, where the plant is considered as a system. Applying systems theory, we need to identify maintenance inputs and outputs, which are necessary to highlight the real cost incurred by performing maintenance work and the value added by these costs, *i.e.*, the economic output of the sub-system. Maintenance planning and performance usually demands resources, which at the end represent maintenance expenses. The cost factors that can be considered representing the input to maintenance, *i.e.*, maintenance direct cost, can be summarised by the expenses of the following inputs:

- maintenance personnel/labour (man-hours);
- maintenance experience and knowledge (training expenses);
- spare parts;
- risk capital investments in maintenance;
- maintenance tools and consumable material;
- organisation for supporting maintenance work;
- data acquisition and analysis systems;
- maintenance managerial tools; and
- miscellaneous.

At the same time, maintenance as any other working area has its outputs. It will not be possible to highlight whether maintenance contributes to the company profit or not without assessing the technical and economic output of maintenance. In this study we found that the values of the following factors represent the vast majority of maintenance technical contribution, *i.e.*, outputs:

- less planned stoppages;
- less failures and shorter stoppage time;
- maintaining high quality production;
- increasing productivity and production rate;
- reducing the short stoppages and disturbances and their stoppage times;
- less failure-related accidents;
- less accidents violating environment; and
- prevention of reoccurrence of the same faults.

These outputs can also be expressed in other technical and economic terms, *e.g.*, cost-effective production, longer production times, more production, less tied up capital, lower assurance premium and less penalty expenses due to less violation of the environment, which leads to more competitive organisation. Identifying the factors that are responsible for generating maintenance costs and income makes it possible to identify maintenance related economic factors.

12.2.1.1 Maintenance Related Economic Factors

One of the major reasons to consider maintenance as a cost-centre is that it is easy to identify the losses due to the lack of or inefficient maintenance, but it is difficult to assess the effect of maintenance on the production process. Consequently, it is easy to assess maintenance cost but difficult to assess the economic impact of maintenance on company profit. In this study, we call the economic factors that are influenced by or influence maintenance: maintenance related economic factors. This is because these factors consist of maintenance costs, savings, risk capital investments and results (profit or losses). Also, we call it so to avoid labelling maintenance as a cost-centre by dividing maintenance related economic factors to direct and indirect costs of maintenance, as usual. Therefore, maintenance related economic factors are divided into four categories (Al-Najjar 2007a):

1. *Maintenance Direct Costs.* The costs that are really incurred by maintenance and are not confused with other activities, *i.e.*,
   - maintenance labour expenses (man-hours)
   - spare parts
   - consumable material
   - instrumentation and software
   - locals and overheads
   - miscellaneous.

2. *Economic Losses, Potential Savings or Potential Income Sources.* These losses represent all the losses that are generated due to unplanned stoppages, such as failures and short stoppages, failure-related accidents, bad quality production because of lack of or inefficient maintenance, *etc.* These economic losses are also considered as *maintenance income resources*, because the more efficient the maintenance the higher the probability of recovering some of these losses in the form of savings, *i.e.*, potential savings. These losses consist of:
   - unavailability cost due to unplanned stoppages, *e.g.*, failures;
   - performance inefficiency costs due to idling, minor stoppages and reduced production speed;
   - bad quality costs due to maintenance deficiencies;
   - delivery delay penalty expenses due to breakdowns;
   - extra energy cost due to disproportional energy consumption;
   - excessive spare parts, buffer and work-in-progress (WIP) inventory costs to avoid the effect of unplanned stoppages on fulfilling delivery schedules;
   - unnecessary equipment redundancy costs to avoid waiting time due to unplanned stoppages;
   - extra investments needed to preserve WIP and redundancies in good conditions;

- expenses due to the absence of professional labour as a result of maintenance-based accidents, compensation labour costs and costs of using less skilled labour;
- penalty expenses for generating more pollution due to bad machine condition and inefficient maintenance related accidents; and
- high insurance premiums due to inefficient maintenance related accidents and their consequences.

3. *Risk Capital Investments for Improving Maintenance Performance.* These investments are required for improving maintenance performance during a specific period of time. Investments in maintenance should be justified by more efficient maintenance performance that increases economic savings. The latter can be achieved through recovering some of the losses, *i.e.*, preventing loss generating events, such as failures, short stoppages, *etc.* Investing in maintenance is usual when the current maintenance policy is not sufficient to handle all the disturbances generated due to machine deterioration and more complexity in the maintenance and production processes. Maintenance performance can be enhanced in several ways, such as effective diagnosis and analysis and maintenance management systems, which is discussed below. The investments can be made in many activities, such as:

   - instrument and software for better maintenance performance
   - training programs to enhance personnel competence.

4. *Maintenance Results.* This concerns maintenance profit or more losses that are achieved after a new investment for improving maintenance performance is made. The income of maintenance is assessed based on:

   - savings due to maintenance more efficient performance
   - profit/losses that are generated.

Most condition-based maintenance policies are based on the fact that most failures can be foreseen, for example when using different types of condition monitoring techniques, and thereby be prevented. Because maintenance is an economic matter this demands access to both economic and technical data. In many situations, there are often several technically applicable maintenance policies to choose from, but the goal is to find the alternative that is the most cost-effective from a holistic perspective.

12.2.1.2 Diagnosis Techniques

A machine state can be evaluated using condition monitoring (CM) systems, such as vibration analysis, which be carried out through identifying active frequencies, *i.e.*, damage causes frequencies and their amplitudes. Amplitude is usually used to indicate and assess the severity of the deterioration. When vibration levels of some frequencies are higher than the normal level, the probability of failure increases if nothing is done and product quality, *e.g.*, in paper mill machines, may also be influenced (see Al-Najjar (1998) and references cited therein). Detecting these frequencies early may lead to maintaining the condition of the component/equipment and prolonging its life and keeping the quality of production within its specified limits for longer, and leads to more time to plan the eventual maintenance actions. In this context, we may say now that we are more certain about identifying damage and deviation in the condition of a component/equipment but less certain in assessing the severity of the damage and time span until a failure is a fact. In other words, better techniques and tools for more reliable assessment of the condition of a component/equipment in the close future is required to reduce production time losses and consequently economic losses toward enhancing company profitability and competitiveness.

The assessment of the condition of a component can be done in several ways. In this study, we focus on intensive utilisation of the available data for more information, but this demands different data at different levels of the work hierarchy. For example, a production manager may need to know how long it will be possible to run a machine or production despite a damaged component. This is necessary to enable him to plan the production reliably in order to accomplish the delivery schedule. The production manager will feel satisfied if he receives more information telling him that it is possible to keep the machine running with an assessed risk of failure and a predicted value of the vibration level in a close period, *e.g.*, at the next planned stoppage or next measuring moment. However, inconsistent data may confuse the assessment of the machine's condition. Overestimation of the fault development rate causes the user to lose some of the machine life and to increase the number of stoppages and spare parts used. The underestimation results in extra production losses because of bad quality products and sometimes to stopping the production at an inconvenient time.

12.2.1.3 Maintenance Management IT-systems

The development of IT systems for maintenance has more or less followed the progress of maintenance strategies. Large companies are likely to have one or more types of computerised support for maintenance administration and/or control. These IT systems can be divided into three main groups (Al-Najjar and Alsyouf 2000, Al-Najjar and Kans 2006, Kans 2008):

Computerised maintenance management systems (CMMS) are independent maintenance specific IT systems that can be integrated partially or fully with other IT systems, *e.g.*, production control or financial system. Examples of CMMS used by Swedish companies are AMOS, Data Stream 7i and API PRO. CMMS usually supports planning and execution of PM activities and handles emergent failures using BDM. A few CMMS also support condition-monitoring and/or reliability-centred maintenance. The emphasis lies on user friendliness with easy to use graphical and flexible user interfaces and support for handheld devices. Support for multimedia files, such as photos of an object in focus, is also common. The functionality is based upon planning and execution of PM, with handling of work orders, spare parts and personnel in focus. Technical and economic data are restricted to maintenance alone. Most CMMS have functions for budgeting, cost control and follow up of maintenance costs. In a few cases failure statistics, capacity measurement and overall equipment effectiveness (OEE) assessment are also included.

Enterprise resource planning (ERP) systems handle the total corporate need of core business information such as finance, production, maintenance and quality. Examples of IT systems within this category are SAP/R3, IFA Applications and Movex. About one tenth of the ERP systems used in Sweden include a maintenance module. These maintenance modules provide about the same functionality as a CMMS, with emphasis on planning and execution of preventive maintenance actions. Some ERP systems include technical and financial follow up of the maintenance activities, but these are normally restricted to cost control, failure statistics and availability analyses. The strength of ERP systems is the data they contain, which are of both technical and financial kind, if all modules are implemented. ERP systems may, in other words, be an excellent source for the data needed for maintenance control and follow up. Unfortunately, it is not so easy in reality; most companies use more than one IT system for information processing, *e.g.*, one for production planning/control, another for financial administration and a third for quality control.

Decision support system (DSS) expert systems are computerised information systems used in decision making processes. They are suitable in situations where:

- the problem is too complex to be solved manually;
- only a part of the variables have defined values and we wish to simulate how different values on the other variables affect the total result;
- interactivity between user and the system is needed; and
- the user understands quantitative and qualitative analysis methods.

An expert system is a variant of DSS. By using data and information within the expert system and a set of rules, the expert system can solve problems too complicated to solve without expertise knowledge. In literature, decision support systems and expert systems are mainly discussed with respect to (or developed for) fault detection and diagnosis (see, *e.g.*, Batanov *et al.* 1993, Groth *et al.* 1996, Molina *et al.* 2000, Kipersztok and Dildy 2002, Yagi *et al.* 2003). Failure detection and diagnosis demands a high level of knowledge about the characteristics of components and machines, *e.g.*, failure causes and consequences, which is suited for expert system support. Another area within DSS is optimising maintenance schedules (see, *e.g.*, Hadzilacos *et al.* 2000, Zhan *et al.* 2001). Some authors like Yam *et al.* (2001) recognise the connection between the different computerised maintenance systems and cost-effectiveness. There are also articles that discuss computer aided design or redesign decision making and how to monitor and extend plant or machine life cycle cost (LCC), (*e.g.*, Hansen *et al.* 1992, Soliveres and Alquier 1997). It is interesting to note that the functionality that maintenance decision support or expert systems provides, such as fault detection, diagnosis and maintenance optimisation is not to be found in commercial CMMS today. Hafkamp and Schutters (2001) recognise this and present the design and implementation of a CMMS with an expert system included.

12.2.2 The Role of Maintenance in Company Business

Managerial concepts and tools belonging to condition-based maintenance do not include a clear theory for treating maintenance in its real context, *i.e.*, as a part of a production or service process. However, if we consider maintenance so, we then have to describe maintenance interactions with all relevant working areas such as operation and production, production logistics, life cycle cost analysis and management, working environment, *etc.* Also, we have to describe the methods and tools required for describing, analysing, quantifying and modelling maintenance technical impact and its conversion to the economic scale for making it understandable. When all that has been done, it will be possible to monitor, control and improve maintenance technical and economic effectiveness. For any production process or enterprise to survive harsh international competition, specific competitive advantages should be created, maintained and improved continuously, *i.e.*, :

- high quality production and machine/process
- competitive price
- delivery on time
- environmentally friendly production process and product
- acceptance by society.

Production process capability as a measure of conformity depends on the elements involved in the production process; these are producing machines, tools, procedures/methods, maintenance policy, operational conditions, competence of operating and maintenance staff, quality system, raw material, managerial functions, disturbances influencing the production process, *etc.* Deviations in the quality of any of the essential input elements may have a major effect on the competitive advantages. In this study, the condition of the production process is defined by the state (quality) of the basic elements constituting it. In general, it is not usual that old and deteriorated equipment/processes can, in the long term, produce quality products in high effectiveness at a competitive price to be delivered on time and are an environmentally friendly process in the sense demanded by society. Deterioration in equipment can be started, or developed, due to external causes, such as misuse, bad raw material, unsuitable lubricants, inefficient maintenance, external shock and harsh operational conditions. For example, these factors constitute about 70% of the total reasons behind failures in rolling element bearings (Bloch and Geitner 1994). Therefore, to prevent or reduce their effect, relevant information parameters should be monitored actively.

In order to provide more accurate real-time mapping and analysis of the condition of a production process, both the product characteristics and the condition of the relevant elements involved in the production process should be monitored and assessed. This is also necessary to prolong the lead-time required for preparing and performing maintenance actions to restore the condition of a component, equipment or process to function according to the specifications. Technical, statistical and economic analyses are effective tools to identify and assess the deviations in the condition of the production process, product quality, working environment and production costs.

Vibration spectrum analysis is a powerful tool for monitoring rotating and reciprocating machines (Al-Najjar 1997). It is also partly suitable for monitoring the working environment and personnel competence in specific areas. Vibration frequencies can be utilised to monitor the machine condition and product quality and to identify and localise a wide range of the causes behind the deviations and, consequently, which elements in the production process that these causes are related to (Al-Najjar 2001). The best output of the production process that fulfils the competitive advantages can be distinguished and achieved through selecting a suitable combination of the process elements, which cannot be achieved without using an easily accessible and special database.

One of the essential forces driving total quality management (TQM) and total productive maintenance (TPM) is the improvement cycle (Deming cycle), *i.e.*, plan–do–check–act. Practically, this cycle has been used in the way that one should act as soon as a failure has occurred. However, it can also be interpreted so that an action is started at an earlier stage, *i.e.*, as soon as a significant deviation in the equipment/process condition is observed. A study done by Bloch and Geitner (1994) reveals that about 99% of mechanical failures are preceded by some detectable indications of condition change. Vibration spectral analysis provides a basis for identification of damage causes, damage development mechanisms and failure modes for most types of faults in rotating and reciprocating machines (Bloch and Geitner 1994, Al-Najjar 1998).

In this study, failure is defined as a termination of a component's ability, to perform its required function, which can be defined on basis of the machine function, capability, production rate, production cost, product quality or personnel/machine safety. Also, total quality maintenance (TQMain) is defined as a philosophy to maintain and improve continuously the technical and economic effectiveness of the production process and its elements. In other words, TQMain is not just a tool to serve or repair failed machines, but rather a means to maintain the quality of all the elements involved in the production process. Thus, TQMain's role can be described as a means for monitoring and controlling deviations in a process, working environments, product quality and production cost, and for detecting damage causes and their developing mechanisms and potential failures in order to interfere (when possible) to "stop" or reduce the machine deterioration rate before the production process and product characteristics are intolerably affected. Also, it is to perform the required action to restore the machine/process to a state as good as it was before. All these should be performed at a continuously reducing cost per unit of high quality product (Al-Najjar 2008). The features characterising TQMain and distinguishing it from other maintenance techniques, such as preventive, condition based, reliability-centred and total productive maintenance, are summarised in the following.

- It covers a wide range of a production process and not just machinery.
- It is based on a new condition-based concept. Actions are planned and performed based on the needs arising due to deviations in the quality of the elements involved in the production process.
- It handles production and maintenance technical and economic problems by integrating tools and methods belonging to both deterministic and probabilistic approaches.
- It advocates the use of a common database that should be updated by real-time data of the essential information parameters for real-time monitoring and assessment of the machine condition and production process technical and economic effectiveness, product quality and the working environment. It then becomes possible to select and improve the most informative CM system and the most cost-effective maintenance policy effectively.
- Consequently, it provides an overall view of the state of the production process, including all the elements involved, and the technical and economic impact of maintenance on a company's business.
- It is based on making intensive use of the real-time data acquisition and analysis to detect at an early stage the causes behind quality and cost factors deviations and machinery malfunctions, and following damage/defect development to prolong the component life.
- It provides tools and methods for proactive-predictive maintenance, *i.e.*, to detect and eliminate the cause behind damage initiation. If it is not possible technologically, detect the deviation at an early stage and predict its development to reduce, or eliminate, the risk of failure.
- It emphasises the systematic maintenance work combining technical, organisational and economic knowledge and experience, where all the theories, tools and methods required are developed and verified.
- It provides the basis for cost-effective and continuous improvement of the production and maintenance process, *e.g.*, even after each renewal when using vibration-based maintenance (VBM) policy through confronting history with the replaced components, *i.e.*, continuous cyclic improvement.

TQMain has been applied partly in about ten case studies. The results of these applications have shown a big beneficial potential and the possibility of applying it as a whole concept (Al-Najjar 2006). Cost-effective maintenance actions cannot be achieved without better coverage high quality data that can be provided by a reliable common database. TQMain provides all the tools and methods required to perform these steps. These tools and methods are usually verified in several case studies and are usually based on the common concept of TQMain. For all these reasons, we have developed MDSS based on the concept of TQMain.

12.3 Development of a Maintenance Decision Support System (MDSS)

Applying TQMain, the objects of study are complex systems and their interactions with other systems. These objects consist of both material entities such as human beings and machines and immaterial entities such as information and strategies. System theory, which studies complex systems by using methods like modelling and simulation (Skyttner 2001) is used to describe MDSS. Figure 12.1 describes a conceptual model, where relevant basic variables and key performance indicators (KPI) are used for mapping, analysis and following up the performance of maintenance and production processes.

The vast majority of the data required for applying the model is probably available at in company's databases. A few additional information parameters may be needed to measure or estimate because they have not been measured before or can be confused with other information. This should be done in a reliable way to avoid any major influence of the certainty of the model. The technical, economic and managerial variables relevant to the model are defined and formulated to be compatible with the MIMOSA database so that it can be used by different users. In order to realise this conception, we have defined and formulised the functions using Du'Pont breakdown charts, which can usually be used to determine what kinds of variables or parameters are needed to carry out the functions. These functions assess the additional costs and profit losses due to, *e.g.*, stoppages, unavailability, spare parts, man-hours, overheads and bad quality. The data needed to apply the model properly are divided into three categories:

- *Raw data:* the data that can be gathered directly from company databases.
- *Processed data:* the data that are a result of processing some raw data collected from the company by using predefined formulas, including both aggregating and de-aggregating data.
- *Estimated data:* these data are usually not available at the company databases and should be estimated based on other technical and economic factors. The data are estimated using the company's experience, database, and mathematical or statistical relations.

The sets of functions, *i.e.*, toolsets, included by MDSS, see also Figure 12.1, contain tools for improving the accuracy of maintenance decisions, simulating alternative technically suitable solutions for a particular problem in a machine/component to select the most cost-effective solution, identifying and prioritising problem areas and problems in a production process and to assess and highlight maintenance, and investments in maintenance, performance and payoff. This is why the raw data include cost factors, losses, investments, stoppage times, production amount, quantity of rejected items, *etc.* For more details we refer the reader to Chapter 13.

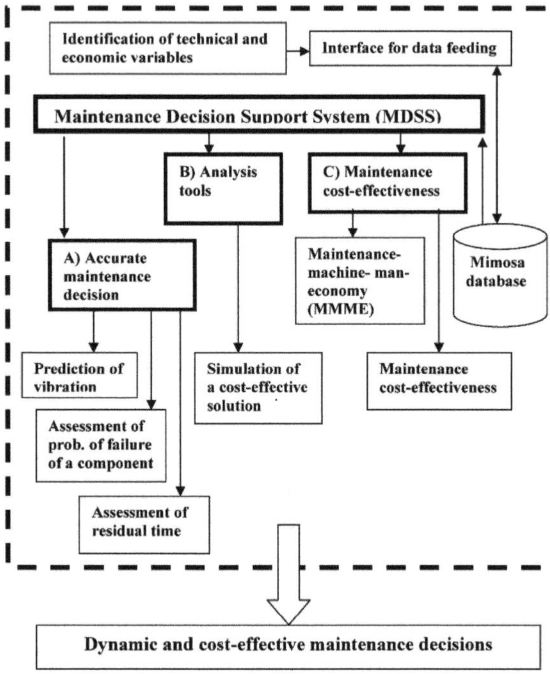

Figure 12.1 Conceptual model of MDSS

12.3.1 Objectives of MDSS

To be able to monitor, analyse, assess, predict and improve the outcome of maintenance actions properly it is necessary to base data gathering on a systematic model for identifying and localising both technical and economic data in the company databases. A computerised means, such as MDSS, for collecting and analysing data reduces problems, disturbance and costs. It allows following up maintenance KPI more frequently, thereby being able to react more quickly on disturbances and avoid unnecessary costs before it is too late. It would help in answering where and why an investment in maintenance may have the best economic payoff in order to achieve cost-effective decisions concerning planning and executing maintenance actions (Al-Najjar *et al.* 2001, 2004). Pintelon (1997) points out the importance of a well functioning maintenance reporting system and also the fact that most systems in this area are limited only to maintenance budget reporting. The major objectives of using MDSS can be summarised as follows:

- To have better data coverage and quality required for monitoring, mapping, analysis, evaluating and controlling the performance of the maintenance process and to enhance the information underlying maintenance decisions.
- To have a more accurate assessment of maintenance performance and its contribution to company profit.
- To identify, localise and prioritise problems.
- To identify the most cost-effective maintenance solution, *i.e.*, maintenance investment, when there are several technically suitable solutions.
- Consequently, the main objective of applying this model is that it is possible to achieve dynamic and cost-effective maintenance decisions.

12.3.2 MDSS Toolsets and Tools

MDSS consists of three toolsets (toolset 1, 2 and 3), see Figure 12.2 (a). Every toolset consists of one or more tools that can be used for different objectives. When applying CBM, usually the level of the observed CM parameter is considered significant when its value exceeds a particular level, *e.g.*, warning level. At that significant level, it is necessary to gather more information concerning the component/equipment in question before making any maintenance decision, *i.e.*, the information required should be more than trend data and the current situation of that component. This is why toolset 1 offers the following additional information:

- prediction of the vibration level at the near future, such as the moment of the next planned measurement or stoppage;
- assessment of the residual time, especially when production manager would like to assure smooth running of the production process until the production is delivered according to the delivery schedule; and
- assessment of the probability of failure based on the previous experience of the same or equivalent components or equipment.

All these measures in addition to the usually available measures, such as trend data and the current condition, will increase the information required to enhance the accuracy of the maintenance decision and its cost-effectiveness.

Detecting problems in a component/equipment increases the need to identify all possible ways of solutions/actions. In many cases these alternative actions/solutions are all technically suitable but not necessarily equally cost-effective. Toolset 2 provides the possibility of simulating relevant alternative maintenance solutions in order to select the most cost-effective one. Toolset 3 makes it possible to follow up, control and assess the losses in production and profit by providing the possibility to:

- assess the losses in production time;
- indentify and prioritise problem areas and basic causes behind losses in the production time; and
- assess cost-effectiveness of maintenance through identifying and assessing maintenance direct cost, savings, investments and results (maintenance losses or profit).

In Table 12.1, a short description of all toolsets and tools constituting MDSS and their acronyms are introduced, see also Figures 12.2 (a) and (b).

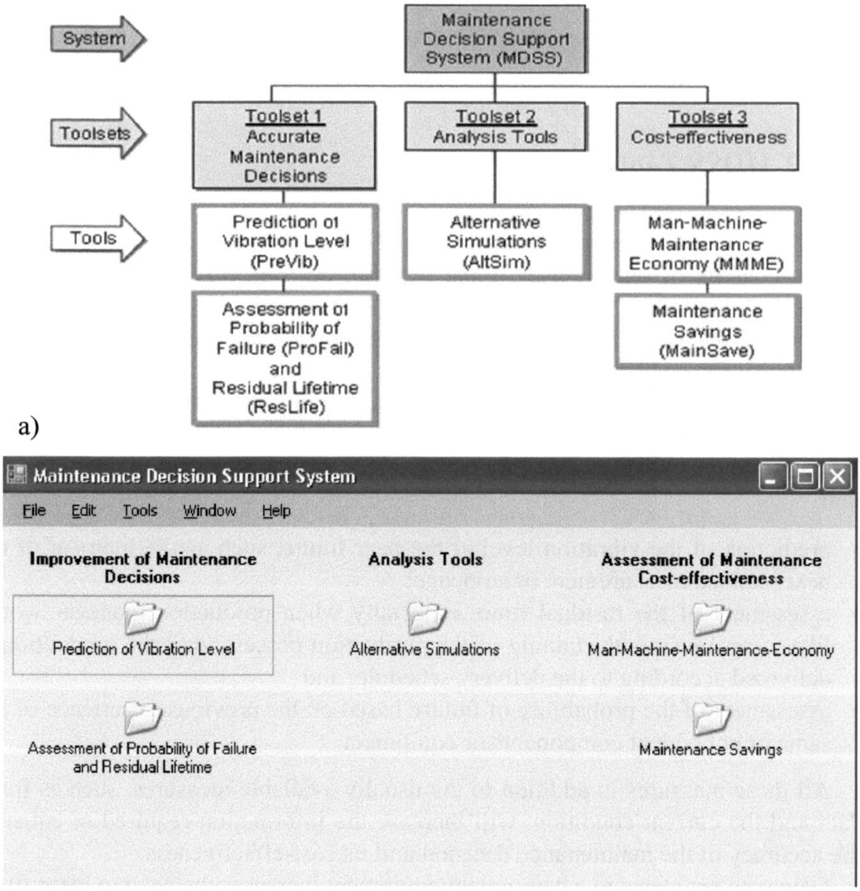

Figure 12.2 (a) Toolsets and tools constituting MDSS, and (b) User-interface of the MDSS software program

Table 12.1 Functions of the toolset and tools included by MDSS

Toolsets	Tools	Features and function
Toolset 1		
To enhance the accuracy of maintenance decisions	PreVib (prediction of vibration level)	It is a mechanistic model for predicting the vibration level of a component/equipment at the next planned maintenance action or measuring moment, when using vibration-based maintenance. Its aim is to predict significant deviations in the condition of a component at an early stage in order to keep the probability of failure of the machine/significant component as low as possible and to avoid sudden and dramatic changes in the vibration level leading to a catastrophic failure until a condition-based replacement is done (Al-Najjar 2001).
	ProFail (probability of failure)	It is a probabilistic approach for assessing the probability of failure of a component/equipment using machine past data at need or when its CM, *e.g.*, vibration, level is significantly high (Al-Najjar 2003).
	ResLife (residual lifetime)	The same tool as in ProFail is utilised for assessing the residual lifetime of a component/equipment when its condition monitoring level, *e.g.*, vibration level is significantly high (*ibid.*). It is used to avoid failures and delivery delays. ResLife can be used to control whether or not it is possible for the production process to proceed according to the production schedule. The probability of failure and residual life are assessed using a modified form of total time on test (TTT-plots)
Toolset 2		
Simulate and select the most cost-effective maintenance solution	AltSim (alternative simulation)	To simulate technically applicable alternative solutions suggested for a particular problem and to select the most cost-effective maintenance solution using an intelligent motor. This tool is important to improve the cost-effectiveness of maintenance investments and planned actions. The selection is done using well-defined criteria, such as the proportion of maintenance savings to the invested capital, total maintenance profit.
Toolset 3		
To map, follow up, analyse and assess maintenance cost-effectiveness	MMME (man-machine-maintenance-economy)	To identify and prioritise problem areas and to assess the losses in the production time. It is beneficial to plan maintenance actions according to a prioritising list.
	MainSave (maintenance savings)	To monitor, map, analyse, follow up and assess maintenance cost-effectiveness, *i.e.*, maintenance contribution in company profit (maintenance savings and profit). It is a reliable tool for securing cost-effective maintenance actions.

12.3.2.1 Accurate Maintenance Decisions: PreVib, ProFail and ResLife

In this section, three services that can be acquired through using two tools for enhancing the accuracy of maintenance decisions are introduced and described.

Prediction of the Vibration Level (PreVib)

Mechanical and electrical systems in power stations, paper mills, hydraulic systems, *etc.*, consist of many sub-systems, components and modules such as rolling element bearings, pumps, motors and gearboxes. Replacing the mechanical parts at the right time, which can be achieved by using CBM or age replacement at cost-optimised intervals, reduces the contribution of each part to the system rate of occurrence of failures (ROCOF) and therefore reduces the ROCOF itself (Sherwin 2000). It is well known that the vibration signal generated by a component/equipment provides detailed information about the condition of that component/equipment in a rotating machine. The vibration level can be measured according to different parameters, such as vibration displacement, velocity or acceleration. Each parameter has its own range of application. When the damage is initiated, the CM parameter gives enough information about a component's condition. The hazard rate can then be kept low as long as the vibration level describing the component condition is still below the warning or replacement level.

VBM policy based on the concept of TQMain increases the chance of detecting damage initiation of significant components/equipment at an early stage to keep the probability of failure of machines very low (Al-Najjar 2000). This effective usage of VBM enables the user to plan and perform cost-effective maintenance actions to reduce the risk of damage initiation. Also, a more accurate assessment of the condition of the machine's significant components means a higher chance of prolonging the mean effective life length of the components. Assessing the component's condition can be done effectively, *i.e.*, with high certainty, through considering both the deterministic and probabilistic approaches simultaneously for modelling the deterioration process, (Al-Najjar 2007b). Issues related to machine function, failures analysis and diagnostics are examples of the deterministic approach. While modelling the time to maintenance action and the probability of failure of the component when damage is detected are examples of the probabilistic approach. When a potential failure (damage under development) of a significant component is detected, predicting the value of the CM parameter, *e.g.*, vibration level, during the interval until the next measurement or planned stoppage, accurately, would enhance the effectiveness of maintenance decision-making process.

A Mechanistic Model to Predict the Vibration Level

In general, the life of a component consists of three different phases: no damage, damage initiation and development (potential failure), and the imminent failure phase, which ends with by condition-based replacement or failure. The second phase represents more than half the total usable life of a component/equipment (Al-Najjar 2000). Technically, the chance that a component will fail depends on its past history, current status and its operating and environmental conditions in the near future. The CM parameter level is usually considered stationary during the interval prior to the initiation of damage and fluctuates about its mean value (x_o). However, after damage initiation and development the CM parameter value, *i.e.*, $x(t)$, may be assumed to be increasing (exponentially) with the operating time (Herraty 1993). A potential failure can be defined as an identifiable physical condition, which indicates that damage is initiated and under development, and an actual failure may eventually occur if no relevant maintenance action is taken. Apart from the case where changes in the component condition have not been detected at an early stage, the component has no chance to fail when there is no damage initiated and as long as the levels of relevant CM parameters are still below the replacement level (x_{th}). Consequently, the probability of failure may be assumed to be very low (approximately zero) during this interval. Denote the level at which a potential failure is considered initiated by x_p, so that $x_o << x_p << x_{th}$.

It is often difficult to tell for certain that the second phase has begun, especially when the number of CM measurements, *e.g.*, vibration, is very small. In many cases, using the cumulative sum (CUSUM) chart of the condition measurements is a better indicator of a potential failure than the measurement itself (Al-Najjar and Ciganovic 2009). This is because of the random fluctuation of the vibration level value, which occurs according to some uncontrolled disturbing factors independent of the component's condition. Also, when using the CUSUM chart to identify a potential failure level, the risk of false alarms can be reduced. Assume that when the vibration level exceeds the potential failure level this insures the initiation of a potential failure and indicates the beginning of the wear-out interval. In this interval, $x(t)$ may be assumed to be a non-decreasing function. The potential failure level should be assessed for each component because it varies even for identical components.

At each measurement, after a potential failure is detected, the time needed by a component to approach the replacement (failure) level is a random variable. Based on the component deterioration rate and the predicted CM parameter level during the near future, it is possible to assess the moment of replacement graphically. However, a probabilistic model is necessary for assessing the component's residual time and for describing the risk of failure or the dispersion (uncertainty) in the predicted CM parameter level after each measurement to enhance the decision making process.

Consider that we use an efficient vibration monitoring system, as long as the relevant vibration frequencies or bands for monitoring the state of a component are identified, the detection of damage can be done with high accuracy (Sherwin and Al-Najjar 1999). Also, an alarm should be released as soon as the level of the vibration frequency or band significantly deviates from x_o approaching or exceeding x_p. Further investigation is then required to make sure that it is not a false warning, which is possible due to randomness in the vibration signals and the existence of noise. When the vibration level exceeds the first alarm level (x_p) this confirms that the damage is under development and there is a significant change.

When the monitored vibration level increases or changes significantly, maintenance engineers need, at each measuring moment, to decide whether or not it is urgent to perform the required maintenance action. This is why it is very important to have as much information as possible concerning the condition of the component/equipment in question to make the decision-making process easier and cost-effective. For example, for any production or maintenance manager it will be much easier to decide whether to proceed running the machine/equipment or stop it if the information below is given (Al-Najjar and Wang 2001).

- Past data such as the vibration trend describing the equipment's behaviour and current condition, including diagnosis of damage causes, development mechanisms and possible failure modes, which are usually available when implementing CBM/VBM policy.
- The predicted CM parameter value during the period after recent measurement, *e.g.*, at the next planned measuring moment or next planned stoppage.
- Assessment of the probability of failure of equipment at that point of time.
- Assessment of the residual lifetime of equipment at the same point of time.
- Whether it is more cost-effective to do maintenance actions now, at the next convenient moment or at failure.

In order to complete the set of information needed for decision-making in the second item above, a model was developed for predicting the vibration level on demand in the near future, *e.g.*, at the time point for the next measurement or planned stoppage (Al-Najjar 2001). The model is illustrated by Equation 12.1. Y is the dependent variable representing the predicted value of the vibration level. It is a function of three independent variables (X, Z and T) and three parameters (a, b and c); i = 1, 2...n and i is the number of measuring opportunities after damage initiation,

$$Y_{i+1} = X_i + a * Exp\left(b_i * T_{i+1} * Z_i^{c_i}\right) + E_i \,. \tag{12.1}$$

Where

Y_{i+1} = predicted value of the vibration level at the next planned measuring time.

T_{i+1} = elapsed time since the damage is initiated and its development is detected.

X_i = current vibration level.

Z_i = deterioration factor, a function of the current and anticipated load and previous deterioration rate. ($Z_i = dx` * Lf/L_c$).

a = gradient (slope) by which the value of the vibration level varied since it started to deviate from its normal state x_o due to initiation of damage until detecting it at x_p.

b_i & c_i = non-linear model's constants.

E_i = model error, which is assumed to be identical, independent and normally distributed with zero mean and constant variance, $N(0, \sigma)$.

The prediction of the vibration level is done when a significant vibration level is approached because of the development of damage. The future vibration level behaviour can be described based on several factors related to the deterioration process, such as previous and current vibration measurements, and previous and future operating conditions, such as loading. In this model, the major focus is on the recent vibration measurements out of the whole vibration history to avoid underestimation or overestimation of the vibration level. In the model, previous vibration measurements that are usually used for illustrating the trend of vibration level development are utilised for identifying the interval when damage is initiated, which is important for the development of a potential failure and the rate of damage development.

Development of the PreVib Software Prototype

For easy use of the model and in order to achieve reliable and faster results, new software based on Equation 12.1 called PreVib has been developed (Al-Najjar and Ciganovic 2009), see Figure 12.3. The user-interface is divided into two halves, with the left half representing input data for prediction and the right half representing the result in the form of a graph. Input data show both database and non-database data in the grey and white boxes, respectively. When a prediction is to be made, the segment (machine), asset (component, *e.g.*, a rolling element bearing) and the location (direction of the vibration measurements) should be specified. Also, we should specify how many measurements in "limit measurements" that we consider, if we just wish to focus only on part of the measurements. Otherwise we should use "none" if we consider all the measurements in the prediction. Then, the measurements can be downloaded by pressing "load", which uploads the data from a MIMOSA database.

The *mean vibration level* (x_0), *prediction time per*iod and *near future load/present load* should all be specified before clicking *predict*. The desired future *prediction time period* can be specified in, *e.g.*, seconds, hours, days, weeks, months, *etc.* When *predict* is clicked, the value of the predict vibration level and the associated date and plot it on the diagram on the right-hand side of the user-interface will be determined. By analogy, the same thing can be repeated any time the user needs to predict the vibration level, for example after the next measurement of the vibration level.

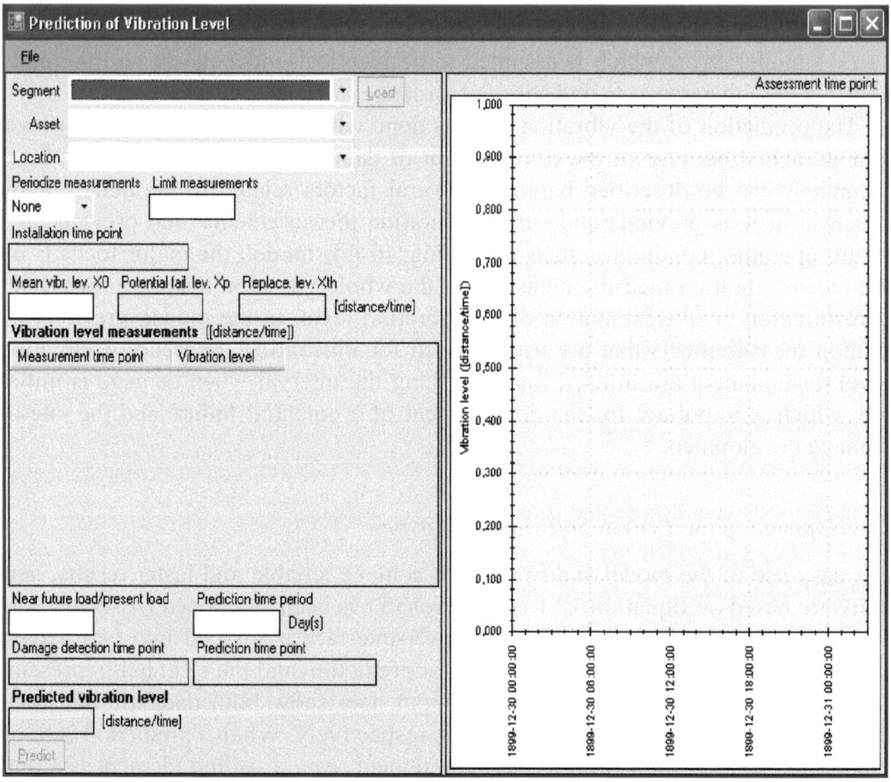

Figure 12.3 User-interface of the software program PreVib

The right half of the user-interface is a graph that will contain two plots, one (which is started earlier) representing the actual vibration level values and second (which is started later) showing the predicted vibration level values. The *y*-axis is the vibration level in, *e.g.*, mm/s and the *x*-axis is the calendar time for measurements and predictions. The graph also has three dashed lines representing the three different levels of the life of a mechanical component *i.e.*, the mean vibration level (x_o), the potential failure level (x_p) and the replacement level. These levels are determined and set based on data from identical components and are automatically retrieved when component data is uploaded. For more details see Chapter 13.

Some of the initial measurements belonging to the first phase will not be shown in the graph, even if they have been considered when estimating the model's constants b and $\check{c}$. The first prediction, as evident, should be done after confirming damage initiation, *i.e.*, the vibration level exceeds or close to x_p, because as long it fluctuates around x_o it means there is no damage is initiated and thereby there is no need to predict the future level. To be able to predict the future vibration level one need at least three measurements. This is why we may need some of the measurements that have been considered as part of the first phase of the component life.

Observe that sometimes the first two predictions are special cases due to the use of the measurements below the potential failure level to assess the model's parameters b and c. The reason is that we want to predict the future state as soon as we have confirmed the initiation of damage, instead of waiting for at least three measurements exceeding (x_p). Waiting may also cause uncertainty in the maintenance decisions. Also, because the deterioration process is a stochastic process, the vibration level can be changed randomly in time. Therefore, some of the level values may acquire a higher level than they should, *i.e.*, exceed x_p, which may be irrelevant to the severity of the deterioration. For each prediction, all previous measurements are used for estimating the model's parameters (b and c) and thus the predicted vibration level itself.

Failure Probability and Residual Lifetime (ProFail and ResLife)

It is usually preferred to base improvements of CM techniques on the techniques' past data. This is because each CM system and its applications may be considered unique. Disturbing factors and noise are the worst enemy of maintenance engineers who attempt to achieve accurate interpretation and analysis of vibration spectrums. These disturbances in many cases lead to over- (or under-) assessment of the equipment condition and consequently to economic losses that may arise due to false signals or failures. TTT-plots for equipment/components are usually obtained from failure data. This graphical method is related to the total time on test (TTT)-plot, as introduced by Barlow and Campo (1975).

In this study, a documented history of the condition-based replacements and failures of the equipment/components in question is used for the purpose of description and assessment. Also, we consider a component, *e.g.*, a rolling element bearing, as the smallest part in the machine and having a particular function. When using CBM, the most important data that are needed are the predetermined levels of the measured CM parameter, such as vibration levels (x_o), (x_p) and (x_{th}), CM measurements and trend. In this section, the theoretical background and the development of the models for assessing the probability of failure of a component and its residual lifetime using past data from the machine and component in question are introduced.

Theoretical Background, Model Development and Failure Model

A failure is defined in the (1993) Standard, the Glossary of Terms used in Terotechnology, BS 3811:1993 as: "The termination of the ability of an item to perform its required function" and it can be defined as: the failure of a machine is assumed to be announced as soon as an unsatisfactory condition is approached, *i.e.*, when the value of the CM parameter (or parameters) exceeds a predetermined level. This level is be defined on the basis of one or more of the elements deciding whether or not the equipment/component is considered to be in a failure situation, such as equipment function, product quality, production cost, *etc.* The failure data needed for statistical analysis are not always available because companies try to carry out replacement before failure, even if this leads to over-maintaining the equipment, due to the consequential costs of failure.

Therefore, special experiments are often required to obtain the required data. These experiments are often expensive, deviate from reality appreciably and are not always easy to conduct. For the given reasons, condition-based replacement data will be considered to obtain TTT-plots as an aid to assess the condition of a bearing. The bearing's history will be utilised to perform this task. In this context, condition-based replacement data mean all the data that are related to the replacements done depending on the analysis and assessments of the measurements collected by the CM system. Suppose that given n observations $t_1, t_2,.., t_n$ from a particular life distribution $F(.)$. Let these observations be ordered according to their sizes, *i.e.*, $t_1 \le t_2 \le .. \le t_n$. Thus $t_1, t_2,.., t_n$ represents n times to failure in increasing order. Let T_i denote the total time generated in ages less than or equal to t_i, *i.e.*, $T_1 = nt_1$, and

$$T_i = \sum_{j=1}^{i} (n - j + 1)(t_j - t_{j-1}) , \tag{12.2}$$

for $i = 1,..,n$ where $t_0 = 0$. Also, let $u_i = T_i/T_n$. The application of TTT-plots is explained in Example 12.1.

Example 12.1. Assume that eight components are observed until failure, their times to failure, ζ_1, ..., ζ_8, and the calculated quantities are shown in Table 12.2. The TTT-plot is obtained by plotting u_i versus i/n, as shown in Figure12.4. U_i is the ratio of the total time on test until the ith unit fails, T_i, and the total time on test until all units fail, T_n. The TTT-plot gives a dimensionless view representing times to failure of the tested components. The deviation of the plot from the diagonal provides information about the deviation of the plotted data (life distribution) from the exponential distribution.

Table 12.2 Failure times and calculated quantities for the TTT-plot shown in Figure 12.4

i	ζ_i	T_i	u_i	i/n
1	8.7	69.6	0.267	0.125
i	ζ_i	T_i	u_i	i/n
2	11.6	89.9	0.345	0.250
3	21.3	148.1	0.568	0.375
4	26.1	172.1	0.660	0.500
5	37.4	217.3	0.834	0.625
6	38.2	219.7	0.843	0.750
7	49.8	242.9	0.930	0.875
8	67.5	260.6	1.000	1.000

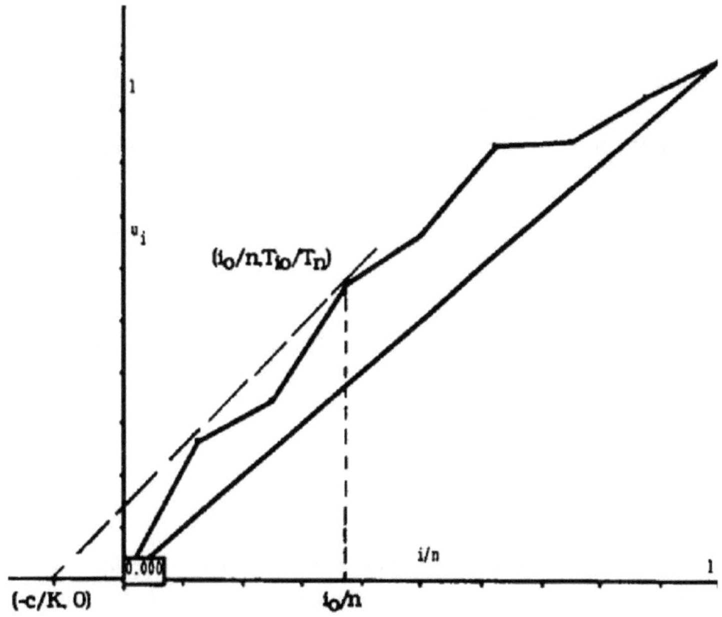

Figure 12.4 TTT-plot of the data given in Table12.2

Also, it is a useful tool for identifying life length models as shown by Bergman (1977). In other words, the TTT-plot is considered as an estimation of the TTT-transform of the underlying distribution of the observed data (Bergman and Klefsjö 1995). If the hazard rate is an increasing function of the component age, then the TTT-plot approaches a concave curve for a large number of observations (see Figure 12.4). Conversely, it approaches a convex curve if the hazard rate decreases. However, the plot of data from a distribution where hazard rate decreases at first and increases later first has a convex curve and later a concave curve.

Vibration-based Maintenance and TTT-plots

In general, damaged components generate specific vibration frequencies. These frequencies can be used to identify the problem and assess the damage severity of the components. For vibration measurements to detect faults in shafts and couplings such as imbalance, bending and misalignment the overall root mean square (RMS) vibration level is a reliable measure to assess the state of the shaft and coupling (Al-Najjar 1997). However, this measure is relatively insensitive to changes in the state of bearings or gears compared with the relevant damage vibration frequencies and bands. When using VBM, the replacement of a component is performed as soon as x_{th} is approached or exceeded. In this context, the TTT may be understood as total time in operation because we are using condition-based replacement (and failure) data instead of just failure or test data.

Suppose that the parameter values, such as the monitored vibration levels of n identical bearings at different locations in the machine but of identical operating conditions are monitored until the respective replacement times, τ_i, $i = 1, 2,...n$, where no failures are experienced. Define $x_j = sup\ x_j(t)$, $t \leq \tau$. That is x_j is the largest value that the monitoring parameter acquired during the monitoring time, *e.g.*, running time (age) t, where $t \leq \tau_j$. Order the indices so that $x_{(i)}$ corresponds to that x_j which is ith in size. Then, the ordered indices are $x_{(1)} \leq x_{(2)} \leq ... \leq x_{(n)}$. Define $S_j(x)$ as the total time on test (operation) generated by the jth bearing before its parameter value exceeds the level $x_{(i)}$ for the first time, *i.e.*,

$$S_j(x) = \inf\{t;\ x_j(t) \geq x_i\}, \tag{12.3}$$

$$Let\ S(x) = \sum_{j=1}^{n} S_j(x). \tag{12.4}$$

Define $T_i = S(x_{(i)})$, which is the total time accumulated by all bearings while their parameter values are less than or equal to $x_{(i)}$, $i = 1,.., n$. The plot of the TTT can then be obtained in the usual way, *i.e.*, the plot of $u_i = T_i/T_n$ against i/n. The ratio i/n is actually an estimation of the probability that the failure occurs if no planned replacement is done, where i is the number of planned (condition-based) replacements done before their parameter value exceeds the level $x_{(i)}$ for the first time. n is the total number of the bearings under consideration and T_n/n is an estimate of the expected time to a planned replacement. Denote by c_1 and c_2 the cost incurred by a planned action and an additional cost that is incurred only at failure. The costs c_1 and c_2 are considered independent of maintenance policy. Denote by $T_{io,\pi}$, the time to planned replacement at the index i_o and when using maintenance policy π. Thus, the estimate of the minimum long run average cost *per* unit time (C_π) when technique π is used can be written as:

$$C_\pi = \frac{c_1 + c_2 \dfrac{i_{o,\pi}}{n}}{\dfrac{i}{n} T_{io,\pi}}. \tag{12.5}$$

The index $i_{o\pi}$ minimising C_π can be estimated graphically (Al-Najjar 1999).

Monitoring Bearing Condition by GTTT-plots

The generalised total test on time (GTTT-plot) is a modified version of a TTT-plot. The GTTT-plot gives the possibility to determine the optimum replacement interval and to distinguish the most cost-effective maintenance policy/strategy when there are several applicable policies/strategies (Al-Najjar 1999). It is also used for assessing the probability of failure of a component at time *t* when damage in the component is confirmed, which can consequently provide information about a component's mean effective life (Al-Najjar 2003). In this section, the use of GTTT-plots for assessing the probability of failure of a bearing and its residual lifetime are discussed.

This assessment is necessary to effectively enhance the information given to describe, *e.g.*, the bearing condition after each vibration measurement, especially when the vibration level increases rapidly or when it is relatively high (significant) and there is a risk of faster deterioration during the time until the next measurement. From practical experience, the level of a CM parameter, *e.g.*, the vibration level, at which a bearing should be replaced (x_{th}) is always treated as the maximum allowable level.

In many cases, a failure occurs soon after the monitoring parameter(s) exceeds a well-selected replacement level. The vibration level at which a replacement should be made can be selected statistically. Sometimes, the value of a CM parameter may increase suddenly to be close to or even higher than x_{th}. In some cases, these increments arise randomly and are not always related to the changes in the state of the component under consideration because they are induced by some temporary external or internal disturbing factors.

Using the databases of the vibration monitoring software program, it is possible to identify the information concerning the times of installation, condition-based replacements and failures and the condition of the bearing during its operating life, or at particular times when using a periodic monitoring system. Most data in the databases of such companies usually consist of condition-based replacements due to the consequential high cost of failure. This, in general, is because the company's policy is to replace a component, such as a bearing, as soon as its vibration level deviates from the norm. By an effective use of VBM policy integrated with the company's IT-system, the machine can run until just before failure as defined by the monitored parameter reaching a predetermined unacceptable value (Al-Najjar 1997). Therefore, these replacements can be treated statistically as failure times. However, to be fully effective, they must all be near-failure times.

Renewals on the basis of the vibration level without taking the vibration history into account and renewal at the first substantial increase in vibration are common reasons for disallowing renewals as near failures. In such cases they must be treated as censored lives. Sometimes, components such as rolling element bearings may be replaced at a higher level than the predetermined level. This may happen due to faster deterioration and faster increase in the vibration level than was anticipated, without exposing the operating safety, machine functions, the production rate and product quality to a bigger risk (Al-Najjar 1999). Prolonging the operating life of a component, *e.g.*, a rolling element bearing, by making its replacement level (x_{th}) higher is important to reduce the number of planned and unplanned stoppages of a machine, which are expensive in many industries due to production losses.

When the predetermined warning level is exceeded, it means that damage is already initiated and getting worse (Al-Najjar 1998). Then, the most interesting question that should be answered before making any decision is: what is the probability of failure of the bearing during the period prior to the next measurement or planned stoppage and how much of the bearing's life is left? Uncertainty in the answers of these two questions costs many companies appreciable losses. In order to use TTT-plots, a laboratory test of identical components at identical operating and environmental conditions is usually needed. Such a test is usually terminated after a predetermined test time or when all, or a predetermined number, of components have failed.

The operating conditions used to test these components usually are not a genuine replication of the real operating conditions. This means that we lose some reality by testing components in laboratory. In companies with expensive downtime, it seems to be impossible to perform real experiments due to the high failure cost. However, the databases of many of the companies using vibration monitoring software and systems include a large number of replacements of identical components. For example, in paper mills it is possible to find a large number of identical components, such as rolling element bearings, installed in different positions of the paper mill machine. In many cases, it is realistic to assume that the operating and environmental conditions are approximately the same for these identical bearings independent of their positions in the machine, because if the conditions were not that similar, bearings of different design would have been installed.

For identical components such as bearings, the bearing life can then be assumed to be independent of the component position in the machine if operating and environmental conditions such as speed, load, lubricant and temperature are approximately the same. A paper mill machine may include about 400 bearings and many of them are often identical. Often during the operating life of the machine, several replacements of bearings can be carried out at different positions of the machine (Al-Najjar 2000). Many positions may undergo several replacements. The replacements of identical bearings may be made at different but adjacent vibration levels where the life length is marginally varied. Then, it would be reasonable to consider that the number of identical bearings that has been replaced during the operating life of a paper mill machine is large, which can easily be verified in companies with well documented databases.

Development of GTTT-plots

We assume that no failure will occur as long as the vibration level of a bearing is still below the replacement level. Such an assumption is not unrealistic because nowadays the number of machines that does not experience failures during long periods, say 2 years, is increasing rapidly due to the use of more effective VBM (Al-Najjar *et al.* 2001). With VBM, the probability that the replacements of a particular bearing will occur precisely at the predetermined vibration level is very low (almost zero).

The replacements are usually performed at either higher or lower levels. These different replacement levels for identical bearings can be utilised to obtain GTTT-plots. Similar bearings can also be used, but we should realise that some of the certainty in the value assessed can be lost. Here, the average time generated in ages is emphasised, *i.e.*, T_i/n, at time t and before $x(i)$ is exceeded, where T_i is the total time accumulated by all running (identical) bearings while their CM parameter values, *e.g.*, vibration level, are less than or equal to $x(i)$ *(for* $i = 1,.., n$), n_i is the number of replacements that have been done until $x(i)$ is exceeded, while n represents the total number of the bearings under consideration and x_{th} is the replacement vibration level. Then the model is

$$U_i = (T_i/T_n), \quad and \qquad (12.6)$$
$$= (T_i/n)/(T_n/n)$$

This plot is actually the usual GTTT-plot but it is plotted on the abscissa of a different definition (Al-Najjar 1999). U_i indicates the proportion of the average exhausted life length of the bearings until their vibration levels approaches or exceeds $x_{(i)}$ divided by the average time generated by n components until their levels equal or exceed $x_{(n)}$, for $x_{(i)}<x_{(n)}$ and $i=1,\ldots,n$. The plot of U_i versus n_i/n can now be used for monitoring the probability of failure of these identical bearings. The replacements of identical bearings at several positions in a paper mill machine are usually performed at different times along the operating life of the machine. Data about n identical (similar) bearings can be collected partly from past data and partly from the running bearing.

The past data of a number of bearing replacements together with the operating time of the running bearing under consideration can be used to estimate T_i at each time when the probability of failure of the bearing needs to be assessed. The average of the total time accumulated by all bearings or mean effective life, *i.e.*, T_n/n, when their vibration levels approaches or just exceeds the replacement level (x_{th}), can be calculated from past data, *e.g.*, vibration measurements, time of installation and replacements, of nominally identical bearings. When x_{th} is high enough, *i.e.*, the bearing's replacement is supposed to be made just before failure, this yields that T_n/n can be assumed constant during monitoring a new installed bearing of the same type.

When the GTTT-plot is obtained, the operating time of the bearing being monitored can be used together with the past data to re-calculate T_i/n at any time t. There are three main situations where the assessment of the probability of failure of a bearing and its residual time are considered very important to highlight the bearing's state:

- At any time $t > 0$ when the vibration level increases, but is not high enough to motivate a maintenance action, and there is a risk of faster deterioration during the time until the next measurement.
- When the vibration level is close to or larger than x_{th} and the production delivery schedule should be fulfilled.
- When the vibration level suddenly exceeds the predetermined warning level, which is lower than the replacement level.

Once T_i/n is assessed, the probability of failure at time t (where t is the time following directly after taking the vibration measurement at the running bearing) can be determined graphically. Also, the residual effective life of the bearing can be predicted when T_n/n is assumed constant for the bearing type in question. This GTTT-plot can be improved continuously after each replacement (or failure), *i.e.*, if the previous GTTT-plot was obtained depending on data from n bearings, the improved plot should be obtained depending on the replacement data from $(n+1)$ bearings, $n = 1, 2,...$ The GTTT-plot can easily be obtained by using software programs.

Development of a Software Program Prototype

Software to easily apply ProFail and ResLife on a daily basis has been developed, see Figure 12.5 (Al-Najjar and Ciganovic 2009). This software tool is supplementary to PreVib for enhancing the accuracy in maintenance decisions. Also, ProFail, ResLife and PreVib can be used jointly and independently. As for PreVib and ProFail the ResLife user-interface is divided into two halves. The left half represents input data required for assessment and in the right half the result is shown in the form of a plot (graph).

When assessing the probability of failure and residual lifetime of a component, the *segment* (machine), *asset* (component, *e.g.*, a rolling element bearing in this case) and the *assessment time point*, *i.e.*, the time at which the assessment should be done followed by clicking the assess-button must be specified. Subsequently the program uploads the table of lifetime data from MIMOSA database and presents the graph on the right half of the user-interface. The y-axis in the graph shows the proportion of the average exhausted lifetime and the x-axis shows the probability of failure for the analysed component. The *probability of failure* of the component in question and its *residual lifetime* are then assessed and displayed on the left-hand side of the user-interface.

The values can be shown by using the cursor on the plot in the right half of the user-interface. If a new assessment is performed after the current assessment time point, then consequently the cross-air pointer will move forward on the curve. Each point on the graph represents each and one of the component lifetimes in relation to each other.

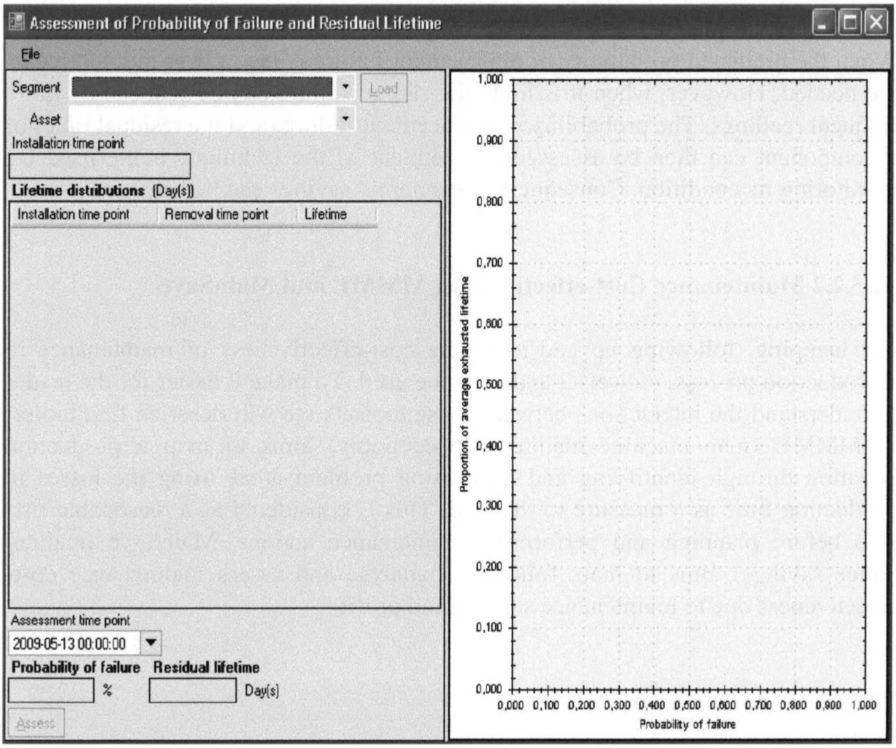

Figure 12.5 User-interface of the software program ProFail and ResLife

Discussions and Conclusions

When using the past history of a large number of identical, or sometimes similar, components, it becomes easier to decide whether or not an increment in the vibration level is serious. However, it does not provide that reliable information required to decide when to perform a maintenance action. Monitoring ProFail and ResLife (GTTT-plot), the probability of failure of a damaged bearing and its residual lifetime at any time $t > 0$ can be assessed, which can be used in addition to the vibration level current value and its historical trend to reliably illustrate the component state and decide when to perform a maintenance action. Monitoring the mean effective life of identical components and their probability density function is useful to detect dramatic changes in the distribution of the time to failure/replacement, which is an indication of whether there are deviations in the quality of these bearings and in their operating/environmental conditions.

When using a periodic measuring system for monitoring vibration, the time between measurements is almost always considered constant. It is usually assessed based on the maintenance staff's experience. However, using the GTTT-plot, maintenance engineers can predict in advance the probability of replacement/failure of the bearing during the period to the next proposed measurement opportunity or planned stoppage, which yields reliable measurement planning. When the predicted probability of replacement/failure is low, fewer measurements are needed. However, when it is high, the risk of failure can be reduced by more frequent readings. The probability of replacement/failure (and the residual time) of a component can then be assessed independent of the technique being used for monitoring its condition. Consequently, economic savings can be achieved.

12.3.2.2 Maintenance Cost-effectiveness: MMME and MainSave

For mapping, following up and assessing cost-effectiveness of maintenance in a production process, toolsets 2 and 3 can be used. To make it easier for the reader to understand the interactions between these toolsets we will describe first toolset 3. MMME (man–machine–maintenance–economy) aims to map a production situation through identifying and prioritising problem areas using the losses in production time as a measure to this end. This is considered as a reasonable first step before planning and performing maintenance actions. MainSave (maintenance savings) aims to map, follow up, analyse and assess maintenance cost-effectiveness due to maintenance savings and profit.

Modelling of MMME Interactions

In this section, the focus is on developing a model that can describe the interactions between man–machine–maintenance and economy so that it will be possible to assess the losses in the production time, identify, localise and prioritise problem areas. Working areas participate in a production process, such as operation, human resources and organisation, quality, working environment, maintenance, production logistics, *etc.*, interact with each other and affect the production process outcome. MMME was developed with respect to a broader maintenance perspective described by TQM to describe the above mentioned interactions. Moreover, it helps to identify causes behind losses in production time and assess these losses. The latter can be utilised easily as a trigger for starting a deeper analysis to justify the solutions suggested when the losses can be assessed (Al-Najjar 2007a). For simplicity, we neglect some variables that might have a marginal effect on the number of failures and production time, such as the quality of the raw material, assuming that the company uses TQM to select and control purchasing of the raw material and the condition of the machine tools, assuming that the tools are of high quality. The development of the model can be described in two stages; first the major categories of losses are identified and then the basic causes/factors (subcategories) behind losses, *i.e.*, the factors constituting theses categories are identified (Al-Najjar *et al.* 2007).

Description of Causes

A classification of causes at the operative level is necessary to easily identify the losses in production time and consequently to analyse and distinguish the root-causes behind them. Therefore, the model is divided into different segments to make it possible to highlight the causes of unplanned stoppages and production-time losses. The operational level is divided into three main categories; human recourses and organisation, the production process and maintenance. A schematic illustration of the classification of causes of failures and the prioritised area for improvement is shown in Figure 12.6.

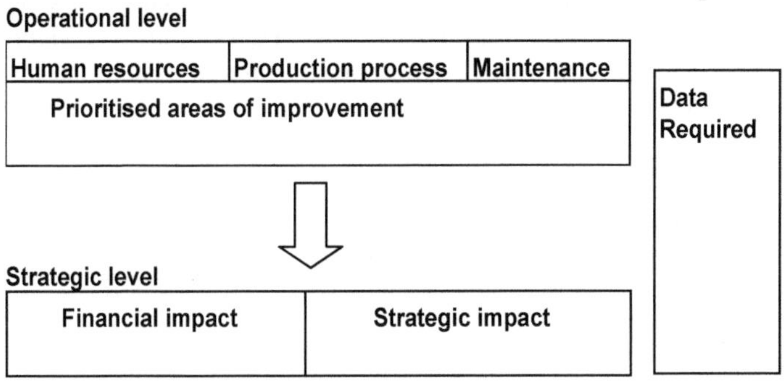

Figure 12.6 Classification of the causes of production time loss

In the strategic level, we translate the technical data from the operative level into economic data. Each category consists of several sub-categories that may potentially lead to losses in the production time. Each sub-category of the production time loss categories on the operative level are described below. Note that for simplicity the lateral interactions between these factors are not considered in the study. The model introduced in Figure 12.7 emphasises the differences between measurements from previous and current periods in order to ease identification and classification of problem areas and quantify the losses. It will not be such a big task to transfer data stated in the model MMME from the operational level to the economic/strategic level, *i.e.*, expressed in money, as long as the losses in production time have already been assessed.

Human recourses and organisation: from everyday experience, one of the important resources required for production is human resources and their organisation. These resources are as much as necessary also a source of problems that may cause losses in production time, such as the case of operator or maintenance technician unavailability when needed, insufficient training, bad communication, lack of (or insufficient) experience and commitment (Al-Najjar *et al.* 2007). It is not always reliable in meeting the requirements of the enterprises, such as the case when it concerns maintenance organisation support (Al-Najjar 2000). This is why some of the failures and other losses in production time that occur in a production process have their root-causes in this area. Human error and misuse may make up about 30% of the total causes behind unplanned stoppages.

A work environment of a production process can be described using ambient temperature, humidity, noise, vibration, pollution, dirt, *etc.* The working environment influences many attributes that affect the production time, such as the health of the personnel, the working moral/ethic, the condition and failure rates of the physical assets, the repair time needed to restore equipment after a failure, *etc.* Maintaining the quality of the working environment would probably increase the probability of eliminating some of the causes behind disturbances and production time losses. For example, production time losses can be reduced through reducing the probability of unavailability of personnel, enhancement of working moral, reducing the asset failure rate, *etc.* This is why TQMain and TPM promote the need for a clean and well organised working environment (Haarman and Delahay 2004, Al-Najjar 2007a).

Changes in the Operative level											
Period	Human Resources and organisation (stoppages because of these factors)				Production process characteristics that can influence and be influenced by maintenance		Maintenance technical impact on these characteristics at the operative level, i.e. impact of maintenance policy				Total
	Working environment	Maintenance organisation and management tools	Competence, i.e. ability to perform maintenance and operation tasks	Personnel Commitments & communication	Machine condition and characteristics	Production procedures and methodology	Failures	Short stoppages	Stability of production quality	Stability of the production speed and production rate	Change in production time

Impact of deviation in the above on the operation time											Total
Previous											
Current											
Difference											
Target level											
Warning level											
Problem area	Priority				Priority		Priority				

Figure 12.7 MMME model

Clear instructions and routines for performing maintenance action are necessary for making it easy for the personnel to know what procedures to take in different situations. This will get the job done quicker and increase the morale among the workers because they know what to do and this will consequently reduce the losses in production time effectively. Maintenance support is the ability of maintenance organisation to meet production requirements, otherwise the losses in the production may increase. If maintenance staff do not possess the right competence to use the methods and tools required for high quality maintenance work, they cannot be expected to perform the right maintenance action on time (Al-Najjar 2000). Personnel also need to be committed to their work, which means that they are committed to following the established routines and instructions effectively, for example to lubricate a machine each Monday.

Lack of, or ineffective communication in the organisation, especially between the operator and maintenance staff, constitutes an additional risk area for misunderstandings, which may lead to more disturbances and failures. This is one of the important reasons behind the traditional conflict between the operator and maintenance staff where every one blame the others for disturbances in production.

Production process characteristics: in general, production process input characteristics, such as the condition and reliability of producing assets and production procedures influence, and are influenced by, maintenance performance and quality. For example, old and deteriorated machinery induces many failures and disturbances and causes production time losses, which in turn demands more maintenance work (Al-Najjar 2007a). The reliability of a machine is one of its characteristics achieved by the manufacturer and affected by maintenance work, the working environment and operating conditions.

Maintenance technical impact: the maintenance performance quality influences the condition of a component/machine appreciably. For example, the installation of a rolling element bearing eccentrically increases the possibility of severe wear and consequently shorter life of that bearing, meaning more frequent stoppages of production and more losses of production time. In other words, the maintenance performance quality influences machine availability, performance efficiency and quality rate, which are expressed in their basic constituents, such as the number of unplanned stoppages, production speed stability, production quality, *etc.*

Maintenance policy is defined in BS 3811:1993 as "a description of the interrelationship between the maintenance echelons, the indenture levels and the levels of maintenance to be applied for the maintenance of an item". In this chapter, maintenance policy role is not only to perform repair/replacement at need, but rather to detect deviations at an early stage in order to prevent the occurrence of failures and even damage initiation if technologically possible. In general, repair time is strongly influenced by the maintenance policy, because the latter decides on the technology required for gathering relevant data to perform maintenance actions cost-effectively. For example, VBM policy provides much more information about the past and current condition of a component/equipment and product quality compared with PM.

In the MMME model, the focus of the model is on the changes or the difference between the amounts of losses in the production time during two periods. These two periods can be selected in a way that one lies before and the second after a particular change/improvement has been done in the production or maintenance process. Also, they can be selected with respect to other criteria. This facilitates the detection of changes without the confusion of the absolute value of each factor, which are irrelevant in the model. Deviations in all the factors mentioned above can be translated at the operative level to production attributes and their impact measured on the production time scale. The model quantifies changes in the process to hours of production time losses. Changes in the production time losses can then be summed and the increase/decrease of production time losses can be assessed. For more details about MMME see Al-Najjar *et al.* (2007).

MMME Software Development

The software prototype of MMME model has been developed for easy daily based application of MMME. Most of the data required for applying MMME are available in the companies' existing databases. Since the model is able to assess deviations in production time, it is most suitable to be linked to an automated measuring-system for easier and more effective detection, assessment and categorising of the production time losses. The computerised model enables unlimited numbers of people to find the same information and share this information electronically and instantly (Figure 12.8).

The information that should be provided to before clicking the button *assess* are *segment* (machine), *production* (to select either all production or a specified one) and the production periods that will used to compare the losses in production time. When clicking *assess* the software retrieves the data required from MIMOSA database automatically. It is obvious that the work load will be far lower than when the system is fully integrated in the existing databases. Thus, accuracy will increase because the amount of data that has to be assessed or estimated is less and the risk of confusion is greatly reduced.

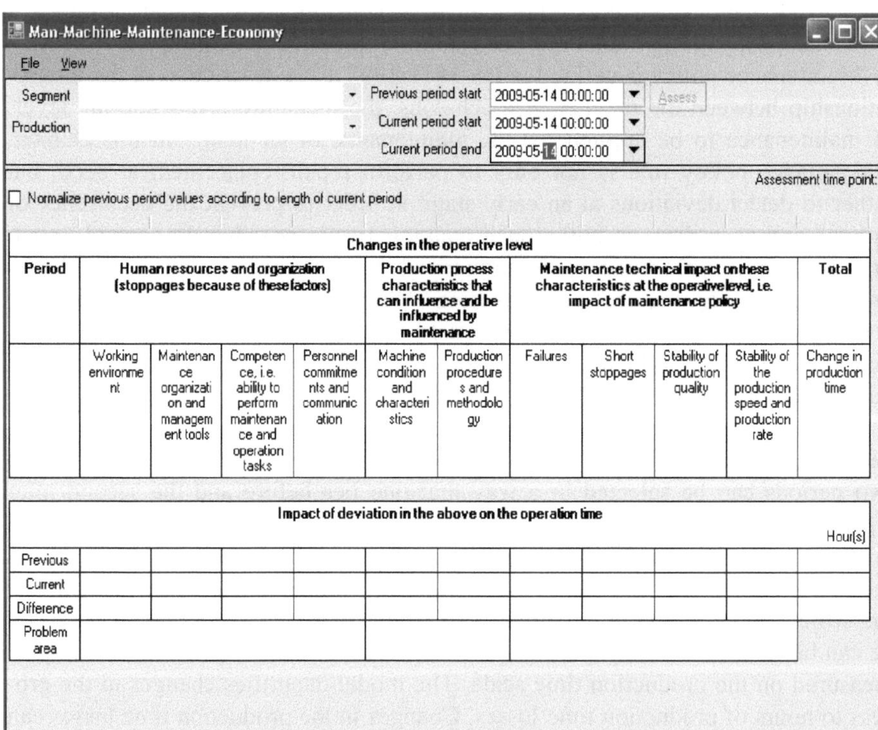

Figure 12.8 User-interface of the MMME software program

Assessment of Maintenance Cost-effectiveness (MainSave)

Manufacturing industries realise the importance of monitoring, analysis and follow up of the performance of production and maintenance processes by using economic and technical key performance indicators (KPIs). These indicators establish a bridge between the operational level in terms of, for example, productivity, performance efficiency, quality rate, availability and production cost, and the strategic level expressed by company profit and competitiveness. In the past, the survival of manufacturing companies was mainly connected to how much a company was able to push into the market.

This situation has changed and today's strategies imply cost minimisation and differentiation and the ability to use available resources in a cost-effective way. The focus on customer needs puts great demands on the production and maintenance systems to meet the goals of high product quality, production safety and delivery on time at a competitive price (Al-Najjar 2007a). Properly identified KPI are required for following up the work done to achieve companies' strategic objectives and daily competition survival. When the profit margin of a plant decreases, the need for a reliable and efficient maintenance policy becomes more important, because it will be more important to reduce the economic losses, production cost *per* high quality item/ton/metre and consequently increase the profit margin. Thus, technical and economic KPIs are important for following a company's performance and for benchmarking.

Theoretical Background

In order to assess the economic importance of an investment in maintenance, it is often necessary to find the life cycle income (LCI) of a machine/equipment, which is usually not an easy task. Also when a company's profit is generally considered, the assessment of the savings achieved is influenced by irrelevant factors, for example the production sold, currency course, wars, crises and product price. Therefore, it is easier to asses the savings, and the return on investment in maintenance (ROIIM), that have been achieved by more efficient maintenance through assessing the reduction in the downtime, the number of rejected items, capital tied in inventories and operating costs, *etc.* (Al-Najjar 2007a). This can be done by breaking down the life cycle cost to its cost factors. Then, analysing and assessing the transactions between maintenance as a system and other maintenance-relevant working areas, *e.g.*, production, quality, inventory, *etc.*, as systems, can be used to highlight the real maintenance role in the internal effectiveness of a producing company.

To be able to monitor, assess and improve the outcome of different maintenance actions it is necessary to develop a model for identifying and localising/retrieving both technical and economic data from company databases. In order to make the process of data gathering and analysis even easier and more cost-effective, the model should be computerised (Al-Najjar and Kans 2006, Kans 2008). Using MIMOSA reduces the technical difficulties and disturbance that may be induced in the current IT-systems of a company (MIMOSA 2006). This would allow follow up of maintenance KPIs more frequently and easily, thereby leading to quicker reaction on disturbances and avoidance of unnecessary costs. It will also be easier to trace the causes behind deviations. The model should also help in interpreting the measurements of relevant basic variables and KPIs in order to achieve cost-effective decisions in planning and executing maintenance actions and to identify where an investment in maintenance may have the best financial payoff (Al-Najjar *et al.* 2004).

Modelling Cost-effectiveness with Respect to Maintenance

A maintenance policy is considered cost-effective if and only if its return on investment (ROIIM) is greater than the capital invested in maintenance. The benefits of the improvements in maintenance are usually collected in other working areas but hardly in maintenance as long as its accountancy system shows just costs. Identifying and relating the benefits generated by a VBM policy is not such an easy task to perform if the mechanisms of transferring maintenance impacts, and technical and economic KPI are not well identified (Al-Najjar 2007a). To justify the investments in maintenance, the cost-effectiveness (C_e) of each investment can be examined by using the proportion of the difference between the average cost *per* high quality product before (B_b) and after (B_a) the improvement to that before (B_b), *i.e.*,

$$C_e = 1 - \frac{B_a}{B_b}. \tag{12.7}$$

This means that all the savings (and possible increments) in the expenses of production, tied-up capital, insurance premiums, *etc.*, including the maintenance cost resulting of maintenance activities (maintenance direct cost) should be assessed.

At the beginning C_e can be ≥ 0 due to the extra expenses incurred by the learning period. This period can be defined on the basis of the nature of each improvement. However, beyond this learning period it should be larger than zero. C_e indicates the percentage of the reduction in the total production cost due to the maintenance impact and can thus be used as a measure of the cost-effectiveness of the improvements. Changes in the production conditions, production and maintenance processes usually lead to appreciable changes in the performance of production and/or maintenance processes.

Therefore, the formulas that have been developed (Al-Najjar 2007a), are used for assessing the impact of maintenance on the company economics during two different periods to highlight changes in production and maintenance performances and results, *i.e.*, savings or losses achieved due to better or worse usage of the available maintenance technologies. These two periods can be selected in the same way that has been described for applying MMME, see above. The formulas for assessing maintenance cost-effectiveness, see below, can be applied independently of the maintenance technique being used.

Denote five of the most popular sources generating savings/losses by S_i, for $i = 1, 2,\ldots, 5$. These are changes in; failures, short stoppages, stoppage time, bad quality production due to inefficient maintenance and additional expenses that can be defined by the user. The total savings or losses can simply be expressed as;

$$\text{total savings} = \sum_{i=1}^{i=5} S_i \,, \tag{12.8}$$

where $S_1, S_2, \ldots, S_5$ are assessed using the following formulas:

1. Failures: the savings or losses in the production cost that have been generated due to less or more failures (S_1) can be expressed by

   S_1 = number of failures avoided * average stoppage time* production rate * profit margin (P_M)

$$S_1 = [(Y - y) * L_1] * \text{Pr} * P_M \,, \tag{12.9}$$

where Y and y are the numbers of failures during the previous and current period, respectively, L_1 is the failure average stoppage time and Pr is the production rate.

2. Average stoppage time: the savings or losses that have been generated due to shorter or longer stoppages (S_2), *i.e.*, longer/shorter production time, is expressed as

S_2 = difference in failure average stoppage time * number of failures * production rate * profit margin

$$S_2 = [(L_1 - l_1) * y] * \text{Pr} * P_M , \quad (12.10)$$

where L_1 and l_1 are failure average stoppage times during the pervious and current period, respectively .

3. Short stoppages: the savings or losses in the production cost that have been generated by less short stoppages (S_3) can be expressed by;

S_3 = [short stoppages in previous period (B) – short stoppages in current period (b)] * average stoppage time (L_2) * production rate * profit margin

$$S_3 = [(B - b) * L_2] * \text{Pr} * P_M . \quad (12.11)$$

4. Quality production. The savings or losses generated due to higher production quality (S_4) can be expressed by

S_4 = [current period high quality production *per* hour – previous period high quality production *per* hour] * number of production hours *per* day (P_h) *number of production days *per* period (Pd) * profit margin

$$S_4 = (p - P) * P_h * P_d * P_M , \quad (12.12)$$

where P and p are the amount (in tons, meters, *etc.*) of high quality product produced *per* hour in the previous and current year, respectively

5. User defined expenses paid by the company to cover, for instance, personnel compensation due to accidents, environmental damage penalty, insurance premium, direct maintenance costs (including labour, spare parts and overheads), tied up capital in spare parts and equipment and penalty expenses of delivery delay. Denote the expenses before and after the improvement, *i.e.*, previous and current period, by E_b and E_a, respectively. Then, the sum of the reduction or increment in these expenses can be expressed by

$$S_5 = \sum_j \left(E_b - E_a\right)_j , \tag{12.13}$$

where j =1, 2,..., n denotes the n types of the expenses that can be expressed by the user. In order to make the model industrially applicable, we tried to make it transparent and avoid using the idea of black box, *i.e.*, the end user feeds in the data required in the software and gets the results by pushing particular buttons without knowing any thing about what has happened inside the software.

Development of the Software Prototype

One of the major reasons behind the lack of techniques for controlling and assessing maintenance economic impact on company profitability and competitiveness is the lack of clear and robust theory, methods and tools required for that difficult task. Also, the difficulties in finding and processing the data required for monitoring, analysing, controlling, following up and assessing maintenance economic impact. This is why a software-tool will make it is possible to perform this task easily and cost-effectively (see Figure 12.9). A software program module (MainSave) for easy and daily use to assess maintenance cost-effectiveness has been developed (Al-Najjar 2009).

The main objective of using MainSave software is to assess production economic losses and maintenance savings/losses due to failures, downtime, short stoppages, quality problems and user defined expenses. In order to run MainSave software, two *production periods*, which will be compared, should be specified. *Segment*, *profit margin*, *total investment* and *depreciation periods* also should be specified. The results of applying MainSave are: assessment of several economic KPI, such as OEE, quality rate, maintenance total saving and profit, total potential saving (economic losses) , rate of total savings to potential savings, rate of savings to investment and rate of savings to investment to potential saving.

MainSave software enables the user to easily and quickly see the contribution of maintenance in company profit as well as the potential for further improvements. Not only does this help the user to assess the current situation, but it also makes it easier to identify problem areas, assess technical and economic losses and motivate investments in maintenance. The latter is important to convince the company's board and executive nanager of the necessity of these investments for enhancing the productivity and effectiveness of the production process. All these results cannot be achieved without high quality and relevant coverage data. Also, the data required for applying MainSave should be easily retrieved by the system.

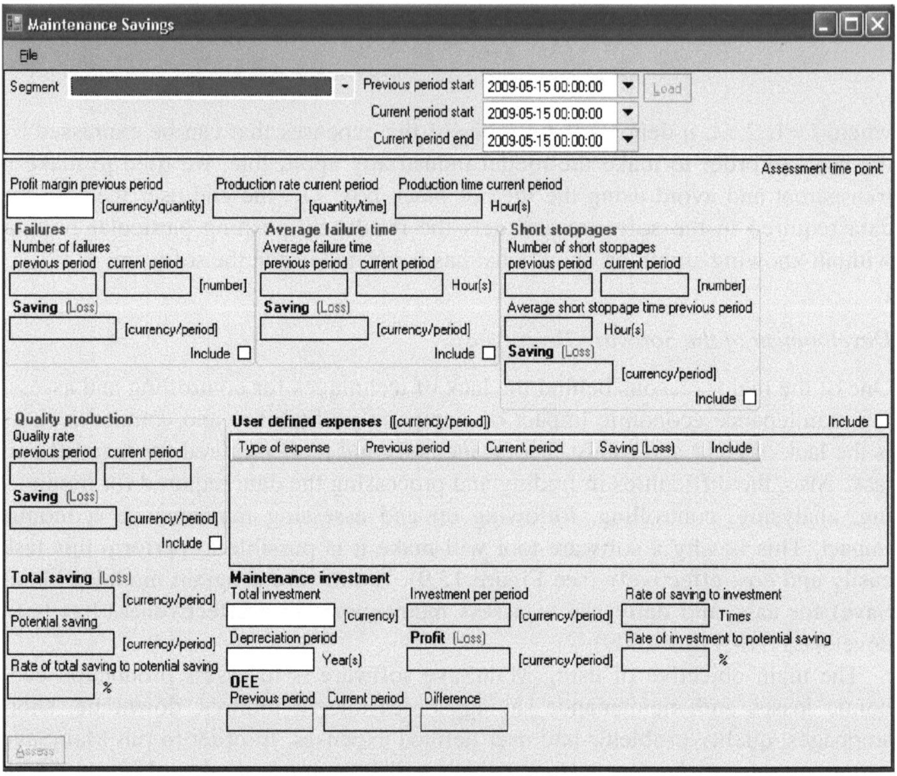

Figure 12.9 User-interface of the software program MainSave

12.3.3.3 Simulation of the Most Cost-effective Solution (AltSim)

With respect to problem analysis and solution, from everyday experience it is always possible to find several alternative solutions for the same problem. All these solutions can be suitable technically. However, they are not necessary equally cost-effective. In other words, different solutions usually mean different investments, probably unequally payoff (ROIIM) and payoff periods, which are very important factors for deciding whether or not an investment should be considered. In many cases, maintenance related economic factors influence a big share of the company income. Mckone and Weiss (1998) cited that the amount of money spent company-wide on maintenance by du Pont (1991) was roughly equal to its net income. This is why selecting the most cost-effective solutions for the problems in producing companies will reduce the economic losses appreciably. These losses may spread over a wide area of the company's activities, such as losses of production time because of unplanned stoppages, bad quality production, an unacceptable working environment, failure related accidents, environmental pollution, *etc.* The economic influence of maintenance in the Swedish industry was estimated in a study (Ahlmann 1998) to be around SEK 190–200 billion annually.

Development of a Simulation Model

In Ljungberg (1998) a case study was conducted at a Swedish car factory, overall equipment effectiveness (OEE) was estimated to be on average around 55%. Also, according to a study by Almström and Kinnander (2007), the productivity of the engineering manufacturing industry in Sweden is about 50%. Therefore, the industry could increase its production capacity without big investments in new machinery if an efficient maintenance policy were implemented. It is not that difficult to recognise the role of maintenance in maintaining and improving OEE by maintaining availability, quality rate, performance efficiency and production rate, machine quality/specifications, as well as fulfilling delivery schedules and reducing violation of the environment. The role of maintenance in a company and its contribution to generating more profit can be clarified if we convert and measure all these factors to just one common well-understood scale, *i.e.*, money. In this way, it will be possible to follow up the whole chain of the investment, *i.e.*,

capital-to-*technical measures in maintenance*-to-*technical measures in production*-to-*economic measures in the strategic level*-to-*capital*

In this section we consider the same theoretical background and formulas used in MainSave to develop a new model that can be used for simulating alternative (AltSim) solutions, which are all supposed to be suitable technically, to distinguish the most cost-effective one. In order to examine several alternative solutions for a particular technical problem to distinguish the most cost-effective one, the model is constructed in the form the nine steps described below.

1. Developing a reference situation mapping the real status of the production system using a minimised version of MainSave software.
2. Identifying and prioritising problem areas in the production and maintenance processes for deeper analysis using MMME software. This can also be done using toolset I (PreVib, ProFail and ResLife) to identify a problem that might be in the potential failure or imminent failure phase.
3. Technical analysis of problems is important to identify the root-causes behind damage initiation and damage developing mechanisms, and for suggesting three alternative solutions to solve the problem. For example, using different CM technologies, a new analysis software programme, training courses for better usage of the available maintenance technologies or to enhance the quality of maintenance actions. We use just three solutions to reduce the number of alternatives and to indicate three different levels of confidence: acceptable, good and very good technically, which are weighted by max. 60%, max. 80% and max. 100%, respectively.

4. Assessing the approximate investment capital (and the depreciation period) required for each alternative solution mentioned above in III, which may have unequal costs.
5. Examination of the effect of each solution on the production and maintenance processes should be done properly through anticipating the effect of every solution on the performance of production and maintenance processes to highlight its ability in, *e.g.*:
   - detecting problem symptoms at an early stage;
   - eliminating unplanned and unnecessary stoppages;
   - prolonging the lead-time;
   - reducing bad quality production; and
   - reducing failure related accidents.
6. Converting all these technical effects to economic measures.
7. Comparing the capital invested for each solution/technique and the payoff (savings).
8. Assessing total savings/profit and ratio of savings to investment for each alternative.
9. The most cost-effective solution is that which acquires higher profit. The ratio of savings to investment can be used to distinguish the most cost-effective solution, but not necessary that with the high profit.

Development of a Software Prototype

Applying the model developed above manually is very labour and time demanding. Therefore, software prototype AltSim has been developed to support the end user for daily application to select the most cost-effective solution when there are several alternatives. The alternative solutions suggested for solving a problem influence all the categories of savings/losses stated in the reference situation, *i.e.*, white coloured boxes. To run the AltSim software, *segment*, *profit margin*, *reference period* and *anticipated period* should be specified. When the button *load* is clicked, the data required to describe the reference situation will be retrieved from the MIMOSA database (see Figure 12.10). Figure 12.11 shows alternative solution 1, where all the data shown in the grey-coloured boxes are similar in all three alternatives and are retrieved as a reference situation from the MIMOSA database on the same basis as the MainSav software module.

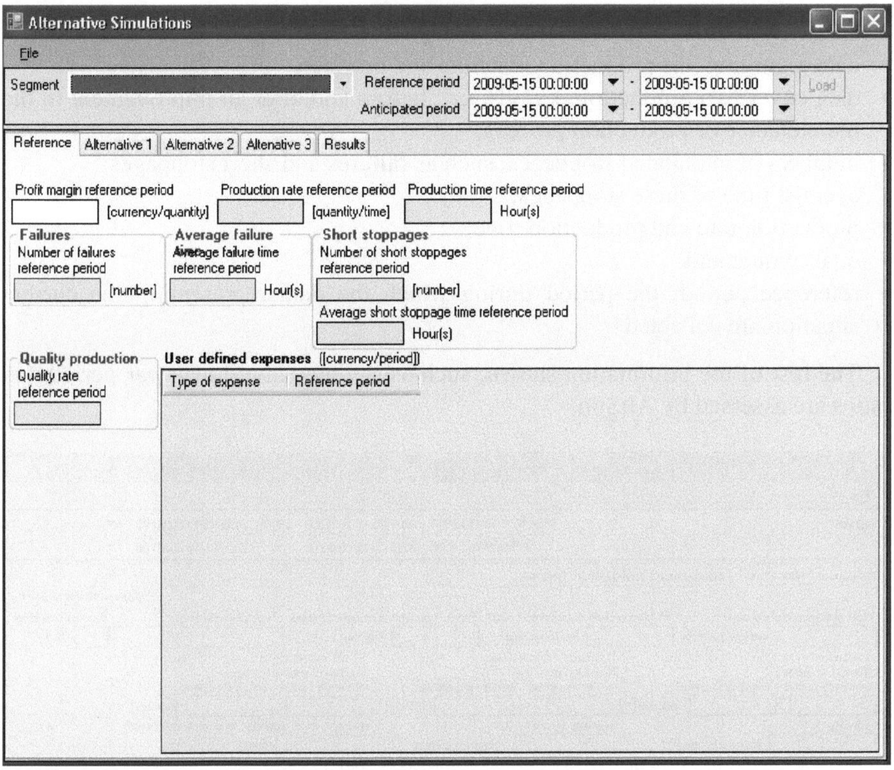

Figure 12.10 User-interface of AltSim software, reference situation

The user-interfaces of alternatives 1, 2 and 3 are identical except that concerning the data that should be anticipated in the white coloured boxes are different for different alternative solutions. When the information required for alternatives 1, 2 and 3 is anticipated, AltSim will automatically suggest the most cost-effective alternative when pushing the button *results* (see Figure 12.12). The data required for running AltSim are: real data representing the reference situation and retrieved from the database (database datasets in the grey boxes), and anticipated data (data that are expected), such as the number of failures when using a particular solution/CM technology during the anticipated period, *e.g.*, 1 year, *i.e.*, non-database datasets in the white coloured boxes (Figures 12.10 and 12.11). Non-database datasets that are even used for MainSave, are:

- profit margin per each item, ton, metre or cubic metre, *etc.*;
- total investment in maintenance for improving its performance;
- depreciation period, period of the investment life length;
- anticipated period for comparison, *e.g.*, 1 year; and
- production time and rate during the anticipated period.

The data described by database datasets are:

- data regarding one production machine and product;
- data cover two production periods, *i.e.*, before and after an improvement in the maintenance or production process;
- numbers of unplanned stoppages, such as failures and short stoppages;
- average time of these stoppages;
- production rate and production time;
- quality rate; and
- reference period; the period during which the data representing the current situation are collected.

The rest of the information shown, such as savings, investment *per* period and ratios are assessed by AltSim.

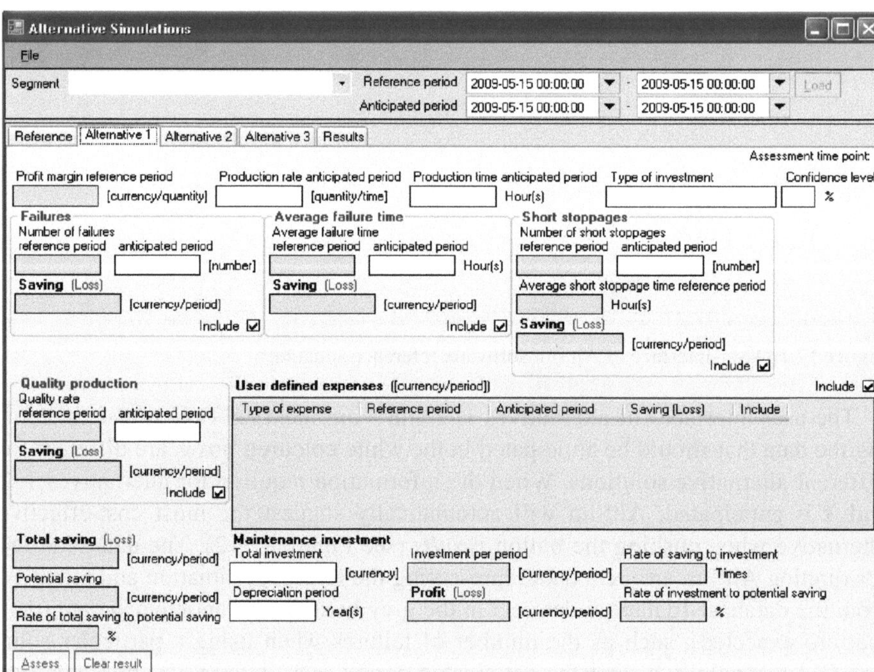

Figure 12.11 User-interface of AltSim software, alternative solution 1

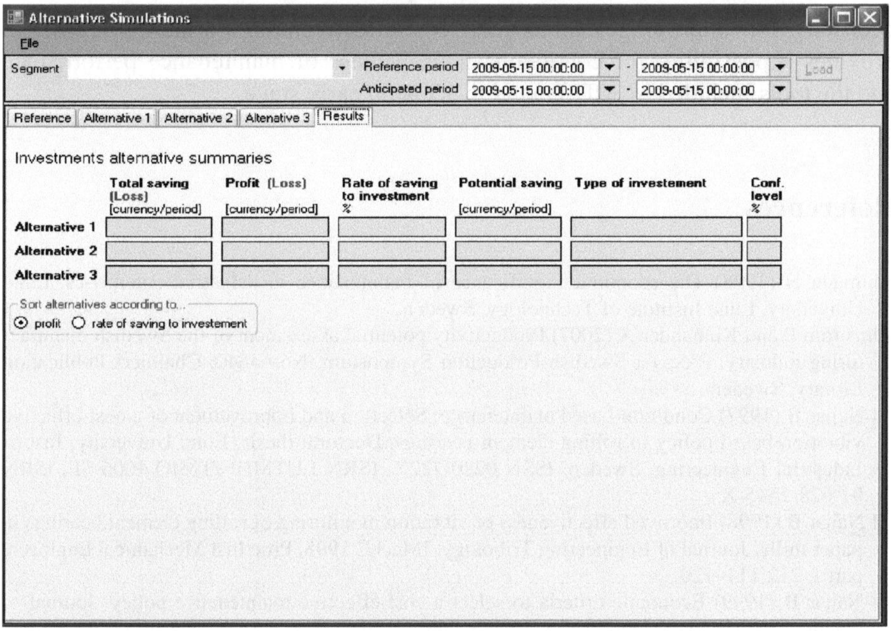

Figure 12.12 User-interface of AltSim software, results

12.4 Conclusions

The main results that have been achieved in this chapter are the development of MDSS, which consists of three strategies and five software modules for enhancing maintenance cost-effectiveness through effective problem detection, analysis and more accurate maintenance action planning. It is also used for mapping, following up, analysis, control, assessment and improvement of the role of maintenance in company business.

Predicting the vibration level at the next planned measuring time or next planned stoppage reduces the risk of an unexpected dramatic deviation in the condition of significant components and keeps the probability of failure of the component close to zero. Applying MainSave provides the possibility of controlling, after a short time of application, whether the cost-effective maintenance solution suggested by AltSim is really cost-effective. The models developed in this chapter are so flexible that they can be adapted for different production processes and problem causes/classification as expressed by MMME.

The main conclusions that can be drawn from this chapter are: application of the tools included by the three strategies (toolsets) implies that maintenance is a profit-centre and not just additional cost. Also, it provides a unique opportunity for gathering more relevant information required for higher accuracy of maintenance decisions for improving maintenance cost-effectiveness. Furthermore, MDSS provides a possibility for continuous improvement of maintenance performance and the tools included by MDSS, and to act at an early stage.

References

Ahlmann H (1998) The economic significance of maintenance in industrial enterprises. Lund University, Lund Institute of Technology, Sweden

Almström P and Kinnander A (2007) Productivity potential assessment of the Swedish manufacturing industry. Proc 1st Swedish Production Symposium, No 64640, Chalmers Publication Library, Sweden

Al-Najjar B (1997) Condition-based maintenance: Selection and improvement of a cost-effective vibration-based policy in rolling element bearings. Doctoral thesis, Lund University, Inst of Industrial Engineering, Sweden, ISSN 0280-722X, ISRN LUTMDN/TMIO-1006-SE, ISBN 91-628-2545-X

Al-Najjar B (1998) Improved effectiveness of vibration monitoring of rolling element bearings in paper mills. Journal of Engineering Tribology, IMechE 1998, Proc Inst Mechanical Engineers part J, 212:111–120

Al-Najjar B (1999) Economic criteria to select a cost-effective maintenance policy. Journal of Quality in Maintenance Engineering 5:236–248

Al-Najjar B (2000) Accuracy, effectiveness and improvement of vibration-based maintenance in paper mills; case studies. Journal of Sound and Vibration 229:389–410

Al-Najjar B (2001) Prediction of the vibration level when monitoring rolling element bearings in paper mill machines. Int Journal of COMADE 4:19–27

Al-Najjar B (2003) Total time on test, TTT-plots for condition monitoring of rolling element bearings in paper mills. Int Journal of COMADEM 6:27–32

Al-Najjar B (2006) Total quality maintenance for assuring continuous improvement of company's profitability and competitiveness: case studies. COMADEM 2006, 12-15.6.2006, Luleå, Sweden, 181–191

Al-Najjar B (2007a) The lack of maintenance and not maintenance which costs: A model to describe and quantify the impact of vibration-based maintenance on company's business. Int J of Production Economics (IJPPM)107, 55:260–73

Al-Najjar B (2007b) Establishing and running a condition-based maintenance policy: Applied example of vibration-based maintenance. WCEAM2007, 12-14.6.2007, Harrogate, UK. NDT, Northampton, 106–115

Al-Najjar B (2008) Maintenance function deployment (MFD) for cost-effective and continuous improvement of company's business: using total quality maintenance (TQMain). Euromaintenance 2008, 8-10.4.2008, Brussels, Belgium

Al-Najjar B (2009) A computerised model for assessing the return on investment in maintenance: following up maintenance contribution in company profit. Proceedings of WCEAM 2009, Athens, Greece. Springer, London

Al-Najjar B and Alsyouf I (2000) Improving effectiveness of manufacturing systems using total quality maintenance. Journal of Integrated Manufacturing Systems 11:267–276

Al-Najjar B, Alsyouf I, Salgado E, Khoshaba S, Faagorg K (2001) The economic importance of maintenance planning when using vibration-based maintenance. Växjö University, Terotechnology, Sweden

Al-Najjar B, Andersson D, Jacobsson M (2007) A model to describe the relationships man-machine-maintenance-economy (MMME). IMECS2007, 21-23 March, Hong Kong II:2127–133

Al-Najjar B and Ciganovic R (2009) A model for more accurate maintenance decisions (MMAMDec). Proceedings of WCEAM 2009, Athens, Greece. Springer, London

Al-Najjar B, Hansson M-O, Sunnegårdh P (2004) Benchmarking of maintenance performance: a case study in two manufacturers of furniture. IMA Journal of Management Mathematics 15:253–270

Al-Najjar B, Kans M (2006) A model to identify relevant data for accurate problem tracing and localisation, and cost-effective decisions: a case study. Int Conference ICME 2006, Oct. 2006, ChengDu, China. Masato Online Enterprises 2010

Al-Najjar B and Wang W (2001) A conceptual model for fault detection and decision-making for rolling element bearings in paper mills. Journal of Quality in Maintenance Engineering 7:192–207

Barlow RE and Campo RA (1975) Total time on test processes and applications to failure data analysis, in reliability and fault tree analysis. Barlow, Fussel and Sigpurwalla. SIAM , Pennsylvania

Batanov D, Nagarur N, Nitikhunkasem P (1993) EXPERT-MM: A knowledge-based system for maintenance management. Artificial Intelligence in Engineering 8:283–91

Bergman B (1977) Some graphical methods for maintenance planning. Annual Reliability and Maintainability Symposium, 467–471

Bergman B and Klefsjö B (1995) Quality from customer needs to customer satisfaction. Studentliteratur Lund, Sweden

Bloch HP and Geitner, FK (1994) Machinery failure, analysis and troubleshooting. Gulf Publishing Company, London

British Standard (1993) Glossary of terms used in terotechnology. BS 3811:1993

Groth T, Grimson W, Allahwerdi N, Baudin J, Duignan F, Gaffney P, Hayes R, Huhtala K, Larsson O, Modén H, Stephens G (1996) OpenLabs advanced instrument workstation services. Computer Methods and Programs in Biomedicine 50:143–159

Haarman M, Delahay G (2004) Value driven maintenance – new faith in maintenance. Mainnovation, Dordrecht

Hadzilacos T, Kalles D, Preston N, Melbourne P, Camarinopoulos L, Eimermacher M, Kallidromitis V, Frondistou-Yannas SS, Saegrov S (2000) UtilNets: a water mains rehabilitation decision-support system. Computers, Environment and Urban Systems 24:215–232

Hafkamp P and Schutters, G (2001) Designing and implementing a maintenance management system with an expert support system: methodology, implementation, expert IT-system, achieved cost reductions. 16th International Conference and Exhibition on Electricity Distribution 1:142–147. Institute of Electrical Engineerning, London

Hansen WA, Edson BN, Larter PC (1992) Reliability, availability, and maintainability expert system (RAMES). Annual Reliability and Maintainability Symposium 478–482

Herraty AG (1993) Bearing vibration-failures and diagnosis. Mining Technology, Feb, 51–53

Kans M (2008) On the utilisation of information technology for the management of profitable maintenance. PhD thesis, Växjö University Press, Sweden

Kipersztok O, Dildy GA (2002) Evidence-based Bayesian networks approach to airplane maintenance. Proc 2002 International Joint Conference on Neural Networks, 2002. IJCNN '02, 3:2887–2892. IEEE, Piscataway, NJ

Ljungberg O (1998) Measurement of overall equipment effectiveness as a basis for TPM activities. International Journal of Operations and Production Management 18:495–507

Mckone K and Weiss E (1998) TPM: planned and autonomous maintenance: bridging the gap between practice and research. Production and Operations Management 7:335–351

MIMOSA (2006) Common Relational Information Schema (CRIS) Version 3.1 Specification. http://www.mimosa.org/.

Molina J M, Isasi P, Berlanga A, Sanchis A (2000) Hydroelectric power plant management relying on neural networks and expert system integration. Engineering Applications of Artificial Intelligence 13:357–369

Pintelon L (1997) Maintenance performance reporting systems: some experiences. Journal of Quality in Maintenance Engineering 3:4–15
Sherwin D and Al-Najjar B (1999) Practical models for condition monitoring inspection intervals. Journal of Quality in Maintenance Engineering 5:203–221
Sherwin D (2000) A critical analysis of reliability-centred maintenance as a management tool. 4th Int Conf Maintenance Societies, 23-26.5.2000, Wollongong, Australia
Skyttner L (2001) General systems theory. World Scientific, Singapore
Soliveres H and Alquier AM (1997) A particular aspect of DECIDE bid decision support system: modelling of life-cycle processes and costs. IEEE Int Conf on Systems, Man, and Cybernetics, 12–15.10. 1997, Computational Cybernetics and Simulation, 4:3609–3614
Yagi Y, Kishi H, Hagihara R, Tanaka T, Kozuma S, Ishida T, Waki M, Tanaka M, Kiyama S (2003) Diagnostic technology and an expert system for photovoltaic systems using the learning method. Solar Energy Materials and Solar Cells 75:655–663
Yam R C M, Tse P W, Li L, Tu P (2001) Intelligent predictive decision support system for condition-based maintenance. Int. Journal of Advanced Manufacturing Technology 17:383–391
Zhan W, Jiwei G, Jingdong X, Guoqing T (2001) An introduction of a condition monitoring system of electrical equipment. Proc. International Symposium on Electrical Insulating Materials (ISEIM 2001), 221–224. Institute of Electrical Engineering of Japan, IEEE (Tokyo)

Chapter 13
Dynamic and Cost-effective Maintenance Decisions

Basim Al-Najjar and Niclas Eberhagen

Abstract. In general, all the elements involved in a production process, such as tools, machinery, methods, competence and working environment are exposed to changes. Thus, to maintain and improve company profitability and competitiveness, it is necessary to reduce losses through maintaining and improving the quality of the elements involved in the production process. Therefore, the maintenance strategy applied should be dynamic to be suited for all these changes cost-effectively. In producing companies, the accuracy of maintenance decisions is essential for reducing economic losses generated due to unnecessary stoppages in companies of intensive capital investment where stoppage time is expensive, especially in paper and pulp mills, refineries, power stations and engineering manufacturing. Therefore, it is necessary for maintenance and production managers to have a system providing the data required to achieve dynamic and cost-effective maintenance decisions.

In this chapter, the new maintenance decision support system (MDSS) for achieving dynamic and cost-effective maintenance decisions developed in Chapter 12 is tested and discussed. The developed system is MIMOSA compatible, consists of three toolsets and five software modules for performing six services to achieve different objectives, *e.g.*:

1. higher accuracy of maintenance decisions;
2. selection of the most cost-effective maintenance solutions;
3. identification and prioritisation of problem areas and assessment of losses in production time; and
4. mapping, follow up and assessment of maintenance cost-effectiveness (maintenance savings and profit) to achieve continuous and cost effective improvement.

13.1 Introduction

Maintenance is generally treated as a necessary cost that gives nothing in return. Maintenance should instead be treated as a profit-generating centre since it is closely related to a company's internal efficiency (Al-Najjar 2007a). Using an efficient maintenance policy, such as vibration-based maintenance (VBM) policy for rotating machines, leads to fewer planned stoppages and failures, lower level of spare part inventory and a smoother production process. It will also lead to a higher quality and more profitable production process especially in the process, chemical, energy and recently in manufacturing engineering industry (Al-Najjar and Alsyouf 2003). Therefore, it becomes essential to use relevant data to be able to control maintenance cost-effectiveness. To be able to monitor, map, analyse, assess, predict and improve the outcome of different maintenance actions properly it is necessary to gather and use the data covering the relevant disciplines, *e.g.*, technical, financial and managerial data. To make the process of collecting and analysing data even more trouble-free and cost-effective, it should be computerised in a decision support system (DSS). This allows following up maintenance performance measures more frequently, thereby being able to react more quickly on disturbances and avoid unnecessary costs. It will also be easier to trace the causes behind deviations. A system developed especially to highlight the role, performance and impact of maintenance on company business would help to achieve cost-effective and dynamic maintenance decisions. Thereby, planning and executing maintenance actions and answering where and why an investment in maintenance may have the best financial payoff can easily be achieved (Al-Najjar *et al.* 2001, Al-Najjar 2007a). Pintelon (1997) points out the importance of a well functioning maintenance reporting system and also the fact that most systems in this area are limited to only budget reporting.

13.2 MDSS for Dynamic and Cost-effective Maintenance Decisions

In this section, the MDSS role in achieving dynamic and cost-effective maintenance decisions by combining the deterministic and probabilistic approaches is introduced and discussed. Furthermore, an application scenario of MDSS is also described and discussed.

13.2.1 Deterministic and Probabilistic Approaches

Condition monitoring (CM) technology as is well-known as a source of information that can be utilised for mapping the state of a component/equipment. However, condition-based maintenance (CBM) is a maintenance strategy and it can be achieved in different policies through utilising different CM parameters or sets of parameters if demanded. CBM is not just a tool for gathering measurements from CM systems, or a tool for diagnosis and prognosis of the condition of a component/equipment that is exposed to a deterioration process. CBM consists of all these activities: data acquisition and management, analysis, interpretation, fault detection, diagnosis, prognosis and prediction, decision-making, planning and performance of maintenance actions (Al-Najjar 2007b).

Maintenance cost-effective decisions cannot be achieved effectively without relevant data of high quality that describes properly production and maintenance processes. Identification of and gathering relevant data demand properly conducted technical analysis of the production and maintenance processes' problems for identifying relevant CM parameters providing the information required for decision making. A systematic technical analysis eases and secures the identification and application of relevant CM systems as described by the eight-step procedures shown in Figure 13.1. These eight steps are classified in three big categories: (1) mapping, classifying and prioritising, (2) technical and economic analysis, and (3) results, identification and work improvement.

In this study, we use CBM to indicate the utilisation of condition-based data for planning maintenance actions instead of what is sometimes called predictive or diagnostic maintenance, because CBM consists of all these activities. CBM usually applies the deterministic approach for data interpretation and analysis, diagnosis, prognosis and decision-making. In many cases, even the prediction of the time to action or the level of the CM parameter value in the near future is done by using an extrapolation technique (Al-Najjar 2000).

Applying only the deterministic approach can be counted as one of the shortages in CBM applications, because real life can hardly be expressed or described effectively by only the deterministic approach. Probabilistic modelling of the residual time or the assessment of the probability of failure of a component by converting CM measurement to the time scale without considering analysis, diagnosis and prognosis cannot give a reliable picture of the condition of the component/equipment either. This is because there is an inbuilt appreciable probability of failure in each statistical model, no matter how robust it is, as long as the spread of the time to failure is large, *i.e.*, the standard deviation is big, especially in mechanical components and systems. In these types of components the spread in the time to failure can be between months and years (Al-Najjar 1998).

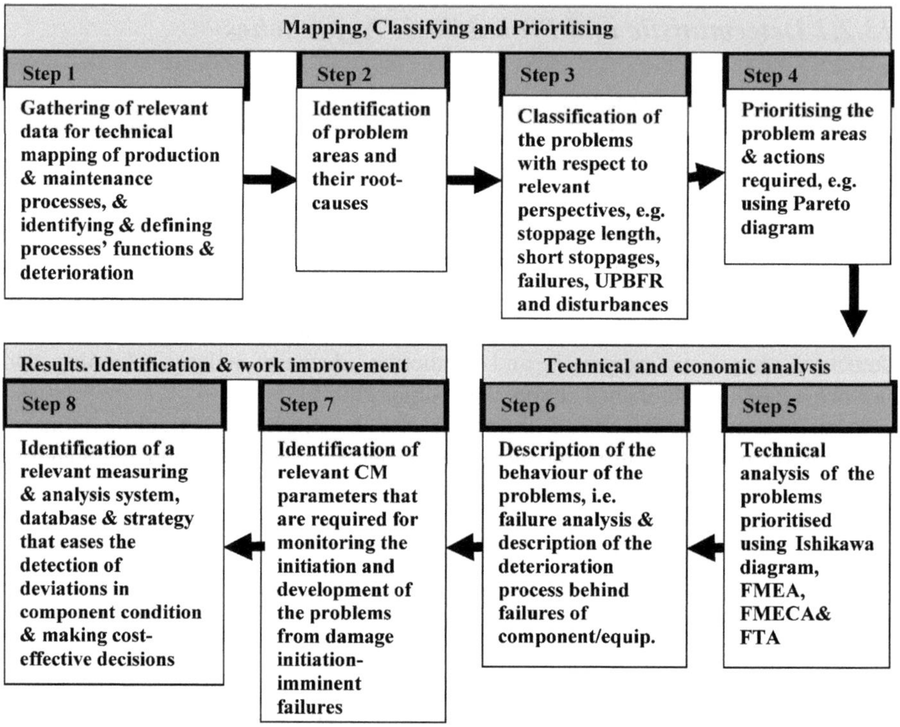

Figure 13.1 A systematic way to identify, analyse and achieve a cost-effective decision-making process

From everyday experience, the more relevant and accurate data concerning the deterioration process under consideration, the lesser the need for probabilistic modeling. Keeping this in mind, the latter cannot be eliminated as long as the deterioration process is a stochastic process. When it is not possible to provide more reliable data for enhancing the certainty of the decision-making process, we believe that probabilistic modeling is a natural complement to the deterministic approach. For example, exact prediction of the time to failure using the information provided by a CM system is very difficult if not impossible because the deterioration process is a stochastic one. When the CM parameter value is increased significantly, the assessment of the probability of failure of a component, its residual time, *etc.*, in addition to the historical and current CM measurements and diagnosis would enhance the description of the previous, current and future state of the deterioration process and component life behaviour (Al-Najjar 2003, Jardine *et al.* 2006).

The activities required for identifying and selecting relevant CM parameters and making the decision making process cost-effective are many and complicated (Moore and Starr 2006). Significant components, damage initiation causes, damage developing mechanisms, imminent failures, failure modes and their consequences should be identified clearly. These identifications help to identify which CM parameters can be used for monitoring one or more phases of the deterioration phenomenon and finally selecting the most informative and cost-effective CM system. All these steps are important for establishing and running a cost-effective CBM policy (Al-Najjar 2007b).

In order to approach an accurate mapping and assessment of the condition of a component/equipment several variables should be considered simultaneously, such as previous measurements and their trends, current measurements, the deterioration rate, the maximum allowable limit for the CM value, diagnosis and prognosis results, the failure rate, residual time, probability of failure, *etc.* Usually, only the first three variables or the last two variables are used (Al-Najjar and Wang 2001, Jardine *et al.* 2006), respectively. Therefore, in MDSS we use additional information parameters to describe the situation of a component/equipment with more detail. In addition to the trend data and the current measurements that reflect the past and current condition of a component/equipment, MDSS predicts the vibration level in the near future, the next planned stoppage or measuring opportunity, the probability of failure utilising past data from the same machine/component and the residual life of the component in question.

13.2.2 Dynamic and Cost-effective Maintenance Decisions

A cost-effective CBM policy influences many working areas in a producing company. Therefore, an estimation of the maintenance impact, *i.e.*, savings generated due to more efficient maintenance, is of interest to many companies because it means lower production costs and highlights the role of maintenance in the company's business, profitability and competitiveness (Al-Najjar 2007a). This is why maintenance is considered as a profit centre in this study. Also, we consider a cost-effective decision that increases savings due to less unplanned and planned stoppages, shorter stoppages, better product quality, *etc.* In economic terms, it is a decision that contributes to more profit for the company. Dynamic and cost-effective maintenance decisions demand a flexible maintenance system that enables the user to follow the changes in the production process and surroundings dynamically and cost-effectively. Also, it demands relevant data and a clear description of the links describing the conversion of money-to-money in the plant, *i.e.*,

money (capital investments in maintenance)- to-*maintenance technical output-* to-p*roduction technical output-* to-*money.*

These three steps are necessary to convert maintenance impact to a measure, such as money, which everyone in a company from the floor workers to the chief manager can easily understand and use for comparison between maintenance inputs and outputs and judging maintenance cost-effectiveness. Thus, maintenance actions can be planned cost-effectively through selecting the most proper moment for the task when all the costs are low, which in turn increases the technical and economic impact of maintenance. Also, it will not be difficult to assess the costs of maintenance activities and how much maintenance has generated income (savings and profit). To highlight this point in Figure 13.2 we introduce two cases showing the logic flow of cause–result.

a) A stable production process --- *leads to* --- Enhance the probability of the delivery on time which in turn --- Fulfils company strategic objectives

b) Fewer failures and shortened repair time --- *result in* --- Prolonged production time which in turn --- Increases production --- *and leads to* --- Better profit margin which means --- More profit --- *and consequently* --- Better competitiveness

Figure 13.2 Logic flow of cause–result of maintenance impact described in two cases

TQMain promotes that the data required should be gathered in a common database without the duplication that usually occurs when each department collects its own data (Al-Najjar 2008). Integrating data from relevant working areas has many benefits. For example, the integration of the data collected from production and VBM policy provides good opportunities for monitoring, measuring and improving reliability, availability and productivity of the producing machines. Integrating the data from VBM database with those from the databases for production and quality control establishes the basis for monitoring, measuring and improving the quality rate and causes behind quality deviations in addition to the latter characteristics. The integration of the data from VBM policy and quality control provides a possibility for monitoring, measuring and improving the quality system, because in this context VBM policy works as a quality assurance tool, see above. A reliable modification of manufacturing equipment can also be achieved.

When the cost accountancy program is integrated with the common database, TQMain offers particular criteria and tools to assess the cost-effectiveness of the technical improvements (Al-Najjar 2007a). Establishing a common database, such as a modified MIMOSA database, provides the information needed. In this database:

- effective diagnosis is possible due to the possibility of avoiding ambiguities;
- past data can be used to assess warning and replacement levels effectively and improve their values continuously;
- damage and failure causes can be identified and followed up with high accuracy;

- reasons behind deviations in the quality of any of the elements involved in the production process, such as working environment, production/operation and quality, can be identified, localised and followed up effectively; and
- product cost factors and maintenance related economic factors can be identified, monitored, analysed and assessed effectively. Also, deviations in these factors can be traced effectively and possibly in real-time.

Many possible sub-optimisations and analysis are currently being abandoned because people simply do not have the time to co-ordinate data from several sources and hunt for missing data only to find ambiguities affecting the values of model parameters.

13.2.3 Application Scenario of MDSS

Dynamic and cost-effective maintenance decisions that generate more profit cannot be achieved without dynamic rules for the decision making process. In this study, a dynamic decision is considered, *i.e.*, the decision that is made using dynamic rules instead of a general rules or general decision model. For example, damage initiation cannot in any way mean termination of the ability of a component/equipment to perform its function measured on an operational, economic or safety basis.

It is given that the data required for mapping, analysis and control of the condition of the component/equipment are available and accessible. The rules for making a decision of stopping a machine to replace damaged component should be dynamic in such a way that a maintenance engineer can follow up the development of the damage development. In other words, a damaged component should not be replaced as soon as the damage is detected because in this context what should be considered is the impact of the damage on the production, cost and safety.

Rules for dynamic decisions will make it possible to reduce failures and utilise as much as possible of the component life without increasing the risk of failure. It also means that the maintenance action required can be carried out at any time the risk of failure has approached a significant level. This is what MDSS offers as a possibility, *i.e.*, to maintain a machine cost-effectively in a dynamic (real) environment without increasing the risk of failure. In Figure 13.3, the sequences of applying MDSS allows the application of one tool, more than one tool or all the tools.

Steps of Applying MDSS

Step 1. Identify the needs before applying MDSS, *i.e.*, enhancement of maintenance decision, identification of problem areas or assessment of maintenance cost-effectiveness.
Step 2. Identify the respective tool to be used from all of the tools provided by MDSS (PreVib, ProFail, ResLife, AltSim, MMME, and MainSave), for more details see Chapter 12.
Step 3. Control and clean the data required according to the guide and instructions for data gathering and the list of variables required for running MDSS.
Step 4. Feed the data required to the database using user-interface.
Step 5. Start MDSS.
Step 6. Connect to the appropriate database.
Step 7. Set up MDSS with the right scales for the units, *e.g.*, time units or currency.
Step 8. Select the tool that you would like to utilise.
Step 9. Select appropriate data selection criteria within the tool, *i.e.*, select the right information for segment, asset, *etc.*
Step 10. Download the required data to the tool.
Step 11. Input non-database data manually what is decided by the tool.
Step 12. Use the tool assessment button to perform the task.

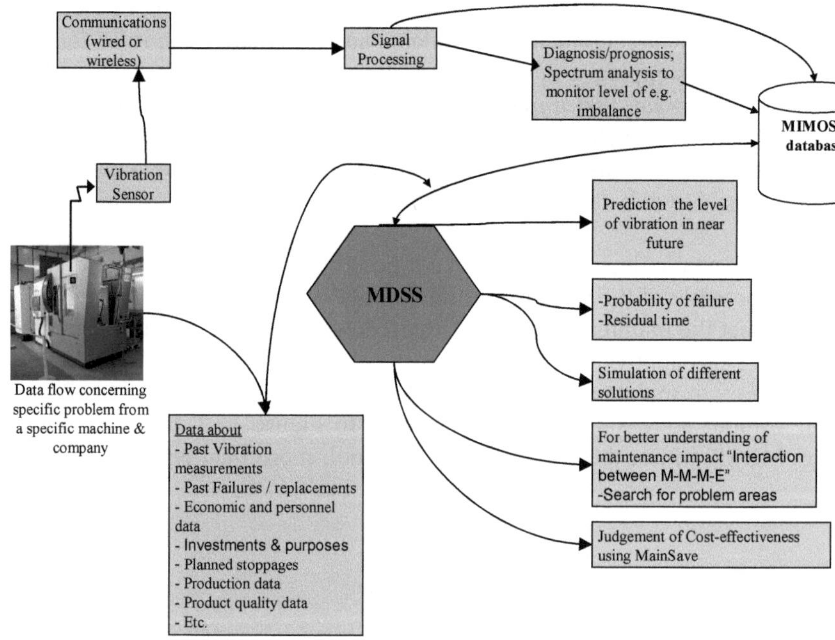

Figure 13.3 Steps of applying MDSS

Assume that the condition of a particular component, such as rolling element bearing, shaft or spindle in a CNC-machine is monitored by using vibration sensors, see Figure 13.3. When the vibration level of interest approaches or exceeds the warning level, it is possible to utilise PreVib to predicting the vibration level in a particular point of time, for example in the next measuring moment or the next planned stoppage. In many cases, this prediction is important for production security and product quality. For the same component, using the past lifetime data of the machine, it is possible to assess the probability of failure and residual life (using ProFail and ResLife) for more reliable planning of the production and securing the delivery schedule to avoid any delay and consequently delay penalties and losses of customers and market shares. Applying ProVib, ProFail and ResLife, means that the already available data, *i.e.*, trend data, current situation data and installation/replacement data are intensively used for adding more information to enhance the underlying information for achieving dynamic and cost-effective maintenance decisions.

ProVib, ProFail and ResLife clarify whether there is a problem in the machine, which can be eliminated and suggest several alternative solutions. For example, the shorter life of a particular rolling element bearing, which leads to more stoppages and shorter production time, which occurs due to occasional dramatic changes in the operating conditions, such as ambient temperature. In order to detect these changes in the ambient temperature we can use:

- a subjective way, *i.e.*, human hand-sense
- temperature sensors
- a regular examination of the lubricant characteristic and temperature.

These alternatives can all be technically suitable but not necessary equally economically effective. Also, they may demand unequal investments of risk capital and different operating costs. Furthermore, they lead to different results and accuracy in detecting the changes in the ambient temperature. For example, the difference in the accuracy of measurements when using the human hand for detecting changes in the temperature compared with that when using a digital sensor for the same purpose. Using AltSim for simulation means selecting the most cost-effective solution, for more details see Chapter 12.

In order to identify and prioritise problem areas and follow up the investment, MMME and MainSave can be used, respectively. MMME helps to identify and prioritise the root-causes behind the losses in the production time, which in turn eases and focuses the work plans for the most urgent problem areas and problems. MainSave assists the user in quantifying the losses or gains generated by maintenance performance through comparing the technical and economic variables belonging to two periods, *i.e.*, previous (before investment) and current (after the investment). The same thing can be done for other periods for a particular producing machine and product. Observe that what has been judged to be the most cost-effective solution using AltSim is know possible to assess more accurately to justify previous investment decisions and also to identify new problem areas for investments, especially those areas where losses can justify the suggested investment.

13.3 Data Required to Run MDSS

In order to utilise all the toolsets and tools and services provided by MDSS, different data sets are necessary, such as:

1. *Technical data*, *e.g.*, times and number of stoppages, type of stoppages, planned and real production, production rate, defective items (production of bad quality), depreciation period.
2. *Economic data*, such as profit margin, maintenance cost factors, user defined expenses, maintenance investments.
3. *Human resources data,* such as the impact of competence, commitments and communication on the production time, *i.e.*, classification of stoppages with respect to the above mentioned factors.

13.3.1 Datasets

The data that are required to run all the tools provided by MDSS are introduced in Table 13.1. It is a compact table without repetition of variables. The table takes into consideration applying all tools individually or jointly. This is why we cannot see some variables that are usually repeated when using more than one tool, such as is the case with AltSim, MMME and MainSave. AltSim and MMME use the same data as MainSave. The major reason why there is no repetition of those variables is to make sure that all the variables required for running every particular tool need only be feed once. In Table 13.1 lists all the variables required for applying all the tools included by MDSS.

Table 13.1 List of variables required for applying MDSS tools

Variable category	Variable (data input)	Value	Actions taken when no data are available	Description of the variable/data	Comments
MIMOSA information sets					
Asset	User identification tag			Unique user identification tag that follows the nomenclature of the plant. This is normally an abbreviation or user-defined alpha-numeric code from which the plant derives meaning, *e.g.*, machine or component serial number, or other means for identification of a specific instance of a physical machine or component.	This refers to a specific instance of a physical machine or component.
Asset	Model			Model of the machine or component	
Asset	Segment			Optional indicator of which segment the machine is associated with. If asset is a component, then it is not associated with a segment.	
Asset type	Name			Descriptive name of the type of component or machine. MIMOSA has predefined names.	
Asset on segment	GMT installed			The time point when the component was installed on a machine (yyyy-mm-dd hh:mm:ss).	A minimum of 2 sets of time point data for identical components on the same machine and equivalent location in addition to time point of the current installed component in question.

Table 13.1 (continued)

Variable category	Variable (data input)	Value	Actions taken when no data are available	Description of the variable/data	Comments
Asset on segment	GMT removed			The time point when the component was removed (due to failure or condition based replacement) from a machine (yyyy-mm-dd hh:mm:ss). It may be undefined if a specific component has not yet been removed, *i.e.*, is still installed.	A minimum of 2 sets of time point data for identical components on the same machine and equivalent location.
Model	User identification tag			Short user-meaningful identification of the type of machine, component, or product, *e.g.*, the name of the machine, component, or product model.	
Segment	User identification tag			Unique user identification tag which follows the nomenclature of the plant. This is normally an abbreviation or user-defined alpha-numeric code from which the plant derives meaning.	
Segment type	Name			Descriptive name of the type of service segment or machinery. MIMOSA has predefined names.	
Measurement location	Asset			The specific component the measurement is associated with.	
Measurement location	User identification tag			Unique user identification tag that follows the nomenclature of the plant. This is normally an abbreviation or user-defined alpha-numeric code from which the plant derives meaning.	

Table 13.1 (continued)

Variable category	Variable (data input)	Value	Actions taken when no data are available	Description of the variable/data	Comments
Measurement location type	Name			Descriptive name of the type of measurement location. MIMOSA has predefined names.	
Measurement event	GMT event			Time point for the vibration measurement (yyyy-mm-dd hh:mm:ss) taken for a specific component and location.	
Measurement event numeric data	Data value			Vibration level measurement of significant level greater than or equal to potential failure level, for a specific measurement event.	A minimum of 3 previous measurements of same component are required.
MIMOSA extended information sets					
Model process model specific	Theoretical cycle time			Theoretical production cycle time (> 0) given in seconds.	It can be different for different machine and product.
Model vibration level specific	Mean vibration level			Mean vibration level during normal operation of a component, *i.e.*, when no damage or problem is initiated.	
Model vibration level specific	Potential failure level			The vibration level indicating that damage development is confirmed with a component, *i.e.*, warning level (> mean vibration level).	
Model vibration level specific	Replacement level			Vibration level that if it is exceeded indicates that the component should be replaced, *i.e.*, replacement/failure level (> potential failure level).	

Table 13.1 (continued)

Variable category	Variable (data input)	Value	Actions taken when no data are available	Description of the variable/data	Comments
Production	GMT production start			Starting time point for the production planned for a particular production order in the machine (yyyy-mm-dd hh:mm:ss).	In order to achieve clear results in maintenance cost-effectiveness we should have data belonging to at least two periods, *i.e.*, two orders of reasonable length such as one month or a couple of weeks.
Production	GMT production end			End time point for the production planned for a particular production order in the machine (yyyy-mm-dd hh:mm:ss).	In order to achieve clear results in maintenance cost-effectiveness we should have data belonging to at least two periods, *i.e.*, two orders of reasonable length such as one month or a couple of weeks.
Production	Planned production time			Planned production time for that particular order during the period of study given in hours.	In order to achieve clear results in maintenance cost-effectiveness we should have data belonging to at least two periods, *i.e.*, two orders of reasonable length such as one month or a couple of weeks.
Production event	GMT event start			Starting time point for the event (deviation in production speed, *e.g.*, production speed reduction/increase, or production stoppage) (yyyy-mm-dd hh:mm:ss).	For each particular start, stoppage or change in the production speed, during the period of the order specified above for a production.
Production event	GMT event end			End time point for the event (deviation in production speed or production stoppage has been corrected) (yyyy-mm-dd hh:mm:ss).	For each particular start, stoppage or change in the production speed, during the period of the order specified above for a production.

Table 13.1 (continued)

Variable category	Variable (data input)	Value	Actions taken when no data are available	Description of the variable/data	Comments
Production event	Production event type			The event type can only be 1 or 2, where 1 = stoppage, 2 = deviation of production rate, *i.e.*, production speed reduction/increase.	
Production event	Actual cycle time			Only used if production event type = 2. Actual production cycle time during the event period (> 0) given in seconds.	
Production event	Actual production quantity			Only used if production event type = 2. Actual production quantity during the event period (>= 0).	
Production event	Event cause subcategory			ID-value linked to each of the ten subcategories that subdividing the major three categories of MMME.	The subcategories are: - Working environment - Maintenance organisation and management - Competence - Personnel commitment and communication - Machine condition and characteristics - Production procedure and methodology - Failures - Short stoppages - Stability of production quality Stability of the production speed and production rate
Production follow-up	GMT follow-up			Time point for production quantity follow-up registration (yyyy-mm-dd hh:mm:ss), *i.e.*, when to register the actual total and the defective production measured according to the company's policy (time schedule).	This will happen during the whole period of the production order.

Table 13.1 (continued)

Variable category	Variable (data input)	Value	Actions taken when no data are available	Description of the variable/data	Comments
Production follow-up	Actual production quantity			Actual production quantity, *i.e.*, production of accepted and rejected quality measured according to the company's policy (time schedule).	This will happen during the whole period of the production order.
Production follow-up	Defective production quantity			Quantity of not accepted production quality out of the actual production quantity measured according to the company's policy (time schedule).	This will happen during the whole period of the production order.
User defined expense type	Name			Unique descriptive name of the type of expense related to machines.	For example: - Insurance premium - Failure related accident fees/penalties/expenses - Environmental related penalties/expenses
User defined expense	GMT expense			Time point for the user defined expense (yyyy-mm-dd hh:mm:ss).	This will be assessed during the whole period of the production order.
User defined expense	Expense			The expense/cost related to a machine.	This will be assessed during the whole period of the production order.
Non-database information sets					
Manual user input	Profit margin			Monetary value *per* quantity	This will be assessed during the whole period of the production order.
Manual user input	Total maintenance investment			Monetary value *per* period	This will be assessed during the whole period of the production order.

Table 13.1 (continued)

Variable category	Variable (data input)	Value	Actions taken when no data are available	Description of the variable/data	Comments
Manual user input	Depreciation period			Periodic value	This will be assessed during the whole period of the production order.
Manual user input	Load ratio			Anticipated future load through present load	

13.3.2 Data Gathering

The data required to run one or more of the tools provided by MDSS can be fed through a user-interface developed specially for this purpose. The user-interface is constructed to be user friendly and easy to handle in order to secure a high quality data gathering process, *i.e.*:

- from relevant data belonging to the working areas under consideration;
- for high quality data;
- to reduce human error;
- with less uncertainty in the information;
- for even real-time measurements; and
- to assist real time analysis, result and maintenance decisions.

The database designed using a relational database as the end product consists of conceptual, logical and physical modelling. The conceptual modelling is in most cases represented in ER-diagrams, while the logical model consists of a database scheme describing the relations of the database and the physical modelling deals with storage issues. The ER-model of MDSS does not describe a traditional planning/execution centred view on maintenance. We have instead chosen to see maintenance as an integrated part of the production process. Furthermore, we have tried to describe the factors of influence that are found in the life cycle of a machine or equipment (in the ER-model called system). The variables that should be measured and fed in for the application of all MDSS functions are described according to MIMOSA. However, these variables are associated with engineering descriptions so that the end user, such as the worker on the floor, can understand what every MIMOSA variable means in order to secure gathering of the right data. The interface is constructed so that it is possible to apply one or more of the tools:

- for the same component and machine;
- for different components in the same machine;
- for different components in different machines; and
- at different point of times for different/similar components in the same machine or in different machines.

More details concerning the structure and test of the database are given below.

13.4 Database Required for MDSS

A successful mapping cannot be achieved without a well-structured database/data model including all relevant data required for the task. Identifying relevant data helps to achieve accurate problem tracing and localisation within maintenance and production processes (Al-Najjar and Kanz 2006). In this section, the MDSS data model and mapping to a company data model and the MIMOSA database user-interface are introduced and discussed. A test of CRIS/MIMOSA database user-interfaces is also conducted.

13.4.1 MDSS Data Model

Supportive datasets/classes are shown in Figure 13.4 and the MDSS model is shown in Figure 13.5. This represents a coherent and consistent view of the datasets and their relations satisfying the data needs of all of the tools of MDSS. Figure 13.5 is made in order to accommodate for the possibility of having several measurement locations for sensors on a single asset. The data model was successfully tested against the conceptual models developed in earlier stages of the project, showing that all minimum relevant data are present in the database model in order to monitor and follow up the production process and maintenance cost-effectiveness. Tests with historical data validated the model technically.

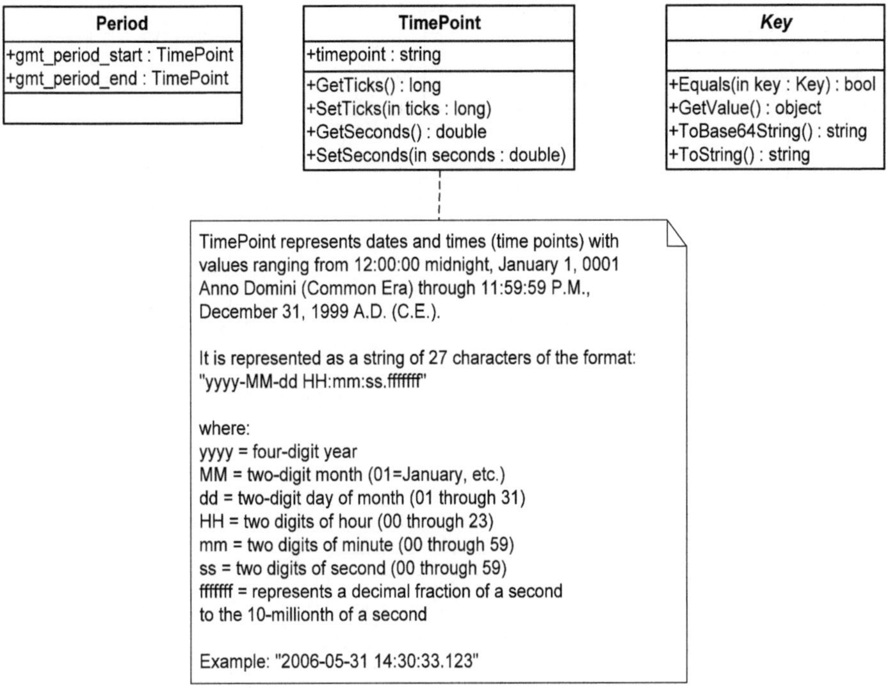

Figure 13.4 Support classes for the MDSS data model

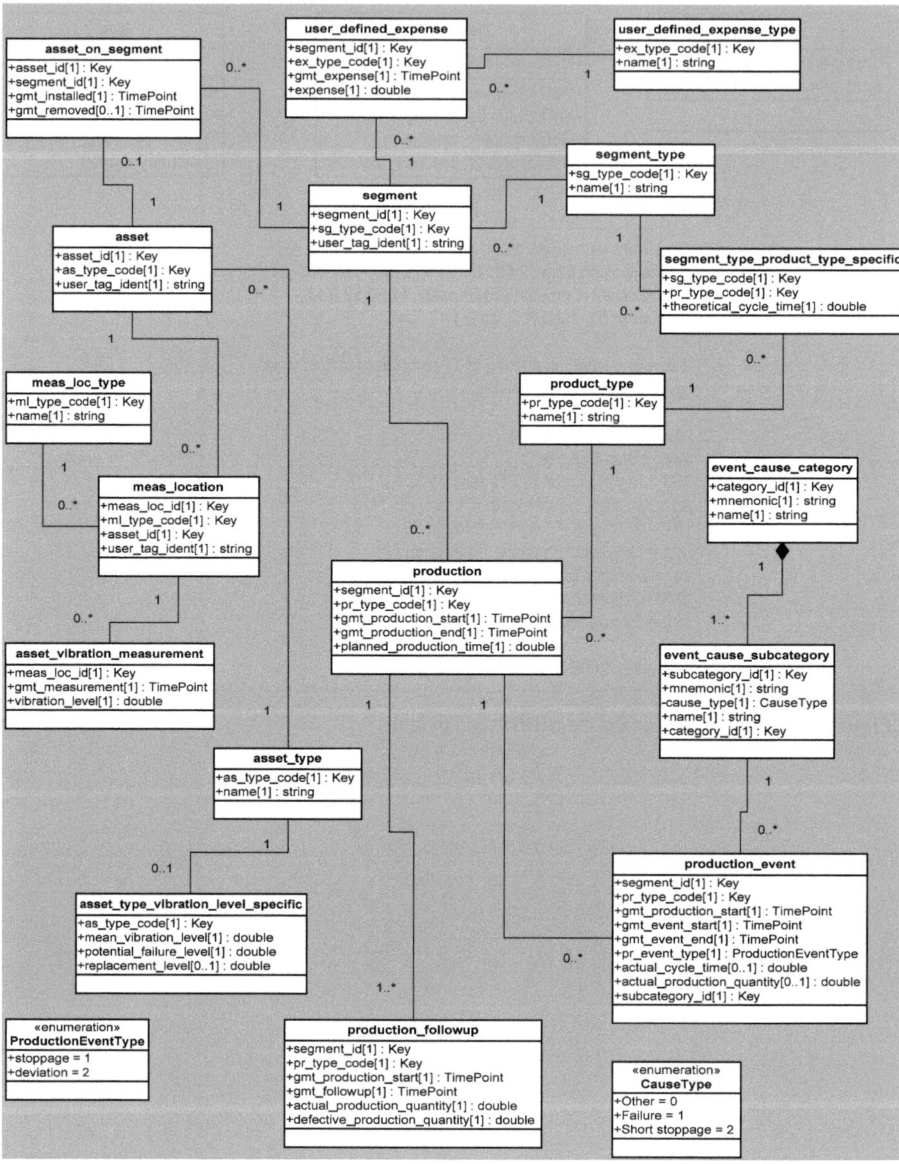

Figure 13.5 The MDSS data model

13.4.2 Mapping to Company Data Models

In order to feed data to the data model of MDSS, thereby satisfying the data needs of the individual tools, an interface to external data sources/models must be established. This is important since, from the point of view of usage of MDSS, no prior knowledge of the surrounding environment, wherein it will function, may be known beforehand.

The system must be able to stand on its own regardless of the characteristics of specific company databases. This was achieved by developing a separate component of the system, a data source agent. The agent acts as the middle man between the internal structure of the system and the outside world. The task of the agent is to map the MDSS data model to the models in the outside environment. The agent is componentised and made separable to the system. This means that as MDSS is targeted for different companies, and hence different external data sources/models, only the agent must be adjusted to accommodate the specific company databases, whilst retaining the integrity of the MDSS data model.

The agent only represents a very small part of the system, and thus only minimum adjustment is needed. This is normal practice in developing any system. Figure 13.6 shows a schematic view of the interaction between the internal data model of MDSS, the data source agent and external data sources/models. To uphold the integrity of the data model of the system in interacting with the data source agent, especially as the agent may be adjusted to different external data sources/models, an interface was established, specifying the interaction between the internal data model and the agent. No matter how the agent is adjusted to accommodate different external demands, as long as it respects the specifications of the interface, MDSS will function independently of the external environment. Figure 13.7 shows the interface that dictates the interaction between the individual tools of MDSS and the data source agent in order to satisfy their respective data needs.

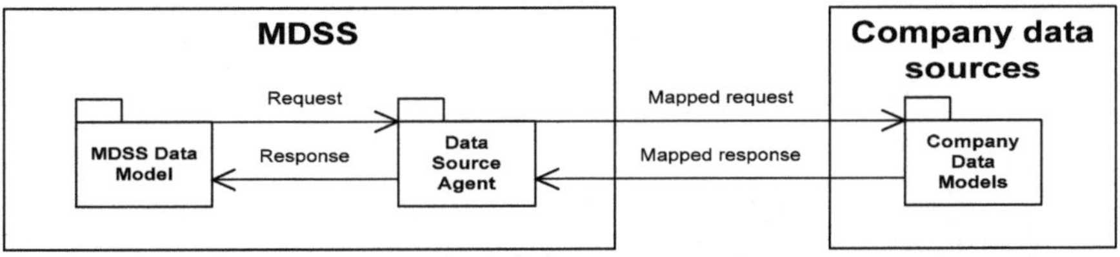

Figure 13.6 Mapping of the MDSS data model to external/company data models

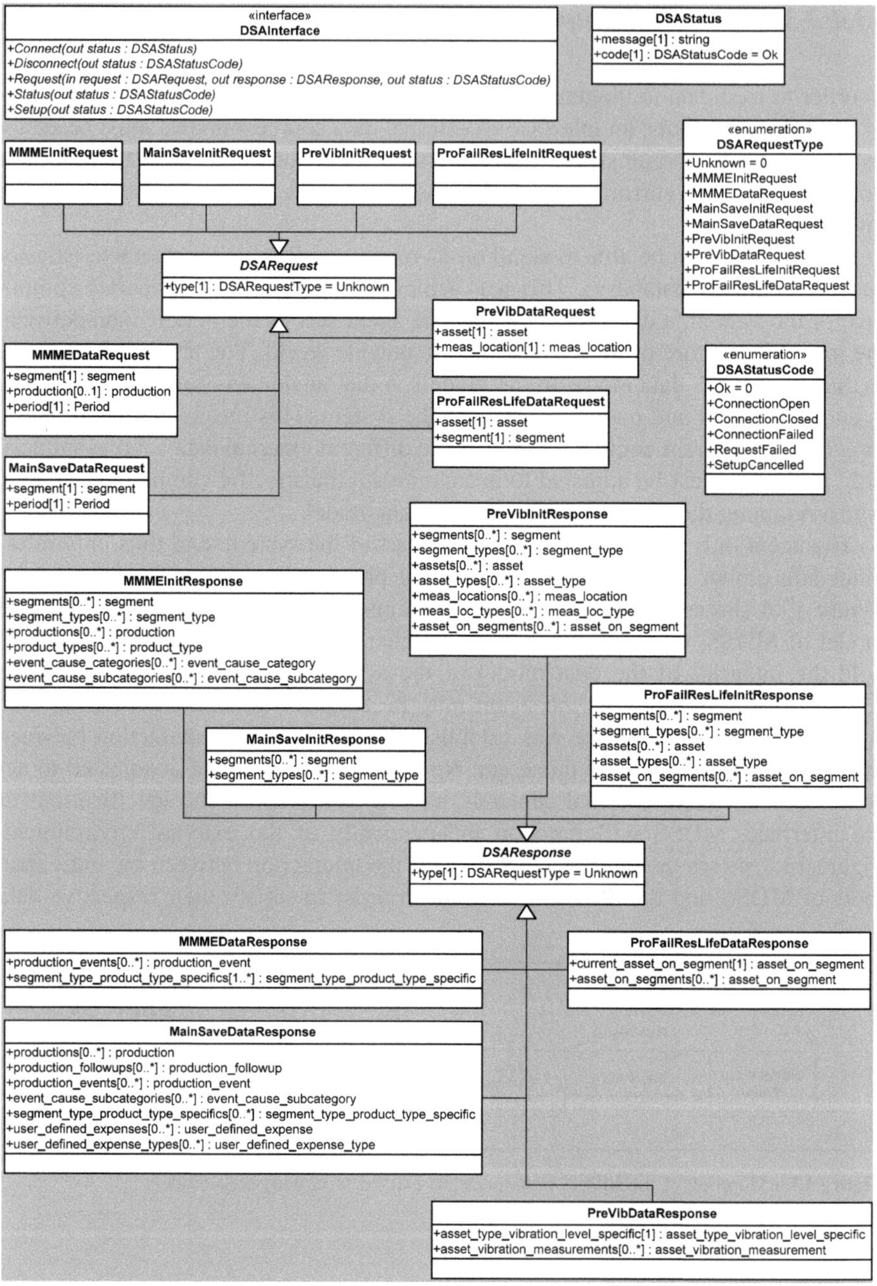

Figure 13.7 Interface between the data source agent and different tools of MDSS

The basic structure of the interface is based on a request – response messaging structure. First individual tools send a request to be initialised with data so the user may make initial selection of available datasets. Then, based upon the selection a criteria request for specific data is made. This is the very basic structure of any interface and is normal practice in developing componentised software as in the case of MDSS and the data source agent. In Figure 13.7, it is possible to distinguish the individual tools data needs both in terms of initialised data and specific data needed to perform their respective function.

13.4.3 Mapping to CRIS/MIMOSA

It was decided early on that CRIS/MIMOSA (MIMOSA 2006) was to be used as the common data model for integration of all software components. A Microsoft SQL server database based upon CRIS/MIMOSA was established. All partners developing software agreed to use this database for storing the data gathered and used within the project. This database came to represent a common company data model.

Thus, CRIS/MIMOSA represents the data model that must be accommodated and respected by the data source agent component of MDSS in satisfying the data needs of the internal data model. Therefore, the data source agent was written to map internal data demands of the individual tools of MDSS, through the interface, to the structure of the data model of CRIS/MIMOSA. Figure 13.8 represents this mapping of internal data demands to the structure of CRIS/MIMOSA. The individual datasets and their relations represent individual tables of the relational Microsoft SQL server database. Not all tables are used directly by the system.

Datasets/classes; enterprise, enterprise_type, site_type, site_database, site, eng_unit_type and ref_unit_type represent tables that needed to be defined in order to uphold relational integrity of the CRIS/MIMOSA data model. Datasets/classes; segment_type, asset_type, asset_on_segment, model, segment, asset, meas_location, meas_loc_type, data_qual_type, meas_event, mevent_num_data represent those tables that are used directly, *i.e.*, they may contain data that satisfy the data need of MDSS.

However, the mapping to CRIS/MIMOSA was not a straightforward process. Even though CRIS/MIMOSA contains many differently defined data tables, to serve many different purposes, it was discovered that not all data needs of MDSS were that easily met. Therefore it was decided that it was allowed to extend the CRIS/MIMOSA data model with new tables that could hold datasets not defined in the original model. Datasets/classes tables; user_defined_expenses_type, user_defined_expenses, event_cause_category, even_cause_subcategory, production, production_event, production_followup, model_process_model_specific, model_vibration_level_specific represent the extended table defining datasets found within CRIS/MIMOSA.

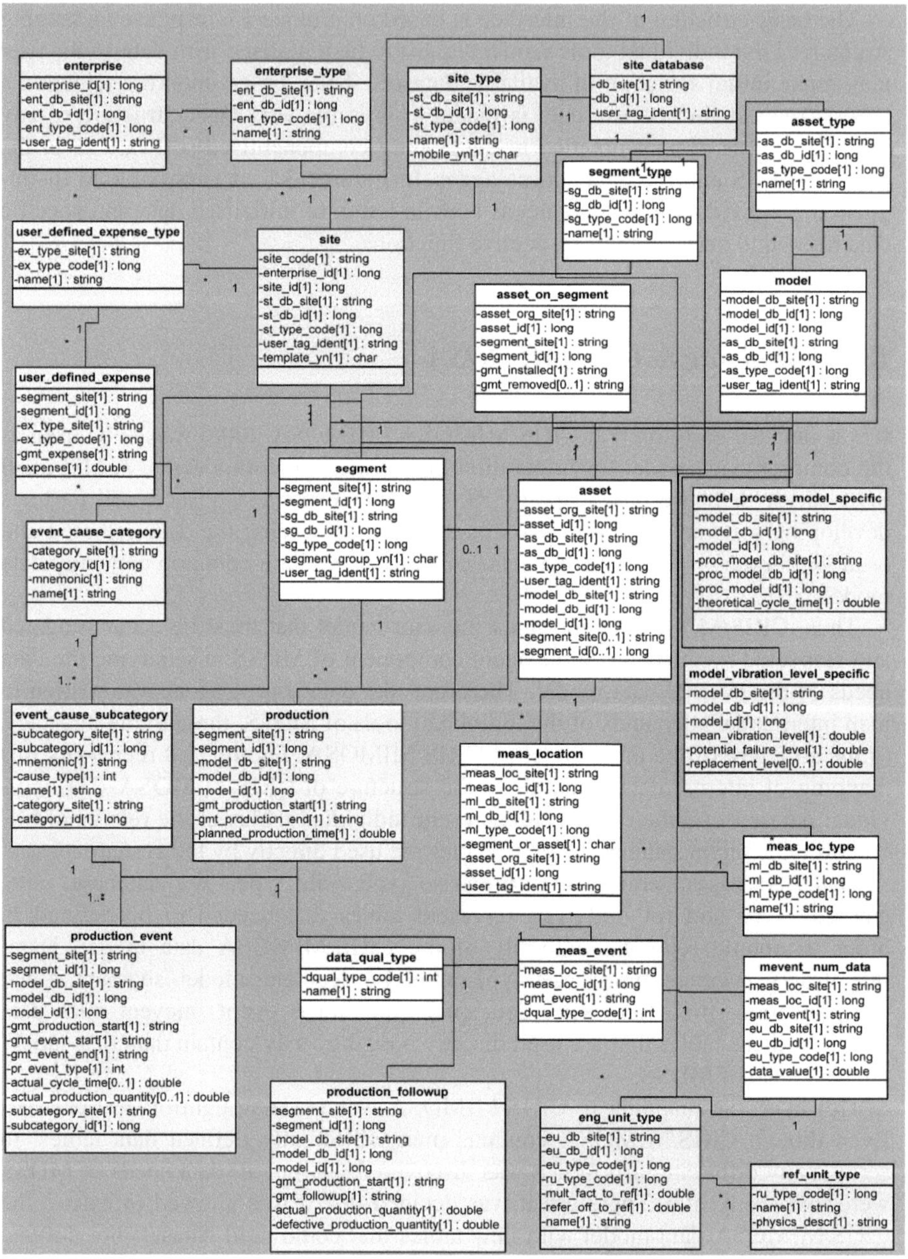

Figure 13.8 MDSS data model mapped to the extended model of CRIS/MIMOSA

13.4.4 CRIS/MIMOSA Database User-interface

To support the data gathering process and for successful application of MDSS a special user-interface application to the CRIS/MIMOSA database was developed (see Figure 13.9). This was absolutely necessary since the CRIS/MIMOSA data structure is not easily accessible or understood, and many users had never encountered it before. The database user-interface application was developed as a small Microsoft Access database linked to the tables in the Microsoft SQL database server. The database user-interface is pedagogically structured to allow choices of which MDSS tools to use to enter or revise data. The structure reflects the way the user works with the actual MDSS. Figure 13.9 shows the main menu window of the database user-interface application. Here it is possible for the user to choose which tool(s) to use to enter or revise data.

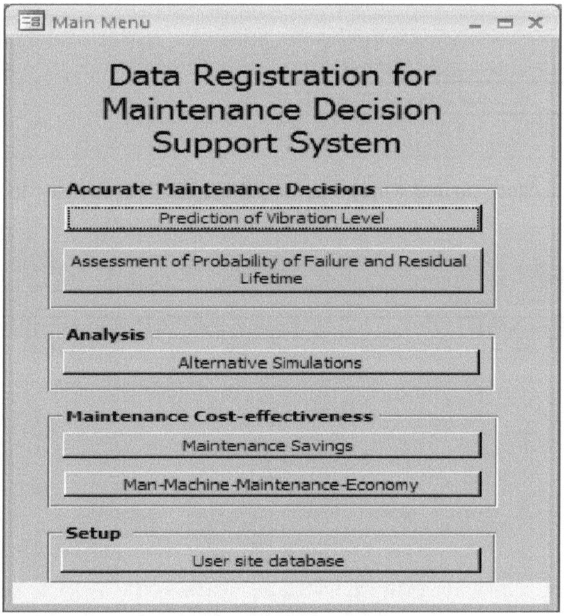

Figure 13.9 Main menu window in CRIS/MIMOSA database user-interface

Choosing one tool, *e.g.*, *maintenance savings*, the user is presented with the next level of menu choices, dependent upon the tool of choice at the previous level (see Figure 13.10). Here, the user is presented with choices of which datasets to enter or revise data. The menu choices represent the complete set of datasets that need to be addressed for successful performance of the specific tool. Not all data sets are necessary to fill in, but precision of analysis within MDSS increases as more data is entered. Choosing one dataset, *e.g.*, *production* or *production follow-*

up, the user is presented with a window that allows a user to easily enter new or revise existing data (see Figures 13.11 and 13.12). This is the lowest level in the menu hierarchy and the user is presented with the actual dataset view, *i.e.*, the table. Here, the user-interface is designed as to hide many technical details that may confuse a user, but nonetheless are important for the integrity of the database, such as related tables and constraints.

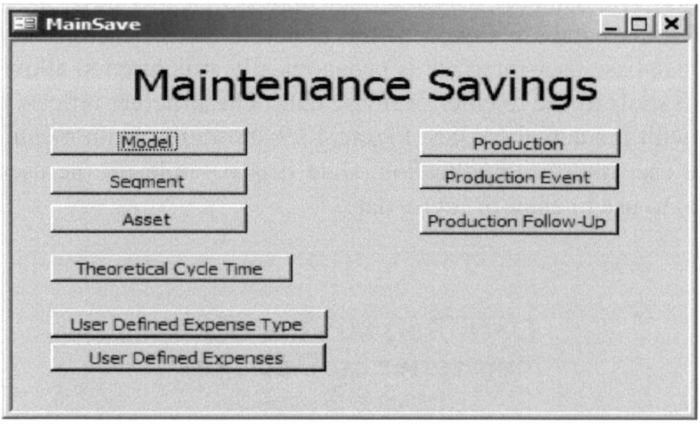

Figure 13.10 Maintenance savings window in the CRIS/MIMOSA database user-interface

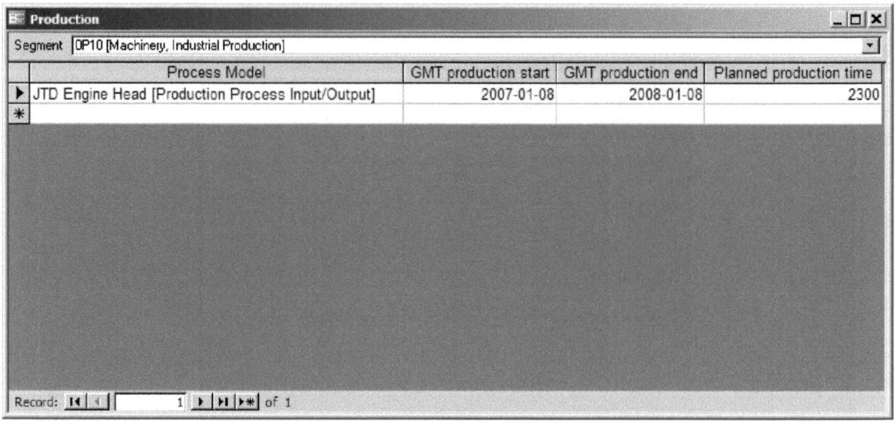

Figure 13.11 Production dataset window in the CRIS/MIMOSA database user-interface

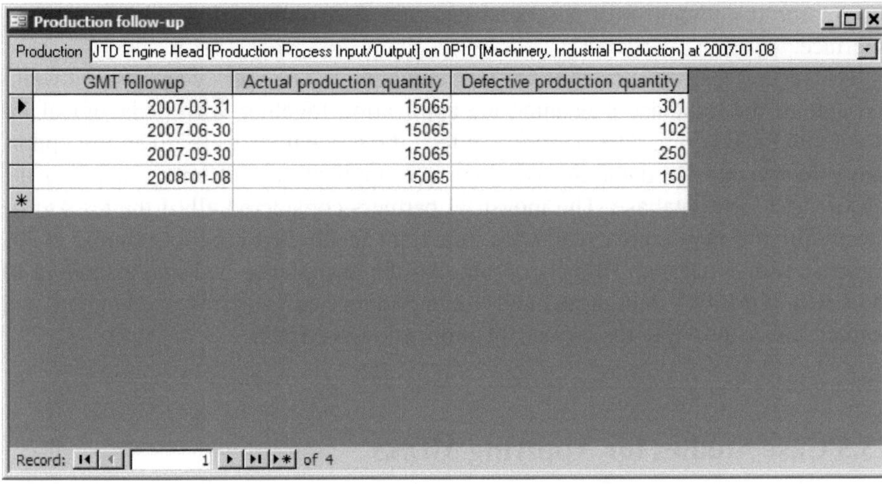

Figure 13.12 Production follow-up dataset window in the CRIS/MIMOSA database user-interface

Here, data to be entered or revised are given names that are as self-explaining as possible, corresponding to the unique list of data variables presented in Table 13.1. To further help the user in entering correct data, format and constraint checks are programmed into the user-interface, *e.g.*, negative planned production time is not allowed and dates are checked so that they are entered in chronological order. The layout of the database user-interface is similarly structured for the rest of the datasets and tools of MDSS. This gives the users a unified and coherent approach in working with the database user-interface application. As a user moves through the different menus of each respective tool within the database user interface he/she will discover that many choices of datasets reoccur. All tools share to a lesser or greater extent the same datasets. The database user-interface was intentionally designed in this manner to remind the user of which datasets of each respective tool were necessary to address should the user choose to only utilise a some of the tools rather than the complete set of tools of MDSS.

13.4.5 Test of CRIS/MIMOSA Database User-interface

The database user-interface application was tested both in laboratory settings, with typical data during development, and by industrial partners with real data. Two tests with the industrial partners were arranged. First, the industrial partners were given the unique data variable list (Table 13.1) and a user guide for data gathering, together with the database user-interface application in order to collect and enter data at their respective sites. Although only one company contributed with real data in this first test, it was met with success and provided important feedback and insight for the planning of a second test and fine tuning of the database user-interface.

The second test was arranged for four companies. The users were given a walk-through of the database user-interface application together with pedagogical instructions for entering data. During the test, the companies were given the opportunity to try out the database user-interface application and enter data into the CRIS/MIMOSA database. The industrial partners considered all of the tests to be successful and they contributed with important feedback to the fine tuning of the database user-interface. This also promoted the acceptance and understanding of the CRIS/MIMOSA data model and the importance in gathering as complete and correct data as possible for successful application of MDSS.

13.5 Case Studies for Applying MDSS

In order to verify the concept, applicability and functionalities of MDSS for dynamic and cost-effective maintenance decisions, sets of real industrial data have been gathered for testing all MDSS tools. In this section, we illustrate these tests, the datasets used therein and the final results.

All the six tools included by toolsets 1, 2 and 3 in MDSS were tested using industrial data from Goratu (CNC-machine manufacturer) and FIAT/CRF (car manufacturer). The data were provided by FIAT/CRF and AltSim, MMME and MainSave (toolsets 2 and 3) were used. The data gathered from Goratu were used for testing the tools (software modules) PreVib, ProFail and ResLife (toolset 1, accurate maintenance decisions). The data gathering was partly done by using the database user-interface. The test results of the tools in toolset 1 are introduced and discussed below.

13.5.1 Toolset 1: PreVib, ProFail and ResLife

MDSS test results using print screen figures are introduced and discussed below (Al-Najjar and Ciganovic 2009).

Prediction of vibration level (PreVib): Goratu provided data related to a specific CNC-machine component (a motor) on which vibration measurements of two different locations/directions (y and z directions) were taken. However, only data for one of the directions were readily available for immediate analysis as the other one was not "clean" in the sense that stochastic disturbances in the measurement dataset had not been discarded, making the analysis impossible. The dataset is considered clean if the values of vibration level measurements are not decreasing or fluctuating about x_o, see Chapter 12. Otherwise the dispersed measurements dispersed are ignored.

At Goratu, the reference levels for the machine type were also recorded. These levels represent the normal vibration level (the vibration level when no damage is initiated, x_o), the potential failure level (a level denoting that damage is initiated and under development, x_p) and the replacement level (the maximum allowable vibration level that should not be exceeded for avoiding failure, x_{th}). These levels are needed for assessing the severity of damage development and for predicting the vibration level in the close future, *i.e.*, at the next planned stoppage or next planned measuring moment.

Additional data are required to be manually gathered, *i.e.*, non-database data were also provided. These data were the load ratio (the ratio of the future to the present loads of the producing machine in question) and the actual time period for predicting the next vibration level. These data are not available in the database, but are supplied by the user when utilising the tool for analysis. For further details behind actual data needed and the algorithm for calculating the damage development curve and predicting the vibration level, the interested reader is advised to consult Chapter 12.

Based upon the data provided for the *y*-axis location of the motor, the recorded vibration level measurements in addition to the manually data input, a first test of the tool and analysis of data could be performed. Figure 13.13 shows how the tool has successfully retrieved the selected data from MIMOSA database and performed an analysis with the given data. In this case, *i.e.*, when the ratio of future to present load is 1 and the vibration level is needed to be predicted after 5 min, the predicted vibration level is 0.035 mm/s. The plot that consists of four measurements and lies in the right side of the right half of the user-interface shown in Figure 13.13 represents the predicted values of the vibration level at different point of times, while the plot consists of five measurements and starts earlier represents the real measurements of the same vibration level and the same point of time. The two curves converge in time, which indicates that the predicted value is gradually getting closer to the real value. The basic reasons behind that are that the model learns of previous measurements, because the model constants are reassessed after each new vibration measurements and at each predicting moment.

In this application, we also observe that the trend of the deterioration increases with time. Note that all previous vibration level measurements concern the condition of the machine before it has become significant, *i.e.*, before the first measurement shown in Figure 13.13 are ignored by the model. In Figure 13.14 we use typical data where the time between measurements (about 1 month, *i.e.*, a usual interval for many users) is longer than that used for the case company Goratu. The major objective of using typical data in this case is to reveal the ability of PreVib to handle any set of data irrespective of the interval between measurements.

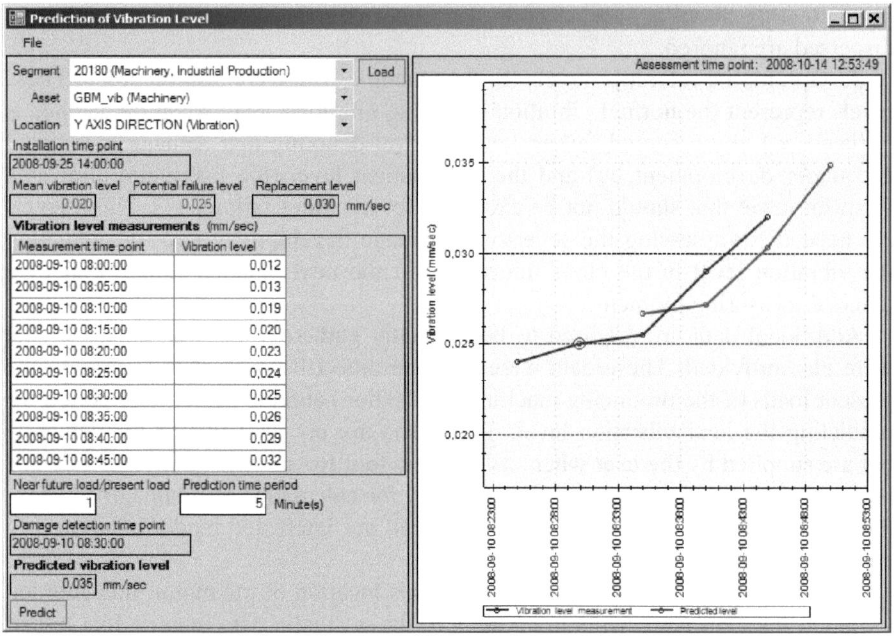

Figure 13.13 User-interface of the software module PreVib and analysis results of motor "GMB_vib"

By means of a probabilistic approach it is possible to model the time to maintenance action and thereby optimise the life and costs related to the significant component in question. This strongly advocates considering both mechanistic (PreVib) and probabilistic (ProFail and ResLife) when modelling time to maintenance action. In the next paragraph we present the second tool, *i.e.*, the probability of failure and residual life time (ProFail and ResLife).

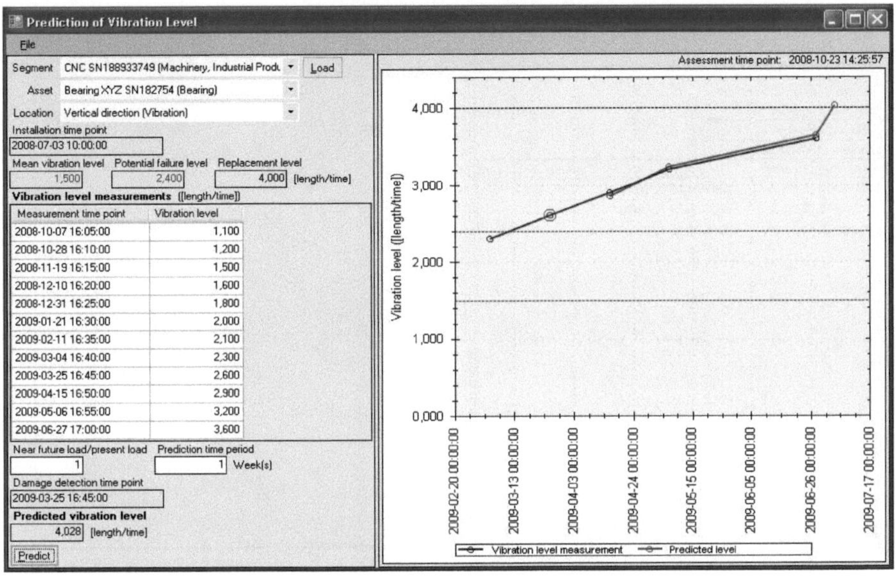

Figure 13.14 User-interface of the software module PreVib, using typical data

Assessment of probability of failure and residual lifetime (ProFail and ResLife): Goratu provided lifetime data, installation times and replacement/removal times of the same motor model as that targeted for analysis. All the data needed to perform the analysis were retrieved from the MIMOSA database. The results of the analysis were that the probability of failure is 100% and there is no more time left for that component. The reason is that the average life length of the previous identical component (it was about 3.5 days) is much shorter than that of the running component, which until the time of performing this analysis was about 20 days (see Figure 13.15).

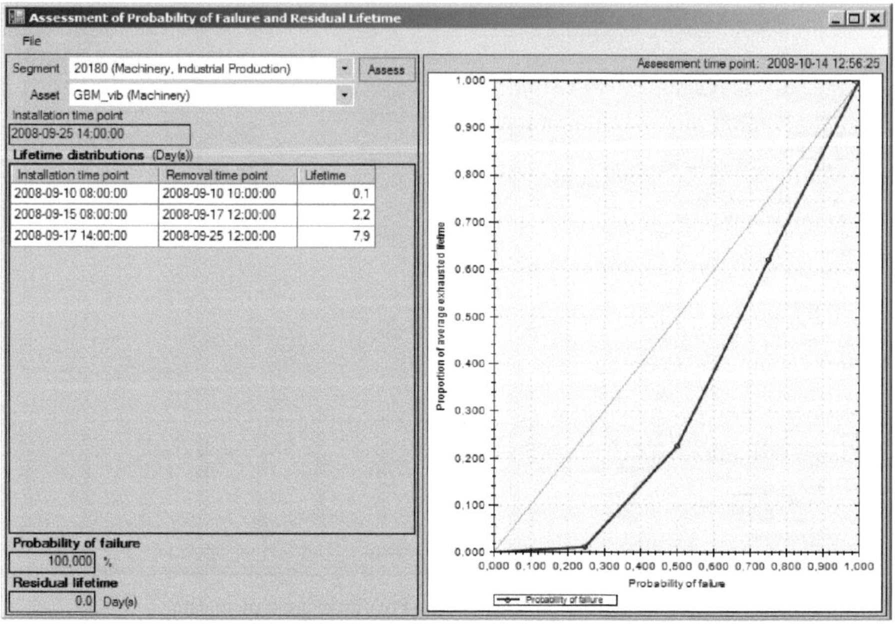

Figure 13.15 User-interface of the software module ProFail and assessment of probability of failure and residual lifetime, motor "GBM_vib"

In Figure 13.16, we use typical data to highlight the differences in the analysis results when different datasets are used. Using typical data the probability of failure is 33.333% and residual lifetime is 32.5 days. As we noticed from the tests of toolset 1, the dataset required is not extensive. A reliable conclusion about the outcome of the analysis is that ProFail can be used to assess the probability of failure of a component and its residual lifetime, and the analysis results are completely based on the input data quality. Also, it can be emphasised that toolset 1 can be applied for datasets of different numbers of measurements and times between measurements, type of component and interval between measurements.

However, these tests show that the tools function as expected and they perform with the data given in a consistent manner. Also, the tests show how the tools have successfully retrieved the selected data from the database and performed an analysis with the given data. The number of replacements (failures or condition-based) that have been used for assessing the ProFail and ResLife is very small (three). More replacements lead to higher accuracy of the assessments of the probability of failure and the residual time. In other words, the accuracy of the assessment can be enhanced continuously by more planned replacements or failure data of identical or similar components/equipment.

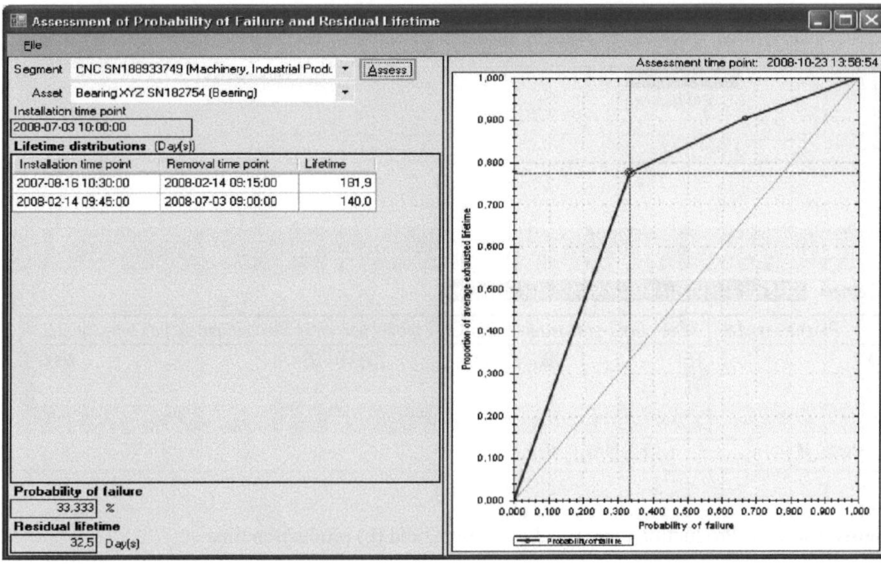

Figure 13.16 User-interface of the software module ProFail and assessment of probability of failure and residual lifetime analysis results for rolling element bearing using typical data

13.5.2 Toolset 2: AltSim

The major objective of applying toolset 2 (software module AltSim) is to simulate different technically applicable alternative maintenance solutions for a particular problem for distinguishing the most cost-effective one. The industrial data required for conducting the AltSim test were gathered from FIAT/CRF and are shown in Figures 13.17 (a) and (b), 13.18 and 13.19. The data are theoretical cycle time, planned production time, stoppages and event cause category, and bad quality production. The only data that have been feed in the white boxes manually are the profit margin. AltSim analysis results of simulating three alternative solutions are shown in Figures 13.20 (a–c) and d. Figure 13.20 (a) shows the reference data that MDSS retrieved from the database. The machine at FIAT/CRF that was considered for the MDSS test is a CNC-milling machine that produces engine heads. The operation performed by the machine is milling. This machine is considered to be a bottleneck in the production line, which makes it critical for the whole production process. The data were collected at two periods (8th January 2007 to 7th June 2007 and (8th June 2007 to 8th of January 2008). Failure data was collected with high precision.

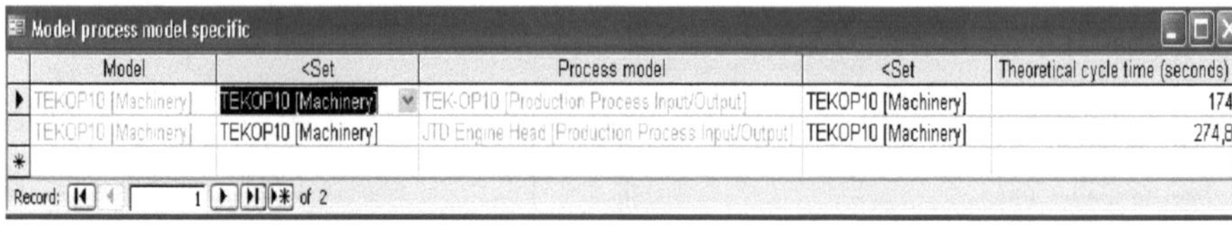

Model	<Set	Process model	<Set	Theoretical cycle time (seconds)
TEKOP10 [Machinery]	TEKOP10 [Machinery]	TEK-OP10 [Production Process Input/Output]	TEKOP10 [Machinery]	174
TEKOP10 [Machinery]	TEKOP10 [Machinery]	JTD Engine Head [Production Process Input/Output]	TEKOP10 [Machinery]	274,8

(a)

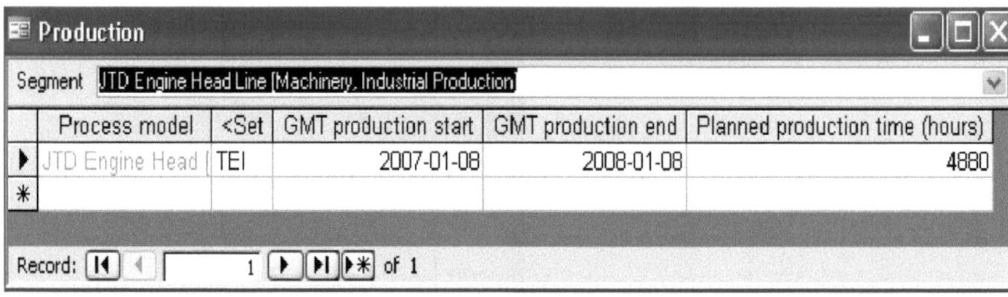

Process model	<Set	GMT production start	GMT production end	Planned production time (hours)
JTD Engine Head [	TEI	2007-01-08	2008-01-08	4880

(b)

Figure 13.17 (a) Production theoretical cycle time, and (b) production time

Losses in production time due to decreased production speed, management problems, human recourses related problems, machine characteristics, *etc.* were initially collected in one huge category called *waste* without any notation of the real causes and the associated losses in the production time. This was because the company had not had any previous necessity to gather these types of data. MMME demands sorting out the different types of losses accumulated in the *waste* category with respect to the sub-categories. This means that the losses in the production time estimated *per* each sub-category is not an exact measurement, because of the deviations in the distribution of *waste* into different sub-categories; measuring error, human error, incompatible measuring systems and techniques, and measuring procedures. FIAT/CRF personnel suggested three alternative solutions using different CM systems for solving the problem of the stiffener that was behind a large part of the failures:

Alternative 1

- 3 × vibration sensor for stiffener system and cast blocking system;
- 1 × (alternative to existing) laser for cast positioning;
- prognostic method for stiffener and cast blocking sensors;
- the investment required for this CM system is estimated to be 30000 units; and
- the depreciation period is estimated to be 4 years.

Alternative 2

- 1 × vibration sensor for stiffener system;
- prognostic method for stiffener and cast blocking sensors;
- the investment required for this CM system is estimated to be 15000 units; and
- the depreciation period is estimated to be 4 years.

Alternative 3

- Installation of 1 × sensoring (alternative to existing) laser for cast positioning and training program;
- the investment required for this CM system is estimated to be 10000 units; and
- the depreciation period is estimated to be 4 years.

Production event

Production JTD Engine Head [Production Process Input/Output] on JTD Engine Head Line [Machinery, Industrial Production] at 2007-01-08

GMT event start	GMT event end	Production event type	Actual cycle time (seconds)	Actual production quantity	Event cause subcategory
2007-01-19 10:59:00	2007-01-19 13:00:00	Stoppage			Failures
2007-01-23 09:49:00	2007-01-23 13:30:00	Stoppage			Failures
2007-01-26 06:00:00	2007-01-26 06:48:00	Stoppage			Failures
2007-01-29 00:40:00	2007-01-29 02:00:00	Stoppage			Failures
2007-02-16 06:57:00	2007-02-16 14:00:00	Stoppage			Failures
2007-02-16 14:00:00	2007-02-16 17:45:00	Stoppage			Failures
2007-02-19 07:27:00	2007-02-19 08:00:00	Stoppage			Failures
2007-03-02 16:09:00	2007-03-02 16:45:00	Stoppage			Personnel commitments and communication
2007-03-26 05:19:00	2007-03-26 06:00:00	Stoppage			Failures
2007-03-29 10:25:00	2007-03-29 11:00:00	Stoppage			Failures
2007-03-29 15:17:00	2007-03-29 16:00:00	Stoppage			Failures
2007-05-21 11:48:00	2007-05-21 13:00:00	Stoppage			Failures
2007-06-01 01:00:00	2007-06-01 02:12:00	Stoppage			Working environment
2007-06-01 03:00:00	2007-06-01 12:36:00	Stoppage			Maintenance organization and management tools
2007-06-01 13:00:00	2007-06-01 17:48:00	Stoppage			Competence, i.e. ability to perform maintenance and
2007-06-02 01:00:00	2007-06-02 09:24:00	Stoppage			Personnel commitments and communication
2007-06-02 10:00:00	2007-06-02 12:30:00	Stoppage			Production procedures and methodology
2007-06-03 01:00:00	2007-06-03 22:30:00	Stoppage			Machine condition and characteristics
2007-06-04 01:00:00	2007-06-04 04:36:00	Stoppage			Failures
2007-06-04 05:00:00	2007-06-05 06:00:00	Stoppage			Short stoppages
2007-06-06 01:00:00	2007-06-06 15:24:00	Stoppage			Stability of production quality
2007-06-06 16:00:00	2007-06-06 19:36:00	Stoppage			Stability of the production speed and production rate
2007-06-07 01:00:00	2007-06-07 03:48:00	Stoppage			Failures
2007-06-08 10:18:00	2007-06-08 12:36:00	Stoppage			Failures
2007-07-09 06:21:00	2007-07-09 09:00:00	Stoppage			Failures
2007-07-12 12:20:00	2007-07-12 13:30:00	Stoppage			Failures

Record: 1 of 50

Production event

Production JTD Engine Head [Production Process Input/Output] on JTD Engine Head Line [Machinery, Industrial Production] at 2007-01-08

GMT event start	GMT event end	Production event type	Actual cycle time (seconds)	Actual production quantity	Event cause subcategory
2007-07-19 18:12:00	2007-07-19 21:25:00	Stoppage			Failures
2007-08-16 06:37:00	2007-08-16 07:14:00	Stoppage			Failures
2007-08-16 15:58:00	2007-08-16 18:30:00	Stoppage			Failures
2007-08-17 01:14:00	2007-08-17 23:00:00	Stoppage			Failures
2007-08-19 07:48:00	2007-08-19 09:30:00	Stoppage			Failures
2007-08-20 01:23:00	2007-08-20 09:00:00	Stoppage			Failures
2007-08-20 23:45:00	2007-08-21 04:00:00	Stoppage			Failures
2007-08-23 08:27:00	2007-08-23 13:00:00	Stoppage			Failures
2007-08-23 21:20:00	2007-08-24 02:30:00	Stoppage			Failures
2007-09-14 11:40:00	2007-09-14 12:15:00	Stoppage			Failures
2007-09-17 07:00:00	2007-09-17 11:00:00	Stoppage			Failures
2007-09-28 04:23:00	2007-09-28 06:45:00	Stoppage			Failures
2007-10-05 04:24:00	2007-10-05 14:24:00	Stoppage			Failures
2007-11-14 21:30:00	2007-11-14 23:00:00	Stoppage			Failures
2007-12-19 01:00:00	2007-12-19 23:12:00	Stoppage			Failures
2007-12-20 01:00:00	2007-12-20 08:33:00	Stoppage			Personnel commitments and communication
2007-12-21 01:00:00	2007-12-21 18:00:00	Stoppage			Machine condition and characteristics
2007-12-22 01:00:00	2007-12-22 03:20:00	Stoppage			Production procedures and methodology
2007-12-24 01:00:00	2007-12-24 08:00:00	Stoppage			Stability of production quality
2007-12-25 01:00:00	2007-12-25 07:24:00	Stoppage			Stability of the production speed and production rate
2007-12-26 01:00:00	2007-12-27 23:12:00	Stoppage			Short stoppages
2007-12-30 01:00:00	2007-12-30 02:54:00	Stoppage			Working environment
2007-12-30 03:00:00	2007-12-30 20:18:00	Stoppage			Maintenance organization and management tools
2007-12-31 01:00:00	2007-12-31 13:00:00	Stoppage			Competence, i.e. ability to perform maintenance and

Record: 1 of 50

Figure 13.18 Production events

Production follow-up

Production JTD Engine Head [Production Process Input/Output] on JTD Engine Head Line [Machinery, Industrial Production] at 2007-01-08

GMT follow-up	Actual production quantity	Defective production quantity
2007-03-31	17203	425
2007-06-30	16401	408
2007-09-30	13702	349
2008-01-08	16622	448

Record: 1 of 4

Figure 13.19 Production follow up data

The data describing the reference situation are shown in Figure 13.20, which is completed by the profit margin (95 units). The impacts anticipated out of all the three investments are specified in Alternative 1, 2 and 3 and shown in Figures 13.21–13.24, respectively. Based on the long term economic results of the anticipated impacts of the three investments (alternatives), AltSim motor has selected alternative 1 as the most cost-effective although it demands bigger investment (Figure 13.24).

The final conclusion that can be drawn is that the size of the investment cannot be used as a criterion to judge the cost-effectiveness of an investment in maintenance. To judge the cost-effectiveness of investments in maintenance we should consider the long term maintenance economic payoff, *i.e.*, return on investment in maintenance. The comparison between these three alternatives can be made with respect to either the potential total profit that will be generated or the potential rate of the savings to the investment. Notice that the most cost effective alternative can be different depending on which criterion is used. In this case we used the profit absolute value to distinguish the most cost-effective maintenance solution. Notice that in the original user-interface of the software module, Total Saving, Profit/Losses are shown in black and red, respectively, to be distinguished easily.

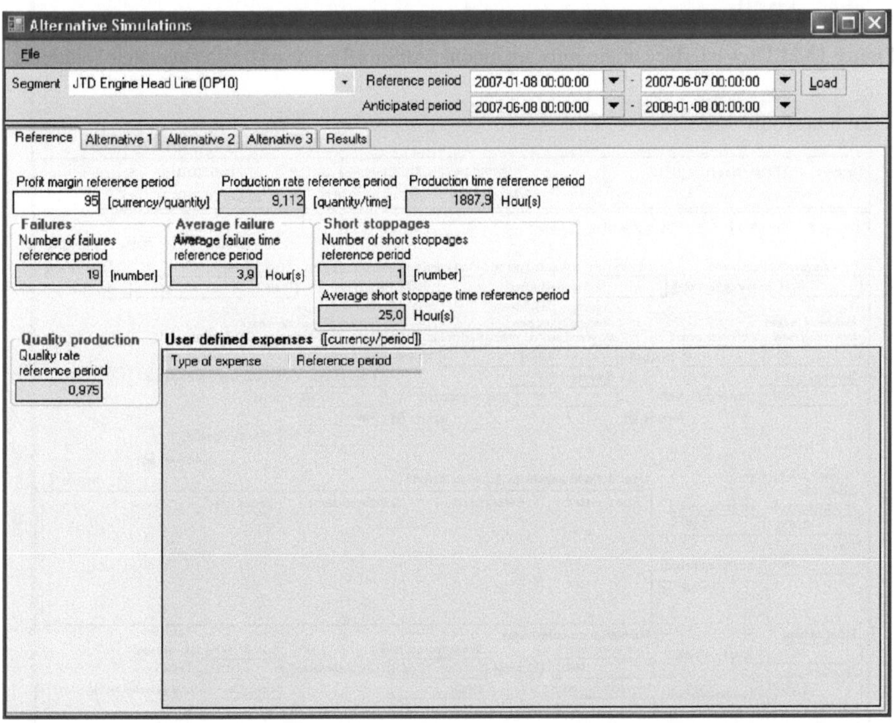

Figure 13.20 User-interface of the software module AltSim. Test results with reference data

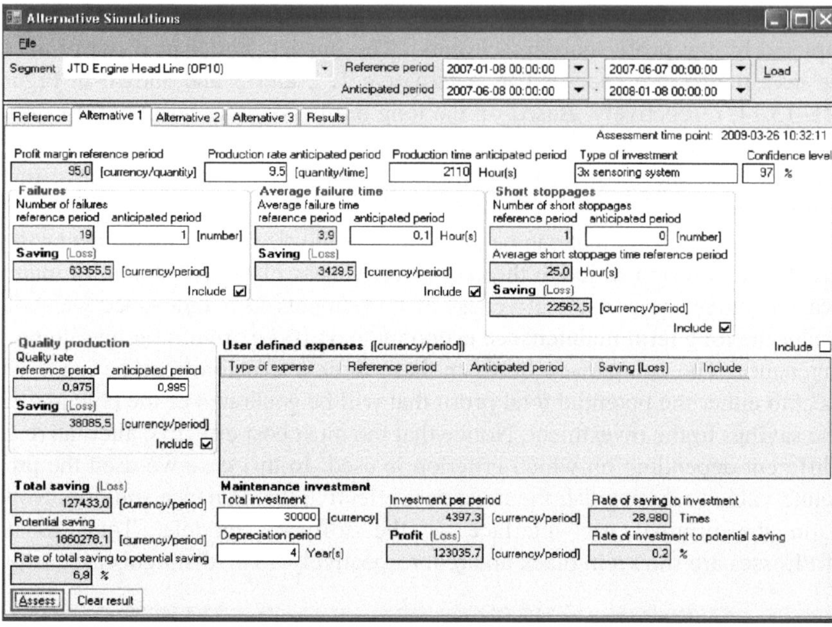

Figure 13.21 User-interface of the software module AltSim. Test results with alternative solution 1

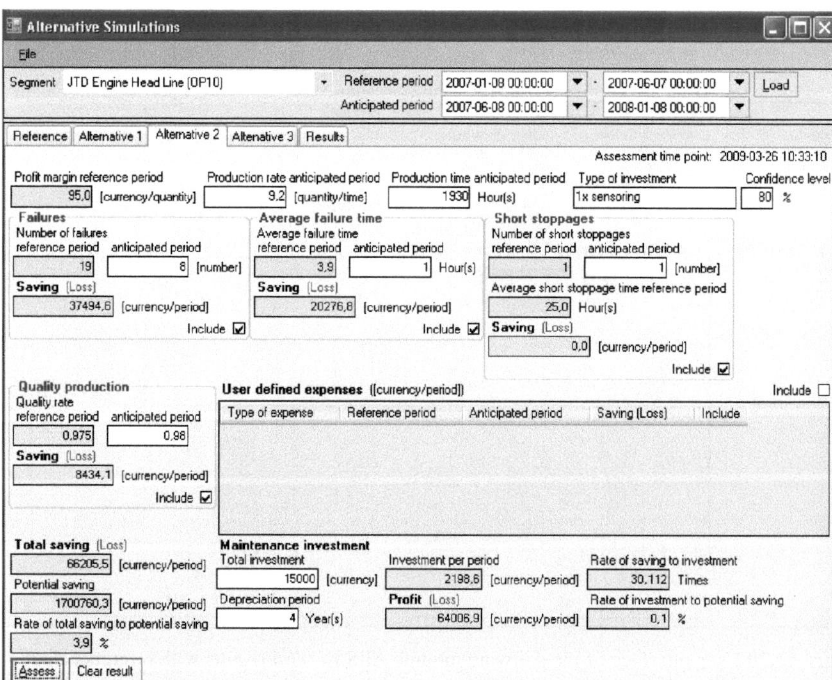

Figure 13.22 User-interface of the software module AltSim. Test results with alternative solution 2

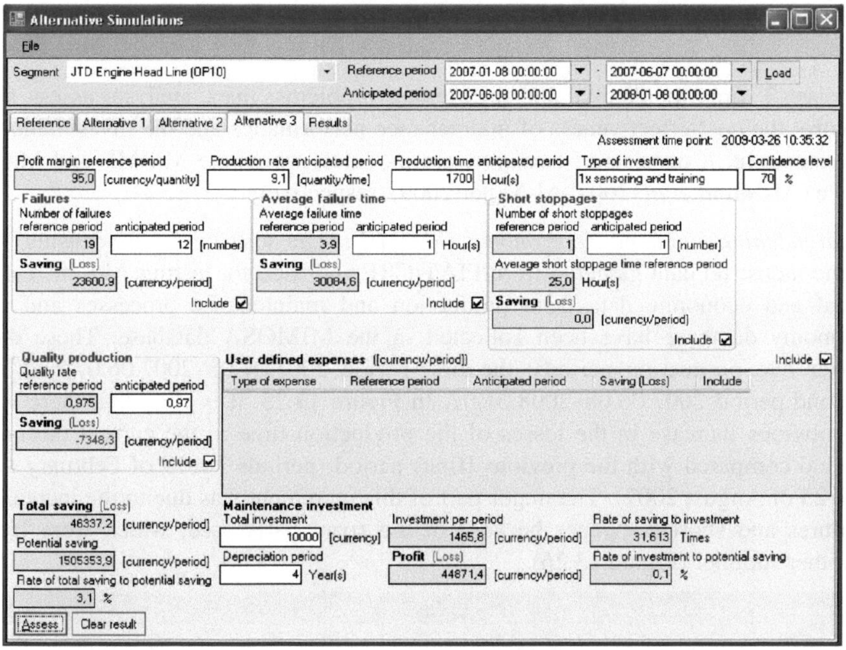

Figure 13.23 User-interface of the software module AltSim. Test results with alternative solution 3

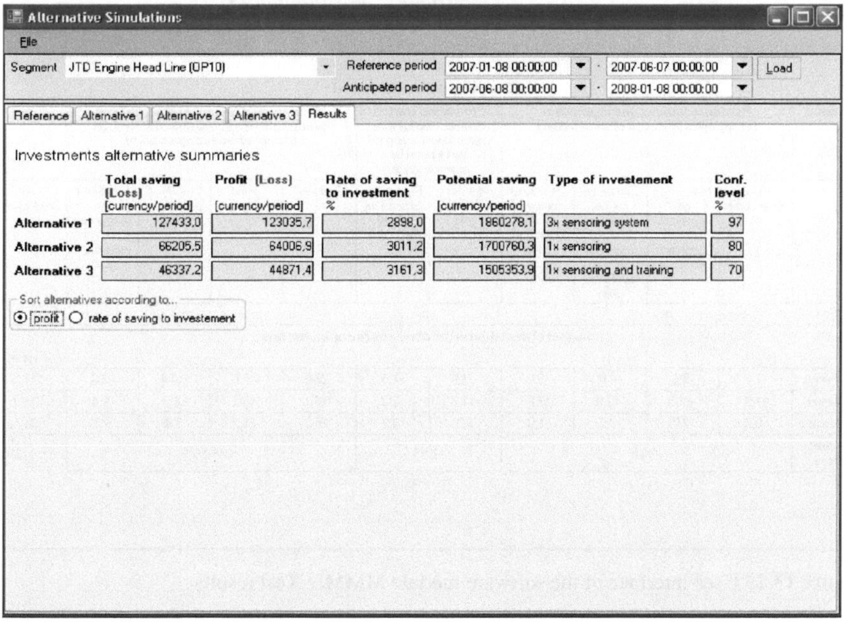

Figure 13.24 User-interface of the software module AltSim. Test results with results of the selection

13.5.3 Toolset 3: MMME and MainSave

Toolset 3 aims to identify and prioritising problems, map, analyse, assess and control the cost-effectiveness of maintenance performance and the investments in maintenance. It consists of two software modules, which are MMME and MainSave (Al-Najjar *et al.* 2007, Al-Najjar 2009), respectively.

Man-machine-maintenance-economy (MMME): The tests were done using the same industrial data gathered from FIAT/CRF and used for testing AltSim. Technical and economic data from production and maintenance processes and the economy database have been collected in the MIMOSA database. These data cover two production periods: the first period 2007.01.08–2007.06.07 and the second period 2007.06.08–2008.01.07. In Figure 13.25, it is clear that there was an obvious increase in the losses of the production time in the current (second) period compared with the previous (first) period (periods 16–19 of February and 16–23 of August 2007). The major part of this increment was due to the increased failures and short stoppages because of the spare parts used, which were from another supplier (Figure 13.26).

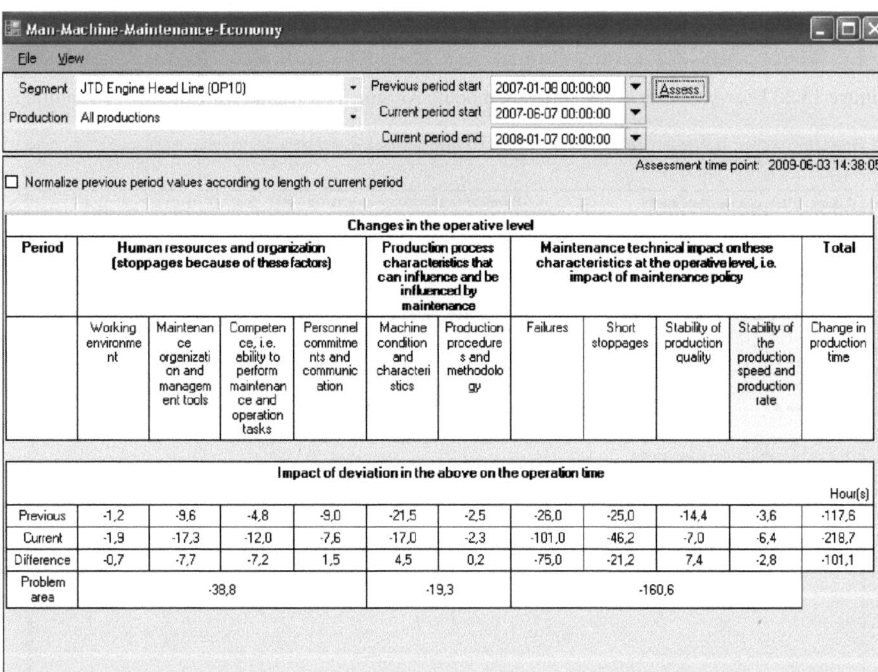

Figure 13.25 User-interface of the software module MMME. Test results

The major conclusion that can be drawn is that by applying MMME it is possible to use losses in production time as a criterion for identifying and prioritising roots of problems and problem areas. Assessing the amount of losses in the production time helps to identify and select the most economically beneficial areas for future investments in maintenance. The losses in production time increased by more than 96 h of production time, and 91 h thereof is due to failures and short stoppages.

During the test period FIAT/CRF did not have any reduction in theoretical production speed. This means that the actual speed of the machine was equal to the theoretical speed during the whole period, which is not usual. The explanation for this is that the losses due to a lower production speed ended up in the *waste* category at FIAT. Therefore, it was unfortunately not possible to analyse and test this part of the data.

The data required for applying MMME cannot always be found in the company's databases because they are not normally needed and collected, but are available in the company's processes and can be gathered. MMME can be used for analysing the losses/gains, *i.e.*, lost (-) or gained (+) hours, in production time and prioritising the root-causes/problem areas behind that when it is needed. It is also easy to compare different machines with the same or similar attributes in order to use previous experiences throughout the company in a cost-effective way. It is obvious that not all the model's categories of losses (see Figure 13.25) can be measured with high accuracy the first time, such as is the case when measuring the impact of personnel commitments and communications on the production time.

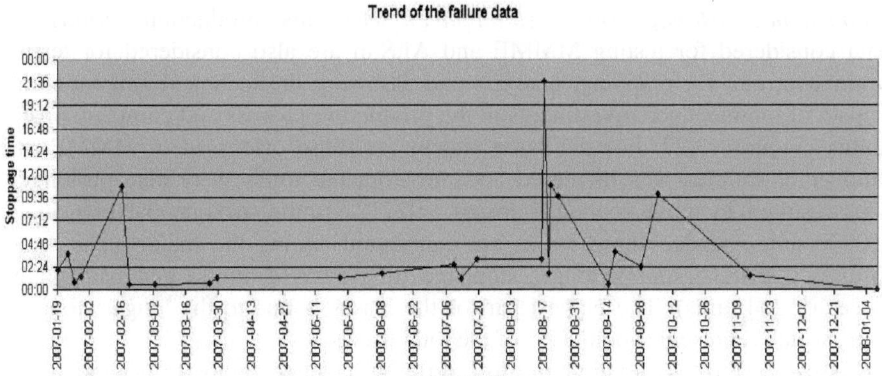

Figure 13.26 Failure data and trend of the failure data from FIAT

To secure collecting the right data, instructions on how to identify and collect right data are used:

1. When collecting the data, one machine in a production station (producing a particular product) should be considered.
2. The reasons for every stoppage or reduction of production rate/production speed that occur should be categorised and specified properly with respect to the sub-categories/factors expressed in MMME.

3. The time points of the start and end of the stoppage or the reduction of production speed should be specified.
4. Only stoppages during the planned production time should be considered, but not planned maintenance activities or time planned for education or meetings, *etc.*
5. External factors that are completely out of the control of the company should be excluded, such as power shortage due to bad weather or power supplier.
6. Stoppages are handled in the first part of the datasheet and reduction of production speed in the second part of the datasheet.

The MMME categories shown are relevant for the type of industry tested and they cover relevant aspects of losses in production time of a production station or machine. During the development of the software prototype it was discovered that it can be useful to add a feature that enables the user to normalise the data. This makes it possible to compare a long period with a short one by comparing an average from a long period to the current situation. To get an overview of the production time, it will be possible to add new functions enabling the user to set levels or reference values for detecting changes/deviations in the process output. A warning level can be used for indicating whether or not the losses in production time are higher than acceptable, and which sub-category contributes with the biggest amount of losses in the production time. The number of different levels is practically unlimited.

Maintenance cost-effectiveness (MainSave): The same production periods and data considered for testing MMME and AltSim are also considered for testing MainSave, *i.e.*, for mapping, analysis and assessing the technical and economic impact of maintenance investments in the production process and company profitability, Figure 13.27. For the same reasons as those discussed in MMME, the number of failures was increased and the stoppage times were also prolonged. This resulted in more economic losses (-92712.8 units) despite the investment (30000 units) made. These losses are distributed among the major areas: more failures (-25483.8 units), longer stoppage time (-65965 units) and bad quality losses (-1264 units). The biggest part of the losses is due to the longer stoppage time, which represents about 71% of the total losses.

Assessing the losses belonging to a different category helps to primarily estimate and judge the size of the risk capital that should be invested for solving the problems. Note that the saving with the (-) sign means losses. Also, the overall equipment effectiveness has been increased due to (may be) more overtime work that has been used for compensating the losses in the production time. In the test of AltSim, MMME and MainSave, it is obvious that periods of 6 months are not the ideal frequency for data gathering.

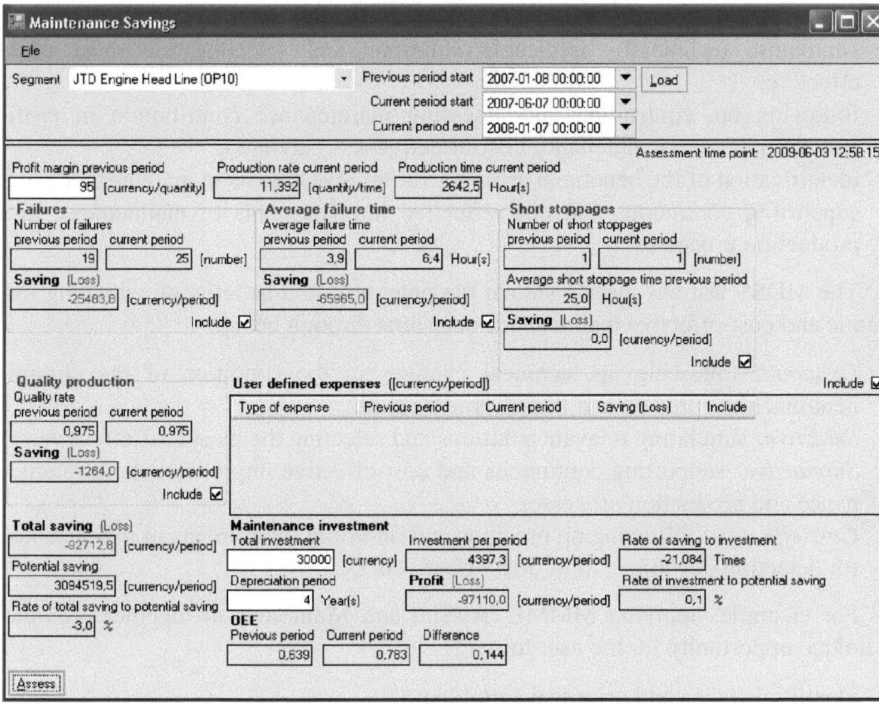

Figure 13.27 User-interface of the software module MainSave. Test results

13.6 Results and Discussions

When using VBM it is common that the component replacement can be done either just lower or higher than the predetermined replacement level. This is because the probability of doing a replacement at an exact predetermined level is negligibly small. If the replacement is done at a higher/lower level than the predetermined one it may lead to failures/loss in residual lifetime of the component, respectively, and consequently it generates economic losses. Higher accuracy in assessing the condition of a component yields a higher probability of avoiding failures and planning maintenance actions at the time when all costs are at minimum, *i.e.*, cost-effective maintenance.

The major result of the study presented in Chapters 12 and 13 is the development of a new decision support systems and software prototype (MDSS). It consists of five software modules for achieving different objectives:

- enhancing maintenance decision accuracy;
- mapping, following up, analysis and improving maintenance and production processes through identifying and prioritising problem areas;

- identification and assessment of losses in production time;
- simulating technically applicable solutions and selecting the most cost-effective;
- following up, controlling and assessing maintenance contribution in profit generation due to maintenance cost-effectiveness (savings);
- identification of the beneficial areas for future investments in maintenance; and
- supporting continuous and cost effective improvements of maintenance and production processes.

The MDSS test has clearly shown the potentials and benefits of achieving dynamic and cost-effective maintenance decisions through being:

- *Dynamic:* following up technical changes in the condition of the component/machine, process and improvement results.
- *Selective:* simulating relevant solutions and selecting the most cost-effective.
- *Supportive:* supporting continuous and cost-effective improvements in maintenance and production processes.
- *Cost-effective:* following up previous and on-going investments in maintenance for detecting deviations in its cost-effectiveness.

For example, applying MMME, AltSim and MainSave all together provides a unique opportunity for the user to:

- identify, localise and prioritise a problem;
- simulate, select and judge in advance the cost-effectiveness of the investments suggested for applying a particular maintenance action/solution; and finally
- follow up the investment results as to whether it contributed to more profit for the company.

The same thing can be said when applying PreVib, ProFail, ResLife, AltSim and MainSave. Using the MIMOSA database reduces technical difficulties and disturbances that may be induced in the current IT-systems of a company.

13.7 Conclusions

The main conclusion of this study is that when applying MDSS it is possible to achieve the above mentioned objectives. Using MDSS, it is possible to act at an early stage in both tactical and strategic levels to fulfil a company's strategic goals of continuous improvement of its profitability and competitiveness. Applying MDSS makes it possible to handle real-time data, analysis and decisions and give necessary information about maintenance and other working areas to the decision-maker.

It provides better data coverage and quality, which are essential for improving knowledge and experience in maintenance and thereby aid in increasing the competitiveness and profitability of a company. Predicting the CM parameter level at the next planned measuring time or next planned stoppage reduces the risk of an unexpected deviation in the condition of significant components. This means that the probability of failure of a component, such as a rolling element bearing, can be kept very low (close to zero) until damage is initiated and under development, given that damage can be detected at an early stage through using an efficient CM system. This would reduce the number and duration of planned and unplanned stoppages, which has a direct impact on the company's productivity, and hence, its competitiveness.

By using the different tools/software modules of MDSS there is no duplication of data gathering. Also, the MDSS modules are flexible and can be used in different companies, for different machines, components and products. It is also possible to show how maintenance affects the profit of a company in a way that would be difficult with the tools and techniques available even if the data required are available. Development of relevant and traceable KPIs for technical and economic maintenance control has made this possible.

All these functionalities are possible when high quality data can be collected using the advanced techniques in data gathering, transmission and analysis that have been introduced in the previous chapters of this book.

References

Al-Najjar B (1998) Improved effectiveness of vibration monitoring of rolling element bearings in paper mills. Journal of Engineering Tribology, IMechE, Proc Inst Mechanical Engineers part J, 212:111–120

Al-Najjar B (2000) Accuracy, effectiveness and improvement of vibration-based maintenance in paper mills: case studies. Journal of Sound and Vibration 229:389–410

Al-Najjar B (2003) Total time on test, TTT-plots for condition monitoring of rolling element bearings in paper mills. International Journal of Condition Monitoring and Diagnostic Engineering Management (COMADEM) 6:27–32

Al-Najjar B (2007a) The lack of maintenance and not maintenance which costs: a model to describe and quantify the impact of vibration-based maintenance on company's business. International Journal of Production Economics (IJPPM)107, 55:260–273

Al-Najjar B (2007b) Establishing and running a condition-based maintenance policy: applied example of vibration-based maintenance. WCEAM2007, 12–14.6.2007, Harrogate, UK. NDT, Northampton, 106–115

Al-Najjar B (2008) Maintenance function deployment (MFD) for cost-effective and continuous improvement of company's business: Using total quality maintenance (TQMain). Euromaintenance 2008, 8-10.4.2008, Brussels, Belgium

Al-Najjar B (2009) A computerised model for assessing the return on investment in maintenance: Following up maintenance contribution in company profit. To appear in the proceedings of WCEAM 2009, Athens, Greece

Al-Najjar B and Alsyouf I (2003) Selecting the most efficient maintenance approach using fuzzy multiple criteria decision making. Journal of Production Economics 84:85–100

Al-Najjar B, Alsyouf I, Salgado E, Khosaba, S, Faaborg K (2001) Economic importance of maintenance planning when using vibration-based maintenance policy. Project report, Växjö University, Sweden

Al-Najjar B, Andersson D, Jacobsson M (2007) A model to describe the relationships man-machine-maintenance-economy (MMME). IMECS2007, 21-23.3.2007, Hong Kong, II:2127–2133

Al-Najjar B and Ciganovic R (2009) A model for more accurate maintenance decisions (MMAMDec). To appear in the proceedings of WCEAM 2009, Athens, Greece

Al-Najjar B and Kans M (2006) A model to identify relevant data for accurate problem tracing and localisation, and cost-effective decisions: a case study. The International Journal of Productivity and Performance Measurement (IJPPM), 55(8): 616–637

Al-Najjar B and Wang W (2001) A conceptual model for fault detection and decision-making for rolling element bearings in paper mills. Journal of Quality in Maintenance Engineering 7:192–207

Jardine AKS, Lin D, Banjevic D (2006) A review on machinery diagnostics and prognostics implementing condition-based maintenance. Mechanical Systems and Signal Processing 20:1483–1510

MIMOSA (2006) Common relational information schema (CRIS) Version 3.1 Specification, http://www.mimosa.org/

Moore W J, Starr A G (2006) An intelligent maintenance system for continuous cost-based prioritisation of maintenance activities. Computers in Industry 57:595–606

Pintelon L (1997) Maintenance performance reporting systems: some experiences. Journal of Quality in Maintenance Engineering 3:4–15

Chapter 14
Industrial Demonstrations of E-maintenance Solutions

Julien Mascolo, Per Nilsson, Benoit Iung, Erik Levrat, Alexandre Voisin, Fernando Garramiola and Jim Bellew

The aims of some industrial demonstrations were to integrate the technological and information technology strands in a business sense and to verify the effectiveness of the complete DynaWeb solution, *i.e.*:

- Assessment of functionality and robustness of the complete hardware and software system developed in the different parts of the project.
- Assessment of the e-maintenance concept and strategy and the cost-effectiveness model developed in the project, which is presented in the previous chapters in this book.
- To test the developed methods for optimising maintenance strategies and technology choices using real data, providing the benchmarks against which the research is measured.
- To demonstrate the effectiveness of new strategies and cost effectiveness on field data.
- Assessment of user interface, *i.e.*, the usability of DynaWeb combined with assessment of training modules.
- Identification of the best practice and meeting the needs in training and education for technical, IT and management in maintenance. The results of research have been integrated in new courses.

The validation activities were performed and at the end of the project the final output is:

- a demonstration performed at four different test sites at FIAT CRF, Volvo, Goratu and Martechnic;

- a technical and economical evaluation of these demonstrators; and
- recommendations for further implementation, development and industrialisation.

Detailed information on the results achieved are reported in the following sections.The following results have been achieved:

- FIAT CRF tested and demonstrated the integration of 25 Dynamite hardware, software components and services. As a summary of the demonstration:
  - The overall result is extremely positive, with technical and economical feasibility proven.
  - The level of the quality of components and adequacy to requirements was high, with people extremely dedicated to enhancing their components and testing them on the demonstrator.
  - As expected, integration was not straightforward and required major effort from all involved partners.
- Volvo tested the Tekniker oil sensor system designed to measure the level of oxidation of the lubricant by spectroscopy of visible light. The demonstration was done in a hydraulic system in a real industrial environment, a production line in the foundry. In conclusion:
  - The oxidation sensor hardware and software worked well in the foundry installation.
  - The environment in the foundry at Volvo was extremely dirty, which was a good test for the sensor but made it impossible to have the computer at the same location.
  - Connection into production line hydraulic systems required pressure reduction (normally by bypass flow).
  - The sensor required a continuous low speed oil flow without air bubbles and at a low oil pressure.
  - The sensor signal jumped up and down depending on, e.g., irregular oil flow, air bubbles, *etc.*, which made the interpretation more difficult and not straightforward.
  - Volvo IT policy made it almost impossible to demonstrate communication with the MIMOSA database using SQL server communication but a one-way web service communication to store data in the Mimosa database was created and included in the software and finally tested.
- For global demonstration, Goratu tested several Dynamite components and their communication to the Mimosa database:
  - The VTT particle scatter lube sensor for a hydraulic system, the Tekniker water content lube sensor for a cooling system and the Wyselec vibration measurement system for spindle vibration were implemented at a Goratu machine, and the data collected sent to the MIMOSA database. All the data

collected provided great information for Goratu, where these was no information related to these issues. Apart from this, the web services provided an important tool for machine reliability. Web services allow us to implement the online diagnosis and condition monitoring, which it was impossible until now.
  - The handheld vibration unit and the PDA maintenance user interface were tested too. These allow inserting new assets on the database and taking measurements using a PDA, which due to its size it is very comfortable for the user.
  - This demonstration gave new feedback to Goratu, where this information will be used for machine improvements and new utilities for the customers. The innovation is very important, but some improvements are needed for a full implementation in an industrial environment, *e.g.*, higher flows and pressure for sensors and better filters for vibration system.

- At Martechnic the demonstration consisted of a simulated application of a stern tube bearing/tail end shaft assembly from an 8000TEU container ship. For logistical and security reasons this demonstration could not take place on board the ship. Using a specially designed test rig conditions on board ship were replicated and the cycled lube oil was progressively contaminated with water and particulate matter.
  - The demonstration evaluated four sensors. Three of them performed satisfactorily communicating their results *via* two separate routes to the MIMOSA database, while one was not enabled for DynaWeb communication.
  - The demonstration was deemed a success and the economic scenario surrounding this application clearly demonstrated considerable benefits from applying the Dynamite concepts.

14.1 Global Demonstration in a Milling Machine Environment

Julien Mascolo

This chapter presents the results of testing the DynaWeb concept and components and the FIAT CRF demonstrator, focussing on one side on both technical and business issues. Possible next steps for implementation are also reported.

On the technical side, it defines and addresses the main results obtained and demonstrated in the FIAT group. In particular the following points have been addressed:

- the FIAT group environment for requirement elicitation and testing;
- the results of testing in the FIAT group; and

- the settings for the final demonstration in CRF (research centre of the FIAT group).

On the business side, it recalls the business analysis and status. On the industrialisation side it sketches a roadmap for development. The analysis has focused on results and has adopted a methodology developed on the in depth analysis of the international literature and the best practices. In particular, the business analysis has addressed the competitive advantages/disadvantages of the Dynamite approach, in order to develop a technology focused on the important technological and business success factors, leading to a high competitive positioning and a relevant market share.

The main objective of CRF for participating in the project was to develop innovation for the FIAT group sectors, with the aim of increasing plant productivity and the availability of machines. In the framework of the project, applications of the project techniques, including diagnosis and prognosis, have been performed at four locations of the FIAT Group.

From the technical side the tests have validated the overall approach and its implemented and available elements.

An economical assessment has been performed, and the approach is deemed to be widely applicable to our production facilities.

Overall the project results have validated the work done by the FIAT group sectors and their partners in the project.

14.1.1 Objectives of the Test and Demonstrations

The main objectives were to develop methodologies and tools for giving early warnings of any incoming failure or deterioration of performances and assessing the impact of the maintenance strategy on the overall availability of the machines.

During the testing and demonstration phases the CRF objectives were to apply these methods and tools on a number of applications inside the FIAT Group. The global demonstration at FIAT was intended to cover as much as possible – actually nearly all – the DynaWeb components. Applying and testing components at different locations was resulted from different facts.

First of all, several FIAT group sites were interested in different elements of the DynaWeb approach; the lube sensors and the predictive maintenance in one case, the e-maintenance and wireless aspect in others. On the other hand, installation at a single workshop would not have reflected the requirements and interests of different users with respect to the variety of tools of the complete DynaWeb platform. Testing in many locations yields different requirements and feedback from different users and this provides interesting viewpoints. There was also a gap in terms of timing between the collection of data for economical analysis, with different phases of analysis, refinement and new requests for data, which lasted 2.5

years, and the sensor implementation and test, performed during the 4 final months of the project.

Taking into account all these considerations, three locations were selected at production sites, with another site for the final demonstration event and integration on a single machine at CRF. A plant was selected for the oil sensor implementation, due to previous experience in the field also in collaboration with oil providers (Castrol). Another plant was selected due to the company interest and experience in applying sensors for monitoring the quality of production. Furthermore, a specific machine in that plant was chosen due to the fact that an identical machine is available at CRF. Tests were performed at CRF using actual production data such as workload, typical speeds, tools used, *etc.*

Altogether 24 out of 27 DynaWeb components were tested in the context of the FIAT Global demonstration and demonstrated at CRF.

Use cases were defined to cover the main objectives for the industrial users:

- Use case 1: strategic decisions, such as whether to implement a new maintenance policy or assess previous ones.
- Use case 2: support in case of failure (stoppage).
- Use case 3: support in case of incoming failure (degradation/loss of production).
- Use case 4: support to periodic maintenance.

Use cases 1, 3 and 4 were given top priority for implementation, while use case 2 was considered as a sub-use case of the second one. In details the steps for each use case were defined using an appropriate system. The DemoSteps are shown as an example in Figure 14.1.

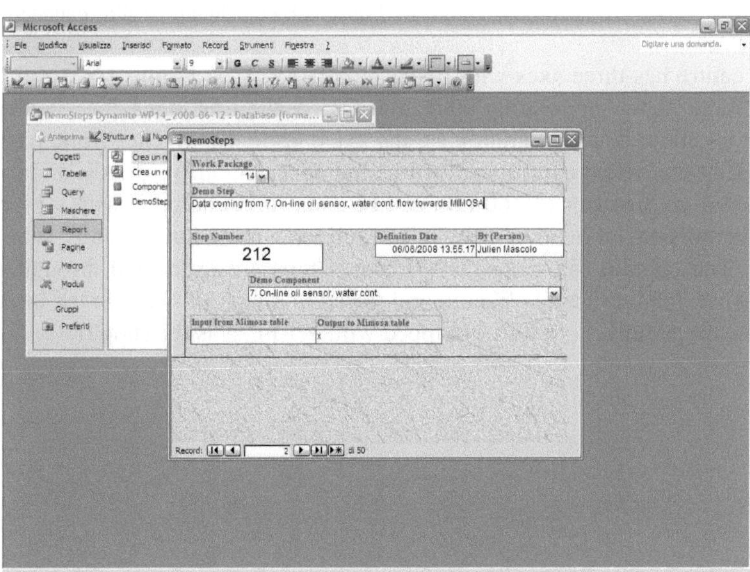

Figure 14.1 A step of the demonstration in the *Demosteps* tool

14.1.2 Description of the Test Platform

The test platform (site 67890002) was a machine located on the CRF premises. This machine mainly works on prototypes and small lots for the main customer FIAT Powertrain Technologies (FPT). The workload on the machine is not even, with peaks due to clients requests. The milling machine is identical to one used in another company of the FIAT group, with the same client and same typology of products.

The machine is a Urane (Figures 14.2–14.4), designed for high speed machining of lightweight alloys, equipped with special safety devices for the machining of magnesium, including hydrogen and flame sensors and an inert gas extinguishing system. The machine is a high speed machine tool, with axes driven by linear motors, which in theory reduces preventive maintenance.

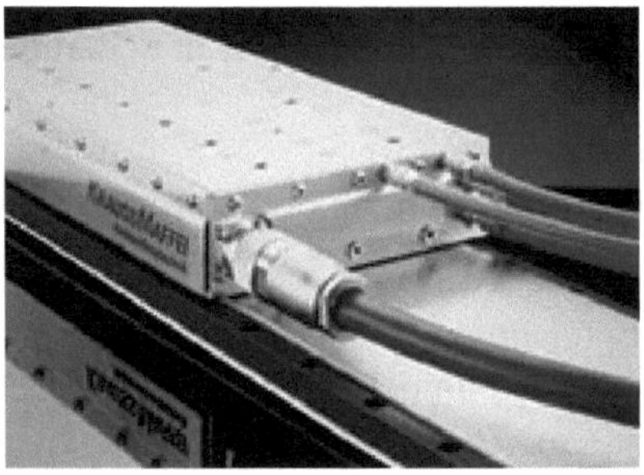

Figure 14.2 Linear motor

Figure 14.3 Machining of alloy cylinder-head

The centre has three axes with linear motors (*X*, *Y*, *Z*, Figure 14.2) and further two rotations B for the rotating table, and Q1 for the tool magazine. Siemens linear motors allow fast acceleration on the axes, up to 1.5 g. The Urane has a Siemens 840D numerical control, as well as Siemens Simodrive drives, and tools up to a maximum diameter of 160 mm can be used. It is equipped with a high pressure oil/emulsion lubrication system, as well as a minimum quantity lubrication system. The spindle can reach 24000 rpm. Chips are evacuated through a conveyor, while the lubricant is filtered and stored in a tank (800 l). On the infrastructure side, the premises are fully equipped with Wi-Fi, as is the shop floor.

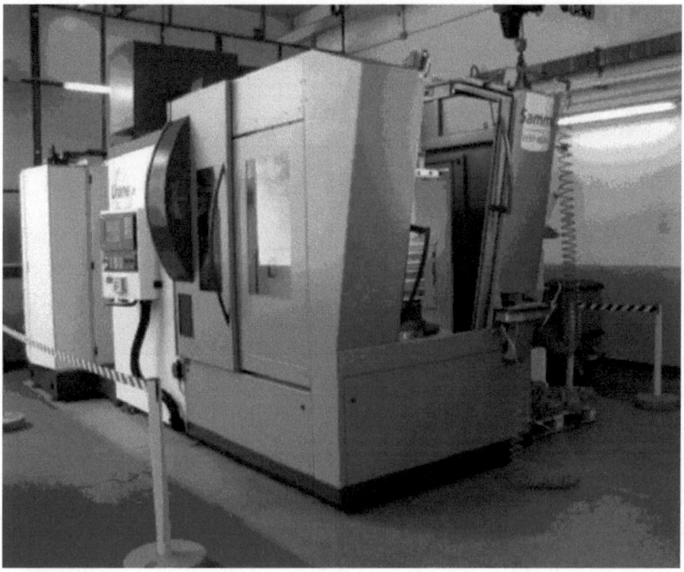

Figure 14.4 CRF Urane 20 machine tool

14.1.3 Description of the DynaWeb Components Tested

The demonstration held at CRF was based on the tests in other companies of the FIAT Group involved in the project, where 25 out of 28 Dynamite results were been assessed. In the following we will report the use of these components, together with the improvements needed, if required.

14.1.3.1 Smart Tags and PDA Support

Smart tags have been used in the CRF setting for both asset and location identification and active mobile asset tracking. The demonstration room was mapped as below (Figure 14.5).

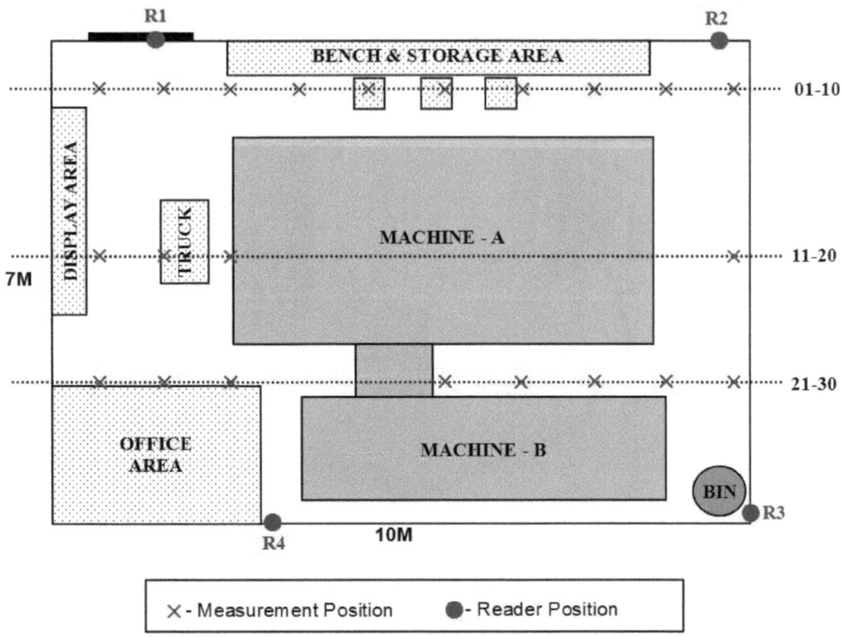

Figure 14.5 Demonstration room at CRF for RFID use

Some new passive smart tags were been brought to tag different assets and locations directly in the demonstration room by using the PDA interface. Then this information was stored in the MIMOSA database to identify items for daily maintenance procedure.

In a second step, the system was demonstrated to detect a new active smart tag when it is moving into the room. Then, the system shows the position/location of the smart tags on the map.

The overall evaluation of the tests was positive. The tests demonstrated the closing of the loop of information: from the demonstration site using the PDA interface to MIMOSA and back to consultation from a PDA or other devices on the shop floor.

Figures 14.6–14.9 shows the relevant information displayed to the maintenance operator accessing the machine: the location, photo, characteristics, design and information on spare parts.

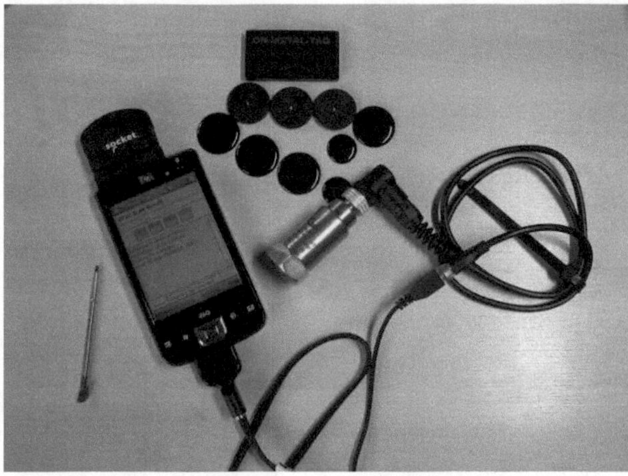

Figure 14.6 PDA used for the test with tags and portable vibration sensor

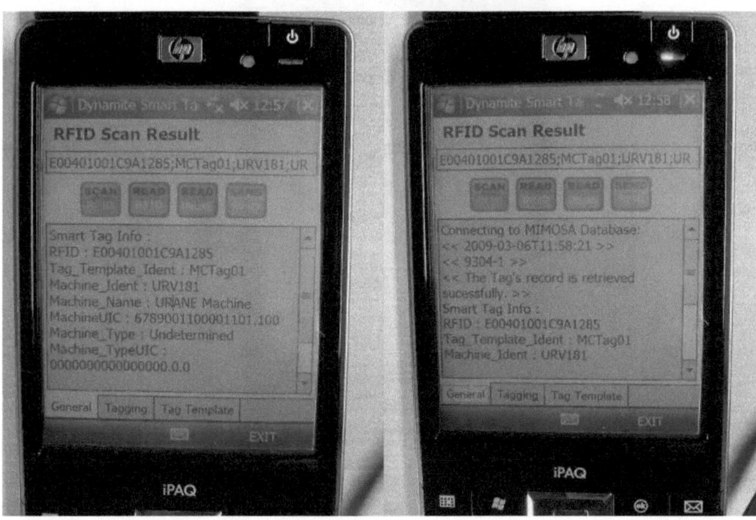

Figure 14.7 Scanning and transmission to MIMOSA

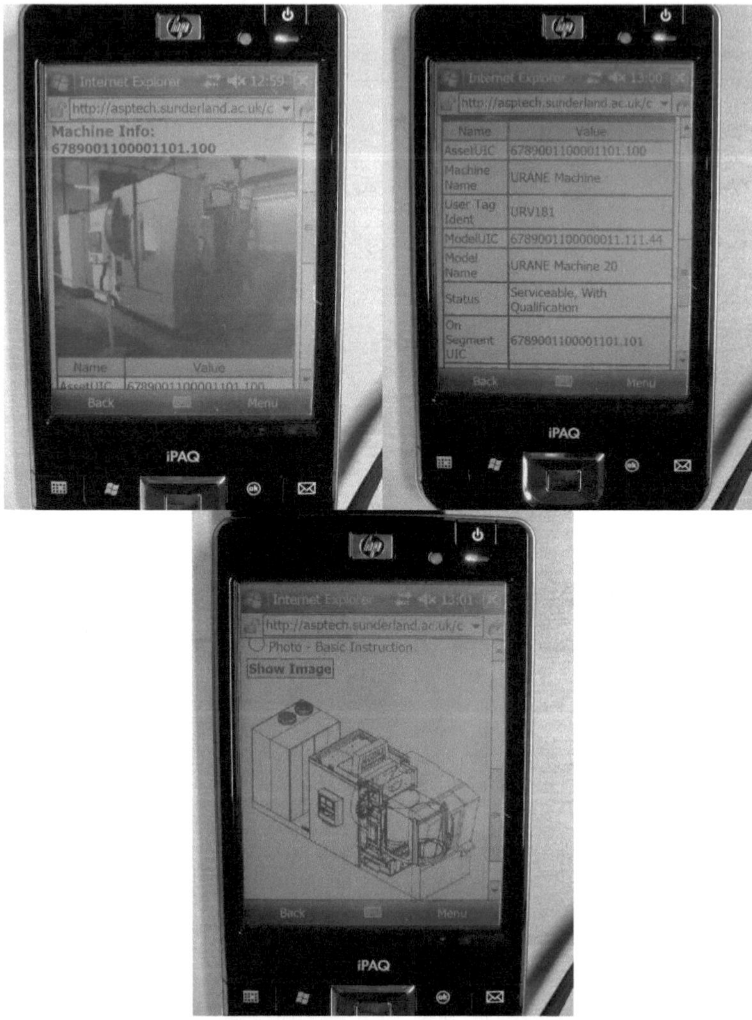

Figure 14.8 Retrieval of machine-relevant information from MIMOSA

Figure 14.9 Parts-relevant information from MIMOSA

Figure 14.10 summarises the substitution procedure and is a guideline for the maintenance operator accessing the machine.

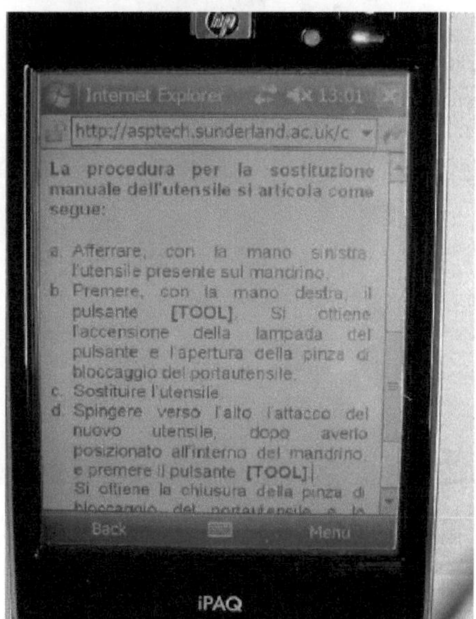

Figure 14.10 Maintenance procedure retrieved from MIMOSA

14.1.3.2 Handheld PDA Vibration Data Collector

A PDA has been selected for tests, able to act as a USB host to enable reading of the vibration sensor data. An application for the PDA has been developed for driving the device.

The tests were performed with the device inserted on the rotating plate. In the picture below, the wire for the handheld device is circled in black. Tests have shown an excellent usability and relevance of the results.

Some examples of the results of analysis are displayed in the Figures 14.11–14.13.

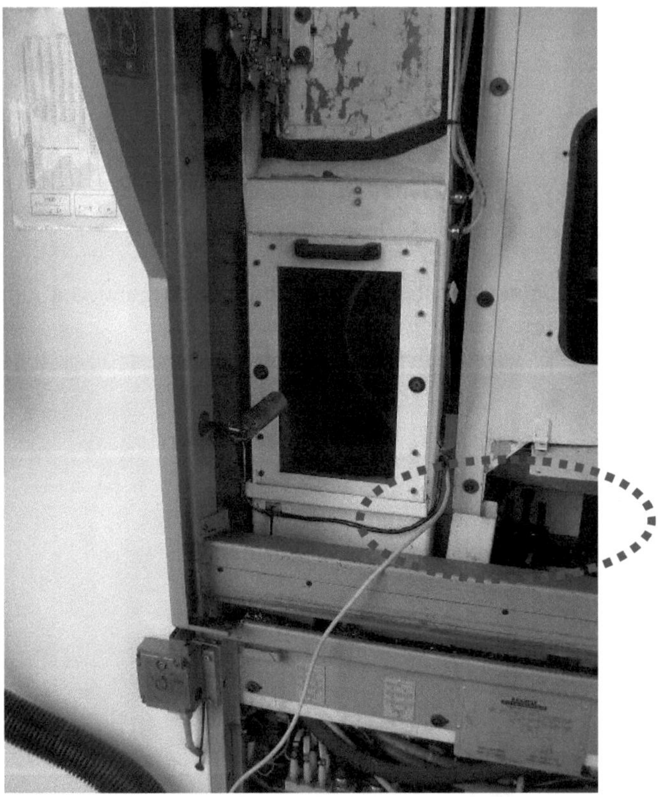

Figure 14.11 Portable PDA-Vib device setting

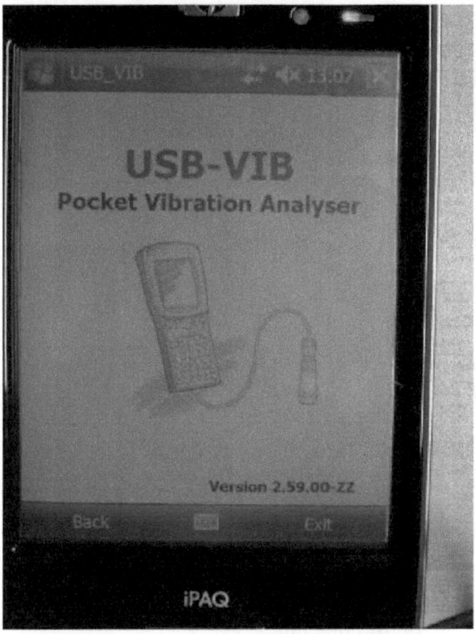

Figure 14.12 Main interface of the PDA-Vib

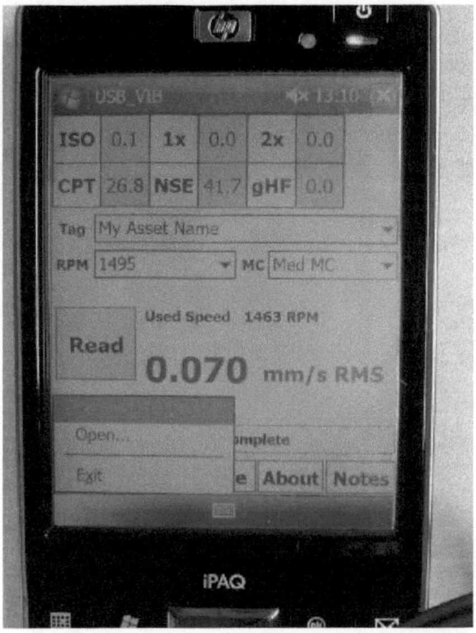

Figure 14.13 Results of analysis (PDA-Vib)

14.1.3.3 Vibration Measurement System

An industrial computer was installed on the shop floor for the testing of the vibration measurement system located near the machine tool. The sensor and the dedicated PC are shown in Figure 14.14.

Measurements were performed on the machine located nearby. For to safety reasons and in order to avoid any interference of the sensor with the tools, the sensor was screwed onto the rotating plate (see Figure 14.15). The vibration sensor can be screwed/unscrewed easily.

Figure 14.14 Vibration sensor and dedicated PC

Figure 14.15 Vibration sensor installed on the machine

The computer on which the sensor was installed had a direct connection to the Internet, the MIMOSA database and the web services from Tekniker.

The tests performed on the shop floor demonstrated the closing of the loop of information and the flow of data from the shop floor to MIMOSA and to the web services (Figures 14.16 and 14.17).

The approach was validated and the results showed a good potential for assessing on-line the health of the machine. The next steps will be to define specific models for wear-out, which was out of scope for this phase of the project.

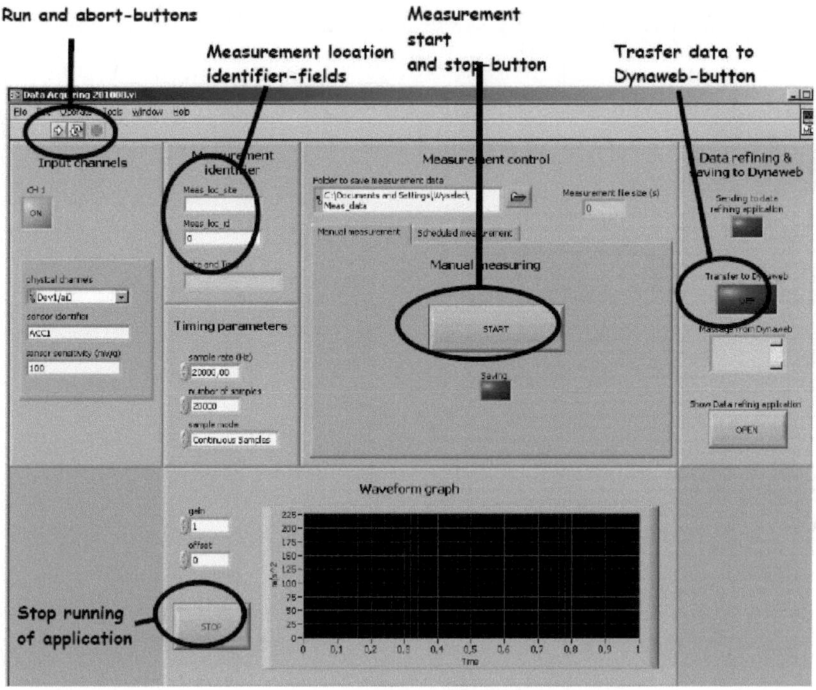

Figure 14.16 Interface of the measurement system with main functionalities

14.1.3.4 Oil Sensors

Regarding oil sensors, the industrial needs of the FIAT Group as concerns metal-working fluids have been divided into three different types of lubricant:

1. *Emulsion:* water-based coolant is used widely in production of power train components, mainly for aluminium components such as cylinder heads and gearboxes. Monitoring of the chemical condition of the emulsion can reduce disposal of coolant and increase the work piece quality.
2. *Net oil lubricants* are used in critical operations like grinding or deep drilling. Machining performances are related with the oil proprieties that can change during the working hours. This modification is not detected easily.
3. *Hydraulic oils* are in a centralised system able to move different axes on transfer lines. The presence of water or other solid particles can stop the whole transfer line or produce axis movement out of tolerance.

In the Dynamite project, the FIAT Group decided to focus its investigation on the last point: hydraulic oil. At this stage it was better to avoid monitoring the emulsion because of the complexity of the task and because the first systems able to monitor the oils that are going to be tested at FIAT are now available on the market.

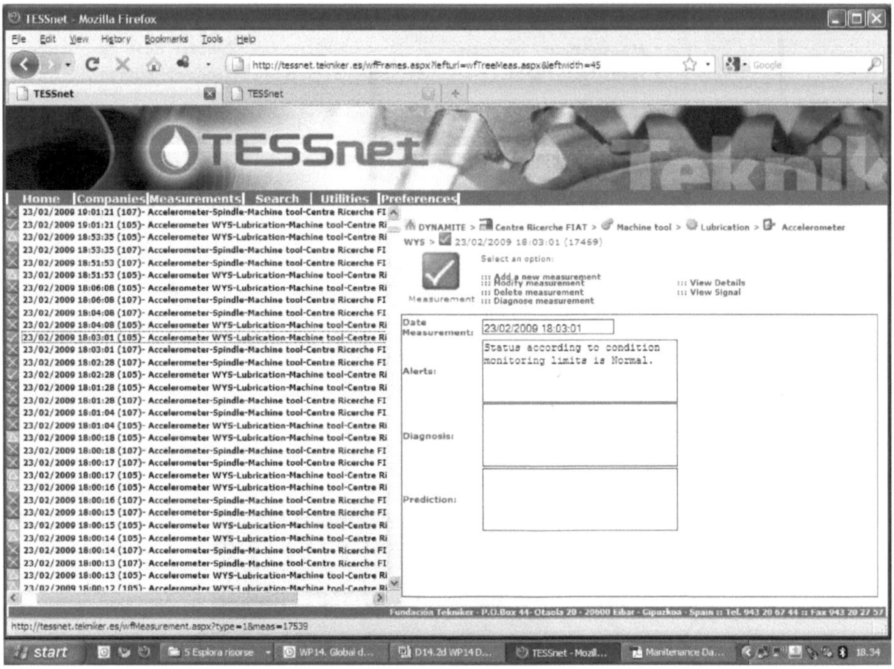

Figure 14.17 Result of the analysis in TESSnet

The idea of CRF was to develop a monitoring system for hydraulic systems based on similar sensors. These systems should monitor the main parameters that can be changed during use on-line, leaving to the determination of the other characteristics to the off-line chemical analysis.

The cooperation between the FIAT Group and the specialists of one of our main oil suppliers and the Tekniker specialists enabled us to find a list of variables that can monitor the performances of the lubricant during its use. Six variables were detected and these are listed below. The letter that follows the number means importance: the first three are the more important, then the other two, followed by than the last one. The final implementation of sensors focussed on the on-line analysis of the variables shown in Table 14.1.

Table 14.1 Parameters evaluated and selected for on-line sensor analysis

Lubricant parameters to be monitored	
Parameter	Description
Cinematic viscosity 40°C	Time measurement [s] of oil flow through a capillary tube by IR detector
	Influence of oil to a rotation speed of two cylinders
Water	Molecular iodine is produced by an electrode and it reacts with the humidity in the oil. When Iodine no longer reacts, the system detects it and stops the measurement.
	Esoteric reaction with $CaCl^2$; measuring of increased temperature
NAS (* only for hydraulic oil)	Laser light scattering of particles distribution in fluid suspension
Acidity	Analysis of KOH
Cinematic viscosity 100°C	Time measurement [s] of oil flow through a capillary tube by IR detector
	Influence of oil to a rotation speed of two cylinders
Density 15°C	Frequency vibration of a U tube
	Density by floating

The installation of the sensor required the following equipment.

- Power source (220 V for each sensor).
- Computer for data acquisition (with USB connection).
- Connection (by-pass) with the lubricating system. An outlet and inlet were needed from the lubricating system for the oil measurements. Hydraulic connectors and fittings were used to connect the sensors to the CRF system. The by-pass was the hardest issue to solve, as the oil tank was hardly accessible. Figure 14.18 illustrates the difficulties for installation and the technical solution implemented.

In Figure 14.18 the dotted arrow indicates the flow of oil towards the tank while the other arrow indicates the by-pass towards the oil sensors. The interface of the water contents' sensor is shown in Figure 14.19.

Figure 14.18 Installation of by-pass plus pressure reducer for oil monitoring

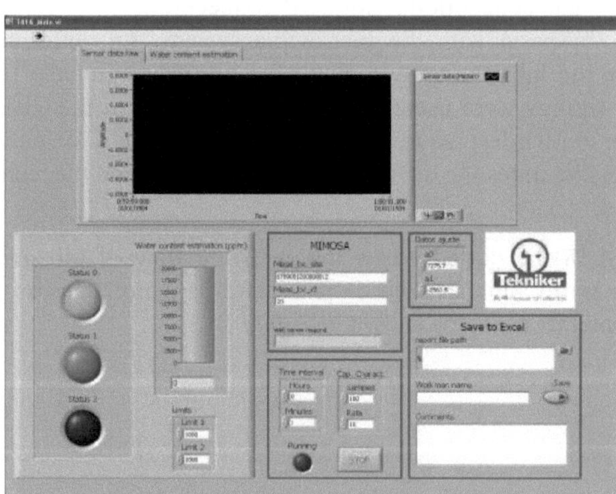

Figure 14.19 Interface of the water contents sensor

The results of testing were highly satisfactory. The installation proved to be feasible in such a difficult context as the demonstration shopfloor. The data flow from the sensors to MIMOSA and the analysis back to the shopfloor PC was successful. An example of the results of analysis from TESSnet is displayed in Figure 14.20.

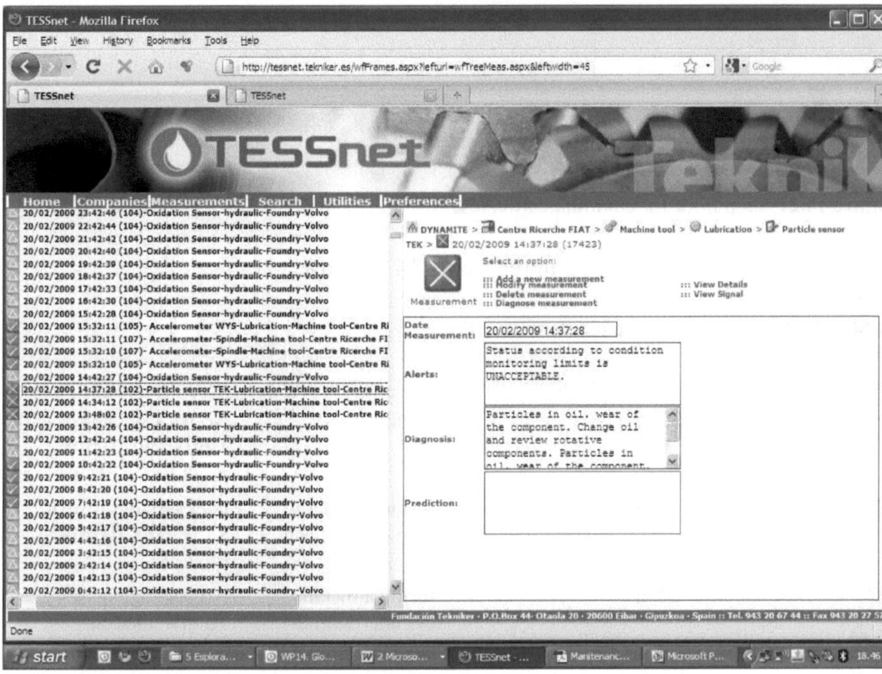

Figure 14.20 Results of analysis by TESSnet

The approach was validated also in this case and the tests proved the approach for assessing on-line the particles in oil as being both feasible and efficient.

14.1.3.5 Communication

As presented in the previous paragraphs, the communication was thoroughly tested. We tested the connection of the sensors with the gateway under different conditions and distances on the shop floor. To increase the relevance of the test and the difficulties for testing, a shop floor located next to the Urane was selected, with many machines working contemporaneously and at a higher distance for transmission. We used two types of wireless modules: 1 mW and 10 mW power transmission. The diagram of the testing area at CRF is shown below, with the positions of where sensors were placed (Figure 14.21).

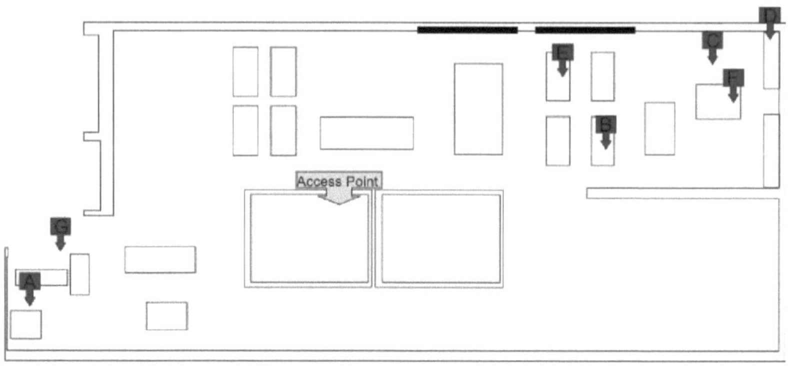

Figure 14.21 Shop floor for communication test

The test results are included in the Table 14.2 below.

Table 14.2 Results of communication tests

Position of testing	Distance from gateway	Special conditions	Results
A	22 m	None	No package losses with 1 mW and 10 mW modules
B	15 m	Inside the metal case of the machine	package losses with 1 mW and 10 mW modules
C	20 m	Inside a metallic cylinder with diameter ~50 mm.	No package losses 10 mW modules
D	30 m	None	No package losses 10 mW modules.
E	10 m	Very close to working machine	No package losses 10 mW modules.
F	25 m	Under a machine	More than 50% package losses 10 mW modules

Upon conclusion of the tests, the maximum distance achieved was tested: a signal was received even about 100 m away from the gateway with 10 mW modules. After that distance there weres a lot of packages losses. The gateway and devices were not visible to each other (no line-of-sight). The measurements were taken inside the plant with walls and working machines between the gateway and the sensors (Figures 14.22–14.24).

Figure 14.22 Location of devices on the shop floor

Figure 14.23 Shop floor with gateway and location of devices

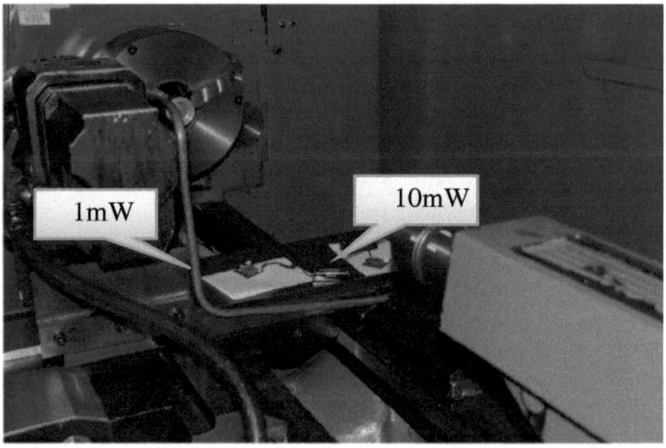

Figure 14.24 Position B setting

14.1.3.6 MDSS

This section focuses on MDSS cost effective maintenance support. The MDSS tool was installed at CRF and tested with data from:

- CRF machine tools (sensors) and
- another plant of the FIAT Group (economical data).

The detailed objectives of the tests were to demonstrate:

- hardware and software requirements and installation;
- connection requirements and problem solving;
- installation of MDSS and database user interface application;
- identification of missing data, preparation and inclusion of additional data in the database application; and
- training and extensive testing of the system.

During the tests and demonstrations, MDSS succeeded to communicate with the MIMOSA database by downloading specific information generated in CRF. The tests also demonstrated the successful integration of PDA, for the generation of machine-specific data, and MDSS, for the retrieval and analysis of data. The test for the tool man–machine–maintenance–economy (MMME) was done using industrial data gathered at the FIAT Group during 2 years in the framework of the Dynamite project. Here, data for a production period on a specific production machine and product, such as planned production and production events, were collected and later entered into the database using the database user-interface application. During the period, with respect to the previous period, the losses in production time increased by 95.5 hrs. The highest priority is the sub-category *maintenance technical impact/failures*. After many technical meetings with the plant staff, the problems behind this loss were finally identified, and linked to the maintenance of two particular systems (Figures 14.25 and 14.26).

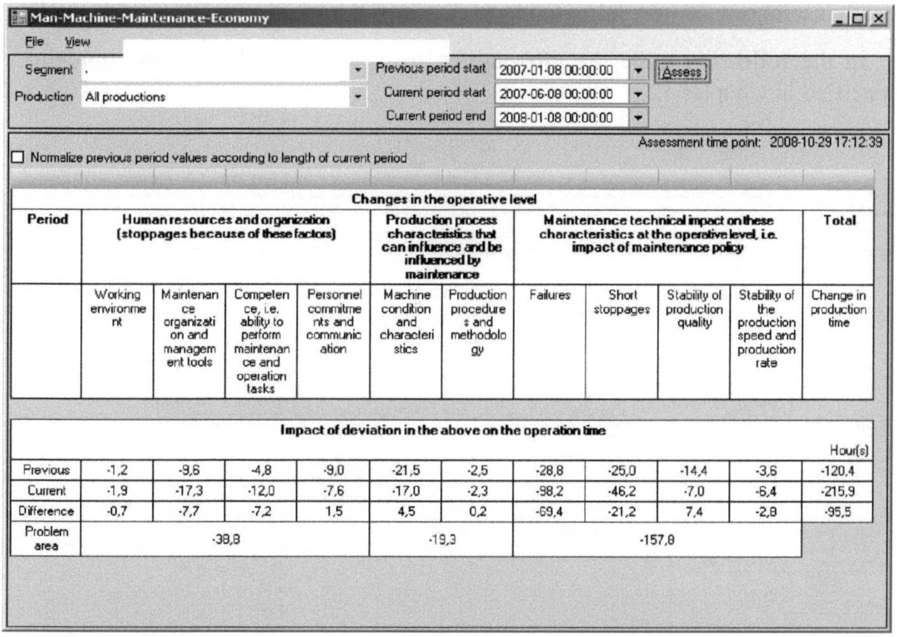

Changes in the operative level											
Period	Human resources and organization (stoppages because of these factors)				Production process characteristics that can influence and be influenced by maintenance		Maintenance technical impact on these characteristics at the operative level, i.e. impact of maintenance policy				Total
	Working environment	Maintenance organization and management tools	Competence, i.e. ability to perform maintenance and operation tasks	Personnel commitments and communication	Machine condition and characteristics	Production procedures and methodology	Failures	Short stoppages	Stability of production quality	Stability of the production speed and production rate	Change in production time

Impact of deviation in the above on the operation time (Hour(s))											
Previous	-1,2	-9,6	-4,8	-9,0	-21,5	-2,5	-28,8	-25,0	-14,4	-3,6	-120,4
Current	-1,9	-17,3	-12,0	-7,6	-17,0	-2,3	-98,2	-46,2	-7,0	-6,4	-215,9
Difference	-0,7	-7,7	-7,2	1,5	4,5	0,2	-69,4	-21,2	7,4	-2,8	-95,5
Problem area	-38,8				-19,3		-157,8				

Figure 14.25 MMME results

It was thus decided to identify and test several alternatives for solving this problem. The alternatives were identified with the final client. The process required a thorough analysis of the causes of function and dysfunction of the system, costs and potential. All of these were tested with the system:

Alternative 1: Installation of sensoring system, including:

- Production decrease
- 2 vibration sensors for systems 1 and 2
- 1 (alternative to existing) laser for system 2
- Prognostic method for system 1 and 2 sensors.

Alternative 2:

- Production increase.
- 1 vibration sensors for system 1
- Prognostic method for systems 1 and 2 sensors.

Alternative 3:

- Production decrease
- Installation of one (alternative to existing) laser for system 1.

In the following table the alternatives are specified following the guidelines specified in Chapter 12.

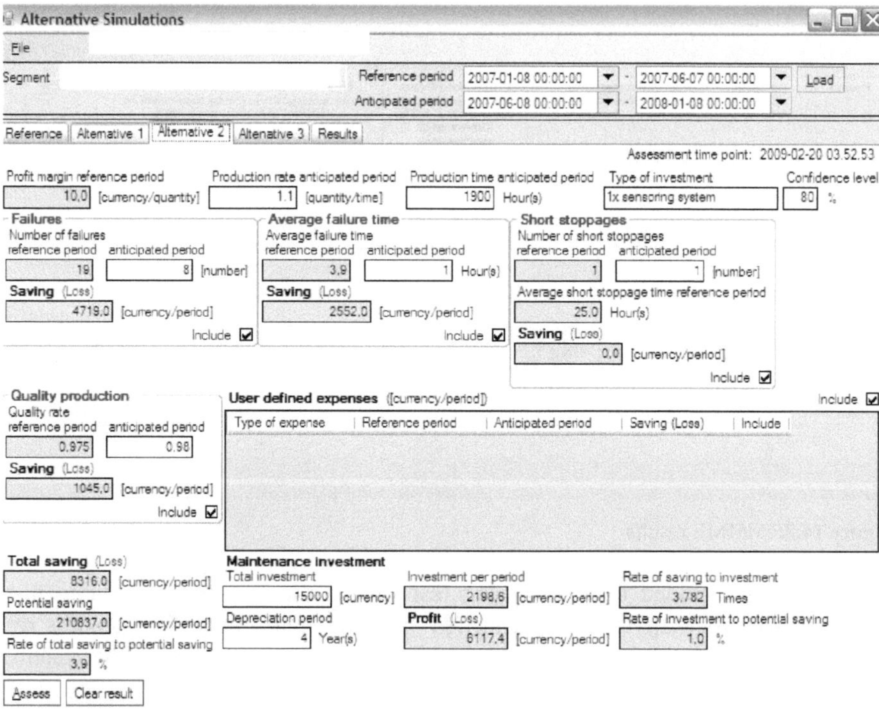

Figure 14.26 Alternative 1 specification

The results of the simulations are included in Figure 14.27.

In synthesis of the table above, the analysis showed higher potential savings for alternative 2 and better ratio saving to investment for alternative 3. Both alternatives will be further assessed internally to the client.

In conclusion the successful integration between MDSS–MIMOSA database–PDA has been proven. The results of the analysis, in particular the economical analysis, have been found excellent. The overall evaluation of the system is thus very promising.

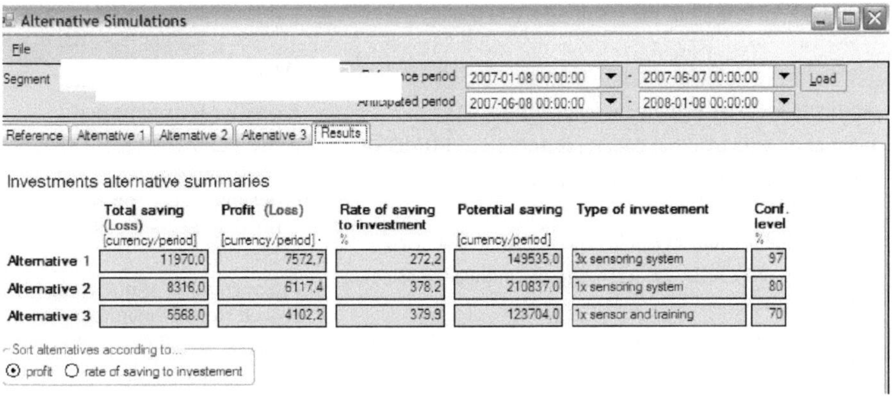

Figure 14.27 Simulation results and selection of the most cost-effective alternative

14.1.4 Economical Evaluation

As the research centre of the FIAT Group, CRF acts as a research department for the FIAT Group sectors: Fiat Auto (including FIAT, Alfa-Romeo, Lancia, Ferrari, Maserati, *etc.*), Case-New Holland (CNH), IVECO, Comau and other companies. CRF, in collaboration with the FIAT Group, has developed, installed and tested prototypes of components and services for two companies with DynaWeb.

In the business model enabled by DynaWeb, we studied how to integrate the functionalities developed into services provided to industrial users, with respect to the existing telematics services provided by the FIAT Group to clients under the name Blue&Me. The exploitation strategy lies in:

- evaluation of a potential market for services to industrial users focussing on the predictive maintenance and integrated with the FIAT Group telematics platform (Blue&Me);
- evaluation of willingness to pay through interviews of final clients;
- refinement of DynaWeb demonstrators for the FIAT Group, based on the existing telematics platforms (Blue&Me); and
- optimisation of synergy between the DynaWeb demonstrator and the FIAT Group internal research projects.

Business Model and Resulting SWOT Analysis

In the framework of the project a business analysis was performed. We give only an extract of the resulting SWOT analysis in Table 14.3 below.

Table 14.3 An extract of the resulting SWOT analysis

Strengths	Weaknesses
Ability to increase equipment availability Integration with external data repository (*e.g.*, MIMOSA) Strong brand images of FIAT Group Companies as innovation providers Open/modular solution (inclusion of new web services) Scalability (capability of extending the number of relevant handled objects) Retrieval and storage of more data about process actors (shop-floor, machine tools, operators...) Good ROI and cost effectiveness Capability of modelling the wear-out of different types of systems Ability to assess residual life of systems	Extensive tests to be performed to assess the lifetime of tags Extensive tests to be performed to assess the power harvesting Fragmentation of market approach from Dynamite partners
Opportunities	**Threats**
May increase legislative compliance (*e.g.*, on oil usage and recycling) Is perceived by the market as innovative and useful Builds on existing infrastructures Added components are relatively few Vector for company-wide optimisation: Blue&Me is used in processes before and after selling of the vehicle Improved safety of operators	Internal clients may not perceive it as an important feature Scarce "willingness to pay" of the clients for the functionalities Increased responsibility of the producer in the case of incident/dysfunction

14.1.5 Conclusions

The area where DynaWeb is active is very promising. The DynaWeb solutions represent a big step forward from the existing state-of-the-art in the FIAT Group. The approach is in any case more wide-spread in the product area (*i.e.*, vehicles) than in the process area (*i.e.*, manufacturing systems).

In synthesis, FIAT tested and demonstrated the integration between 25 DynaWeb hardware components, software components and services. The tests were carried out in three different locations. The final demonstration was done in an industrial machining centre similar to those used in vehicle production. As a summary of the demonstration:

- The overall results are extremely positive, with technical and economical feasibility proven.

- The level of quality of components and adequacy to requirements was high, with people extremely dedicated to enhancing their components and testing them on the demonstrator.
- As expected, integration was not straightforward and required a major effort from all partners involved.
- Some components were not delivered on-time, which resulted in a unit test, *i.e.*, not integrated. This did not reduce the value of the demonstration, which proved to be of high value.

14.2 Foundry Hydraulic System Demonstrator

Per H. Nilsson

The DynaWeb components were tested in various industrial environments. The chosen place for the system consisting of an oil sensor and computer with DynaWeb software was a foundry. The sensor managed to operate well in the very tough environment but special care had to be taken when collecting, storing and communicating data. Especially communicating data caused problems. The demonstration of the DynaWeb components from the foundry opened a new discussion in the company on the benefits of predictive maintenance.

As a part of the global application demonstration one oil sensor system was tested in hydraulic systems in real industrial environments, the production line in the Volvo foundry. The total test time for the oil sensor was from June 2008 to February 2009. The tested system was from Tekniker and measured the level of oxidation of the lubricant by spectroscopy of visible light, see Chapter 7.

The sensor system saved data in a local log-file and database. The values were compared with analysis in the laboratory of oil samples taken during the test period. The trend for the oxidation level of the oil was according to expected trends and variations. The hardware and software worked well during the test period. The sensor module being installed in the hydraulic system in the foundry was exposed to an extremely dirty environment, *e.g.*, black dust.

Communication with the DynaWeb was done through a specific web service developed for this sensor, which is described in Chapter 3. The communication was one-way only but was the only possibility to get data through the Volvo firewall.

The data is now available at the global MIMOSA database for testing web services and also at Tekniker with the TESSnet platform. The oil sensor signal was analysed by the TESSnet platform and the status displayed according to the levels set for normal, monitor and alarm.

14.2.1 Objectives of the Test and Demonstrations

The aim of the Volvo foundry sensor demonstration was to test the effectiveness and robustness of an oil sensor when exposed to a very harsh industrial environment and to explore the effectiveness of data transfer and processing.

14.2.2 Description of the Test Platform

The first step of the demonstration in the Volvo Powertrain foundry in Skövde, Sweden, was to identify hydraulic systems suitable for demonstration of the sensor measuring level of oxidation in a hydraulic lubricant. A requirement was that the machine using the hydraulic system should be a part of Volvo's existing e-condition monitoring system named Duga. A hydraulic system in the foundry was chosen. The sensor had to be mounted in a side flow of the hydraulic system due to the limited ability of the sensors to withstand the high pressure, 70–80 bar, in the primary hydraulic circuit.

A foundry is a very warm and dirty place, which causes extra problems in terms of finding a suitable place for a computer. The place for the computer should have access to Internet (Intranet) and be fairly clean. Such a place is normally hard to find in a foundry.

The second step was to prepare Volvo Powertrain's IT architecture for the requirements of the DynaWeb global demonstration. From previous work it was known that an open communication with the DynaWeb components was troublesome. All data from the global demonstration was at first stored both in a local Microsoft access database and as an excel-file, and Tekniker created a solution to at least put the sensor data to the global MIMOSA database through a web service.

Discussion of DynaWeb Concept at EUCAR

The Dynamite project was approved as a EUCAR project. EUCAR is the European Council for Automotive R&D and most of the European automotive producers are members of EUCAR. Every EUCAR project has to report twice a year to EUCAR. At the EUCAR meeting on 24 September 2008 the Dynamite project was presented and discussed. Especially the DynaWeb concept was focused. All EUCAR members thought that they preferred a local version of the DynaWeb concept inside the company firewall. At the meeting it was also ascertained that smaller companies probably would prefer the on-line DynaWeb.

14.2.3 Description of the DynaWeb Components Tested

The demonstration focused on the following DynaWeb components: a lubricant oxidation sensor, data collection and sensor control software and communication to the MIMOSA database.

14.2.3.1 Sensor Measuring Oxidation of the Lubricant by Spectroscopy of Visible Light

The location chosen for the industrial validation of the sensor measuring the oxidation of the lubricant by spectroscopy of visible light developed by Tekniker was a hydraulic pump and tank serving the giant shaker in the Volvo foundry. The environment at the testing place was very tough, 45°C, and extremely sooty. The principle of the oxidation sensor is described in Chapter 7.

Communication is limited in the foundry. Volvo has a global WiFi system but this was not an alternative in the foundry. Instead local LAN was used and a 30 m long RS422 cable had to connect the sensor with the control unit.

The test platform can be seen in Figures 14.28 and 14.29.

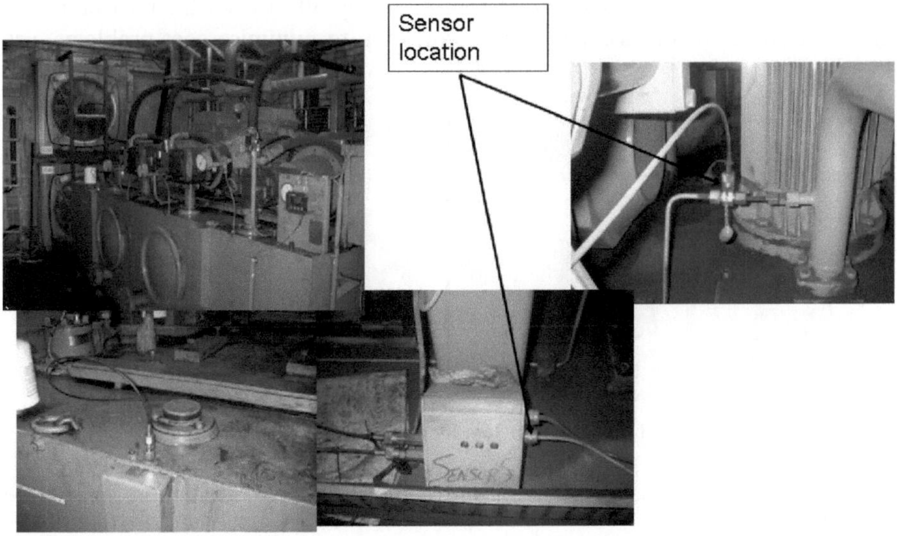

Figure 14.28 Oxidation measurement system installed on-line in the foundry at Volvo Powertrain in Skövde, Sweden

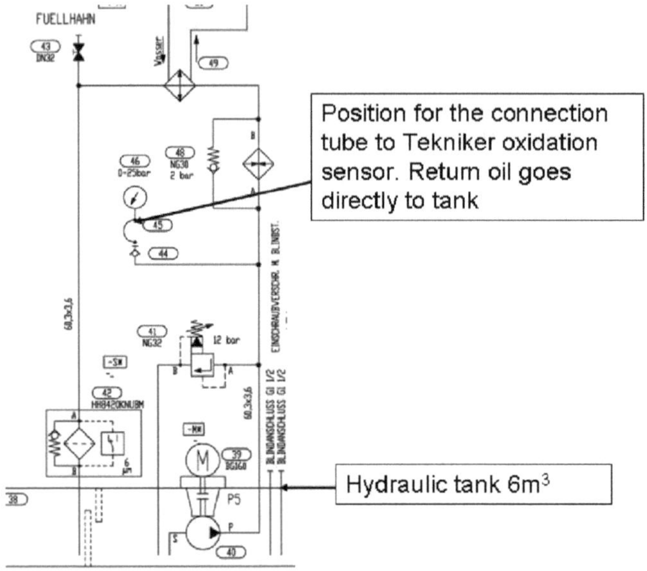

Figure 14.29 Position of the oxidation sensor in the hydraulic system

The control system was installed approximately 30 m from the measurements unit in a corner of the welding repair workshop to minimise soot problems, see Figure 14.30.

Figure 14.30 Installation of the control system for the oxidation sensor

Information and the parameters for the demonstration of oxidation sensor are listed below.

- Asset no.: 283-9043 hydraulic system in foundry, line 3.
- Sensor: oxidation sensor from Tekniker (visible light).
- Hydraulic system with three pumps above the tank (6000 l). The system has mainly been running two-shift (06.30–23.00) and is closed down at night. During the weekends (Fridays 23.00–Mondays 06.00) it is also closed down. Also

the circulation through the filters and cooler (and sensor) is stopped. When needed oil is filled into the system just before the filters.

- There is a problem with oxidation caused by implosion of oil (valve closing). There are rapid pressure changes due to valves closing and opening. The implosion causes local high temperatures that lead to oxidation. Cavitation problems can also occur because the pumps are placed above the tank.
- Fluid: Bartran 46 BP.
- Sensor location: in the circulation loop from big tank through pump, filter, cooler and back to tank. In this loop, the sensor is mounted after the pump but before filters and cooler using a Minimess tube (1.5–2 mm inner diameter). Return pipe directly to tank.
- The sensor controller is located close to the sensor and powered by 230 V.
- The flow rate through the sensor is: » 11 ml/min.
- Oil samples are taken after the sensor by loosening the return pipe.
- Pressure: about 2.8 bar.
- Temperature in tank: 42.4°C, 41–43°C.
- The colour of the used oil is typically dark brown (like thin coffee) and has a sharp smell.
- There is an electrostatic filter connected to the tank (ELC clean tech).
- Computer location: in office of the welding workshop, line 3 connected to Volvo intranet (LAN). A power supply of 230 V is needed for the computer.
- The connection between sensor and computer is by a RS422-cable of about 25–30 m.
- Logged values from the online oxidation sensor were stored every 60 min.

Some additional information and comments for the hydraulic system:

- A complete change of oil is made every third year.
- Current oil has been used 2.5 years (changed during summer 2006).
- Normal check of oil status is made visually (and by smelling).

14.2.3.2 TESSnet Platform

The data stored from the oxidation sensor were also available for analysis through the TESSnet platform described in Chapter 3. The TESSnet platform can be used as a predictive maintenance management system, displaying information about the machine status and the measurements done as seen in Figure 14.31 below.

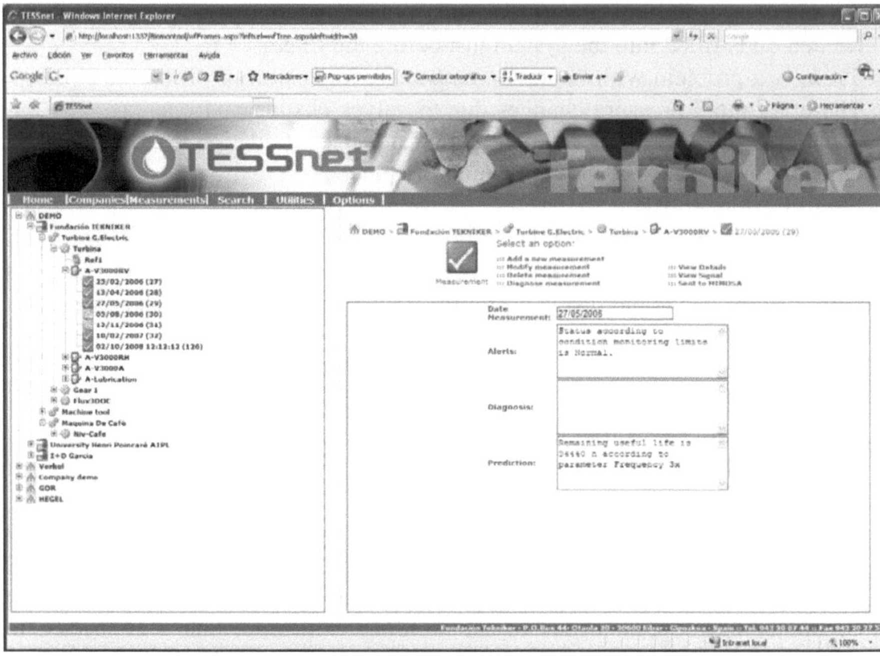

Figure 14.31 Oxidation sensor measurements in TESSnet

14.2.3.3 Data Storage in the Global MIMOSA Database

The software for oxidation oil sensor stores the measurement in three different places:

- excel file, on the local PC
- access file, also on the local PC
- MIMOSA database, through a specific web service for this sensor.

One of the problems to solve in the foundry was to connect to the MIMOSA database to its location at IB Krates in Talinn. The SQL server uses specific ports that are closed in the Volvo plant for to security reasons, according to Volvo IT policy. However, it was possible to connect to a web service since this does not require opening a specific port. Internet web pages, http protocol, as well as web services use the standard port 80. When the web service was installed outside Volvo in Tekniker's server, it was possible to configure the firewall at Tekniker, in such way that connection to MIMOSA database was enabled (Figure 14.32). In this way, the data is successfully stored into the MIMOSA database.

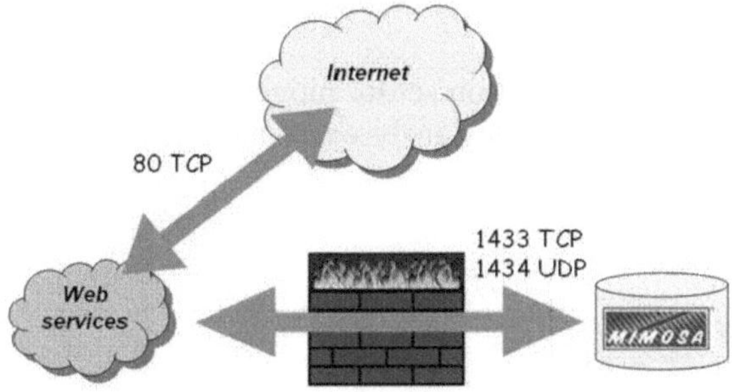

Figure 14.32 Connection through firewall

In return, after importing the data from MIMOSA, it is possible to see the measurement, and perform condition monitoring and diagnosis using TESSnet as seen in Figures 14.33 and 14.34 below.

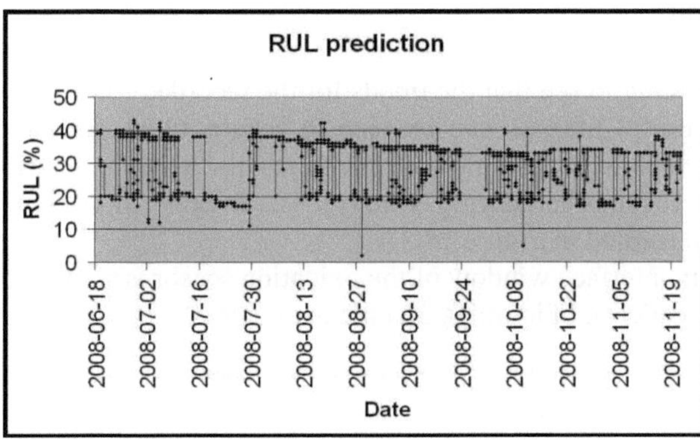

Figure 14.33 Plotted log file values for predicted RUL in percent from the oxidation sensor mounted in a hydraulic system in the Volvo foundry

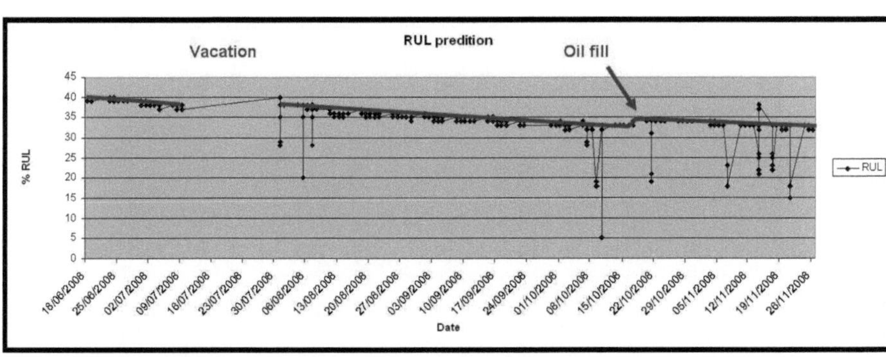

Figure 14.34 Plotted log file values as in Figure 14.33 but with expected outliers removed. The general trend downwards as expected from oil degeneration is clearly seen. A small jump upwards can be explained by the filling of new oil (230 l)

14.2.4 Reference Measurements and Software

The logged data from the oxidation sensor mounted in the foundry was compared with Volvo hydraulic oil samples analysed in the laboratory at Tekniker (using ASTM D 664-07).

14.2.5 Results

The evaluation of results was done in two categories: sensor performance and data storage and communication.

14.2.5.1 Sensor Measuring Oxidation of the Lubricant by Spectroscopy of Visible Light

It is very interesting to see that the trends for the recorded sensor signal in percent of RUL (rest useful life), as can be seen in Figure 14.34, follows the expected trends for oil very well. The flow in the hydraulic system is turned off during nights and at weekends and during holidays. The irregular flow complicates measurements, as illustrated in Figure 14.33.

The program interface window of the oxidation sensor is shown in Figure 14.35 (status normal/dark) and Figure 14.36 (status monitor).

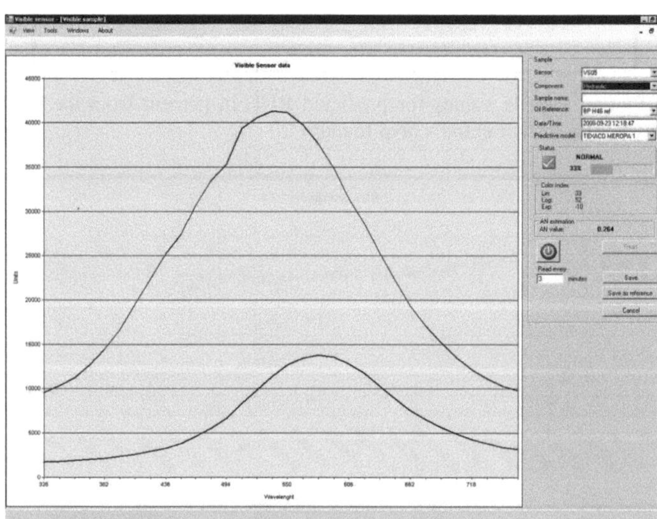

Figure 14.35 The program interface window for the oxidation sensor showing status normal (dark) for the estimated % RUL value of 34% (Lin) and the corresponding AN value of 0.259

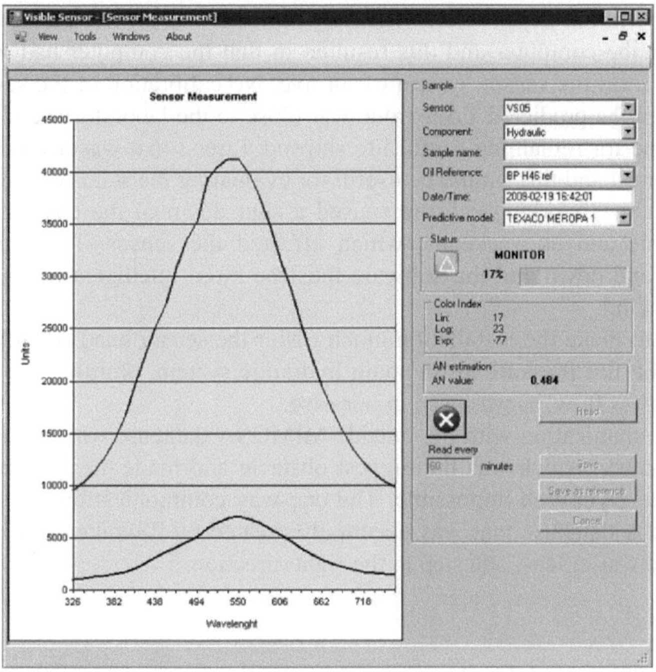

Figure 14.36 The program interface window for the oxidation sensor showing status monitor for the estimated % RUL value of 17% (Lin) and the corresponding AN value of 0.484

Volvo has no other online sensor measuring the level of oxidation in hydraulic oil systems in the factory so this result is very interesting. The lower limit for percent of RUL should be evaluated in order to know when the oil needs to be changed based on the sensor measurement and not based only on the time elapsed from the earlier oil change, normally 3 years.

14.2.5.2 Data Storage and Communication

Data storage in the global MIMOSA database was demonstrated through a one-way communication through a web service set up by Tekniker. The sensor data was also demonstrated on the TESSnet platform where the data could be analysed and the status displayed for all logged parameters.

14.2.6 Technical Evaluation

The overall experience with the oxidation sensor system was good and the hardware and software worked well. They survived the extremely dirty environment in the foundry without any problem. The only problem was the need for a clean environment for the computer and this resulted in that the computer had to be about 30 m away from the sensor system to survive. No calibration of the sensor signal was done but the predicted TAN value was close to the laboratory results. The estimated value for remaining useful life showed a trend that was according to that expected for oil and this should be useful for evaluating the status of the oil.

The industrial working scheme caused a shut down of the circulation circuit during nights and at weekends, which affected the sensor. The sensor signal jumped up and down and the software must be more intelligent to suppress outliers of this kind.

In order to make the installation much easier the sensor head should be able to withstand the full pressure in the main hydraulic system, omitting the need for a separate bypass flow, as was used in our case.

Data communication with the outside MIMOSA database was due to the strict Volvo IT policy, which was the biggest obstacle and made the demonstration of the complete DynaWeb impossible. The one-way communication to store data in the MIMOSA database that was eventually created by Tekniker and included in the software was at least one step in the right direction.

14.2.7 Economical Evaluation

Because of lack of experience at Volvo with non-DynaWeb state-of-the-art solutions for monitoring the oxidation level of hydraulic oil systems it is very hard to make any relevant economical evaluation with respect to a DynaWeb commercial solution. If the DynaWeb solution is to be a commercial success it must be cheap enough, easily installed, easily handled, robust, *etc.*, so that it can be fitted in all hydraulic systems in a factory.

If it is possible to increase the oil change interval from 3 to 4 years for our hydraulic system in the test above by monitoring the status of the oil some money can be saved. The tank contains 6000 l and going from a yearly rate of 2000 l to 1500 l saves 500 l on average *per* year, which should be in the order of 1000 € *per* year.

14.2.8 Conclusions and Recommendations

14.2.8.1 Conclusions

The oxidation sensor hardware and software worked well in the foundry installation and the sensor signal showed a trend according to the expected one. The predicted TAN value was close to the laboratory results.

The sensor signal jumped up and down depending on, for example, irregular oil flow, air bubbles, *etc.*, which made the interpretation more difficult and not straightforward. The software needs to be intelligent to suppress outliers when the oil flow is off. The sensor required a continuous low speed oil flow without air bubbles and at a low oil pressure. Connection into production line hydraulic systems requires pressure reduction, normally by bypass flow.

The environment in the foundry at Volvo was extremely dirty, which was a good test for the sensor but made it impossible to have the computer at the same location. Volvo IT policy made it almost impossible to demonstrate communication with the MIMOSA database but a one-way communication to store data in the MIMOSA database was created and included in the software and finally tested.

The use of PDA with wireless connection to the Volvo intranet is not currently allowed because of a strict IT policy and this made it impossible to include PDA wireless communication in the testing.

14.2.8.2 Recommendations

Based on the global demonstration at Volvo Skövde recommendations for the future use of Dynamite results can be made.

There are four main concerns when demonstrating DynaWeb in industry:

1. Access to the global MIMOSA database. As pointed out by the EUCAR Manufacturing board, Section 14.2.2.1, all EUCAR members, in practice the entire European automotive industry, prefer a local version of the DynaWeb concept inside the company's firewall. Many companies have restricted data communication outside the company. Smaller companies normally do not have these strict restrictions.
2. Local DynaWeb. If industry prefers a local version of DynaWeb it must be clear how this will be implemented.
3. Sensor development. The sensors must be able to handle industrial outliers better, *e.g.*, pressure and flow variations. Air bubbles are always a problem.
4. Industrial environment. An industrial environment is normally much more demanding than a laboratory environment. Sensors have to withstand large variations in parameters like humidity, temperature, pressures, vibrations, flooding of different liquids and soot/dirt levels. This normally means that electronic equipment has to be kept in a sheltered environment. In practice this means that the AD converter and the first level of processing may be more that 30 m away from the sensor head and that WiFi is not always possible.

14.3 Automatic Strip Stamping and Cutting Machine Demonstrator

Benoit Iung, Eric Levrat, Alexandre Voisin and Nicolas Krommenacker

The feasibility of integrating the DynaWeb components to form the e-maintenance architecture has been tested on the TELMA Platform (Levrat and Iung 2007). TELMA (Figure 14.37) is based on a physical process and is as a test bed relevant to both an automation architecture and a maintenance architecture. It was developed from market components to be as close as possible to an industrial context. In this way TELMA implements a physical process dedicated to unwinding metal strips. This process is similar to industrial applications such as sheet metal cutting and paper bobbin cutting. The physical process is divided into four parts: bobbin changing, strip accumulation, punch cutting and advance system (Figures 14.38 and 39). Each part is composed of several components, such as pneumatic cylinders, chucks, marking systems and motors.

The TELMA platform is located at Nancy University and was developed mainly for supporting e-maintenance purposes, from both educational and research points of view, by integrating:

- the engineering and deployment of CBM and proactive maintenance strategies in a way consistent with OSA/CBM proposals (Lebold and Thurston 2001);
- integration of these strategies within enterprise strategies context; and
- assessment of the strategy impact on the performances of a global manufacturing system, such as productivity (availability, maintainability), quality and costs.

Thus, the platform is designed for:

- local use in the framework of conventional training activities;
- remote use *via* Internet for operation on industrial e-services and for accessing information, such as production data and performance data, *via* a virtual private network;
- use for e-teaching and e-learning as application support of courses in the e-maintenance domains; and
- experimentation on research issues, such as prognostic and decision support processes and for making demonstrations on e-maintenance components, software and web-services as needed for Dynamite (Arnaiz *et al.* 2006).

In its conventional structure, the TELMA platform is based on three levels: (1) the physical layer, (2) the automation layer and (3) the business layer.

Figure 14.37 The TELMA platform

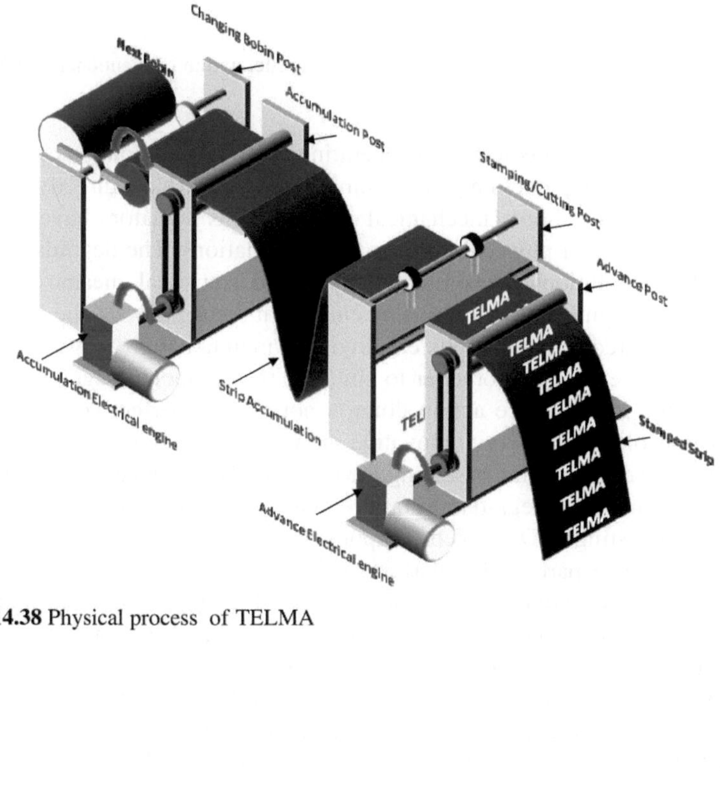

Figure 14.38 Physical process of TELMA

At the physical layer several different sensors and actuators are linked with the automation system, which is composed of control screens, control boards, PLCs (TSX Premium with Web Interface), Altivar for Engine control with web interface, web-cam and remote input/output.

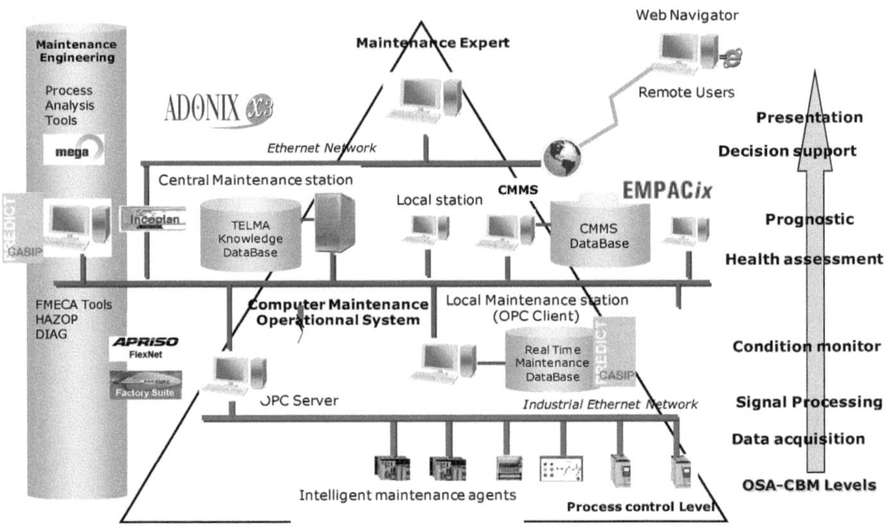

Figure 14.39 Integration of all the maintenance components within conventional TELMA platform

Another PLC is fully dedicated to generating artificial degradations and failures by employing signal processing algorithms or by modifying directly the input and/or output signals. Some mechanical parts such as actuators have also been added to simulate other physical failures and degradations. The degradations, failures and their evolutions are simulated by sound statistical means, employing Weibull law, random events and Markov chains. In the same way, the failed components are repaired in simulations employing a maintainability law, exponential law or by concrete action. Moreover to simulate the component ageing, it is considered that the maintenance action done is not always perfect (*i.e.*, as good as new, as bad as old). It is a very innovative view for the platform and offers an efficient way for validating and assessing the e-maintenance capacities by emulating technical *vs.* functional degradation/failure. These capacities are used when implementing the testing of DynaWeb components.

The maintenance part of the platform (Figure 14.39) is built on the CASIP-KASEM (http://www.predict.fr) product, used for the engineering and implementation of the e-maintenance services. It supports a local real-time maintenance system, a centralised maintenance system, sitting on top of an Oracle database and some remote stations. These systems are integrated through databases. The maintenance part is also integrated with an ERP system called ADONIX, a CMMS system called OPTIMAINT and a MES system called FLEXNET.

With this conventional structure as a starting point, several modifications of TELMA have been performed since 2005 in order (a) to support the deployment of different components and (b) to develop all the DynaWeb test procedures.

14.3.1 Objectives of the Test and Demonstrations

The demonstration objectives of the DynaWeb components done on the TELMA platform can be structured in two complementary ways:

- A demonstration of the feasibility of the connection of the DynaWeb ICT components to form an e-maintenance architecture. These main components are PDA, wireless communications, web-service and the MIMOSA database.
- A validation of the e-maintenance architecture according to technical and interest criterions but not to quantitative end-user utility indicators, which is the scope of demonstration on site.

In relation to the feasibility item, it consists more precisely in testing:

- Key advanced functionality of *smart PDAs* in the field within e-maintenance architecture. Assessing the PDA software concepts.
- The capabilities of tailored *wireless communication features* in the field. The emphasis is on the demonstration of capability as an enabling technology for advanced function.

In relation to the architecture item, it consists more precisely in testing:

- A structure for *standardised connectivity* through the MIMOSA database, Gateway and Collector. The design is fundamental to the effectiveness of the global package. It is the backbone of interactive e-maintenance and must be shown to work in accordance with user requirements and to present an effective standardisation across the spectrum of managerial and technical activities.
- *Demonstration* on automated machine to machine communication, as well as personalised/customised human to machine communication.

All the previous tests according to feasibility and architecture aspects can be summarised into two categories:

1. *Internal test:* to consider each ICT component of the DynaWeb platform and test its function in order to validate its ability to tackle data issued from a process other than the one used by the component developer.
2. *Integration test:* to integrate the set of components in order to validate the ability of the Dynaweb platform to tackle an e-maintenance problem from early stages, *i.e.*, sensor detection, to high level decision and scheduling of a maintenance work order. Because of this second level of testing, in the first one the

function required by each component in order to carry through the second level was included.

These two types of tests led to considering moving from isolated validations related to each component to integrated validations related to the component assembly as a whole.

14.3.1.1 Isolated Validation

Isolated validation aims at testing the adequacy of the component within the DynaWeb e-maintenance framework. The tests are conducted first, by people who are not the component developers and then on data that are not issued from component developers either. Two dimensions are used in order to assess the adequacy of the component to the test context: functional and informational dimensions.

The functional dimension is considered through the use of different sets of functional tests:

1. *Demostep test:* this test is defined by the developer and shows a typical use of the component. It describes how the function should be used from the developer's point of view.
2. *Empirical test:* this test corresponds to an end-user user operation of the component. Hence, the developer's logical sequence of use may not be followed. Such tests are helpful when several functions are available and may be called separately.
3. *Use cases:* this test puts the function required by the component in order to fulfil its mission in the integrated validation scenarios. The use cases are similar to those available for the demonstrations on site.

The informational dimension for integration is based on the use of the MIMOSA standard (http://www.mimosa.org) for information repository. Considering the components, the data on which it is based are noticed. The possible sets of data are:

1. Other databases than MIMOSA: this is the worst situation. It means that the component is working with a proprietary database and is not able to read/write information in a MIMOSA database.
2. dyn_mimosa database: this is the first MIMOSA database proposed on a remote database server (located in the IB Krates company, Estonia) within the Dynamite context. Its current status is that the management of primary and foreign keys of tables is no more active. This leads to the possibility of integration of data that are not MIMOSA compliant because the references between tables are not ensured.
3. Another MIMOSA database: such a database is a developer's/partner's own MIMOSA database. Its use allows validating MIMOSA compliancy as far as the developer/partner has fully implemented MIMOSA package but with a lack

of communication to the remote database server (server located at IB Krates, Talinn, Estonia) in order to fully integrate DynaWeb.
4. dyn_mimosa3 database: this is the current operational MIMOSA database, which is fully compliant with the MIMOSA standard. The use of this database ensures compliancy with other DynaWeb components.

14.3.1.2 Integrated Validation

Integrated validation must lead to validation of the DynaWeb e-maintenance framework functions from an end-user point of view. In such tests, two scenarios are selected in consistence with end-user scenarios proposed by FIAT within the framework of the final demonstration sites. These two end-user scenarios have been adapted to be executable with the TELMA platform. The aim is to have a test scenario where all the agents *vs.* actors are involved (from degradation to service result). Such a scenario results in demonstrating first the interoperability of single components and second ability of DynaWeb to tackle the end-user problem but without consideration of cost effectiveness.

14.3.2 Description of the Test Platform

In order to be able to develop the two types of tests and validations, the TELMA platform has been improved mainly to integrate wireless functionalities supported by e-technologies (Levrat *et al.* 2008, Iung *et al.* 2009). This up-grading is based on the implementation of the Dynamite components to be tested (Figure 14.40). While keeping the integration facilities the improvements are related to:

- The change of wired communications to wireless communications. Collectors at field level are introduced with ZigBee protocols (Dynamite product developed by PRISMA, MicaZ), gateway for transferring data from the ZigBee area to the WiFi area (PRISMA product, StarGate SPB400), and WiFi access point (Emmanouilidis *et al.* 2008). The Wi-Fi access point (Cisco Aironet 1200) is classically configured with a pre-shared security key of 128 bits. The device and configuration done such that communication is possible, although insecure but working with every wireless (802.11) device. This access point is also configured with a DHCP relay to let the main network configure the wireless device's network configuration.
- The use of wireless sensors is more adapted to provide plant information for decision-making in maintenance. On TELMA, these wireless sensors are MEMS to measure vibration on the accumulation system. An electrical connection has also been added to the angular position oscillating arm to enable the plug in of a wireless sensor. The angular position sensor is used as a testing

sensor for wireless sensor communication. Smart tags are also plugged in on some physical components such as an engine to store static maintenance information of the last maintenance action done on the component.

- The use of PDA to aid the maintenance crew on site. This is an HP iPAQ hx4700 (Pocket PC) in which the maintenance functionalities developed in Dynamite to access to the MIMOSA database, to retrieve information and to access the web browser are implemented.
- The use of a MIMOSA database in form of an SQL server able to integrate all the data produced and consumed by the different components. This allows solving interoperability issues. However, some maintenance actors such as components and software are not able to address the MIMOSA database directly (*i.e.*, components that are not able to send SQL requests through the network). For this case, a common interface called MIMOSA translator has been developed. The MIMOSA database was not located on the TELMA platform but on a remote database server to show the feasibility of remote connections to the database.
- The implementation of web-services and the way to the access to them. The web services were prognosis, diagnosis, condition monitoring, scheduling web-services. For example, the prognosis developed for the platform (Muller *et al.* 2008) aims at computing the remaining useful life of the asset/segment. The same prognosis web-service addresses reliability based prognosis (taking into account influences variable through the use of PHM) and condition based prognosis (use of parameters to make the choice) (Voisin *et al.*, 2008). The proportional hazards models (PHM), also called the Cox model, is a classical semi-parameter method. It relates the time of an event, usually death or failure, to a number of explanatory variables known as covariates. It has the form of

$$L(t|x_1, x_2, \ldots, x_n) = h(t) \exp(b_1x_1 + \ldots + b_nx_n),$$

where L is the hazard function or hazard rate, $\{x_i\}$ are covariates, $\{b_i\}$ are coefficients of the model-effects of the corresponding covariates, and $h(t)$ gives the effect of duration on the hazard rate.

As the communication between the different items for fulfilling web service is performed by means of XML (client, agent), a software agent is in charge of collecting data from the MIMOSA database and sending them to the web service. The use of TCP/IP allows locating the client and the agent on the same PC/PDA or on different PC/PDAs.

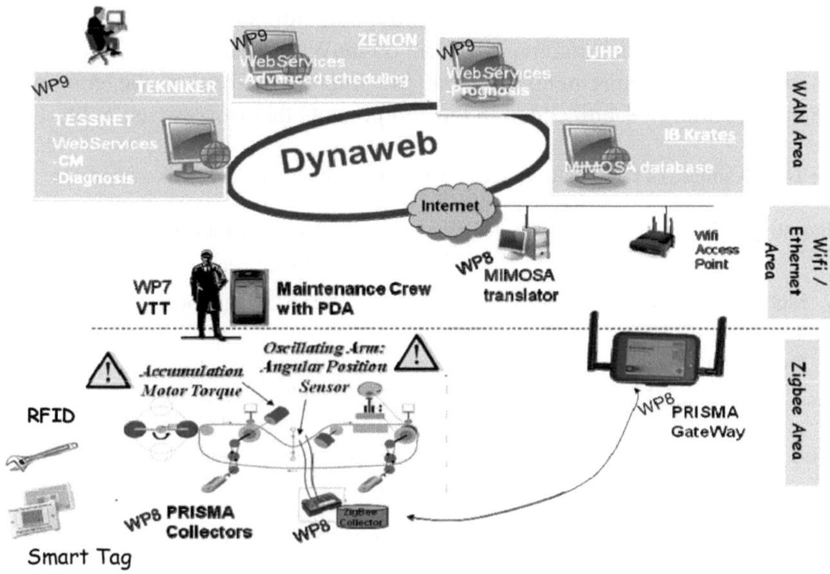

Figure 14.40 New e-maintenance TELMA platform for testing DynaWeb components

14.3.3 Description of the DynaWeb Components Tested

All the DynaWeb components previously explained and tested on the TELMA platform are summarised in Table 14.4 (15 components).

Table 14.4 The demonstrated DynaWeb components

Demonstrated DynaWeb components
9. Mobile maintenance PDA user interface
10. Smart tag PDA support
13. PDA scheduling support
14. Smart PDA maintenance user interface
15. Communication SW module
16. Mimosa translator
17. Collector (= Gateway)
18. Wireless communication system for e-maintenance
19. Condition monitoring web service
20. Diagnosis web service
21. Prognosis web service
22. DynaWeb e-maintenance platform (TESSnet)
23. Scheduling web service
25. DynaWeb platform
26. Mimosa database

For each component tested, the requirements for supporting its integration within the TELMA architecture to form DynaWeb architecture were defined. Main technology requirements concern the interoperability of the components since the technology is embedded in the components themselves (see Table 14.5).

Table 14.5 Examples of communication and hardware requirements for two components

Demonstrated DynaWeb components		Type of software	Communication requirement	Hardware requirement
9.	Mobile maintenance PDA user interface	Exe file	SQL server port	PDA, wireless access point
10.	Smart tag PDA support	Web page	Internet connection	PDA

The interoperability between the components is mainly supported by:

- a MIMOSA database as a common data repository; and
- services to interoperate by defining the right to pass through the firewalls and communication checking points.

The second point requires a lot of communication ports to be opened in order to make the SQL server communication allowed to the IB Krates server and to make the web services call possible.

The first point required deployment of a MIMOSA database (Dyn_Mimosa3) fulfilled in a consistent way with the e-maintenance platform information. This means that the MIMOSA structure must be understood and the new TELMA-DynaWeb topology must be translated to the MIMOSA structure and finally specific TELMA information must be used such as sensor measures, prognosis data, *etc.*, as occurrences of MIMOSA tables. Thus the MIMOSA database stores all the platform information allowing the different DynaWeb components to share the same information with the same meaning for supporting interoperability constraints.

The full procedure to configure one MIMOSA database with the e-maintenance structure is the following (only the first step is illustrated Figure 14.41):

- Declare the enterprise, the sites and the database since the three entities has a cyclic reference. Two sites have been declared for this demonstration since the TELMA platform is in the AIPL building (at the campus) and the DynaWeb service server is in the undergraduate building,
- Declare the process architecture of TELMA. The tables to be fulfilled are segment and segment child.
- For process architecture define the real components: the assets. An asset can be composed of several assets and linked to a segment.
- As one asset is a sensor (*e.g.*, angular position) it is necessary to associate a transducer to this asset.
- Define the measurement locations that are related to the devices and able to produce measurement data.
- Declare for each asset the possible degradation mode (hypothetical event) that will be related to the observed degradation mode (proposed event). For TELMA, two degradation modes for two assets were defined: a seizure for the accumulation motor and a loosening of the V-belt.

For each component tested, a list of information was provided before launching the test procedure in order to check the consistency of the material put at the disposal of demonstration step.

The list documents the component status:

1. A short description is extracted from the user documentation given with the component.
2. A technical check (see Table 14.6) summarises information on the component type and in link with MIMOSA and software database.
3. A documentation check gives information on the component module, its demostep (see Table 14.7 below with an example of Demostep), *etc.*

The generic frame used to structure the information is presented in Table 14.6. One table has been developed *per* component leading one to consider the component in the right status for this demonstration phase. This means it is to be accepted as delivered: “preliminary” test.

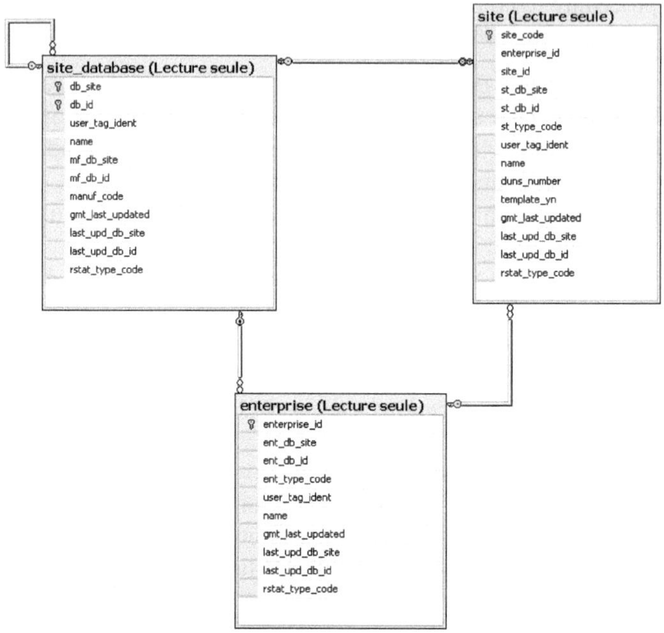

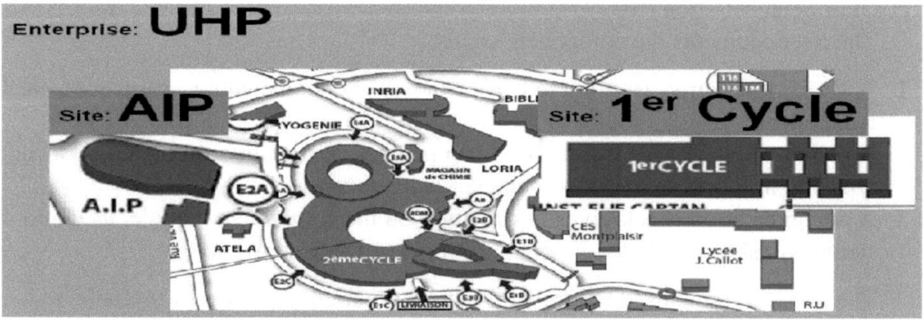

Table - dbo.enterprise* Summary

enterprise_id	ent_db_site	ent_db_id	ent_type_code	user_tag_ident	name	gmt_last_updated	last_upd_db_site	last_upd_db_id	rstat_type_code
67890013	0000000000000...	2	3	UHP	UHP	NULL	NULL	NULL	1

Table - dbo.site_database* Summary

db_site	db_id	user_tag_ident	name	mf_db_site	mf_db_id	manuf_code	gmt_last_updated	last_upd_db_site	last_upd_db_id	rstat_type_code
789001300000013	113	UHP First TELMA	UHP First TELMA	NULL	NULL	NULL	03/06/2008 21:...	NULL	NULL	1

Table - dbo.site* Summary

site_code	enterprise_id	site_id	st_db_site	st_db_id	st_type_code	user_tag_ident	name
6789001300000013	67890013	13	0000000000000...	2	1	AIPL	UHP AIPL
6789001300000019	67890013	19	0000000000000...	2	1	Faculté des Sciences	UHP Faculté des sciences

Figure 14.41 Enterprise site: conceptual tables in MIMOSA,corresponding TELMA-DynaWeb data, storing -DynaWeb data in tables

Table 14.6 Test protocol

Documentation check

Has the component's module(s) been defined in the software module database?	Yes	No
Has the approach been documented in a deliverable report?	Yes	No
Which report:		
Has a DemoStep query file for testing been provided by partner in charge?	Yes	No
Have the modules been documented with UML?	Yes	No
Has a user guide been delivered with the component?	Yes	No

Technical check

Kind of software component:	☐ exe file ☐ DLL ☐ Web page ☐ Web service	
Are the web resources available to all partners?	Yes	No
In the case of web resource, web address:		
Does the module provide data into Mimosa DB at EB Krates server?	Yes	No
Does the module rely on data from tags, sensors or Mimosa DB?	Yes	No

Component's software module list (extracted from software module database):

Software module	Software deliverable

14.3.4 *Reference Testing Procedure*

After checking the component status (preliminary test) and as has already been explained in Section 14.3.1, tests have been developed according to two aspects: internal tests and integration tests.

14.3.4.1 Procedure for Internal Tests

Internal tests aim at verifying that the DynaWeb component is working. Hence it is mainly based on the developer's view for testing the initialisation, parameterisation and functioning of the component. With this test, both informational and functional views have to be checked.

The informational view considers the data the component is working with. Thus the informational view has to check the compliancy/use with MIMOSA database.

The functional view has to check that the component is working from the developer's point of view. This view is described in a Dynamite Demostep file that explains step by step how to make the component work. Hence, such a Demostep file is expected for every component. An example of the Demostep file for the component 19 condition monitoring web service is presented below.

The internal tests were developed with these informational and functional objectives to assess the degree of internal capacities for each component.

Table 14.7 An example of the Demostep file for the component 19 condition monitoring web service

Work package	Demostep	Step number	Demo component	Definition date	By (person)	Input from Mimosa table	Output to Mimosa table
10	Display web service interface for vibration data acquisition web service	1500	19	12/09/2008 09:51	Eduardo Gilabert	NULL	NULL
10	Fill the form with information related to a vibration measurement and a press invoke button	1505	19	12/09/2008 09:56	Eduardo Gilabert	NULL	meas_signal
10	Display web service interface for vibration data manipulation web service	1510	19	12/09/2008 10:14	Eduardo Gilabert	NULL	NULL
10	Complete the form with measurement identification and press invoke button	1515	19	12/09/2008 10:15	Eduardo Gilabert	meas_signal	meas_param
10	Display web service interface for condition monitoring	1520	19	12/09/2008 10:17	Eduardo Gilabert	NULL	NULL
10	Complete the form with asset identification and press invoke button	1525	19	12/09/2008 10:18	Eduardo Gilabert	asset, meas_location, loc_config, loc_limits	meas_diagnosis

14.3.4.2 Procedure for Integration Tests

Integration tests consist in playing two scenarios issued from FIAT end-user user scenarios used in final demonstration phases. Indeed within Dynamite, FIAT has defined four use cases based on real industrial process demand. The four use cases are:

- Use case 1: support decisions
- Use case 2: support in the case of failure (stoppage)
- Use case 3: support in the case of incoming failure (degradation/ loss of production)
- Use case 4: support to periodic maintenance.

On TELMA only use cases 2 and 3 are used for testing DynaWeb components.

Use case 2: Support to Failure (Stoppage)

The first use case represents the case when a sudden failure occurs in the TELMA physical process. The failure leads to a stoppage of the process. The following steps must be supported well by the components to be tested in an integrated way to implement this scenario:

- Asset failure occurs and is recorded in MIMOSA.
- A maintenance work order is requested and scheduled.
- Maintenance crew (dedicated to the machine) is activated; operator gets a work order on his PDA from MIMOSA.
- Data regarding work order and previous maintenance actions on the asset is downloaded onto the PDA.
- PDA supports location/access instructions to the asset.
- Maintenance crew go on site.
- Check of identity of asset with RFIDs.
- Visualisation past data of asset.
- Call to web service remote diagnostics.
- Exchanges knowledge with remote experts.
- VoIP communication between operator and expert on data in the knowledge system (MIMOSA).
- Action is performed and case is stored (MIMOSA/tag: time of action, type of action, ID of operator, spare part used).

For example, step 3 addresses functionalities normally supported by the PDA scheduling support (Figure 14.42) and the scheduling web service.

Figure 14.42 Asset menu of the PDA to give all the TELMA assets

Use Case 3: Support to Incoming Failure (Degradation/Loss of Production)

The second use case represents the case when the TELMA physical process derives from its nominal functioning point meaning that a degradation or production loss is running. Hence, contrary to the previous scenario, no process trigger exists. The following steps have to be supported well by the components to be tested in an integrated way to implement this scenario:

- Data coming from sensors flow towards MIMOSA.
- Maintenance crew performs a round and discovers something unusual (*e.g.*, abnormal sound from an asset):
  - maintenance crew request asset information
  - call condition monitoring for assessing the current situation
  - if needed, call diagnosis
  - if needed, call prognosis
  - if needed, call scheduling for a maintenance work order.
- Day of scheduled maintenance work order:
  - maintenance crew is informed of maintenance action to be performed
  - maintenance crew go on asset
  - maintenance crew verify asset information
  - exchanges knowledge with remote experts
  - VoIP communication between operator and expert, on data in the knowledge system (MIMOSA).

- Action is performed and case is stored (MIMOSA/ tag: time of action, type of action, ID of operator, spare part used).

The scenarios are concretely carried out through the use of a Dynamite Demostep file that lists all the steps to be performed. Every step corresponds to a single function that is either supported by a DynaWeb component or by an agent of the process. Table 14.8 shows a part of the Demostep defined for the detailed scenario 2 corresponding to asset degradation.

Table 14.8 Part of the Demostep defined for the detailed scenario 2 corresponding to asset degradation

[Work package]	[Demo step]	[Step number]	[Demo component]	[Definition date]	[By (person)]	[Input from Mimosa table]	[Output to Mimosa table]
10	Record failure in MIMOSA asset_event table	10100		14/10/2008	Alexandre Voisin		Asset_id, Site ID, sg-as-event_type(ev_db_site, ev_db_id, event_type_code), gmt_event_start
10	Define MWO	10101		14/10/2008	Alexandre Voisin		Asset_id, site id
10	Schedule corrective MWO	10102	23	14/10/2008	Alexandre Voisin		Asset ID Site ID, start Date, Start time, End Date, End Time, wm_type_code , Priority,

These tests steps are expressed from an end-user point of view, providing details of each step in relation to offered component functionalities. In other work, each step of the use case is explained by answers to generic questions (what, where, who), thus leading to the definition of *detailed scenarios* (Tables 14.9 and 14.10).

Each detailed scenario is structured by columns as follows:

- Step: number of the event in the temporal sequence.
- Trigger: describes the trigger of the function, attention must be paid to the field in order to know if it comes from a user interface or a software interface, *i.e.*, call to and/or return from a software module.
- What: describes the expected function.
- Where: defines the site where the function is required. The sites are:
  - asset site
  - maintenance crew site
  - DynaWeb
  - MIMOSA site.

- Who: gives the name of the DynaWeb component.
- How: gives, in case a non-dynamite component is used, how the function could be implemented.
- N° dyn: gives the number of the component according to the table entitled Dynamite-DynaWeb demonstration component delivery schedule.
- Data required: gives the list of data require prior to the function.

Table 14.9 Part of the detailed scenario related to use case 1: support in the case of failure (stoppage)

#	Trigger	What ?	Where ? Site	Who ?	How ?	N° dyn	Data required
1	Asset Failure	Record failure in MIMOSA asset_event table	Asset Site /MIMOSA Site	CASIP or process crew	SQL request		Asset_id, Site ID, sg-as-event_type(ev_db_site, ev_db_id, event_type_code), gmt_event_start
2	Record created in MIMOSA asset_event table	Define MWO		SQL Server or process crew	SQL server alert or manually		Asset_id, site id
3	Ask for scheduling corrective MWO	Schedule corrective MWO	DynaWeb	WS Scheduling		23	Asset ID Site ID, start Date, Start time, End Date, End Time, wm_type_code, Priority
4	Record created in MIMOSA MWO table	Send Alert to PDA of Maintenance crew	MIMOSA or DynaWeb	??	SQL Server Alert		Maintenance crew PDA address
5		Receive Alert on PDA	Maintenance Crew Site	PDA maintenance user interface		9/14	

These detailed scenarios were completed with the developer's point of view in order to achieve the technical functioning of the component. These final scenarios were those used to develop the integration tests to test how the components work with their own functionalities but together to implement the scenario as it is expected by the user.

Table 14.10 Part of the final scenario related to use case 1: support in case of failure (stoppage)

[Work package]	[Demo step]	[Step number]	[Demo component]	[Definition date]	[By (person)]	[Input from Mimosa table]	[Output to Mimosa table]
10	Record failure in MIMOSA asset_event table	10100		14/10/2008	Alexandre Voisin		Asset_id, Site ID, sg-as-event_type(ev_db_site, ev_db_id, event_type_code), gmt_event_start
10	Define MWO	10101		14/10/2008	Alexandre Voisin		Asset_id, site id
10	Schedule corrective MWO	10102	23	14/10/2008	Alexandre Voisin		Asset ID Site ID, start Date, Start time, End Date, End Time, wm_type_code , Priority,
10	Send alert to PDA of maintenance crew	10103		14/10/2008	Alexandre Voisin		Maintenance crew PDA address
10	Receive alert on PDA	10104	9	14/10/2008	Alexandre Voisin		

14.3.5 Results

By using the reference procedures previously explained, technical tests were conducted between July and November 2008. According to several criteria (delivery time, state of work, user documentation available, response time to question, *etc.*) components were tested once or many times. All the components (15) were tested and the results are representative on the preliminary tests (is the component material consistent enough to be used for this demonstration phase?), internal tests and integration tests.

The main results of these tests show that the percentage is calculated on the basis of 15 components (equal to 100%).

14.3.5.1 Preliminary Tests

First the preliminary tests are referred to the component status (see Section 14.3.3.). There are mainly based on developer's point of view and as such on Demostep files. Figure 14.43 presents the result of delivery of Demostep files.

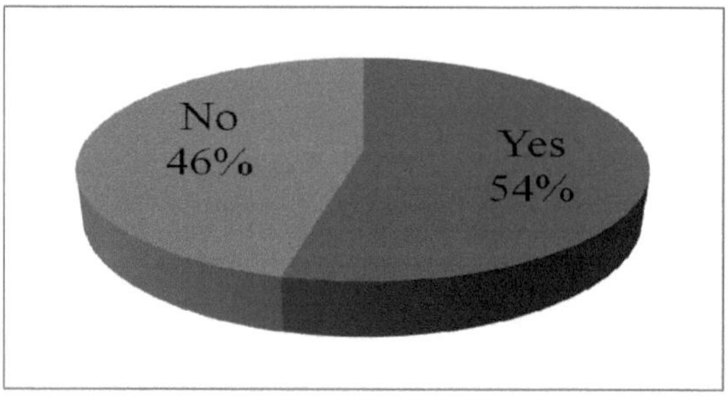

Figure 14.43 Demostep file delivered for testing the 15 components

Then, three situations have been underlined in relation to the component declaration in the software database:

1. No modules are declared by the partner in charge of the component. Hence, the component has not been declared.
2. Some modules have been declared by the partner in charge of the component but no clear link with the component exists.
3. Some modules have been declared and the link with component is clear.

After checking initially this situation, more explanations where requested from partners when it was needed in order to retrieve new versions of software modules. Figure 14.44 presents the result of the correct software module declaration. The question mark means that the component partner declares some module but the declarations do not correspond exactly to components.

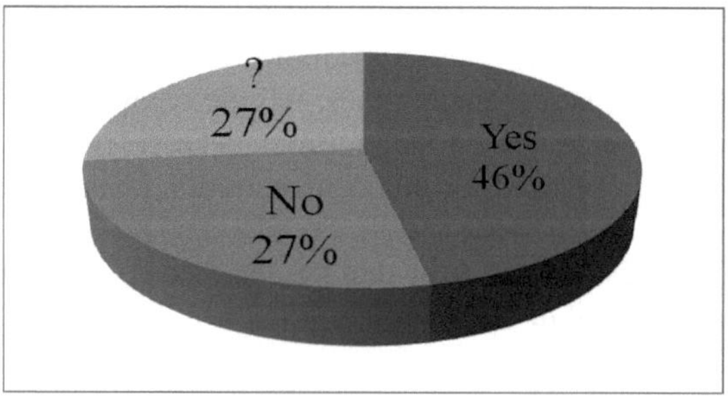

Figure 14.44 Software module declaration

14.3.5.2 Internal Tests

Concerning the functional view, Figure 14.45 shows four situations for the components:

1. ? (20%): this means that some components were not tested in their last version until end of November 2008. A new slot of time may be investigated to test the components again.
2. N (0%): no delivered component is not working at all.
3. P (7%): (1/15) component are partially working regarding internal tests. Some of their developer's functions are not working properly.
4. Y (73%): (11/15) components have succeed to the internal test.

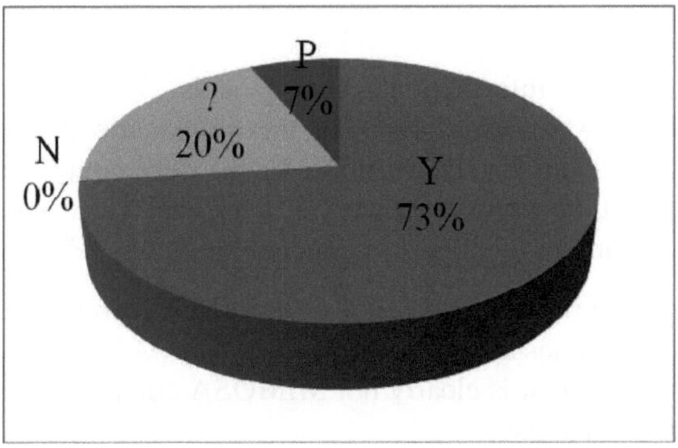

Figure 14.45 Functional view for internal tests (Y→ fully working, P→ partially working, N→ not working, ?→ not tested)

14.3.5.3 Integration Tests

In relation to the integration tests, the functional aspect, which is based on the evaluation of use case functional requirements, was developed first. Figure 14.46 shows four different situations:

1. U (20%): 3/15 components were not tested in their last version until end of November, the same ones as for the internal tests.
2. N (7%): 1/15 components did not fulfil the functional requirements for the use cases.
3. P (13%): 2/15 component partially fulfilled the requirements for the use cases.
4. Y (60%): 9/15 components fulfilled the requirements for the use cases.

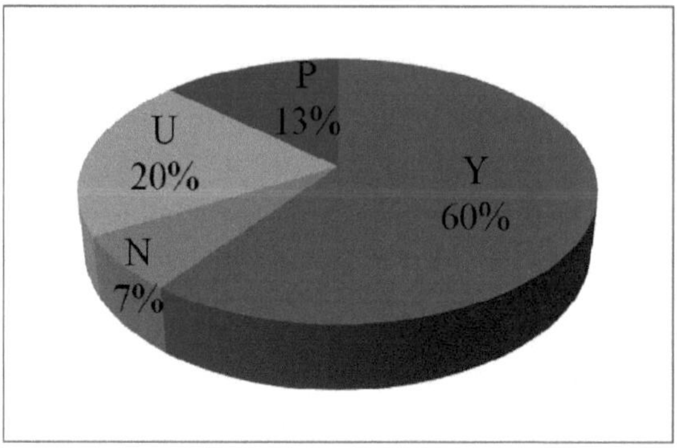

Figure 14.46 Functional view for integration tests (U→ untested, N→ not working, P→ partially working, Y→ working)

Then the tests on the informational aspect, which is based on the use of the MIMOSA database, were developed. For this informational view, four components (DWC15, 17, 18 and 25, see Section 14.3.3.) were not relevant since they do not have data interaction with the MIMOSA database. Thus only 11 components were considered leading to identification of four situations (Figure 14.47):

1. U (27%): 3/11 components were not tested. This corresponds to the components that were not assessed previously.
2. N (9%): 1/11 component is clearly not MIMOSA compliant since it works on a proprietary Oracle database.
3. M (9%): 1/11 component is partially MIMOSA compliant according to the test. The upgrading of the component to dyn_mimosa3 (remote MIMOSA database) has been performed but not tested.
4. M3 (55%): 6/11 components are fully MIMOSA compliant and work with dyn_mimosa3.

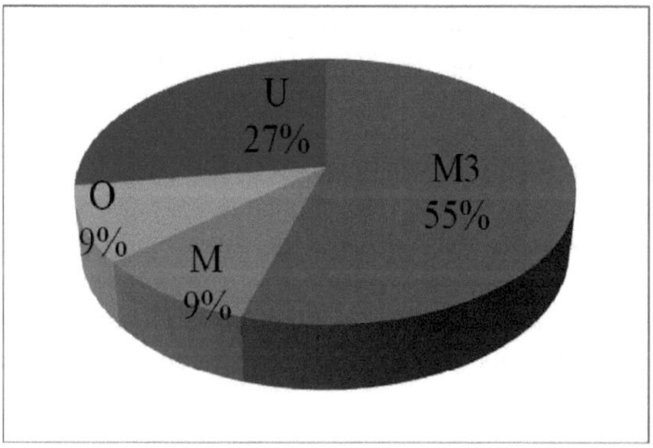

Figure 14.47 Informational view for integration tests (U→ untested, O→ not MIMOSA compliant database, M→ partially MIMOSA compliant database, M3→ fully MIMOSA compliant database)

14.3.6 Conclusions

The tests conducted on the TELMA platform aimed to validate the DynaWeb ICT components to assess their functionalities (internal tests) but mainly their capacities to be interconnected to form an e-maintenance platform (integration tests).

This required first to adapt the TELMA platform in order to support an efficient implementation of the DynaWeb components. The main improvements (see Figure 14.40) concerned adding (1) communication features from a physical point of view (*e.g.*, Wi-Fi access point) and from logical point of view (*e.g.*, by granting access through port authorisation), and (2) the translation of TELMA architecture information into the MIMOSA standard and database.

Then, according to the end-user requirements described in the FIAT use case, two test scenarios were developed in order to point out the component function expected by the end user and to execute scenarios for demonstrating components' interoperation leading to their integration.

The first comments on the tests developed on TELMA for assessing DynaWeb components are:

- Many tests have been performed leading to many improvements of the components; for example several versions of PDA are delivered to the demonstration step in a chronological way.
- The testing procedures are well established and recognised by all partners. The testing procedures (internal and integration) will be used by the partners in order to make their own tests in order to pursue the integration tests.

Concerning the internal tests, the results are positive, meaning that the components make several functionalities available.

Nevertheless, the results of the integration tests are less positive. This is mainly due to the following:

1. MIMOSA compliancy was not ensured for all of the components. This issue should be resolved quickly because it depends only on the ability of providers to be well in phase with MIMOSA.
2. Missing functionalities according to use case requirements. The developer's functionalities do not fit completely the use case requirements and some adjustments are required from the developer's or the end-user's point of view to assign the components with the right capacities needed for the final implementation on site.

According to this previous report on the integration aspect, mainly based on MIMOSA compliance, Dynamite should continue its interoperability effort to form a consistent e-maintenance platform according to two aspects:

- MIMOSA-based initiatives to develop a standard in the maintenance area to normalise integration issues. Indeed, today MIMOSA is not really a standard approved by all, but initiatives advocated by international organisation are ref-

erenced to MIMOSA for normalising interoperability profiles between all maintenance processes such as diagnosis, prognosis, scheduling *etc.*, and between maintenance processes and production processes. Two main initiatives are: (a) ISO TC 18435-1 (ISO TC 184/SC 5 N897–ISO/CD 18435-1) industrial automation systems and integration – diagnostics, capability assessment and maintenance applications integration, Parts 1–5; (b) OpenO&M (http://www.openadm.org/), which is an initiative of multiple industry standards organisations to provide a harmonised set of standards for the exchange of operations and maintenance data and associated context.
- Wireless-based initiatives to focus on wireless monitoring and alerting needs for the industries. Such initiatives are intended to provide reliable and secure operation for non-critical monitoring, alerting, supervisory control, open loop control and "soft" closed loop control applications. The main initiative is the ISA 100 Wireless Networking Committee, which is progressing toward its first standard ISA100.11a.

14.4 Machine Tool Demonstrator

Fernando Garramiola

Goratu builds high technology NC machines for a wide variety of industries, such as the paper, railway, car and petroleum industries. Due to wide experience and a dedicated R&D department, Goratu satisfies customer needs by building machines fulfilling their machining requirements. Goratu provides machines to customers spread around the world. Moreover, developing countries have increased their orders and they have a difficult market where the lack of spares parts is a hard constraint. These reasons make maintenance very important, in order to avoid that a machine failure stops production. Diagnosis and prognosis are becoming more important for manufacturers, in order for the customer to have confidence in the machine. Machine reliability is the key for Goratu marketing.

Maintenance travel and personnel expenses are considered high for Goratu, which has led Goratu to be involved in e-maintenance, which allows maintenance for many customers even remote from Goratu. E-maintenance is an asset for full maintenance with the lowest cost.

14.4.1 Objectives of the Test and Demonstrations

The main objective for Goratu is to avoid stoppages, providing an e-maintenance system. Machine variables should be able to be controlled to avoid machine failure (*e.g.*, vibration may be related to bearing damage). Moreover, the customer may be far away, and the collected data must be transferred to Goratu. In order to do this, an internet connection for transmitting the collected information is needed.

Wired sensors and a wireless sensor network make a great asset for monitoring the machine, providing a useful platform for other sensors. The information collected from sensors will be sent to the MIMOSA database, so it can be analysed and Goratu can take the maintenance steps needed.

14.4.2 Description of the Test Platform

The test machine platform, as can be seen in Figures 14.48 and 14.49, is a machine located at Goratu factory. This machine is not for internal production, so it machine will be sent to a customer later for his application. As a machine tool manufacturer, the test machine platform will be the model for later maintenance service of our customers, who will be located away from Goratu.

The Goratu shop floor has WiFi radio frequency available for the connection between the local PC and the MIMOSA database. CMOpS and the expert will be located at the Goratu office, connected to internet through Ethernet or WiFi, so the CMOpS agent can ask for the different web services and the MDSS.

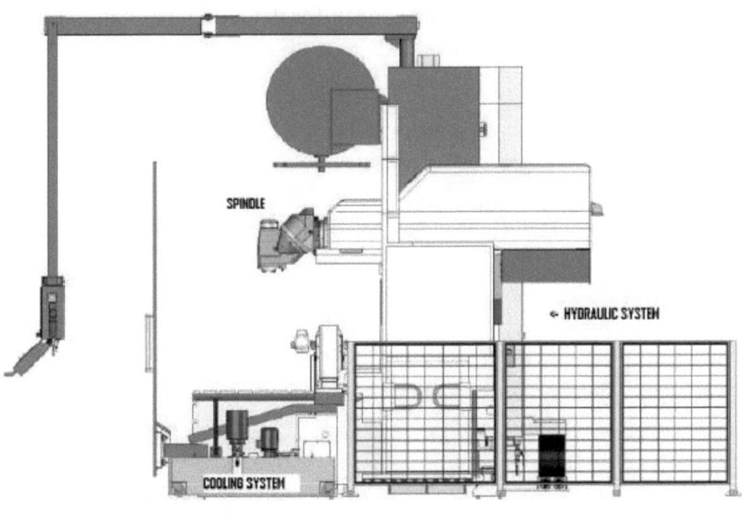

Figure 14.48 The machine used as the test platform at Goratu

Different sensors were connected to the machine, such as a water content lube sensor, a particle scatter sensor and a vibration sensor.

The particle scatter oil sensor analyses the machine hydraulics system oil. The hydraulics system is responsible for the tool opening, counterweight for the vertical Z axis and the blocking of rotating tables, so it is very important for the machine performance that the oil is not degraded. The hydraulics system has a 12 l tank, and it is recommended to check and top this up every 200 h and to completely change it every 2000 h to assure that the machine will work under the specified conditions. Using the particle scatter oil sensor will probably allow an increased time between oil changes by changing the oil only when the oil is degraded. The hydraulics system provides a 60 bar pressure, which needs to be reduced to comply with the sensor requirements.

The water content oil sensor is located in the cooling system. The coolant is usually a mix of oil and water, where the water content may be around 4%. The cooling system is used to avoid excessive heating of the work piece and the milling tool, which can reduce the quality on the work piece's surface finishing. The cooling system tank contains 100 l of coolant. The percentage of water in the coolant needs to be controlled to avoid rust on the work piece. A coolant with an excessive quantity of water can damage the work piece and the machine components (*e.g.*, the table).

A vibration sensor is connected to the machine. The vibration sensor is located on the machine head, close to the spindle. The spindle vibration measurement is very helpful to clarify if the machine is working under the specified conditions, *e.g.*, spindle speed, axes feed, tool wear and cutting thickness. The spindle reaches 3000 rpm. Moreover, the vibration analysis allows obtaining the natural frequencies of the machines with some sensor measurements, which can provide a better knowledge of the machine, as well as a mean for detecting mechanical failures.

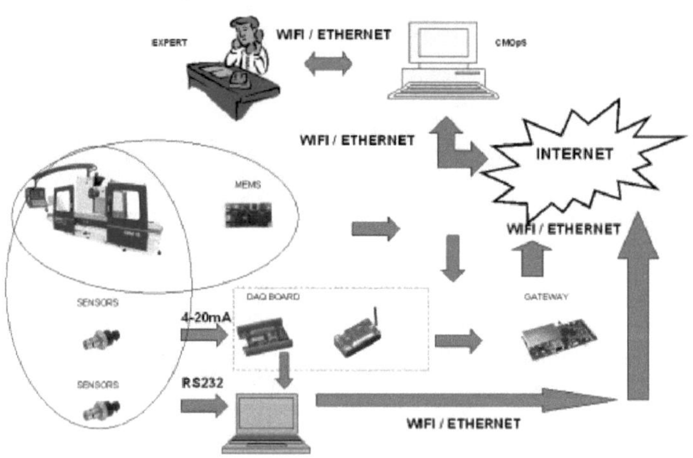

Figure 14.49 Goratu platform for global demonstration

14.4.3 Description of the DynaWeb Components Tested

The Wyselec vibration measurement system is utilised for vibration data collection and analysis. The vibration measurement system utilises a market vibration sensor (Wilconox Model 784A). The vibration sensor is wired to a signal conditioning board, which provides excitation voltage for the accelerometer. The signal conditioning board is wired to the measurement station (industrial PC with 1GHz CPU, 512 MB RAM and 40GB HDD). The measurement station can hold up to four DAQ units with eight ports each. The measurement rate is up to 200 kS/s with 16 bit accuracy. All channels can be measured simultaneously. The measurement station is connected to the local network through Ethernet TCP/IP. Thus, all the measurements are sent to the Mimosa database, located in an external data server, Figures 14.50 and 14.51.

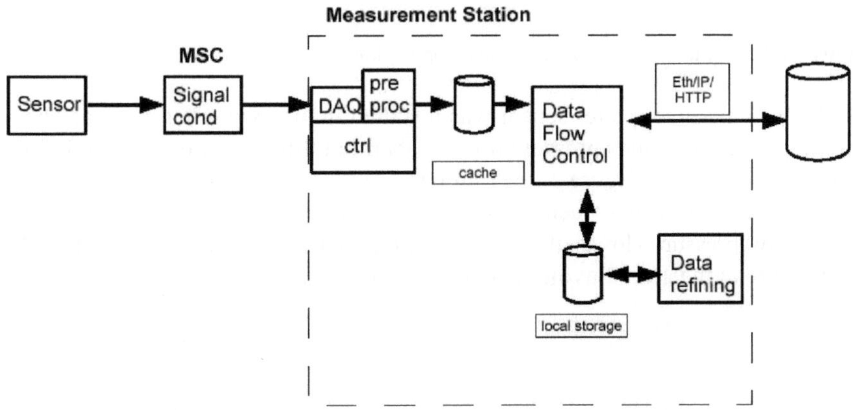

Figure 14.50 Vibration measurement system diagram

As the measurement station is available at the local network, or from an external PC using a VPN, the machine data acquisition (MDAQ) application can be started on line using a remote desktop. This allows running the application either during a test time doing a manual measuring or running a scheduled measurement. The data collected can be sent to the MIMOSA database if it is selected on the MDAQ application.

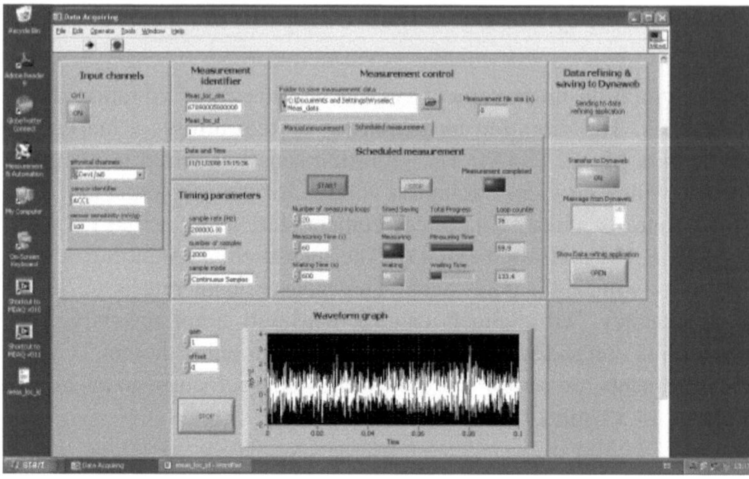

Figure 14.51 Vibration measurement system application

The VTT particle scatter sensor was installed in the hydraulics system. The hydraulics system provides pressure up to 80 bar, but this pressure is too high for the sensor requirements (Figure 14.52). In parallel with the machine hydraulic circuit a pressure reduction has been built, so the sensor could be provided with an oil flow at low pressure (lower than 2 bar). A manometer has been installed to show the pressure at the sensor avoiding damage to the sensor.

Figure 14.52 Pressure adaptation for sensor requirements

The particle scatter oil sensor is connected to a measurement unit, which is connected to a PC through the USB port, and the PC should be connected to internet either wired (local network) or wireless (WiFi). Once the connection to the lube service is ready, new data can be collected or see the previous collected data can be used, as can be seen in Figure 14.53. All the data collected is directly sent to the MIMOSA database.

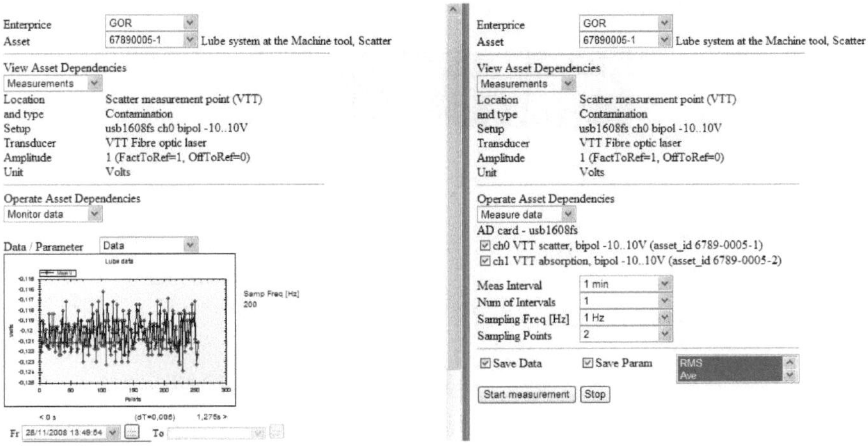

Figure 14.53 Lube service application

The Tekniker water content oil sensor is installed in the cooling system, so it can measure the water percentage in the coolant. This is of great importance, in order to avoid rust on the work piece and on the machine table.

The water content oil sensor is installed in series with the cooling pump (Figure 14.54). The main problem that has been found was the pipe diameter of the sensor; this diameter is too low for the pump flow, inducing an increase in pressure quickly, producing coolant leaks.

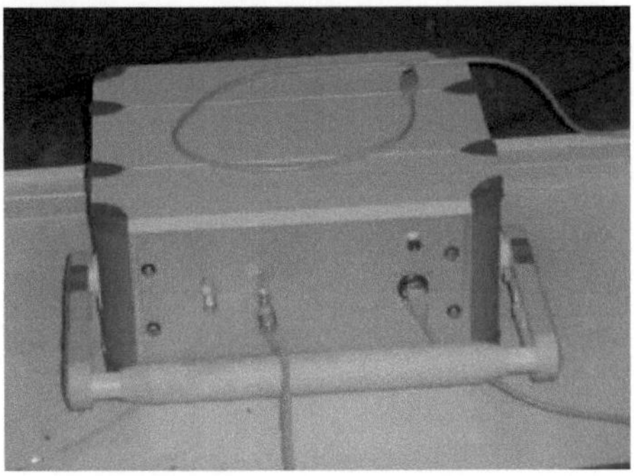

Figure 14.54 Tekniker water content sensor

The sensor is connected to a PC through a USB port. The application is installed on the PC; it shows the water content and has a traffic-lights style interface to alert if the water content is over limit 1 or limit 2; the first one can be a warning and the second one an alarm. The limits can be selected after measuring the reference coolant, which will be used to calibrate the application, as shown in Figure 14.55.

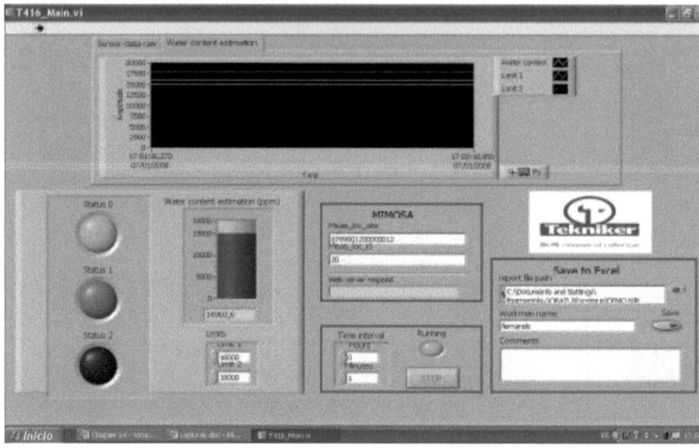

Figure 14.55 Water content sensor application to show the water content related to the allowable limits

The information collected from different sensors will be stored in the MIMOSA database. After that the TESSnet platform is used as a predictive maintenance management system, displaying information about the machine status and measurements taken. Figure 14.56 shows the vibration and water content sensor data collected in the database, which can be displayed using the TESSnet application.

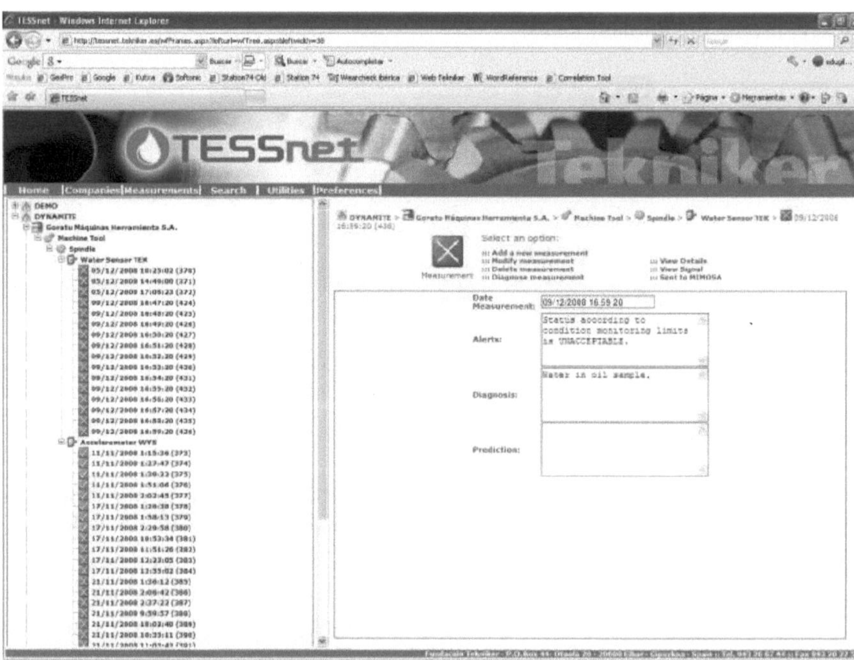

Figure 14.56 Water sensor measurement in TESSnet

The software is synchronised to the MIMOSA database in such a way that the measurements are imported to TESSnet. Moreover, TESSnet is connected to web services for vibration data manipulation, condition monitoring and diagnosis, so that the user can display information about current machine status.

On the other hand, the maintenance expert collects the data from the MIMOSA database to validate cost-effective maintenance support and then he performs the cost-effectiveness strategy. Figure 14.57 shows the vibration data collected for cost-effective maintenance support. The data are collected from sensor measurements and will be utilised for determining the vibration level before the breakdown.

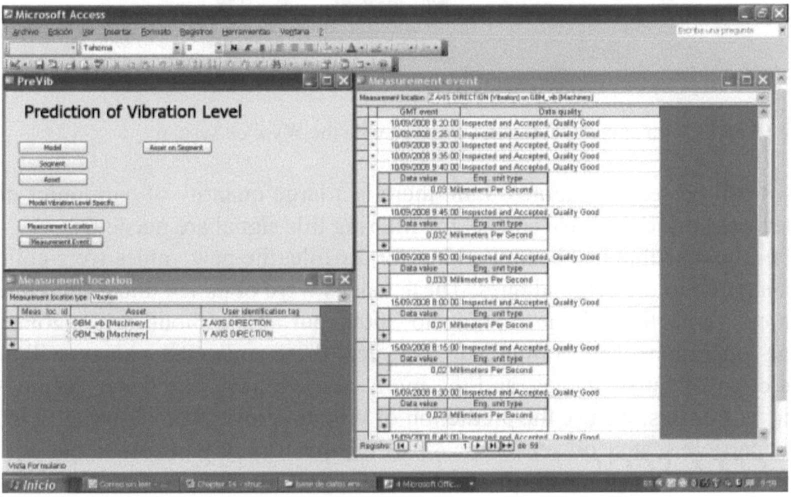

Figure 14.57 Vibration data collected for MDSS

14.4.4 Reference Measurements/Software

The vibration sensor was connected to the machine through a threaded hole. A 2000 rpm spindle speed was programmed. The test was done for a new machine, which had never done any machining. Using the Wyselec vibration measurement system or the vibration handheld unit, the values collected will be the reference values for the spindle vibration. The value of the RMS was collected, and the FFT graphics were also captured.

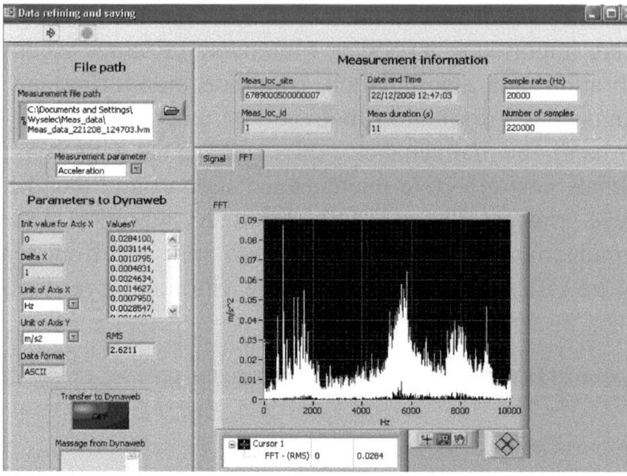

Figure 14.58 FFT for spindle vibration measured with the Wyselec system

As it can be seen in Figure 14.58, there is a large quantity of harmonics due to the electrical drives. Improvements for filtering this signal are needed.

After the machine had been working for a while, the new values were collected and compared to the original ones. These values were obtained under the same test conditions, with the spindle running to 2000 rpm and without any ongoing machining operation. The comparison between these collected values will allow a prediction of future values, and the more data we have, the better accuracy we will have. For this, we use the prediction of vibration level from the MDSS tool.

The particle scatter oil sensor measures the quantity of particles in the oil. Data is first collected from clean oil and then the machine is operated. After the machine has been working for some time, new data are collected and compared to the original ones. As long as the machine is working the number of particles increases, so it is important to determine the quantity limit at which it is necessary to change the oil. Currently the running hours determine the change time.

The water content oil sensor is connected to the cooling system. Until now, Goratu did not have any reference about the percentage of water in the coolant and it was only apparent when the problem of rust arose. Different measurements were taken to see the percentage of water in the coolant. The software application allows warning and alarm limits to be set, so the limits were set measuring the water percentage of the new coolant. Due to the different coolant types it is difficult to do an open calibration, so it is necessary to fix the limits for the coolant used by the customer uses, otherwise failures can occur. Diagnosis web services allow showing the state of coolant as normal, warning or alarm.

14.4.5 Results

All the components provide new assets for improving the quality of the product. The market has a lot of components to measure vibration and water content in the oil, but there are few products that allow storing that information into a database that is accessible from different places.

At the moment, the vibration sensors in the machine were only used for monitoring alarms and stopping the machine, but the data were never stored and, hence, the problem could not be identified. The new components developed in the Dynamite project allow data to be collected and analysed, so not only is machine damage avoided but the source of the problem can also be established. This information can allow us to detect wrong use of the machine or mechanical problems that may arise in the future.

Goratu does not use oil sensors at the moment, so the new sensors are a big improvement, in particular the water content sensor, which can help to avoid many problems due to the wrong coolant being used by customers. At the moment we need to ask the customer for a coolant sample and send it to an expert company to make any conclusions. Thus, the water content sensor may be a great asset for the customer who can control the coolant by himself. The diagnosis web service running with the lube application provides a visual way to activate the steps needed.

The particle scatter oil sensor can prolong the life of the hydraulics system oil, so the oil change will be based on the oil quality and not on the machine running hours. This is a great advantage because we can see that a clean environment allows a life longer than 2000 h of machine running. Until now, the oil change was done through experience, but this more scientific way may be a saving for the customers.

All the communication with the MIMOSA database was quite fluent, although there were doubts due to security rules, as some ports needed to be open for data transmission. The communication through both local network and WiFi connection worked satisfactorily.

14.4.6 Technical Evaluation

There are some technical issues that need to be improved for an industrial environment. For the vibration measurements system, the vibration needs to be separated from the harmonics of the drives. Probably a better filtering is needed, otherwise the bearing vibration could be underestimated due to additional electrical noise. It was tried to isolate the vibration sensor from the machine, sticking it with carton (not conducting component), but the results are not very different from the original ones.

The water content oil sensor needs to increase the coolant flow rate that can pass through it. As currently the flow through the sensor is very small compared to industrial cooling applications a parallel circuit is needed, but this will not work properly due to the large differences between the pipe sections.

14.4.7 Economical Evaluation

Breakdown of electro spindle bearings generates a very high cost. Firstly, because the spindle reparation normally costs over 1000 €, and then, because the repairs should be done by the electro spindle manufacturer, which takes a long time. Sometimes a customer's machine may be out of order for weeks. In this case, the customer may lose manufacturing contracts due to the unavailability of the machine and the incapacity to deliver. Using a vibration system, diagnosis and prediction tools, bearing breakdown can be avoided, and the electro spindle bearing change can be planned, so the customer continues his production without unexpected stops.

A market oil for hydraulics costs around 2 € *per* litre, and we recommend the customer to change the oil tank (12 l) every 2000 h of machine running. Using the particle scatter oil sensor, the oil life can be measured. Moreover, the oil measurement should avoid machine problems due to dirty oil, which could damage the machine and also stop the production.

A market coolant costs around 5 € *per* litre. The cooling system has a 100 l tank in which the coolant needs to be changed if the water content is too high. Otherwise, rust can arise on the machine table or the work piece, which could take hard work to clean, or even grinding works if the damage is more severe. For security, it is recommended to empty the tank and change the coolant at least once *per* year, but if water content measurements are taken it is certain that this period can be longer.

14.4.8 Conclusions and Recommendations

Some improvements are needed but DynaWeb is a new step for e-maintenance in machine tools. The e-maintenance for production is largely used around the world, but the most machine tools manufacturers are now entering in this world.

E-maintenance will increase the efficiency of machine tools, reduce down time and breakdown. This reliability will be a compulsory condition for marketing departments.

A new investment is needed to change the machine tools, and DynaWeb is one step of that change. DynaWeb adds new functions to Goratu machines, which make the maintenance easier to the machine customer. The customer can control the vibration of the bearings continuously and avoid damaging them; he can control the quality of coolant and hydraulics oil and optimise the life of both. Apart from these contributions, the vibration measurement system provides a lot of information, such as FFT, that can better show the possible improvements for machine tools, for example, reduction of the machine vibration.

14.5 Maritime Lubrication System Demonstrator

Jim Bellew

The significant aspects of lubrication systems aboard a ship are that they operate under the most extreme conditions while required to deliver a performance standard far beyond that of equivalent systems ashore. By definition a ship is a remote transportation asset operating in extremes of temperature; its motion is akin to that of an earthquake and there are many other factors that expose its vulnerability. Onboard there are a number of critical lubrication systems upon which the vessel is dependent. Any interruption of service of these systems could leave a ship dead in the water and at the mercy of the elements. The survival of a ship's crew and the security of an asset and its cargo, worth possibly hundreds of millions of Euros, can be dependent upon lubrication systems that are minimally supervised and exposed to a high contamination potential. Because of the asset value any interruption of service is an expensive event. There can be a number of critical lube/hydraulic oil systems aboard a ship:

- the main propulsion engine(s) and power generating engine(s);
- the common fuel rail hydraulic system;
- stern tube/tail end shaft lubricant system; and
- steering engine (rudder) hydraulic system.

The one chosen for this exercise is the *stern tube/tail end shaft lubrication system* on an 8000TEU container ship.

Lube oil circulates around the revolving propeller shaft reducing friction and maintaining in serviceable condition the shaft metal and bearing surfaces and returning to a sump that holds up to 10,000 l of lube oil. Typically for a large container ship, the power of a 12 cylinder, 70,000 kW engine is transmitted through a shaft turning a 120 ton, 10 m diameter propeller at 102 rev/min. The massive forces that surround a ship of 150,000 metric deadweight tons, particularly in heavy seas, creates stress and vibration at the tail-end shaft. The extreme forces acting upon this system can result in sea water ingress and lube oil leakage into the sea, which is a major concern. Corrosion and wear of the shaft can cause deterioration and the system is regularly subjected to mandatory removal and inspection surveys – a process that requires dry-docking the ship at considerable cost and interruption to service!

Lube oil quality. The initial charge of fresh lube oil includes appropriate base oil with an additive pack specifically designed for each application. During the period of use temperature, pressure and pollution cause deterioration of the base oil and depletion of the additive pack reducing the utility of the lubricant. Various materials such as sea and fresh water, wear debris and other contaminants can pollute the lube oil significantly reducing its performance. Changes in viscosity, base number (BN) and other properties occur. Monitoring the quality of lube oil can provide an early warning of deteriorating conditions triggering intervention by the crew. Remedial action can involve changing or cleaning the lube oil – in itself an expensive commodity. In addition, continuous monitoring of the lube oil condition can provide an authoritive record of lube oil conditions over an extended period that classification bodies and other regulatory organisations will accept as evidence of good operating conditions. Water, viscosity, BN, density, impurities and particulate matter can be identified as evidence of deteriorating condition. However, if the condition can be shown to be predominately water free and of optimal standard, then in such circumstances mandatory surveys may be waived with considerable economic and operational benefit to operators.

14.5.1 Objectives of the Test and Demonstrations

For logistic and legal reasons it was not possible to conduct a demonstration on board an operating vessel, therefore a close-to-reality simulation was arranged on a specially designed test rig. The objectives were as follows:

- to demonstrate the operation of the sensor;
- to demonstrate the communication of results locally;
- to demonstrate the communication of results *via* DynaWeb assets to the MIMOSA database.

The objectives of installing such a system aboard a ship are:

- to protect critical systems in order to avoid an interruption of service that could endanger lives, property and environment;
- to extend the life of shipboard machinery for economic benefit by avoiding repairs, unscheduled downtime, and off hire circumstance; and
- to provide input for condition-based monitoring to enable the extension of service time between remove/inspection surveys.

14.5.2 Description of the Test Platform

The test rig was constructed to replicate the conditions of a lubrication oil circulation system of a stern tube/tail end shaft lubrication system. The means of adjusting lube oil flow velocity, pressure and temperature (plus viscosity) were provided to create conditions likely to prevail in a typical stern tube/tail end shaft lubrication system operation. Figure 14.59 provides an outline of this system.

The overall dimensions of approximately 1 $m^2 \times 1.8$ m high provided a convenient working arrangement and the oil capacity of 10 l enabled sufficient volume for accurate measurement while facilitating instant adjustment and oil changes that were both affordable and manageable. The circulation system was mounted on a horizontal framework and the oil circulated by an adjustable gear pump to determine flow velocity. A thermostatically controlled heater provided the desired operating temperature, while the pressure was controlled by partially opening or closing the pressure regulating valve. Suitable instrumentation displayed these parameters. A manifold with a range of fittings allowed access in various ways to connect sensors to the circulating oil and to draw manual samples from the flowing stream. At the lowest point in the circulation system a valve allowed the oil to be drained into a sump to facilitate oil changes. At two points in the system a facility was provided for venting air from the system. The above describes the primary circuit replicating the stern tube/tail end shaft lubrication system. Figure 14.60 is a photograph of the arrangement with the test rig in the foreground and the various sensors and components surrounding it.

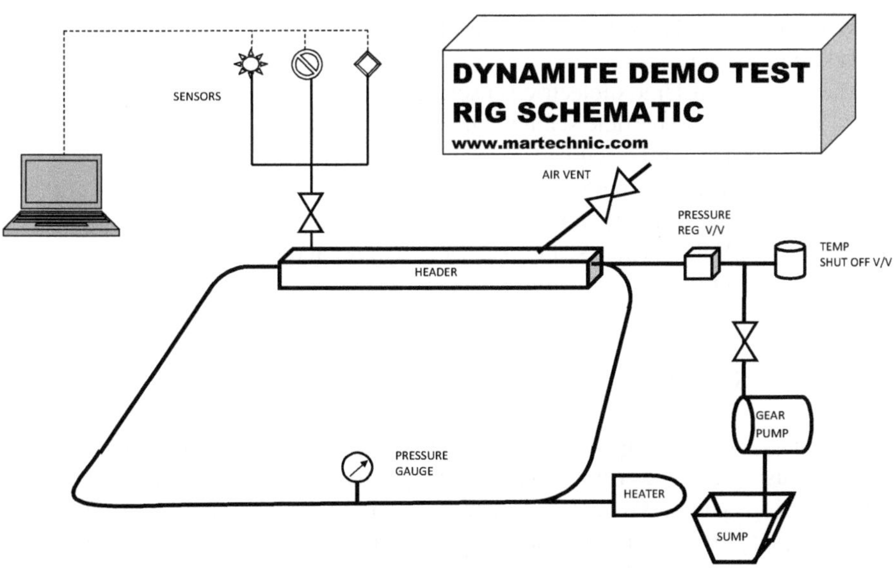

Figure 14.59 Schematic of test rig

Figure 14.60 Photos of test rig, sensors and monitors show readings

The test rig was connected with 3 mm plastic tubing for sample delivery to the four sensors with PCs or laptop and a PDA to provide the interface to the MIMOSA database.

1. Martechnic AHHOI water-in-oil sensor;
2. Tekniker particle sensor;
3. Martechnic AuLUmo dielectric sensor;
4. VTT fibre optical particle scatter sensor;
5. test rig – primary oil circuit;
6. monitor for Tekniker particle sensor;
7. monitor for the AHHOI and AuLUmo sensors;
8. laptop with VTT particle scatter sensor software; and
9. PCs for monitors.

14.5.2.1 The Sampling System

A secondary system designed to simulate the arrangement for monitoring the quality of oil using a bleed from a bypass through the sensors was set up using 3mm plastic tubing. The pressure from the primary circuit provided a constant flow into open sampling bypass circuits. Separate sampling lines feed the four sensors with solenoid valves to open and close the flow. Details of the sensors are given in Section 14.5.3.

The sensors had either integral operating software or were served by software loaded on a computer/PC/laptop that determined the sequencing. The IR sensor was set up for batch processing and the visible sensor operated continuously. After passing through the sensors the samples are returned to the system *via* the sump.

14.5.2.2 Testing Method

The test rig was custom designed to provide the appropriate sample flow equally to each sensor. Clean flushing oil was charged to the system and circulated for two days to ensure that any debris and fouling would be removed and to test the integrity of the plumbing and accuracy of the system components (temperature, pressure, filters, *etc.*). The system was then charged with clean new lubricating oil of the specification used in the stern tube/tail end shaft application.

With all of the sensors fitted and connected to the various electronic and recording systems the oil was circulated and brought to its operating temperature and pressure. These changes were noted on the monitors as the system progressively settled in to a stable operating condition. The testing period commenced with the instruments recording zero water and zero particular contamination. The testing period although not preset lasted for nine days. At regular intervals small known quantities of water where injected into the primary oil flow. The changes in water-in-oil content were detected by the sensors and recorded. Manual samples were periodically taken appropriate to the dosing of water for analysis by field and laboratory tests as a reference check against the measurements. This process was continued until a series of measurements was achieved to demonstrate the system.

Following this, a similar process followed for the particular matter with similar results. All of the sensors identified the changes producing data that could be observed on the various monitors. The objective of the demonstration was not to assess the accuracy or reproducibility of the sensors but to show that the data generated in an operational situation could be managed and communicated. This was achieved.

14.5.2.3 Technical Specifications of the Test Rig

Dimensions:	100 × 100 × 180 cm
Weight (empty):	35 kg
Oil capacity:	10 l
Operating temperature range:	20–70°C
Operating pressure range:	0–3 bar

14.5.3 Description of the DynaWeb Components Tested

Tekniker particle sensor (see Section 14.4.3)

VTT sensor (see Section 14.4.3)

Martechnic AHHOI water-in-oil sensor (see Figure 14.63)
Water exists in lube oil in three different ways:

1. saturated – *i.e.*, dissolved in the water;
2. emulsion – small droplets held in a stable suspension; and
3. free – larger quantities of water mixed without any connections.

The infrared process recognises all three states and consequently has significant advantages over other water measurement methods. Up until 2008 the only means of using infrared analytical processes for this application was to ship a sample to a land based laboratory and wait for the results. As a result of participation in a previous EC project Martechnic developed AHHOI, the industry's first infrared sensor for detecting water-in-lube-oil capable of operating in such a hostile environment. AHHOI has been fitted to all the applications mentioned earlier and as of 2010 has registered over 30 man/years of faultless operation in maritime environments. It can detect total water content in lube oil up to 1% volume (10,000ppm). The sensor is designed for oil operating pressures from 3–10 bar and has built-in self defence mechanisms that filters the oil and controls the pressure to approximately 1 bar (see Figure 14.61 for an overview of the installation). Figure 14.62 shows the sensoring system in the ship.

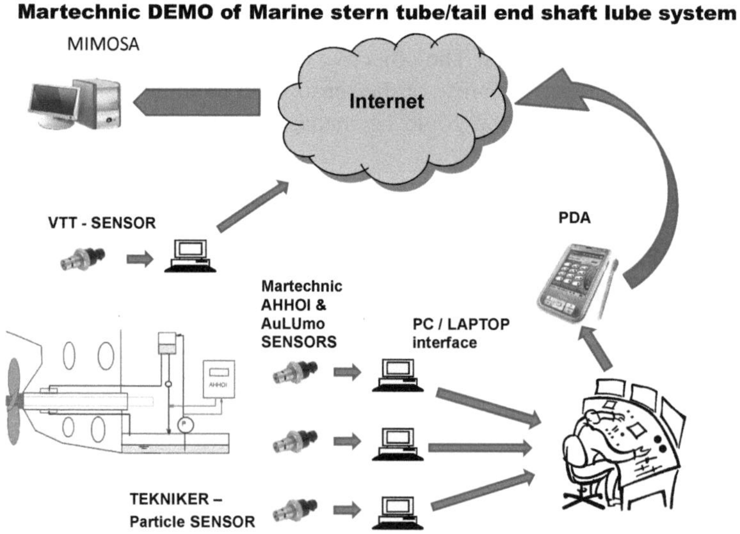

Figure 14.61 DynaWeb arrangement

Technical data

Housing protection class:	IP54
Dimensions of the housing (L × W × H)	355 × 300 × 125 (mm)
Overall dimensions (L × W × H)	400 × 300 × 125 (mm)
Operating voltage:	100–240 V AC, 50–60 Hz
Power consumption:	max. 30 W
Weight:	approx 7 kg
Pressure (primary side):	max. 10 bar
Pressure (measuring cell):	max. 1.5 bar
Ambient temperature range:	0 –70°C.
Temperature range of the oil:	10–50°C.
Measuring range:	0 to 10,000ppm/0 to 1%
Display accuracy:	± 5% of the displayed value
Analogue outputs:	0–20/4–20 mA

Figure 14.62 Martechnic's AHHOI water-in-oil sensor system *in situ* aboard ship

14.5.4 Reference Measurements/Software

The current practice in the maritime industry is for lubrication oil samples to be taken periodically by a ship's crew and dispatched to the lube oil supplier for analysis. After the samples have been analysed at some distant laboratory the results are passed to a technician for interpretation, advice and recommendation to then be communicated back to the ship or its operating company. There are a number of deficiencies in this situation:

- The sample may be incorrectly drawn by the shipboard engineer.
- The container could be contaminated, damaged or lost.
- Damage could occur to the system in the period between the sample being taken and the remedial advice reaching the ship's crew.
- The communication from the laboratory to the ship (sometimes *via* the company) could go astray or be misunderstood.
- Significantly, even if everything within this system goes perfectly a sudden undetected contamination of sea water into the lube oil could seriously deteriorate the component conditions – subsequently placing the ship, its crew, cargo and the environment in jeopardy.

To try to counter this vulnerability manual onboard (field) tests were developed to enable shipboard staff to perform on-the-spot analyses for parameters such as water-in-oil, viscosity, base number and total impurities. However, this practice is still subject to human failure where inadequately trained and motivated staff are required to take and test samples and interpret the results. The location of the stern tube/tail end shaft assembly at the bottom of the ship all the way astern also tends to be a disincentive for crew members to attend to this critical application.

Significantly, classification societies and other regulatory bodies charged with inspection duties require valid/authoritative test data showing a continuous condition of lubricating oil to enable them to determine the time period for safety inspections. The current state of the art does not facilitate this concept.

14.5.5 Results of the Demonstration

All of the sensors tested demonstrated their ability to monitor the changing state of oil quality and transmit *via* various DynaWeb routes these data to the MIMOSA database. Figures 14.63 and 14.64 show some data results.

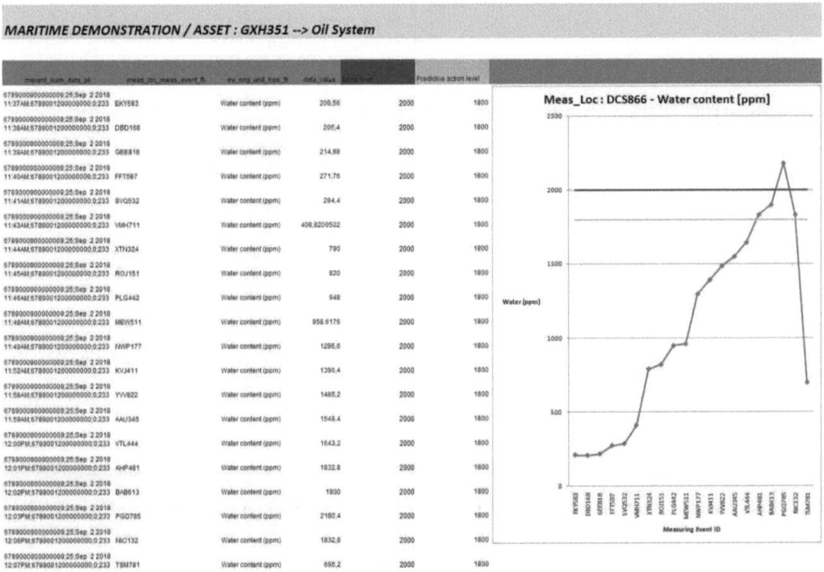

MARITIME DEMONSTRATION / ASSET : GXH351 --> Oil System

mevent_num_data_pk	meas_loc_meas_event_fk	eu_org_unit_type_fk	data_value	[illegible]	Predictive action level
6789000900000009;25;Sep 2 2018 11:37AM;6789001200000000;0;233	EKY583	Water content (ppm)	208,56	2000	1800
6789000900000009;25;Sep 2 2018 11:38AM;6789001200000000;0;233	DBD168	Water content (ppm)	205,4	2000	1800
6789000900000009;25;Sep 2 2018 11:39AM;6789001200000000;0;233	GEE818	Water content (ppm)	214,88	2000	1800
6789000900000009;25;Sep 2 2018 11:40AM;6789001200000000;0;233	FFT587	Water content (ppm)	271,76	2000	1800
6789000900000009;25;Sep 2 2018 11:41AM;6789001200000000;0;233	SVQ532	Water content (ppm)	284,4	2000	1800
6789000900000009;25;Sep 2 2018 11:43AM;6789001200000000;0;233	VMH711	Water content (ppm)	408,8200502	2000	1800
6789000900000009;25;Sep 2 2018 11:44AM;6789001200000000;0;233	XTN324	Water content (ppm)	790	2000	1800
6789000900000009;25;Sep 2 2018 11:45AM;6789001200000000;0;233	ROJ151	Water content (ppm)	820	2000	1800
6789000900000009;25;Sep 2 2018 11:46AM;6789001200000000;0;233	PLG442	Water content (ppm)	948	2000	1800
6789000900000009;25;Sep 2 2018 11:48AM;6789001200000000;0;233	MEW511	Water content (ppm)	958,6176	2000	1800
6789000900000009;25;Sep 2 2018 11:49AM;6789001200000000;0;233	NWP177	Water content (ppm)	1295,6	2000	1800
6789000900000009;25;Sep 2 2018 11:52AM;6789001200000000;0;233	KVJ411	Water content (ppm)	1390,4	2000	1800
6789000900000009;25;Sep 2 2018 11:58AM;6789001200000000;0;233	YVV822	Water content (ppm)	1485,2	2000	1800
6789000900000009;25;Sep 2 2018 11:59AM;6789001200000000;0;233	AAU345	Water content (ppm)	1548,4	2000	1800
6789000900000009;25;Sep 2 2018 12:00PM;6789001200000000;0;233	VTL444	Water content (ppm)	1643,2	2000	1800
6789000900000009;25;Sep 2 2018 12:01PM;6789001200000000;0;233	AHP481	Water content (ppm)	1832,8	2000	1800
6789000900000009;25;Sep 2 2018 12:02PM;6789001200000000;0;233	BAB613	Water content (ppm)	1900	2000	1800
6789000900000009;25;Sep 2 2018 12:03PM;6789001200000000;0;233	PGO785	Water content (ppm)	2180,4	2000	1800
6789000900000009;25;Sep 2 2018 12:06PM;6789001200000000;0;233	NIC132	Water content (ppm)	1832,8	2000	1800
6789000900000009;25;Sep 2 2018 12:07PM;6789001200000000;0;233	TSM781	Water content (ppm)	695,2	2000	1800

Figure 14.63 Results of the AHHOI water monitoring

This example data table shows the graph tracking the progressive dosing of water into the system with the monitored conditions recorded and communicated to the MIMOSA database by the AHHOI sensor system.

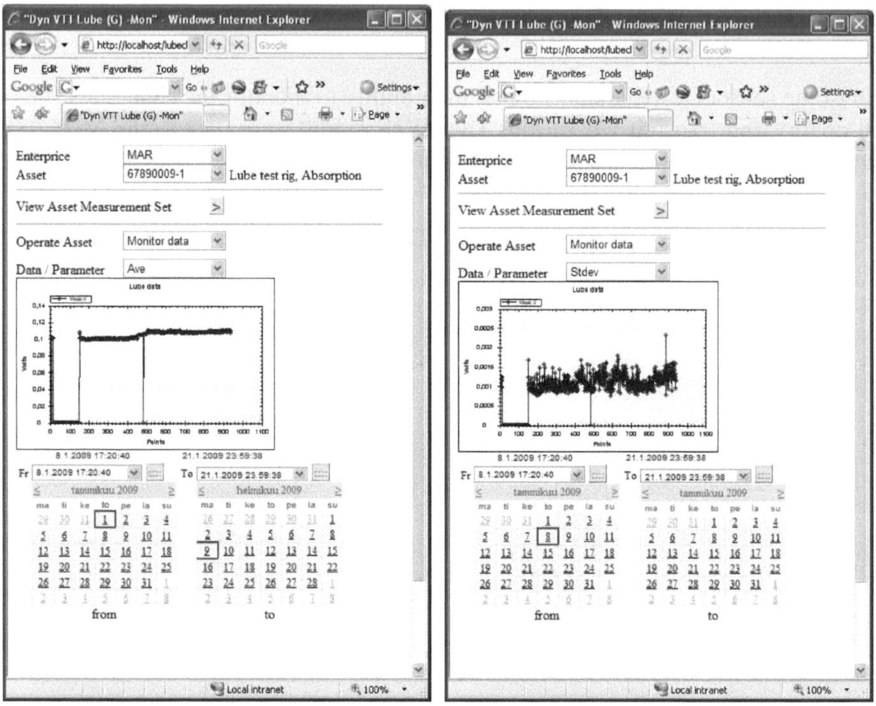

Figure 14.64 Two graphics showing the PDA input to the MIMOSA database *via* DynaWeb

These charts show analysis results from the VTT sensor being communicated to the MIMOSA database *via* a PDA.

14.5.6 Technical Evaluation

At all times the operation, function and condition of the monitored machinery/oil system was supervised beside the standard control mechanism! In this specific case the demonstration underlines the use and need of such a technology for future maintenance management.

14.5.7 Economical Evaluation

The demonstrated system does not replace or compete with any similar system. What it will replace is a logistically heavy ineffective process subject to considerable human error that by definition delivers results too late to be effective. The consequence of this is that the ship operator, crew, vessel and cargo are left vulnerable to the extreme elements of the environment in which they work. At best this causes delays and significant economic penalties, at worst catastrophic consequences. Prior to the launch of the AHHOI the system of manual sampling and shoreside testing represented the state of the art. This circumstance prevails across all of the applications previously listed in the summary.

Demonstrated scenario. Stern tube lube oil costs around $2000/ton with an initial charge of 10,000 l = US $20,000. Replenishment costs are dependent on the degree of contamination and losses. Sample analysis charges have traditionally been included by the supplier in the price of oil, however there is a recent trend to limit this and charge for the shoreside analysis service. The total cost of lubricating the stern tube/tail end shaft *per annum* could be around $10,000 to $20,000, however, included in this equation should be the amortised cost of surveys, repairs and replacements as shown below. It is the practice to drain water contamination on a daily basis; however, the latest lube oils are biodegradable and absorb water into an emulsion that is more difficult to extract. It is common practice for classification societies to require a "remove/inspect" survey at two-year intervals. With modern antifouling hull coatings it is normal for a large container ship to dry dock every five years. It is therefore an expensive, disruptive and possibly an unnecessary event to place a vessel in dry dock for a tail shaft inspection. Classification societies will consider extending the inspection periods if sufficient reliable data is produced as evidence of satisfactory conditions. This DynaWeb demonstration provides a persuasive argument of such evidence.

The potential economic impact of the DynaWeb concept.

Based on a container vessel (8.000 TEU)

Lubrication only cost – say	$15,000 *per annum* (only stern tube)
Dry docking cost	~ $ 500.000
Repair and replacement costs	$500.000–$1.500.000
(shaft repair, pulling, steel works, class surveyor, *etc.*)	
Off-hire/demurrage costs	$10,000 up to 80,000/day
(dependent upon charter rates)	
Dislocation, fines, penalties, costs	$ inestimable
Insurance loading	$ inestimable

Precise figures are difficult to quantify and the nature of the global maritime market delivers massive variation. However, the value of a remotely managed condition-based monitoring system for this specific application is beyond question. The benefits are clearly in *magnitudes* rather than *percentages*. Similar scenarios exist in the other critical maritime lubrication applications listed in the Summary.

Figure 14.65 clearly illustrates the scale of the application being demonstrated and the direct and subsequent economics related to maritime assets can be appreciated.

Figure 14.65 A typical view of a vessel dry docked for removal of its tail-end shaft for stern tube maintenance

A secondary beneficiary of the adoption of this real-time analysis approach would be oil suppliers who currently carry the cost and investment of shoreside laboratories and extensive logistical systems that collect, ship and analyse tens of thousands of lube oil samples every year. Conservatively 90% of these samples do not need to be analysed as the oil being sampled is fit for purpose. The remaining 10% are at risk of being analysed too late for practical benefit, leaving the customers exposed to damage, failure and downtime and the oil supplier vulnerable to charges of failure to deliver a duty of care. While the transition from the traditional manual sampling and testing system to a real-time shipboard installation will take many years, the economic impact of this change on the oil supply industry will be significant. In addition, the adoption of quality monitoring technology could have the effect of democratising the marketplace by liberating customers that are presently tied to a single oil supplier due to the service requirements.

14.5.8 Conclusions

The DynaWeb demonstration delivered all of the anticipated results achieving communication with the MIMOSA database by two separate DynaWeb routes.

The application chosen for simulation is a real-life event where the Martechnic AHHOI sensor system is installed in ships similar to the one described in the above scenario. The only difference is that the sensor output feeds a locally stationed control panel with alarms and recording facility. Extended data communication, both wired and wireless, was not considered by the customer, however considering the remote aspects of the ship's operation such a facility could in the future offer an attractive option.

Based on Martechnic's direct experience in the marketplace and its participation over the past 42 months in Dynamite the company is convinced that the application of this technology will have significant impact operationally and economically in the maritime markets. It is impossible to accurately quantify this without undertaking an evaluation that would be larger and more expensive than the total Dynamite project. The marine industry by definition is a global event and cannot be easily geographically segmented. However, a ballpark assessment based on sound intuitive experience of vessels sized/powered sufficiently to qualify for the installation of such technology number between 40 and 50,000. If only 5% adopted the systems tested across the applications listed in the above summary this would represent a market of 45 million €. More significantly, the benefits that would accrue across the range of consequential activities such as repair, insurance, avoidance of outages, and the avoidance of environmental damage could be measured in the hundreds of millions.

The facility that this technology offers in the area of condition-based maintenance and distant monitoring is yet to be understood but will undoubtedly deliver fundamental change to the maintenance arena.

References

Arnaiz A, Emmanouilidis C, Iung B, Jantunen E (2006) Mobile maintenance management, Journal of International Technology and Information Management 15:11-22.12.2006

Emmanouilidis C, Katsikas S, Giordamlis C (2008) Wireless condition monitoring and maintenance management: A review and a novel application development platform, Proc 3rd World Congress on Engineering Asset Management and Intelligent Maintenance Systems Conference (WCEAM-IMS 2008) 27-30.10.2008, Beijing, China, 2030–2041, Springer, Berlin

Iung B, Levrat E, Crespo Marquez A, Erbe H (2009) Conceptual framework for e-maintenance: Illustration by e-maintenance technologies and platform. Annual Review in Control (2009), doi:10.1016/j.arcontrol.2009.05.05

Lebold M, Thurston M (2001) Open standards for condition-based maintenance and prognostic systems. In: Proc MARCON 2001-5th Annual Maintenance and Reliability Conference, Gatlinburg, USA

Levrat E, Iung B (2007) TELMA: A full e-maintenance platform. WCEAM CM 2007 2nd World congress on Engineering Asset Management, Harrogate UK

Levrat E, Iung B, Crespo Marquez A (2008) e-maintenance: review and conceptual framework. Production Planning and Control, 19:408–429

Muller A, Suhner MC, Iung B (2008) Proactive maintenance for industrial system operation based on a formalised prognosis process. Reliability Engineering and System Safety 93: 234–253

Voisin A, Levrat E, Cocheteux P, Iung B, (2008) Generic prognosis model for proactive maintenance decision support: application to pre-industrial e-maintenance test bed. Journal of Intelligent Manufacturing Special Issue on Intelligent and Cooperative Maintenance in Manufacturing Systems, available on line; DOI 10.1007/s10845-008-0196

Chapter 15
E-training in Maintenance

Christos Emmanouilidis, Vasilis Spais

Abstract. Although maintenance engineering has been in demand for very long time, maintenance training has yet to fully exploit the recent wave of technological advances in information and communication technologies. Besides formal education and theoretical knowledge, on-the-job training and informal education are recognised to have great importance in developing the maintenance profession. Yet, high costs and inflexible training schedules often put off professionals and organisations. As a consequence there are often gaps between existing personnel skills and those needed for the maintenance function. These gaps are further exacerbated as maintenance services are increasingly equipped with innovative technological solutions and personnel also need to get to grips with them.

The Dynamite project has implemented a vertical integration of a stream of novel technologies for maintenance operations, using wireless sensing devices, RFIDs, handheld computers and decision support tools, as well as back office computing infrastructure in order to streamline the maintenance engineering process and make maintenance data transparently available at multiple levels of operation. Desktop and web-based e-learning applications offer academics and industrialists new tools to raise maintenance-related knowledge and competence. This chapter discusses related work in the field and presents dedicated e-learning tools for e-maintenance training.

15.1 Introduction

The rapid technological advances in industrial production and manufacturing processes, as well as the ever increasing global competition are fuelling a growing demand for ensuring adequate personnel competences in the maintenance function. In an era when sustainable business operation and development is a paramount strategic goal, the human factor plays a crucial role in implementing the organisation maintenance strategy, but it can only do so insofar as personnel are adequately trained (Starr and Bevis 2009).

This chapter summarises related work in the field of e-training in maintenance. First a discussion on the need and benefits of maintenance training is provided. Next the focus shifts to current e-learning technologies and requirements for vocational training in maintenance. An outline of current application of advanced learning technologies in maintenance training is provided. The next section presents the e-learning tools offered for training in the use of the innovative technologies for maintenance, introduced by the Dynamite project. These have not been developed with the intention to complete a maintenance training curriculum but they are specialised e-courses to train users so that they can efficiently use e-maintenance technology and tools. Finally, the chapter concludes with a summary of the achieved results and a discussion on current requirements and challenges for training in e-maintenance.

15.2 The Need for Maintenance E-training

Modern enterprises cannot afford to under-utilise their assets, being material or human. As they increasingly need to rationalise their function in a way that reduces costs, increase availability and overall equipment efficiency and enhance safety and quality procedures, the efficient lifecycle management of the engineering assets becomes a key factor to support sustainable operation. This implies that personnel have the necessary skills to perform their intended function. Yet companies and professionals do not share common criteria for the required maintenance-related competences and that has detrimental effect on the capacity of an enterprise to efficiently implement a chosen maintenance strategy. At the same time, this lack of clarity in competence requirements affects personnel mobility and employability.

Substantial work is currently under way in an effort to bring maintenance vocational education and training (VET) training in line with competence requirements for the maintenance function (Franlund 2008, Roe 2003). In Europe the EFNMS has specified competence requirements both for maintenance management, as well as for maintenance technician specialists (Franlund 2008). As engineering assets are of varying nature, form and function, the need for specialised training in specific aspects of the maintenance function is also often highlighted (Starr and Bevis 2009). In particular, the need for the certification in more targeted maintenance-related topics, such as condition monitoring, has led to the drafting of dedicated requirement specifications (Roe 2003) leading to their standardisation (ISO 18436-1:2004).

There is a clear need to establish accredited maintenance-related qualifications as that would be of great benefit to the market, both from the employee as well as from the employer perspective. Employees would enjoy easier mobility and more transparent recognition of their skills, making them employable in wider markets. Organisations would be made more confident to invest in qualified personnel and in that way to more efficiently implement their maintenance policies. On the basis of this need, cross-continent discussions are currently under way to prepare the ground for a global framework for maintenance competences requirements specification. Furthermore, recent surveys of European industry have revealed a clear gap between available competences and required skills and have identified areas that a real improvement in maintenance training can be made (Bakouros and Panagiotidou 2008, Emmanouilidis *et al.* 2009).

With the debate on formalising maintenance competences going on in earnest, there is also an interest in how training and certification should be pursued. Maintenance training is not regularly part of formal education, with the exception of some dedicated courses (Macchi and Ierache 2009). Instead, it is mostly included in vocational education and training curricula. Trainees are often individuals who have already entered their working life and in most cases cannot take a maintenance training course under pressing time constraints.

To mitigate such pressures, while avoiding the high costs of on-the job training, e-training is considered well suited to the specific needs for enhancing maintenance employable skills and competences (Emmanouilidis *et al.* 2008). Although the cost of developing the e-training solution is higher than that of conventional training, the costs associated with running the training, the flexibility offered to the trainees and the fact that e-learning is by design an interactive and engaging training experience, makes it appropriate for maintenance management training.

Considering the potential impact of e-learning on maintenance-related training, a recent survey of 70 professionals involved in the maintenance function in five EU countries (UK, Sweden, Greece, Latvia and Romania) has found that the likely adoption prospects of e-training in maintenance are quite positive (Papathanassiou and Emmanouilidis 2009). It is worth looking at the computer literacy of the interviewees. The majority (94.5%) of interviewees uses computers on a daily basis and many believe themselves to be "very much" familiar with computers (81.8%). Furthermore, 100% of the interviewees responded that they expect to benefit "much" (40%) or "too much" (60%) from a computer-based automated learning platform. There were no negative or indifferent replies on this question. These responses bond well with potential future use of e-training in maintenance.

The next sections discuss current practice in advanced learning technologies that are applicable to maintenance training.

15.3 E-learning Technologies

E-learning has redefined the way education is provided in schools, academia and industry. It is defined as a technological, organisational and management system that enables and facilitates web-based learning. E-learning users, both teachers/trainers and students/learners are offered integrated solutions that facilitate authoring, structuring and delivering educational content, as well as assessing the educational outcome. Such solutions are termed learning management systems (LMS). Most current LMS include functionality to handle lesson content, often in the form of learning objects (LO). Any entity that can be used, reused or referenced in e-learning is called an LO. These systems are more accurately called learning content management systems (LCMS) but both terms are used interchangeably; thus the term LMS will be employed for both LMS and LCMS in this context.

A number of recent studies indicate that e-learning has an exponential impact on learning practice in every level or aspect of education (Huddlestone and Pike 2007, Jones and O'Shea 2004). The number of universities, tutors and professional training providers including e-learning in their curricula demonstrates the maturity that e-learning has reached. There are many obvious advantages of web-based learning, compared with conventional training. Typically, e-learning enables anyone with authorised access, anytime and anywhere to participate in the learning process (Papathanassiou and Emmanouilidis 2009). Especially in ages over compulsory education and in VET teaching, mitigation of time and space limitations is an indisputable benefit. Apart from this flexibility, e-learning is usually associated with lower running costs, compared to engaging a qualified teacher (Ma 2005). On the other hand, e-learning involves extra costs in producing the actual educational content, as well as in customising it for the e-learning environment. A further significant advantage of e-learning is that the whole training and assessment process is computerised. Thus, e-learning lends itself for more transparent, streamlined and standardised knowledge delivery and competence assessment, both sought at a premium in modern industry.

Nevertheless, just presenting information over the web does not provide education and is not considered e-learning in the real sense. A robust and usable way of communication between the teacher and the students and between students themselves is necessary (Papathanassiou and Emmanouilidis 2009). Conferencing, mailing, bulletin boards, chat and forum are among the additional features that are typically integrated with the learning environment and content. Furthermore, e-learning systems greatly benefit by adopting artificial intelligence theories and cognitive science practices, in order to approach and mimic the human tutor. The learning outcome must also be assessed and adequate competence assessment tools and content linked with the learning content and the targeted learner groups are needed for that.

Moving beyond simple presentation of information relevant to the learning subject, the e-learning experience needs to be enhanced to engage the learner groups in a stimulating and practical learning process. Irrespective of the specific conditions of use of any developed learning material, its added value is increased if the learning content can be employed in different systems or varying contexts. Therefore, the re-usability of the learning content should not be ignored. With the internet's ubiquity and the plethora of the offered learning solutions, being proprietary or open-source, re-usability remains an open issue.

15.3.1 Adaptive Learning

A key challenge is to develop flexible e-learning systems, which can be adapted and deployed to serve diverse communities of teacher/learner groups and usage environments. Intelligent tutoring systems (ITSs) have been around for several years, claiming to possess the ability to present the teaching material adapted specifically for each user (Woolf 2009). The main focus is on the development of web-based intelligent learning environments (WILEs). Such an environment can keep track of students' current knowledge on the subject matter by storing it in a student model database. In doing so, the learning environment can adapt the presented content according to the student learning needs, by keeping track of the individual learner's learning path. Typically, the learning content is stored in a standardised format and can be dynamic in nature (Kazi 2004). The benefit of doing so is that the developed content can be ported to another learning environment, which follows the same standard in handling learning objects.

An ITS is typically built around three main concepts: how to handle expert knowledge, learner knowledge and educational principles. The model tracks the learner progression and responses, compares his knowledge with the expert knowledge and employs some form of intelligent reasoning to dynamically generate the best sequence of instructions, learning content and tests for the learner (Karampiperis and Sampson 2006, Hatzilygeroudis *et al.* 2005).

Much of the adaptation capacity of learning management solutions, draw inspiration from an understanding of the way humans learn. Two key principles involved in such a process are (Papathanassiou and Emmanouilidis 2009):

- Learners should be motivated and actively involved in learning.
- People learn in different ways and at a different pace; learning content delivery adapted to their learning styles leads to more efficient learning.

Reflecting on the way humans learn, three main learning theories influence e-learning (Woolf 2009):

- Behaviourism that treats learning as a set of changes on the learner, as he responds to environmental events. The key concept is learning by demonstration,

memorising and imitation. The computer or the teacher acts as the source of knowledge.
- Cognitive science that stipulates that learning is influenced by unobservable internal constructs, such as memory, motivation, perception, attention and meta-cognition skills, thus learning must be based on individual learning needs. The computational focus is on screen design and on human–computer interaction, where the teacher usually has the role of the facilitator or partner.
- Constructivism that claims that learners construct their knowledge as they react with and interpret their environment. The most important differentiation from other theories is that focus is on the learner and his actions and not the teacher or the teaching methods. Thus, the aim is to provide stimuli and support for the users to construct their knowledge.

Personalisation is a key adaptation element in e-learning. It refers to the ability of an e-learning system to tailor the delivery of training content to each individual trainee, according to specified criteria (Woolf 2009). A learning system that supports personalisation should be able to identify learner training needs, based on a specified or inferred learner profile, as well as on trainee-LMS interaction. Based on these identified needs, LMS should adapt training delivery from a multitude of redundant resources to offer what is best suited to the learner. In general, adaptation is the ability of an e-learning system to change the delivered training material, both at navigational and content levels, so that it best fits each user's needs (Dolog *et al.* 2004).

Four prime types of adaptation can be identified (Popescu *et al.* 2007):

- Navigation adaptation, by customising links by generating, hiding, annotating and ranking them.
- Content adaptation, by customising content by hiding or providing extra or different versions of content.
- Presentation adaptation that changes the way content is delivered, by highlighting, adding, removing or sorting parts of it.
- Collaboration adaptation that bases customisation on collaboration preferences between the learners and supports cooperative problem solving.

Personalisation and adaptation base their efficiency on correct identification of learners' learning needs (Papathanassiou and Emmanouilidis 2009). This identification is closely intertwined with the concept of *learner profiles*. There is no unique definition of what a learner profile constitutes but it largely comprises all the important information about a learner in one convenient and searchable location. This information usually includes his preferences, goals, previous knowledge about a subject, general knowledge, achievements, performance, and everything else that could be useful for an automated system or a tutor to decide the appropriate learning material for a learner (Dolog *et al.* 2004). Another research area on improving e-learning experience deals with learning styles. The identification of learning styles is based on cognitive theoretical approaches that attempt to classify users into different categories depending on the most efficient way for them to learn. In this way, a learning style can be defined as a set of characteristic cognitive, affective and physiological factors that indicate how a person learns and interacts in a learning environment (Popescu *et al.* 2007). Categorisation of users according to their learning styles is an efficient way to implement customisation without an unmanageable number of different parameters for each user.

15.3.2 Learning Objects, Standards and Interoperability

The rapid progress in e-learning technologies has lead to the recognition of the need to draft, promote, adopt and implement e-learning standards (Emmanouilidis *et al.* 2008). The motivation for this is that these standards will facilitate e-learning to reach some key goals (Varlamis and Apostolakis 2006):

- *User flexibility:* users will be able to work with different e-learning platforms, following the same standards.
- *Content development flexibility:* content providers will focus more on content production, rather than application development
- *E-learning development flexibility:* vendors will focus more on developing interfaces linking their tools to learning management system platforms, rather developing original complete learning toolkits.
- *Platform development flexibility:* LMS developers will focus on facilitating re-using learning content of standardised format, rather than developing content in new format.

The key expected benefits are (Varlamis and Apostolakis 2006):

- *Interoperability:* easier integration of content from a multitude of sources and providers, with adequate handling of translation and information exchange.
- *Re-usability:* content can be easily reused in different context than the one originally intended for.
- *Manageability:* easier to handle learning process information (profile of learners, history track record, educational target, *etc.*).
- *Accessibility:* easier access to learning content by users on different devices.
- *Durability:* content can be used over long time, irrespective of the original learning platform that was intended for, minimising obsolescence risks.
- *Scalability:* facilitating expansion of e-learning applications and systems to serve wider target groups or organisations.

Achieving goals and benefits such as the above is difficult and costly in the absence of e-learning standards. The IEEE Learning Standards committee has produced LOM (Learning Object Metadata, IEEE 1484.12.1). LOM has been adopted by IMS Global Learning Consortium, the Advanced Distributed Learning initiative (ADL), the Alliance of Remote Instructional and Distribution Networks for

Europe and many other organisations and has become the leading standard for describing learning objects. A milestone in the direction of adopting common standards was 1997, when ADL compiled the first SCORM (sharable content object reference model) component specification based on previously proposed standards, leading up to the 2004 specification (Dodds and Thropp 2006a). Other metadata standards following the Dublin ore metadata initiative or more specialised educational activities to connect high-level pedagogical methodologies with low-level machine interpretable descriptions can also be employed. SCORM in particular has incorporated the best parts of pre-existing e-learning standard groups, like IMS, AICC, ARIADNE and IEEE-LTSC into a new model. In SCORM, sharable content objects (SCO) are defined as the smallest logical, instruction units and considered as the building blocks for learning content. The most recent SCORM version is the 2004 4th edition, available since April 2009. SCORM compliance is intended to lead to fully independent, interoperable LO, which will contain all the information necessary to be deployed in different setups.

Learning objects can be developed for any type of learning curricula or learning objectives but the way this is done need not follow different specifications. The educational technology standards have been developed to define the way educational content is being packaged, sequenced and delivered. Sharing and reusing existing learning objects can reduce costs and speed up the process of course content construction. This reuse is ensured if the learning objects themselves follow specified educational technology standards. If such interoperability is ensured, the bulk of the effort will then need to be devoted to the quality of the implementation and content presentation. Educational content consumers would be able to choose a product that would suit their needs without the fear of sudden obsolescence that is so common in proprietary software solutions. It is through the definition and adoption of such standards that e-learning practitioners can exploit existing information and communication technologies to deliver e-learning modules that can enhance the learning progress, ease the delivery of learning services and streamline the development, delivery and assessment of training.

E-learning interoperability standards deal with the following issues (Collier and Robson 2002):

Metadata definition. Metadata is a convenient way to label each learning object. In this way, it is possible to index, store, discover and retrieve learning objects from multiple educational repositories by employing different tools. Information stored about learning objects is called learning object metadata. Different metadata standards can be employed with the IEEE LOM and its derivative SCORM being the most frequently referenced ones. The LMS learning design team should be concerned with the metadata definition choices.

Content packaging. Content packaging standards deal with how learning objects are stored and delivered to the users. They ensure that learning objects and packages thereof (learning units) will be reusable and editable in any conformant software tool or environment. Content packaging standards are included in:

- The IMS Content Packaging Specification
- The IMS Simple Sequencing Specification
- The IMS Question and Test Interoperability specification (QTI), which is a set of specifications governing assessments and their questions
- The Aviation Industry CBT Committee
- The SCORM Reference Model
- The IEEE Learning Technology Standards committee.

Learner profiles. Learner profiles definitions allow sharing of information about learners. A learner's profile can include personal data, learning plans, learning history, degrees and current e-learning status. Relevant to learner profiles standards are:

- The IMS learner information package (LIP) specification
- The personal and private information (PAPI) specification.

Learner registration. Learner registration contains information about the selection of which learning content and administration components should be delivered to the users. Relevant initiatives include:

- Specification for exchanging offering and registration data among learning systems by the IMS Enterprise working group
- Specification for exchanging the above data in K-12 environment by the Schools Interoperability Framework.

Content communication. Content communication refers to standards about how learners data are linked with the content and the learning activities. Single assessment question answers, course grades and completion status are all included in here. Existing standards include:

- Communication component in computer managed instruction (CMI) by the Aviation Industry CBT committee
- Communication JavaScript API in SCORM

SCORM

A discussion on the main characteristics of the SCORM model, which is the most widely employed educational technology standard, is included here. SCORM was established in 1997 as an ADL initiative. It was developed with the view to incorporate the best parts of the existing e-learning standard groups, like IMS, AICC, ARIADNE and IEEE-LTSC and provide a new, more complete model (Kazi 2004). These standards regulate aspects like metadata, content tracking and content sequencing. SCORM defines sharable content objects as the smallest logical units of instruction and considers them as the smallest building blocks for content. SCOs contain assets packaged for a learning context and they are delivered through a SCORM-conformant runtime environment. Any type of educational digital media, such as text, video, sounds or any form deliverable through the web can be characterised as educational asset (Kazi 2004).

The most recent SCORM version is the 2004 3rd Edition, available since 2006. It is described in three separate volumes; their content is briefly described as follows:

- *SCORM content aggregation model* (CAM). The SCORM content aggregation model contains information on metadata and content packages. The main use of metadata is for searching and discovering learning objects. The content package is a collection of content objects, about a course, a module, a lesson or even a generic collection of related learning objects. In every content package, an imsmanifest.xml file describes the contents of the package and sometimes also the package structure. Information about how an LMS should handle a content package may also exist (Dodds and Thropp 2006a).
- *SCORM sequencing and navigation* (SN). This describes all the requirements of an LMS, which enable it to sequence content objects at run-time. Moreover, it contains information on how the content object should accept and handle user navigation requests. Sequencing and navigation is in turn defined by user choices and achievements at run-time. It comprises sequencing terminology, navigation requirements and models for navigation model, sequencing definition and sequencing behaviour (Dodds and Panar 2006). It deals with the way to respond to the learner choices and activities, so as to define which content object will be delivered next.
- *SCORM run-time environment* (RTE). The actual delivery of the content object is the subject of the SCORM run-time environment. The SCORM RTE defines what an LMS must do in order to be able to deliver SCORM content. SCORM content is divided into two distinct categories SCOs, which are objects with the ability to communicate during run-time and assets that are content objects that are not capable to communicate. Based on these, the way of launching content objects is defined, as well as the way to establish communication between SCO and LMS. RTE also presents a model to track user interaction with the content objects (Dodds and Thropp, 2006b, Qu and Nejdl 2001).

Based on the above, SCORM combines metadata, content packaging and sequencing and navigation to aggregate SCO. Several LMS currently exist, many of them providing SCORM compliance support. A brief survey of the key features of LMS follows.

When considering LMS conformance to SCORM and ease of use, it seems that there is still a conflict between these two. SCORM conformance demands a set of extra variables and descriptions to be used for each learning object. Furthermore, as SCORM appears to be neutral from an educational perspective, it has attracted criticism that it constitutes a technical standard with limited relevance to education (Friesen 2004). Furthermore, it has been suggested that SCORM is primarily concerned with learning content reusability and tracking learners' progress through a learning management system. It does so by working within a single-learner instructive learning approach, that cannot easily exploit other educational elements, which are prevalent in a constructive learning framework, such as those related to the capacity of trainees to learn by building on their own learning process through iterative work with the LMS and system interaction cycles.

The single-learner centric approach of the SCORM philosophy largely ignores collaborative learning schemes and tools (Witthaus 2009). In particular, this criticism is directed not only at SCORM but also at the learning object approaches, stating that they are based on a software engineering philosophy more in line with object-oriented programming, thus the term 'object'. On the other hand, more recently, software engineering has shifted towards the service-oriented architecture (SOA) approach (Gonzalez *et al.* 2009). In line with this, several recent e-learning R&D efforts have sought to establish e-learning interoperability by establishing accessibility *via* web services, rather than relying on standardised learning objects. This stream of research is blended with work on Web 2.0 technologies emphasising more the social aspects of learning and employing rich interactive and collaborative tools and technologies, under the term 'learning 2.0' (Downs 2005), coined to draw attention to the Web 2.0 features (Gütl 2008). As stated in the HELIOS project final report, this trend involves the following:

- Learners create content, collaborate with peers through mechanisms such as blogs, wikis, threaded discussions, RSS and others means to form learning.
- The learning experiences are learner-centred, taking advantage of many sources of content aggregated together into learning experiences.
- Teachers (if any) and learners (students) are peers within social networking environment.
- Learning experiences are increasingly featured by knowledge management, collaboration and search.
- We are moving "from communities of practice to social-networking".
- Finally, there is shift from traditional learning applications and systems managing learning objects within a pre-defined learning architecture to an open learning environment composed of interoperable loosely coupled open-source platforms and tools aimed to support the social interactions of peers.

15.3.3 Learning Management Systems

There are several LMS surveys in the literature. When comparing LMS, there is no single set of criteria to focus on and therefore it is easy to end up with subjective results. Each comparison puts weight on specific factors, directly relative to the context and time that the LMS were tested. The timing of performing the tests can also make a difference as typically LMS releases are becoming available at different frequencies; so a survey can offer a snapshot of LMS features at a specific time. Comparisons have been made against a single criterion, such as adaptivity (Graf and List 2005), functional assessment (Botturi 2004) or SCORM conformance (García and Jorge 2006) or against multiple criteria (Kljun *et al.* 2007, Itmazi and Gea 2006). Evaluations consider both the viewpoint of those concerned with producing an LMS solution for a specific application, as well as for specific targeted teacher and learner user groups. The LMS developers are concerned with the LMS offered features for customising, extending, deploying, upgrading, or content migration from one platform to another. Users on the other hand are more concerned with the offered features to support the learning process, as well as with the system usability. Here we summarise the main factors considered when reviewing LMS and we focus on the features offered by the best known among them.

- *Adaptability.* This refers to the ability of LMS to be modified according to the needs of each installation case.
- *Affordability.* This is broadly related to the return of investment for each LMS solution. Specifically, it examines for each LMS case, cost and time required and the benefits of its use, like improved productivity and efficiency.
- *Interoperability.* The ability of an LMS to be independent of specific software tools and platforms.
- *Reusability.* The ability of LMS to accept and use learning objects from different LMS platforms. This will have also a direct impact on the cost of application of each LMS.
- *Durability.* For how long an LMS will be useful without the need for redesign or a complete change.
- *Accessibility.* This examines the ease to find and use learning objects from anywhere.

The above factors are all among SCORM objectives and are all important from the LMS developer's point of view. It therefore makes sense to examine existing LMS for SCORM-conformance (García and Jorge 2006). Nonetheless, employing e-learning is ultimately about serving user needs and it is therefore important to focus on whether the LMS offers enhanced teaching and learning experience to the teacher and learner user groups. What is important from the user perspective is to have a number of learning-related features present in the LMS and offered in a user-friendly manner.

- *Usability.* This deals with difficulties involved in the installation and consequent use of an LMS. Intuitive environment, integrated authoring tools and visual enhancements are all constituent parts of a user-friendly platform.
- *Features.* Every LMS presents a series of features that can be used in the development and delivery of courses. The extent to which an LMS offers such features can by itself be an important factor for the adoption of an LMS platform.

A major issue when considering adopting a specific LMS platform is whether this is a *proprietary* software tool or an *open source* solution. Many organisations and practitioners would opt for a commercially supported solution justifying their choice by the availability of commercially-driven support in developing and deploying the e-learning solution. Others would prefer to avoid the costs associated with a proprietary ('licensed') solution and work with open source platforms. The latter are distributed together with their source code, making it possible to study the code, run the program for any purpose, customise it according to specified needs and have a much more flexible distribution policy. Open source software is often associated with non-professional, research-only efforts, not applicable for professional deployment. Although this may appear to hold in several cases, there are also examples of open-source software employed for professional use. This is especially true for e-learning, where open source solutions are known to offer (Coppola and Neelley 2004):

- Fast software development due to the parallel work of many programming groups *versus* a single team in proprietary software. The collective work may lead to satisfying the user community needs in shorter time.
- Wide basis for testing as new versions, as test are performed by a multitude of users on much more environments and platforms than any proprietary software.
- Frequent release of new versions and immediate testing of them by the community of the involved users.
- Peer reviewing, testing and improving the code can lead to enhanced transparency, security and better quality in code, as all the weak points or bugs will be most definitely pointed out by the user community.

Among the various LMS solutions, Blackboard/WebCT and Moodle are among the most popular, with SAKAI also enjoying a large user community. Blackboard is perhaps the most widespread commercial e-learning solution, with SumTotal, GeoLearning, Oracle/PeopleSoft, Plateau and Saba being LMS platforms with corporate customer base. Some other commercial e-learning platforms are actually live networking/conferencing software extensions, offering synchronous learning, such as WebEx, Centra, Live Meeting, Interwise and Acrobat Connect Professional (formerly Macromedia Breeze Live). WebCT, acquired by Blackboard, was originally developed at the University of British Columbia in 1995 and it is considered as the first virtual learning environment with commercial success. The SAKAI open source collaboration and learning environment, encapsulates many useful features: SCORM support, shared displays and whiteboard, blogs, *etc.* Moodle is another open source content management system (CMS) solution, with a very large active supporting community and several features making it a collaborative environment for teacher and learner communities.

Between WebCT/Blackboard and Moodle there are many common features. Moodle provides more support for different course formats and grading systems, discussion fora and wiki, while WebCT offers a whiteboard tool enabling learners to share drawings in real time. Both include several aspects of SCORM functionality. Moodle supports SCORM packages in courses and collaborative creation of course content. Moodle is both open source and popular and it constitutes a strong

LMS choice for developing e-learning for e-maintenance. Before presenting the e-maintenance training development, the next section takes a closer look at Moodle.

15.3.4 Moodle LMS

Moodle is an acronym for modular object-oriented dynamic learning environment. Moodle is an open-source learning management system. It offers the typical functionality expected by an LMS and it further aims at stimulating users to explore its pages, interact with the learning material and communicate with the teachers and other learners. Moodle was considered as the chosen platform for developing e-learning for e-maintenance for several reasons (Papathanassiou and Emmanouilidis 2009):

- Moodle has been built on top of a relatively new educational theory, called *social constructivism*. While most other LMS have been designed around technological rather than educational concepts, Moodle has pedagogy in its core (Cole 2008). Constructivism claims that we learn more efficiently when we construct knowledge artefacts for others. Social denotes that this construction of knowledge is even more efficient when it is performed in a collaborative way. Widespread acceptance and use by a large and growing community is evidence of the perception that this has been a very successful LMS project.
- Learners are *free to navigate* through the material and choose the courses that they need. The learning environment can guide them by providing feedback and making recommendations on the courses that may be necessary for them to take. This, in conjunction with a learner enrolment system, gives the course administrator or the tutor adequate control on which learners are attending specific courses.
- *Open-source.* The fact that Moodle is an open source project results in exceptional adaptability and transferability (Emmanouilidis *et al.* 2008). This is a significant advantage and enables easy expansion, amendment and porting of courses to different environments. A system built on proprietary software is always bound to the decisions of the LMS software vendor.
- *Huge user base.* Registered Moodle-based sites and users have grown exponentially, thus indicating a platform with positive outlook. It also designates the existence of an active support community. Furthermore it ensures multilingual support, which is necessary for the wider reach of the training. Moodle already includes language packs for more than 70 languages.
- *Minimal requirements.* Moodle has very limited resource requirements, thus supporting transferability, as lessons can be delivered to platform-independent computers with few hardware restrictions.

Moodle classifies all learning content into two large and distinct categories (Rice 2006):

- *Resources.* These are the static material that learners can read or attend but cannot interact with. They include labels, Web pages, text pages, file directories and links to files.
- *Activities.* This is the interactive course material, where learners can answer questions, upload files and communicate. They include assignments, choices, journal, lessons, quizzes, surveys, wikis, workshops, chats, forums and glossaries. Some of the above further encourage collaboration between users, such as forums, wikis and chats.

Course construction is relatively simple, enabling easy construction of taught curricula. Overall, Moodle provides a highly customisable environment. This customisation is based on self contained, customisable entities called blocks. Blocks can be enabled, disabled and moved around the user screen by an administrator. Most commonly used standard provided blocks are activities, administration, calendar, course/site description, courses, latest news, main menu, online users, people, recent activity, site administration and search forums (Rice 2006). It is worth mentioning here that in spite of the fact that Moodle basic installation contains a full set of blocks, there is a vast amount of additional blocks provided by other members of the open-source Moodle community.

Learners' track records and performance data can also be stored and handled by authorised users. This can be done by handling detailed activity logs. Moodle keeps analytic logs for every type of action that occurs inside the LMS. The track data related to a particular user can shape up an individual learning profile. While it is possible to define a learning path to a user, social constructivism indicates that it is preferable to provide feedback and guidance, thus influence, rather than force the learning pathways.

Moodle manages users by defining user roles. Every user is member of one or more roles. Roles can be defined dynamically and determine exactly the rights of the user inside every piece of the environment. There are a number of predefined roles to facilitate quick setup of courses, like administrator, teacher, course creator and learner. Furthermore, administrators can define new roles with more refined rights. It is worth noting that these roles are completely dynamic, meaning that a learner in one course can be teacher to another, or have the right to create courses for a topic he is well aware of, if that is considered suitable.

Regarding the interoperability support, course content in Moodle is pure HTML, which means that core material is kept in a form completely independent of hardware and software limitations. However, other pieces of information, such as user personal data and qualifications can be stored inside the environment's database. From the e-learning perspective, an important consideration is conformance to learning object standards, such as SCORM. Partial SCORM conformance is already available but full compliance with SCORM is a primary goal for Moodle version 2. Nonetheless, it should be mentioned that there is an open debate at the moment as to the extent that SCORM is in line with social constructivism educational theories, as it is more in line with instructional educational practices. On the other hand, the HTML nature of content delivery makes Moodle content universally available, thus ensuring platform interoperability.

Before turning our attention to the LMS implementation for e-maintenance training, we first briefly discuss advanced learning technologies that are also applicable to maintenance training.

15.3.5 Advanced Learning Technologies

One significant R&D direction for e-learning has been the creation of larger scale educational environment infrastructures and content repositories. GRID technology and services are enabling this shift, but there are two major hindering factors in the procedure (Pairot *et al.* 2005): first the creation of educational hierarchies and organised directory services, so as to make it possible to effectively search for information, based on metadata and second the content standardisation itself. Grid educational infrastructure and technology can essentially support collaborative construction of learning objects with the involvement of many disparate experts, teachers and learners (Rosatelli *et al.* 2006, Yang and Ho 2005).

A different stream of work is related to the penetration of mobile and wireless technologies, which have opened up new opportunities for educational technology in the form of mobile learning (m-learning) (Emmanouilidis *et al.* 2008). While e-learning enabled the delivery of web-based training to remote desktop users, m-learning extends this by offering ubiquitous services to mobile users. In this way it becomes possible to deliver educational content anytime, anywhere and to anyone authorised to access it. Although m-learning can be useful in school and academic education (Mifsud and Marc 2007, Cobcroft 2006), it is lifelong learning that is likely to benefit most from it (Sharples 2000). Indeed a key driving factor for lifelong learning is the fact that school and academic education cannot possibly equip learners with all the skills needed for a professional carrier. On the other hand, lifelong learning participants need to follow a self-paced learning process depended on their available time. Through the use of mobile devices in industrial settings, users can gain access to training material from multiple locations, being at home, office or even at the shop floor, next to the machinery and production processes. In this way, the gap between text and theoretical knowledge and practice is narrowed and the users may engage in a problem-based learning process, actually taking place next the operating environment. The immediacy and ubiquity of the learning process may enhance the learning experience and positively influence the learning outcome through a learn-by-example educational program.

Maintenance training can be made more efficient if accompanied by case studies in a form of on-the-job training. In several domains, such as in aircraft maintenance to name one, this is prohibitively expensive. Augmented reality (AR) technology offers the means to provide problem-based maintenance training, without the costs associated with going through the real case study. AR combines real-world objects with computer-generated data, as opposed to virtual reality (VR) that deals only with computer generated environment. Based on sensors and motion detectors, AR can provide object recognition and learner's motion tracking. Head mounted displays (HMDs), cameras and special clothes usually co-exist in an AR-based training environment (Nakajima and Itho 2003). Industrial maintenance (Li *et al.* 2003), power systems maintenance (Nakajima and Itho 2003), aerospace maintenance (Christian 2007, Haritos and Macchiarella 2005) and medical training (Székely and Satava 1999) are just a few of the sectors involved that are greatly benefited by augmented reality in their teaching procedures.

15.3.6 Vocational Training in Maintenance

Developing maintenance vocational education and training (VET) is of critical importance for industry in pursue to establish adequate maintenance practice, exercised by qualified maintenance personnel. Although focusing on VET and creating a high skilled personnel is a matter of great importance, its implementation relies on public and private funding. Focusing on the private sector, larger companies have established internally funded vocational training policies for most of their employees. The situation is different for SMEs, who are struggling to offer such opportunities and when they do, this is often through participation in public-financed training programs. Therefore, attracting smaller enterprises to VET is a difficult task. Moreover, employees should be persuaded to get involved in VET courses. This group of stakeholders may be unwilling to consume time on VET programmes and as a result it should be informed and persuaded about maintenance training benefits.

Current advances in e-learning and advanced learning technologies can be of great benefit to the delivery of maintenance training. In designing an efficient LMS for that purpose, several questions need to be addressed:

- How will training in the LMS be in line with the general EU recommendations on VET and the specific EFNMS (European Federation of National Maintenance Societies) recommendations on maintenance competences?
- What sorts of maintenance qualifications are envisaged?
- How will the LMS provide and accommodate the transfer of learning outcomes?
- What type of learning paths will be crafted and who will assure that the competence assessment tool keeps step with the recommendations?
- How will maintenance qualifications be divided into units of learning outcomes?

- What is the best way of connecting each maintenance lesson with its expected learning outcomes?
- Would it be recommended to define units, attach credit points to units and connect them to a European Maintenance Quality Framework level?
- What knowledge, skills and competences will constitute the learning outcomes of each unit?

Any maintenance VET training must be quality-assured or validated. According to the recommendations of the Technical Working Group on quality, this should be defined as a context-dependent term. A validated maintenance training system should ideally be accompanied by a carefully designed credit transfer mechanism. Such a mechanism would provide a way of measuring and comparing learning achievements and transferring them from one institution to another, using credits validated in training programmes (Tissot 2004). With this in place, it would become possible to transfer maintenance learning achievements between different countries, organisations and educational systems. A maintenance credit transfer system should support the transparency of proceedings and the comparability of learning outcomes. It should ultimately facilitate the mobility of learners and their qualifications. The availability of e-training systems for maintenance can facilitate formalising qualifications mobility and in particular:

- the transfer of learning outcomes between national systems;
- the transfer of learning outcomes between formal, non-formal and informal pathways;
- the accumulation of learning outcomes;
- transparency of processes;
- mobility of people between countries, professional levels and learning pathways; and
- mutual trust and cooperation in the area of maintenance education and training.

The above discussion highlights important considerations that need to be taken into account when designing an LMS-based training system for maintenance training and competence assessment. However, maintenance training needs are related to both generic maintenance knowledge and skills, as well as to dedicated and focused maintenance-related subjects (Starr and Bevis 2009). For the generic maintenance competences the above matters are important to be addressed to some extent. On the other hand, more focused maintenance-related e-training, such as training for e-maintenance need not follow global design criteria or be included within standard maintenance curricula. Instead, e-training for e-maintenance can be offered as separate and additional maintenance training. The next section provides a brief overview of such a development within the context of the Dynamite e-maintenance architecture. As this architecture comprises technological elements and modules that are likely unfamiliar to a significant proportion of potential end users, designing, developing and offering e-training to familiarise intended user groups with the developed technology and tools was considered as an important additional tool that should be offered.

15.4 E-training for E-maintenance

The execution of maintenance engineering and management tasks require that certain skills need to be possessed by the personnel carrying out the tasks. The nature of the competences needed is in many cases multidisciplinary and many organisations find it hard to define the exact professional requirements and skills needed to efficiently plan and execute the maintenance functions. Recently, the European Federation of National Maintenance Societies has adopted competence requirement specifications for maintenance management, for maintenance supervision and for maintenance technician specialists. These EFNMS specifications have consequently been employed by CEN to draft the Technical Report (TR 15628) on Qualification of Maintenance Personnel (Franlund 2008).

The Dynamite project advocates the use of innovative ICT technologies to manage and execute a range of maintenance-related actions in an integrated, efficient and seamless manner under an e-maintenance framework. A range of innovative enabling technologies are employed, including:

- smart sensing devices;
- wireless communications;
- portable computing devices; and
- web-based e-maintenance services (DynaWeb).

Such technologies are poised to make headways into the maintenance engineering practice. Yet they are not in most cases covered by employed maintenance educational programs. In order to secure acceptance and adoption in industry, e-maintenance technologies need to be incorporated into dedicated maintenance training curricula.

15.4.1 Dynamite E-training: the DynaTrain Platform

Incorporating e-maintenance technologies into maintenance training can benefit much from employing innovative technologies for delivering customised training. The very nature of the ICT technologies employed in e-maintenance lends itself to the implementation of e-learning technologies to deliver the needed training. The Dynamite project sought to exploit such opportunities by employing the popular open source learning management system development platform Moodle, in order to design, implement and deliver e-maintenance training. The Dynamite e-learning platform, DynaTrain has been designed and implemented to deliver training dedicated to the project employed technology and practice.

The training content includes basic knowledge for developing skills to efficiently plan and execute the Dynamite-implemented e-maintenance functions. The training content involves both underlying theoretical background as well as practical examples in the form of 'how to' cases to help personnel carry out certain tasks. The training is divided in training sections, which are segments of educational content that can be studied by viewing a single web page, without or with limited scrolling involved.

The next sections provide DynaTrain examples of e-maintenance training, namely:

- vibration sensing, by Diagnostic Solutions Ltd (UK);
- data acquisition, by Wyselec Oy (Finland);
- PDA and RFID inventory tracking, by the University of Sunderland (UK);
- prognosis web services, by University Henri Poincaré (France); and
- MIMOSA translator, by University Henri Poincaré (France).

15.4.2 Vibration Sensing

This course provides theoretical and practical knowledge and step by step instructions on how to install the device and its software, as well as on how to perform vibration measurement with a USB vibration sensing device. It also includes training content for analysing the vibration data and linking them to common faults, employing the provided software. The course also includes tips on unit conversion and multimedia demo support. The front page with the course outline is seen in Figure 15.1.

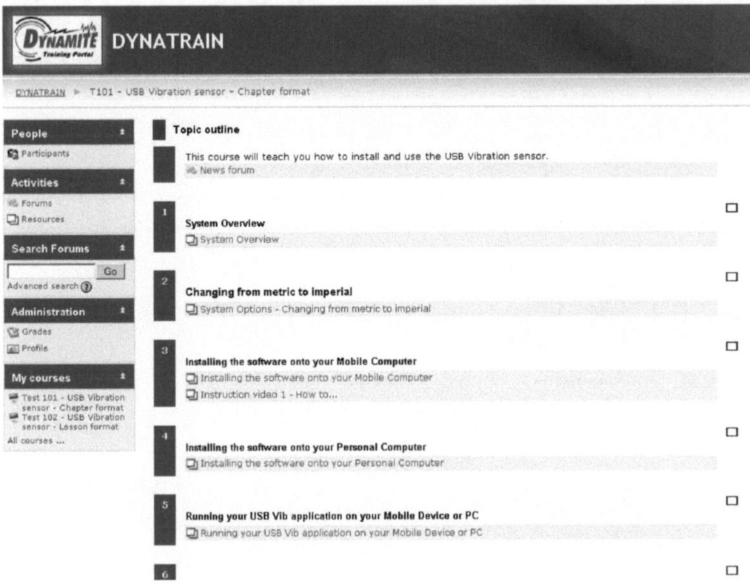

Figure 15.1 Overview of vibration sensor usage training course

The user can control by clicking on a check box on the right-hand side of the course topics to see all sub-course headings or focus on a single one. Focusing on a single one at a time is appropriate for a well-controlled learning path learning pattern, wherein the trainee has to complete one step in a sequence in order to move to the next. Alternatively, it may be appropriate to have a complete overview of each course sub-headings and move quickly and freely from one step to another. The first learning style may be more appropriate for first-time learning while the second is more appropriate for trainees having either a first look at the overall content or attempting a course review. These patterns of learning sequence apply to all developed training courses.

Next, the trainee is provided with instructions on how to install the device and software, including examples of software usage (Figure 15.2), coupled with vibration analysis terms definitions (Figure 15.3).

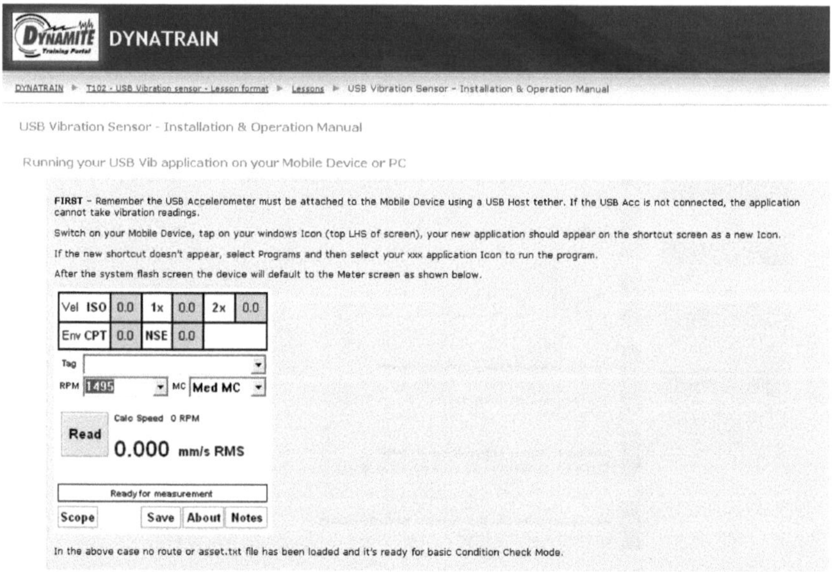

Figure 15.2 Vibration sensor application software usage

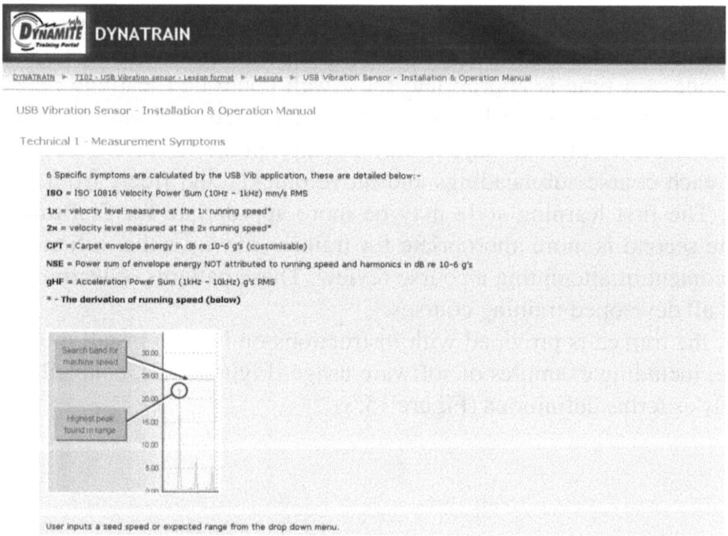

Figure 15.3 Definitions of vibration analysis terms

The software makes available vibration-based condition assessment tips to help the user make a correct diagnosis on the basis of the collected data. Such tips are provided in the form of diagnostic tables, linking observed vibration signal features to known machinery malfunction conditions (Figure 15.4).

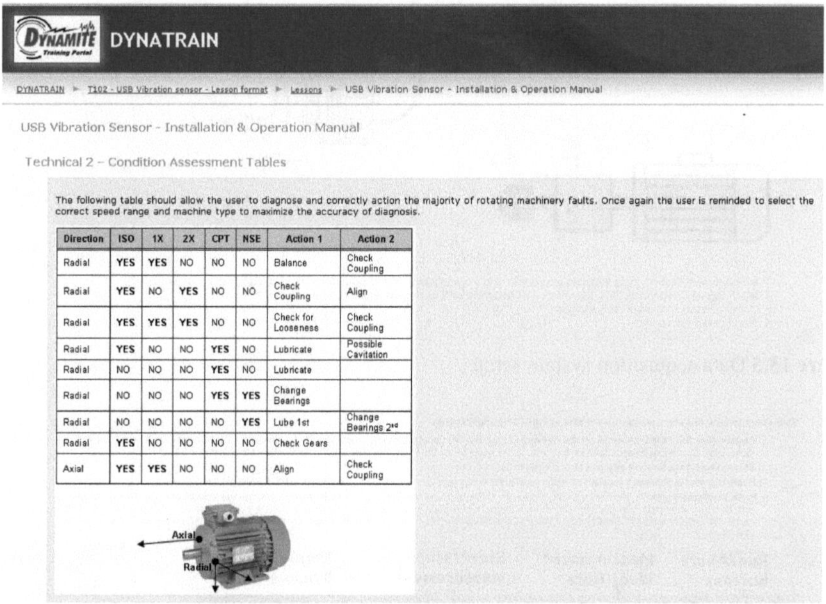

Direction	ISO	1X	2X	CPT	NSE	Action 1	Action 2
Radial	YES	YES	NO	NO	NO	Balance	Check Coupling
Radial	YES	NO	YES	NO	NO	Check Coupling	Align
Radial	YES	YES	YES	NO	NO	Check for Looseness	Check Coupling
Radial	YES	NO	NO	YES	NO	Lubricate	Possible Cavitation
Radial	NO	NO	NO	YES	NO	Lubricate	
Radial	NO	NO	NO	YES	YES	Change Bearings	
Radial	NO	NO	NO	NO	YES	Lube 1st	Change Bearings 2nd
Radial	YES	NO	NO	NO	NO	Check Gears	
Axial	YES	YES	NO	NO	NO	Align	Check Coupling

Figure 15.4 Vibration-based condition assessment

15.4.3 Data Acquisition

The data acquisition training course offers a structured introduction of how to set up a data acquisition system (Figure 15.5) and operate it to take and visualise condition monitoring data (Figure 15.6). It includes step by step instructions for system setup and connectivity, as well as additional information on data flow and usage (Figure 15.7), to make the whole process easier for the user.

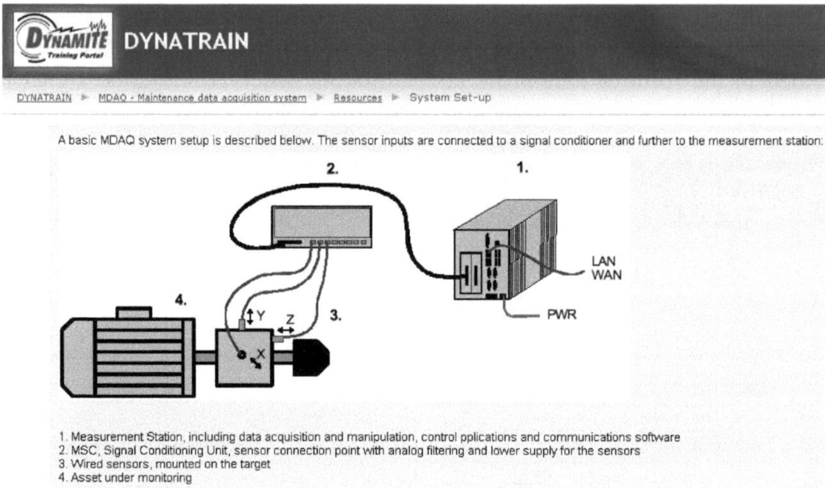

Figure 15.5 Data acquisition system setup

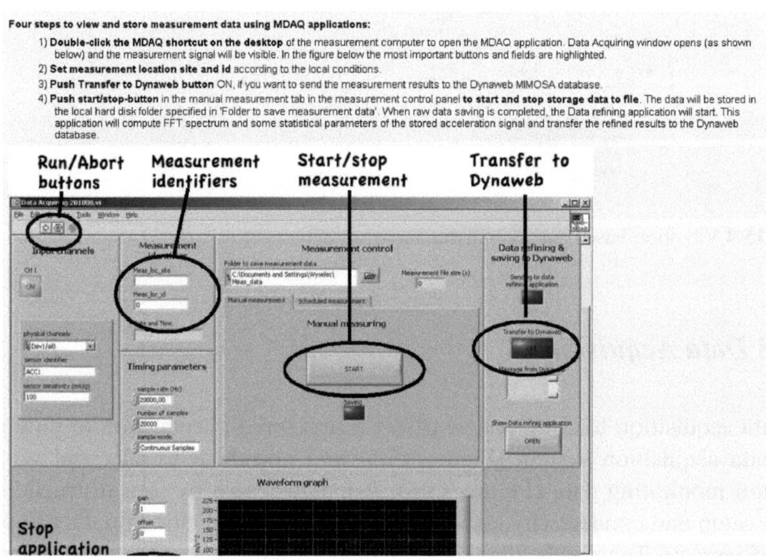

Figure 15.6 Taking condition monitoring readings

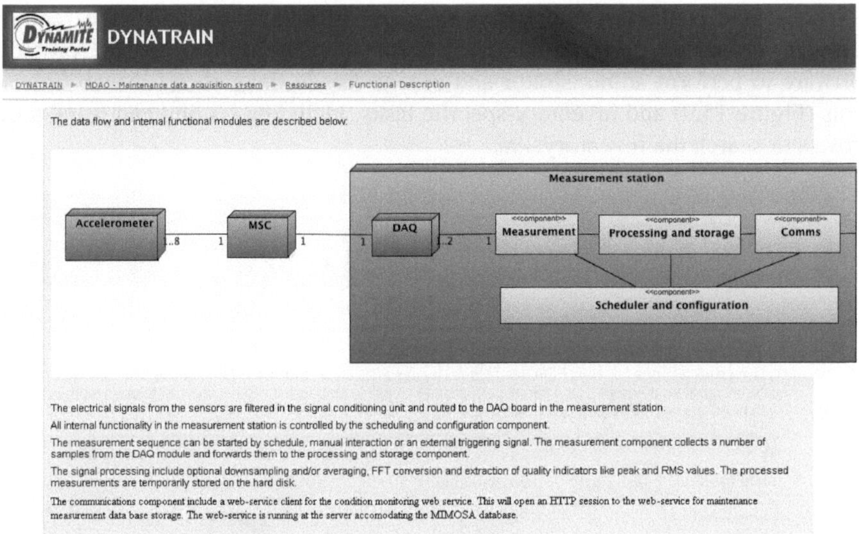

Figure 15.7 Overview of data flow in the data acquisition system

15.4.4 Inventory Tracking System

The inventory tracking system contains software and hardware tools to perform maintenance-related automated inventory tracking, using RFID tags, with the help of a PDA.

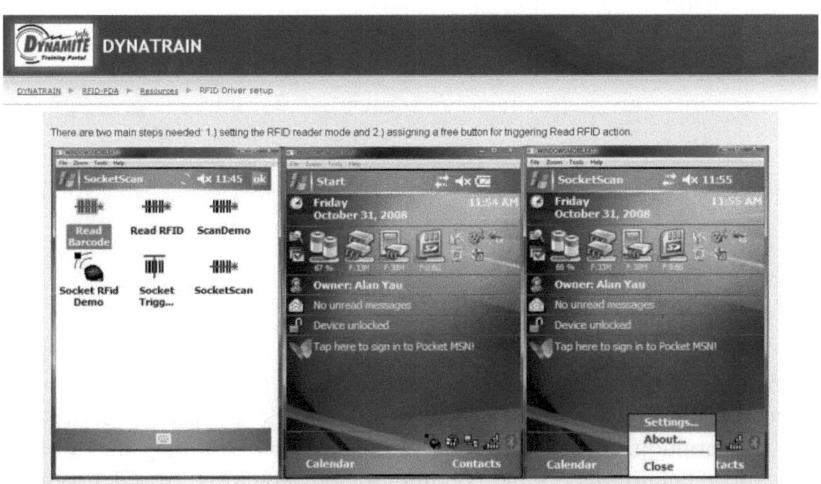

Figure 15.8 RFID driver setup

The course contains content step by step instructions to install the required software and RFID driver (Figure 15.8) and help the trainee use the provided software to perform administrator and user tasks *via* the embedded web browser tools (Figure 15.9) and inventory-specific tasks, including issuing and using tools, browse or search the inventory, *etc.*

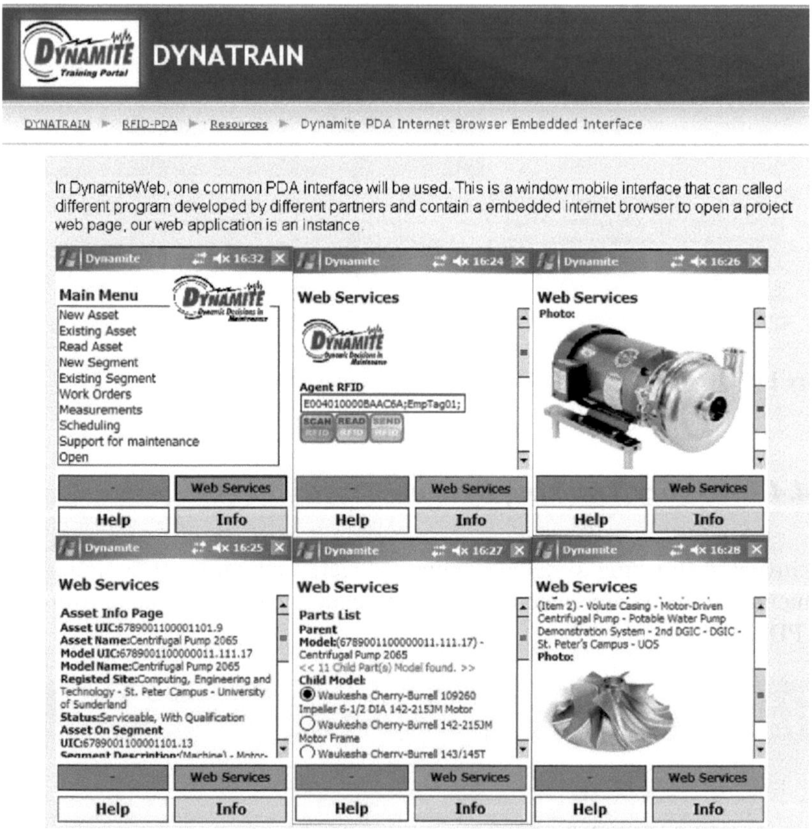

Figure 15.9 Browser embedded interface

15.4.5 Prognosis Web Services

The prognosis web services training course includes the theoretical and practical knowledge necessary to use the provided prognostics application services. It includes a description of the underlying principles for performing diagnosis, with reference to reliability-based and condition-based prognosis. In order to help the trainee (a prospective system integrator or service provider) to better understand the way the prognostic services communicate with the MIMOSA database, the

DynaWeb platform and the user devices, the course provide adequate diagrams to illustrate how this is performed (Figure 15.10).

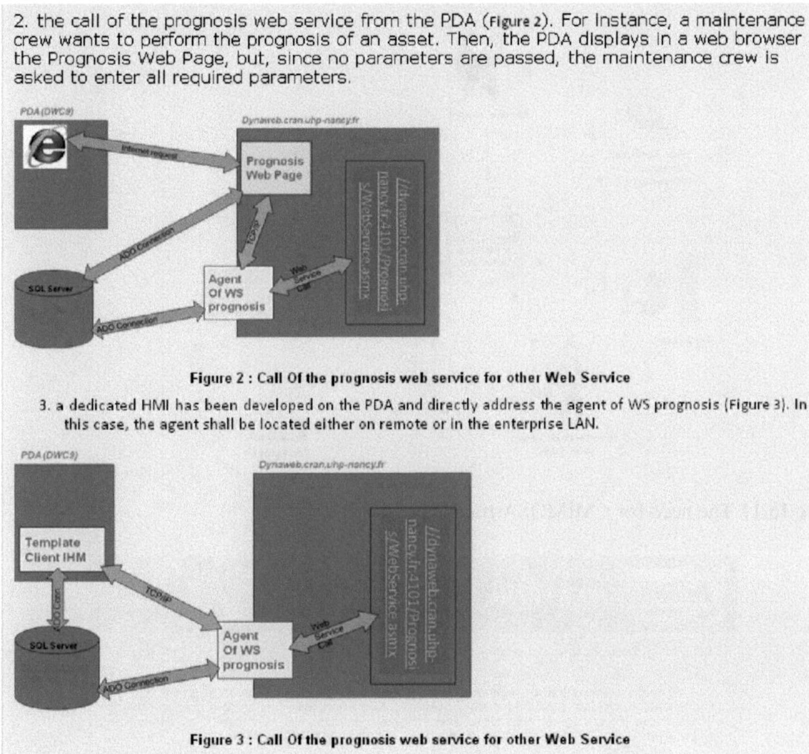

Figure 15.10 Technical specification of the prognosis web service

15.4.6 MIMOSA Translator

In e-maintenance a central issue of concern is maintenance data integration. In Dynamite and in the DynaWeb platform, this is dealt with by adopting the MIMOSA data exchange standard. A dedicated course has been created to guide users on the MIMOSA database usage. A key important functionality is to 'translate' data provided by independent software module to MIMOSA-compliant format, *via* a dedicated tool, called MIMOSA translator (Figures 15.11–15.13).

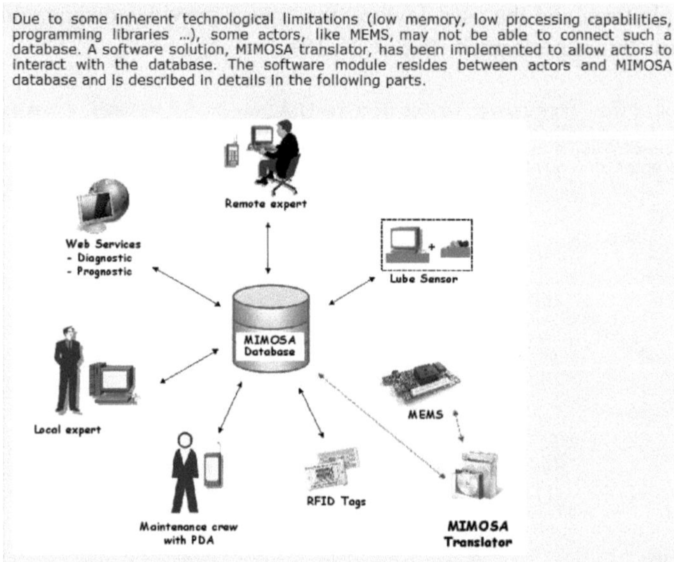

Figure 15.11 The need for a MIMOSA translator

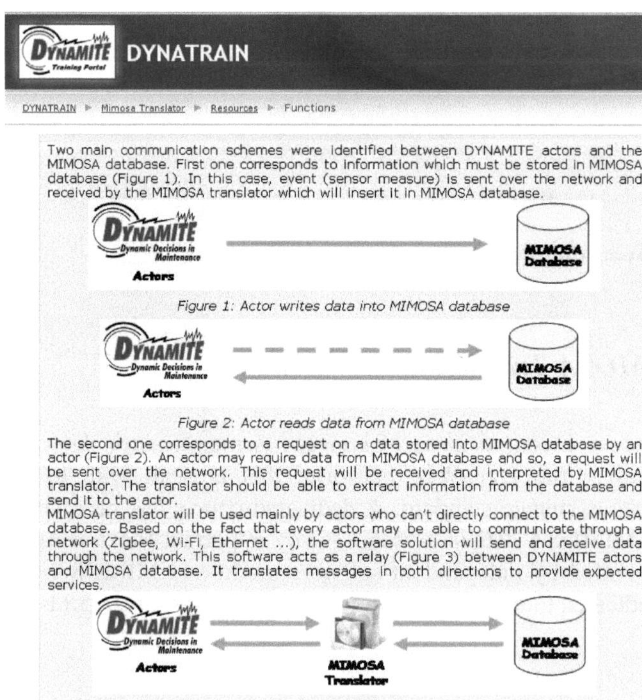

Figure 15.12 MIMOSA translator functionality

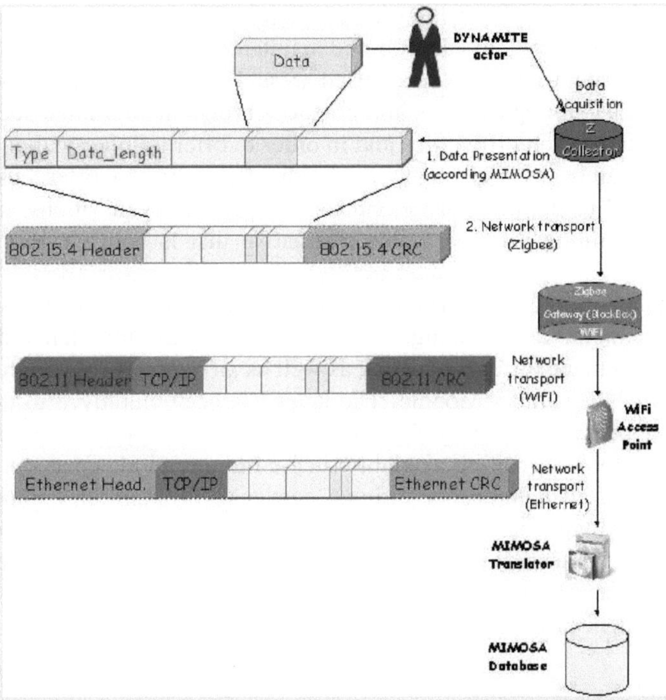

Figure 5: MIMOSA translator communication

Technical Specification

Features	Benefits
MIMOSA connection SQL connection	Able to connect and exchange data with a MIMOSA database
TCP/IP connection	Able to receive and send data through a TCP/IP connection. TCP/IP is a main protocol for network communication.
Modular listening modules	Able to add any type of incoming listener to receive message from different components (COM port, USB, ...)

Figure 15.13 MIMOSA translator communication & specification

15.5 Conclusions

This chapter has looked into the usage of e-learning for maintenance training. As the latter is in most cases not included in formal education and trainees are often people who have already entered their working life, maintenance training is usually part of vocational education and training. Professionals and organisations often cannot afford to engage in conventional courses that are inflexible in terms of time and place scheduling.

Furthermore, on-the job training that is sought at a premium in maintenance training incurs high costs. In addition, the maintenance engineering field often has needs for delivering training in specialised areas. For example, the introduction of e-maintenance technologies, tools and practices brings in additional such training needs. To mitigate such difficulties and in order to offer training without time and location constraints, as well as to retain the flexibility to offer customisable courses, tailored to the need of individuals or specific organisations, it has been argued that e-training offers a viable alternative that is both flexible and cost-efficient.

This chapter offered an overview of such issues, including needs and requirements for maintenance e-training, e-learning technology and standards, current learning management systems benefits, as well as an overview of a highly flexible open source LMS, namely Moodle. The latter has been employed to design, implement and deliver e-training for specialised e-maintenance training. Further work needs to look at making available such e-training modules as optional add-ons to standardised maintenance training curricula. In addition, competence assessment in the maintenance engineering filed can considerable benefit from the introduction of e-assessment tools and more work needs to be directed at meeting demands for streamlining and standardising maintenance management and engineering competence assessment.

References

Bakouros Y, Panagiotidou S (2008) An analysis of maintenance education and training needs in European SMEs and an IT platform enriching maintenance curricula with industrial expertise. Proc of CM-MFPT 2008, 5th Int Conf on Condition Monitoring and Machinery Failure Prevention Technologies, 15-18 July 2008, Edinburgh, UK

Botturi L (2004) Functional assessment of some open-source LMS. University of Italian Switzerland, eLab report

Christian J (2007) Virtual and mixed reality interfaces for e-training: examples of applications in light aircraft maintenance, universal access in HCI. Part III, HCII 2007, In: C. Stephanidis, (ed), Springer, Berlin, 520–529

Cobcroft R (2006) Literature review into mobile learning in the university context. Technical Report, Queensland University of Technology – Creative Industries Faculty

Cole JR (2008) Using Moodle. 2nd ed, O'Reilly, Farnham, UK, 266 p

Collier G, Robson R (2002) E-learning interoperability standards. White Paper, Sun Microsystems

Coppola C, Neelley E (2004) Open source – opens learning: Why open source makes sense for education. The RSmart Group. http://www.opensourcesummit.org/open-source-200408.pdf

Dodds P, Panar A (2006) SCORM® 2004 3rd Edition Sequencing and Navigation (SN). Version 1.0, ADL

Dodds P, Thropp SE (2006a) SCORM® 2004 3rd edition content aggregation model (CAM). Version 1.0., 2006, ADL

Dodds P, Thropp SE (2006b) SCORM® 2004 3rd edition run-time environment (RTE). Version 1.0, ADL

Dolog P, Henze N, Nejdl W, Sintek M (2004) Student tracking and personalisation: personalisation in distributed e-learning environments. Proc of the 13th Int World Wide Web conference. New York, USA, ACM

EFNMS, www.efnms.org, European Certification Committee

Emmanouilidis C, Papathanassiou N, Papakonstantinou A (2008) Current trends in e-training and prospects for maintenance vocational training. In Proc of CM-MFPT 2008, 5th Int Conf on Condition Monitoring and Machinery Failure Prevention Technologies. 2008, Edinburgh, UK. BINDT-Coxmoor Publishing, 466–471

Franlund J (2008) Some European initiatives in requirements of competence in maintenance, Proc. of CM-MFPT 2008, 5th Int. Conf. on Condition Monitoring and Machinery Failure Prevention Technologies, 15-18.7.2008, Edinburgh, UK. BINDT-Coxmoor Publishing, 406–416

Friesen N (2004) Three objections to learning objects and e-learning standards. In: McGreal R (Ed), Online education using learning objects. Routledge, London, 59–70

Garcia, FB and Jorge AH (2006) Evaluating e-learning platforms through SCORM specifications. IADIS Virtual Multi Conference on Computer Science and Information Systems (MCCSIS 2006), IADIS

Gonzalez MAC, Peñalvo FJG, Guerrero MJC, Forment MA (2009) Adapting LMS architecture to the SOA: an architectural approach. 2009 4th Int Conf on Internet and Web Applications and Services. IEEE, 322–327

Graf S and List B (2005) An evaluation of open source e-learning platforms stressing adaptation issues. Proceedings of the Fifth IEEE International Conference on Advanced Learning Technologies (ICALT 2005). Kaohsiung, Taiwan, IEEE:163–165

Gütl C (2008) Moving towards a generic, service-based architecture for flexible teaching and learning activities. In: Pahl C (Ed) Architecture solutions for e-learning systems. Peer-reviewed book chapter, Idea Group, Hershey PA, pp 1–24, IGI Global

Haritos T and Macchiarella D (2005) A mobile application of augmented reality for aerospace maintenance training. Proc The 24th Digital Avionics Systems Conference, IEEE 5.1.3-5.1.9

Hatzilygeroudis I, Giannoulis C, Koutsojannis C (2005) Combining expert systems and adaptive hypermedia technologies in a web based educational system. Proc of the 5th IEEE International Conference on Advanced Learning Technoloies (ICALT'05), 2005

Huddlestone J and Pike J (2007) Seven key decision factors for selecting e-learning cognition. Technology and Work 10:237–247

ISO 18436-1: 2004 Condition monitoring and diagnostics of machines – requirements for training and certification of personnel – Part 1: requirements for certifying bodies and the certification process

Itmazi JA and Gea MM (2005) Survey: comparison and evaluation studies of learning content management systems. MICROLEARNING2005: Learning and Working in New Media Environments. International Conference, Innsbruck, Austria. IADIS

Jones N and O'Shea J (2004) Challenging hierarchies: The impact of e-learning. Higher Education 48(3): 379–395

Karampiperis P, Sampson D (2006) Adaptive learning objects sequencing for competence-based learning. Proc 6th Int Conf on Advanced Learning Technologies (ICALT'06). IEEE, 136–138

Kazi SA (2004) A conceptual framework for web-based intelligent learning environments using SCORM-2004. Proc IEEE Int Conf on Advanced Learning Technologies, ICALT 2004, Joensuu, Finland

Kljun M, Vicic J, Kavsek B, Kavcic A (2007) Evaluating comparisons and evaluations of learning management systems. Proc ITI 2007 29th Int Conf on Information Technology Interfaces, Cavtat, Croatia, IEEE

Li JR, Khoo LP, Tor SB (2003) Desktop virtual reality for maintenance training: an object oriented prototype system (V-REALISM). Computers in Industry 52:109–125

Ma Z (2005) Web-based intelligent e-learning systems. Idea Group, Hershey PA

Macchi M, Ierace S (2009) Education in industrial maintenance management: feedback from an Italian experience. Proc. 4th World Congress on Engineering Asset Management, 28-30.9.2009, Athens, Greece. Springer, London, 531–538

Mifsud L, Marc AI (2007) ‘That's my PDA!’ The role of personalisation for handhelds in the classroom. Proc 5th Ann IEEE Int Conf Pervasive Computing and Communications Workshops (PerComW'07)

Nakajima C, Itho N (2003), A support system for maintenance training by augmented reality. Proc 12th Int Conf Image Analysis and Processing (ICIAP’03), Mantova, Italy, IEEE

Pairot C, Garcia P, Rallo R, Blat J, Gomez Skarmeta AF (2005) The planet project: collaborative educational content repositories on structured peer-to-peer grids. IEEE Int Symp Cluster Computing and the Grid. Cardiff, UK, IEEE 1:35-42

Papathanassiou N, Emmanouilidis C (2009), E-learning for maintenance management training and competence assessment: development and demonstration. Proc 4th World Congress on Engineering Asset Management, 28-30 September 2009, Athens, Greece. Springer, London, 892–901

Popescu E, Trigano P, Badica C (2007) Adaptive educational hypermedia systems: a focus on learning styles. In EUROCON, 2007. The International Conference on “Computer as a Tool”. 2007, IEEE, Warsaw, Poland. 2473–2478

Qu C, Nejdl W (2001) Towards interoperability and reusability of learning resource: a SCORM-conformant courseware for computer science education. In: Proc of 2nd IEEE International Conference on Advanced Learning Technologies, ICALT 2001

Rice WH (2006) Moodle E-learning course development: a complete guide to successful learning using Moodle. Packt, Birmingham, UK, 236 p

Roe S (2003) Condition monitoring certification of personnel in the UK. Insight 45:764–765, ISSN 1354-2575

Rosatelli M, Senger H, Silva F, Stanzani S, Nunes S (2006) Supporting the collaborative construction of learning objects using the grid. Proc 6th IEEE Int Symp Cluster Computing and the Grid (CCGRID’06), Singapore, IEEE

Sharples M (2000) The design of personal mobile technologies for lifelong learning. Computers and Education Volume, 177–193.doi: 10.1016/S0360-1315(99)00044-5

Starr A, Bevis K (2009) The role of education in industrial maintenance: the pathway to a sustainable future. Proc 4th World Congress on Engineering Asset Management, 28-30.9.2009, Athens, Greece

Székely G, Satava RM (1999) Virtual reality in medicine. BMJ – British Medical Journal 13(319):7220

Tissot P (2004) Terminology of vocational training policy, A multilingual glossary for an enlarged Europe. Luxembourg, Cedefop, European Centre for the Development of Vocational Training

Varlamis I, Apostolakis I (2006) The present and future of standards for e-learning technologies. Interdisciplinary Journal of Knowledge and Learning Objects 2:59–76

Witthaus G (2009) The implications of SCORM conformance for workplace e-learning. Electronic Journal of e-Learning 7:183–190

Woolf B P (2009) Building intelligent interactive tutors. Morgan Kauffman, New York

Yang CT, Ho HC (2005) A shareable e-learning platform using data grid technology. The 2005 IEEE Int Conf on e-Technology, e-Commerce and e-Service, 2005. EEE’05, Hong Kong, IEEE

Chapter 16
Conclusions and Future Perspectives

Kenneth Holmberg

Maintenance has developed during the last half century from a low-tech "repair and lubrication man" circulating on the shop floor, to a truly high-tech, scientifically based, multidisciplinary field of technology that is of strategic importance in industrial plants, transportation and power production. Maintenance today means the technological-economical understanding and skills to keep machinery and devices running as required with a high level of reliability, efficiently and without cost-effective disturbances.

It has been possible to improve the reliability and availability in production plants by combining new and advanced technologies like failure analysis, sensors, signal processing, diagnostics, prognostics, modelling, control engineering, reliability design, system analysis, expert systems, risk analysis, *etc.*

The rapid development in the fields of electronics, software design, wireless communication, miniaturisation and hardware computing capacity have brought new possibilities to design devices for high reliability and availability, and continuous condition monitoring wherever they are located worldwide. In case of problems occurring on any system, the best assistance can be rapidly accessed and the device can quickly be brought back to required performance. This new approach is called *e-maintenance* and has been largely described and demonstrated in this book (Figure 16.1).

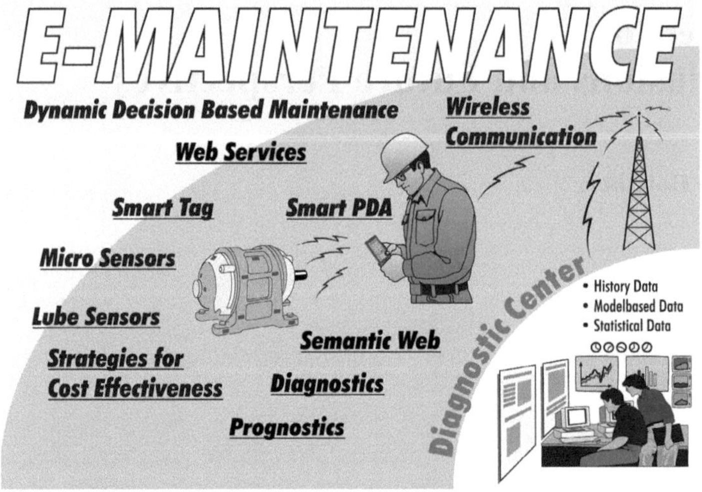

Figure 16.1 E-maintenance is a future flexible, mobile and global solution towards improved cost-effectiveness, safety and sustainability in the modern society

After four years of intensive research and development work of the pioneering DynaWeb e-maintenance solution, the authors of this book came to the following conclusions:

- E-maintenance has today reached the level of real systems that can be implemented in industrial environments. This was demonstrated by the DynaWeb concept comprising 28 software and hardware components including MEMS sensors with energy harvesting, online lubrication sensors, smart tags for identification and location of components, maintenance actions supported by handheld mobile computers (PDAs), wireless communication, diagnostics, prognostics, web services, the MIMOSA database, vibration measurement support, scheduling support, cost-effectiveness and strategic decision support based on technical and economical considerations, supported by adequate e-training.
- The DynaWeb platform, composed of numerous components, was developed as a layered architecture of business processes according to the open system architecture of condition based maintenance (OSA-CBM). This supported synergetic effects as the different components were designed to analyse overlapping data from different angles. The common relation interface schema was implemented as close to standard as possible.
- The e-maintenance modules were integrated with a MIMOSA database and all data were both communicated to and from the database and between the modules according to MIMOSA protocol. This was first found to be heavy and complex but after some practice all users agreed about the benefit of using this protocol. It was experienced as a good solution and actually as the only working solution available today for this kind of large e-maintenance integration.

Moreover, this solution can be easily extended and customised with new and upgraded MIMOSA compliant components.

- MEMS micro-sensors, online lubrication sensors and smart tags or RFIDs are some of the key-components for distributed condition data collection. They have been developed to a level where detailed machinery condition data from components and systems can be cost effectively collected by multiple sensor systems, locally analysed and when necessary transferred to more extensive signal and condition analysis, diagnosis and prognosis. The reduction of sensor energy consumption and highly efficient power scavenging possibilities are future challenges.
- An advanced interface for the maintenance engineer to the machinery system is the handheld mobile computer (PDA) that today is getting closer to smart phones. The PDA can communicate with the different components in the e-maintenance system, check status and local condition, monitor changes, perform diagnosis and prognosis, make work orders and communicate with high level production and business systems. All this can be done either locally close to the equipment or remotely over the web.
- The development of Internet based web services for condition analysis, diagnosis, prognosis and both operational and strategic decision support play a key role in implementing e-maintenance. The web services offer interoperability between independent software applications over the Internet as plug-in components in a distributed environment in a company. Semantic web solutions offer new advantages over previous technologies that make web applications more useful, improve the web search capability and offer better integration interfaces and all without manual intervention.
- Flexible e-maintenance requires good wireless communication, which today is a challenge due to the diversity of applications and the heterogeneous devices that need to be integrated inside the communication system. Experience has shown that early solutions available today have shortcomings in efficient support for all the functional and technological requirements. They are likely to rapidly be succeeded by more flexible solutions. For a particular application, the most efficient technology at a specific level is perhaps not adapted to support at another level of operations. Thus it seems unnecessarily risky to rely on only one technology. It seems as a better solution to develop an e-maintenance concept of wireless gateways connecting areas, in which the most efficient wired and wireless technologies are implemented.
- The MIMOSA database enabled the system to use the collected data for technical-economical considerations, as well as decision support. Enhanced maintenance cost-effectiveness could be demonstrated as an integrated part of the e-maintenance system through effective problem detection, analysis and more accurate maintenance action planning. This provides a unique opportunity for gathering more relevant information required for higher accuracy of maintenance decisions for improving cost-effectiveness as well as the possibility for

continuous improvement of maintenance performance and acting at an early stage.

- E-maintenance is a new working environment for the maintenance engineer with several maintenance functions and actions carried out in a new way with new tools and access interfaces. The adoption of this innovative technology and tools requires effective and continuous training of the personnel involved and this was demonstrated to be achieved by e-training that offers a viable option that is both flexible and cost-efficient.

The DynaWeb e-maintenance concept was demonstrated at three sites in the manufacturing industry and the overall results were very positive, with proven technical and economical feasibility. The quality of the separate components and their functionality was high. Certainly, their integration was not straightforward and required a major effort. When implementing an e-maintenance concept such as DynaWeb, in, for example, the automotive industry, companies may prefer a local version of the system inside the company firewall for security reasons. Many existing software systems in most companies, starting from CMMS and SCADA systems, have to be taken into account when upgrading to e-maintenance. Here, the use of web-based plug-in distributed components is the key to provide the necessary upgrade.

Future implementation of maintenance systems will see greater integration of business and technical systems, with more intelligent use of collected data. They will protect users against change of personnel, with the inherent loss of their learning, and allow better informed choices for decision makers. Technological data collection, with its attendant signal processing for the extraction of information from raw data, will embed an increasing amount of intelligent processing at source, while increasing the speed of communication through wearable computing and robust mesh networking. Sophisticated strategies are under development for mobile plants, vehicles and aircraft, to allow for independent local processing with intermittent communication to a central system for parts ordering and work scheduling.

Limitations to progress include the lack of standardisation of system and communication components, and the need for training. Some hardware, *e.g.*, mobile computing, has a shelf life considerably shorter than high capital engineering equipment. It rapidly becomes obsolete and cannot be replaced without upgrading the software as well as the hardware. Many of the communication networks required exist, but they are not universal, and the business models needed to access them are still under development; *e.g.*, access to cellular telephones and WiFi hotspots. Some businesses are sensitive to security arrangements for public networks.

The use of such wide-ranging systems, from detailed technical programming of smart sensors, through to management of information leading to business-critical maintenance decisions, requires some exceptional people to run it; their mobility will cause them to change employer and job function, so the capture of their knowledge, and the training of new people, will continue to be essential for the exploitation of advanced maintenance. The data from embedded, smart sensors is likely to grow – whether we want it or not – and the problems of managing the data will also grow.

Despite the many challenges related to the development of technology, integration of technical systems, data transfer and user interfaces that exist today, the general conclusion is that e-maintenance has reached a level where it can be successfully implemented in real industrial systems, and already offers improved controllability, availability and cost-efficiency. The future holds a great opportunity for integrated e-maintenance to make a major contribution to profitability and sustainability.

MIX
Papier aus verantwortungsvollen Quellen
Paper from responsible sources
FSC® C105338
www.fsc.org

If you have any concerns about our products,
you can contact us on
ProductSafety@springernature.com

In case Publisher is established outside the EU,
the EU authorized representative is:
Springer Nature Customer Service Center GmbH
Europaplatz 3, 69115 Heidelberg, Germany

Printed by Libri Plureos GmbH
in Hamburg, Germany